The State of Humanity

B

PREVIOUS BOOKS BY JULIAN SIMON

Population Economics

The Effects of Income on Fertility (1974)
The Economics of Population Growth (1977)
The Ultimate Resource (1981)
Theory of Population and Economic Growth (1986)
The Economic Consequences of Immigration (1989)
Population Matters: People, Resources, Environment, and Immigration (1990)
Population and Development in Poor Countries: Selected Essays
Scarcity or Abundance? A Debate on the Environment, with Norman Myers (New York: Norton, 1994).

Other Subjects

Patterns of Use of Books in Large Research Libraries (with Herman H. Fussler, 1969)
Basic Research Methods in Social Science (1969; third edn. with Paul Burstein, 1985)
How to Start and Operate a Mail-Order Business (1965; fifth edn., 1993)
Issues in the Economics of Advertising (1970)
The Management of Advertising (1971)
Applied Managerial Economics (1975)
Effort, Opportunity, and Wealth (1987)
Good Mood: The New Psychology of Overcoming Depression (1993)
Resampling Statistics (Boston: Wadsworth, 1993).

Edited Books

Research in Population Economics: Vol. I (1978); Vol. II (1980) (with Julie daVanzo); Vols. III (1981) and IV (1982) (with Peter Lindert)
The Resourceful Earth (with Herman Kahn, 1984).

The State of Humanity

Edited by

Julian L. Simon
University of Maryland

Managing Editors
E. Calvin Beisner
John Phelps

BLACKWELL
Oxford UK & Cambridge USA
in association with
the Cato Institute

Copyright © Julian L. Simon 1995

First published 1995

Blackwell Publishers Inc
238 Main Street
Cambridge, Massachusetts 02142, USA

Blackwell Publishers Ltd
108 Cowley Road
Oxford OX4 1JF, UK

Library of Congress Cataloging-in-Publication Data has been applied for

ISBN 1 55786 119 6
1 55786 585 X (pbk.)

British Library Cataloguing in Publication Data
A CIP catalogue record for this book is available from the British Library.

Typeset in 10 on 12 pt Plantin
by Pure Tech Corporation, Pondicherry, India
Printed in Great Britain by T.J. Press Ltd, Padstow, Cornwall

This book is printed on acid-free paper

Contents

		Page
Acknowledgments		ix

1 Introduction   1
    Julian L. Simon

PART I: Life, Death, and Health   29

2 Human Mortality Throughout History and Prehistory   30
    Samuel H. Preston
3 The Decline of Childhood Mortality   37
    Kenneth Hill
4 Disease and Health through the Ages   51
    Michael R. Haines
5 The Contribution of Improved Nutrition to the Decline in
    Mortality Rates in Europe and America   61
    Robert William Fogel
6 Trends in Health of the US Population: 1957–89   72
    Eileen M. Crimmins and Dominique G. Ingegneri
7 Mortality and Health in the former Soviet Union   85
    Murray Feshbach
8 Worldwide Historical Trends in Murder and Suicide   91
    Jean-Claude Chesnais
9 The History of Accident Rates in the United States   98
    Arlene Holen
10 World Trends in Smoking   106
    Allan M. Brandt
11 Long-Term Trends in the Consumption of Alcoholic Beverages   114
    James S. Roberts

PART II: Standard of Living, Productivity, and Poverty   123

12 Trends in the Agricultural Labor Force   124
    Richard J. Sullivan

13   The Standard of Living Through the Ages   135
     Joyce Burnette and Joel Mokyr
14   Long-Term Trends in the US Standard of Living   149
     Stanley Lebergott
15   Long-Term Trends in Productivity   161
     Jeremy Atack
16   The Extent of Slavery and Freedom Throughout the
     Ages, in the World as a Whole and in Major Subareas   171
     Stanley L. Engerman
17   Black Americans: Income and Standard of Living from
     the Days of Slavery to the Present   178
     Robert Higgs and Robert A. Margo
18   The Long-Term Course of American Inequality:
     1647–1969   188
     Peter H. Lindert and Jeffrey G. Williamson
19   Trends in Unemployment in the United States   196
     Alexander Keyssar
20   Trends in Costs and Quality of Housing   203
     Richard F. Muth
21   Trends in the Quantities of Education–USA and Elsewhere   208
     Julian L. Simon and Rebecca Boggs
22   Trends in Free Time   224
     John P. Robinson
23   Trends in Poverty in the United States   231
     Rebecca M. Blank
24   How "Poor" Are America's Poor?   241
     Robert Rector
25   Homelessness in America   257
     Randall K. Filer
26   The Recent US Economy   271
     Alan Reynolds

PART III: Natural Resources   **279**

27   Long-Term Trends in Energy Prices   280
     William J. Hausman
28   Trends in the Price and Supply of Oil   287
     Morris A. Adelman
29   The Costs of Nuclear Power   294
     Bernard L. Cohen
30   Trends in Availability of Non-Fuel Minerals   303
     John G. Myers, Stephen Moore, and Julian L. Simon
31   Trends in Nonrenewable Resources   313
     H. E. Goeller
32   Trends in Availability and Usage of Outdoor Recreation   323
     Robert H. Nelson
33   Global Forests Revisited   328
     Roger A. Sedjo and Marion Clawson

34 Species Loss Revisited   346
    Julian L. Simon and Aaron Wildavsky

PART IV: Agriculture, Food, Land, and Water   363

35 Agricultural Productivity Before the Green Revolution   364
    George W. Grantham
36 The World's Rising Food Productivity   376
    Dennis Avery
37 Recent Trends in Food Availability and Nutritional
    Well-Being   394
    Thomas T. Poleman
38 Trends in Grain Stocks   405
    William J. Hudson
39 Trends in Food from the Sea   411
    John P. Wise
40 Trends in Soil Erosion and Farmland Quality   416
    Bruce L. Gardner and Theodore W. Schultz
41 Water, Water Everywhere But Not a Drop to Sell   425
    Terry L. Anderson
42 Trends in Land Use in the United States   434
    H. Thomas Frey

PART V: Pollution and the Environment   443

43 Long-Run Trends in Environmental Quality   444
    William J. Baumol and Wallace E. Oates
44 Atmospheric Pollution Trends in the United Kingdom   476
    Derek M. Elsom
45 Trends in Air Pollution in the United States   491
    Hugh W. Ellsaesser
46 Comparative Trends in Resource Use and Pollution in
    Market and Socialist Economies   503
    Mikhail S. Bernstam
47 Acid Rain   523
    J. Laurence Kulp
48 Stratospheric Ozone: Science and Policy   536
    S. Fred Singer
49 The Greenhouse Effect and Global Change: Review and
    Reappraisal   544
    Patrick J. Michaels
50 Greenhouse Scenarios to Inform Decision Makers   565
    Lester Lave
51 The Hazards of Nuclear Power   576
    Bernard L. Cohen
52 Pesticides, Cancer, and Misconceptions   588
    Bruce N. Ames

53 The Carcinogen or Toxin of the Week Phenomenon:
    The Facts Behind the Scares   595
    Elizabeth M. Whelan

PART VI: Thinking about the Issues   609

54 American Public Opinion: Environment and Energy   610
    William M. Lunch and Stanley Rothman
55 Public Opinion About, and Media Coverage of,
    Population Growth   619
    Rita J. Simon
56 Risk Within Reason   628
    Richard J. Zeckhauser and W. Kip Viscusi
57 Natural Ecology Today and in the Future:
    A Personal View   637
    Kenneth Mellanby

PART VII: Conclusion: From the Past to the Future   641

58 What Does the Future Hold? The Forecast in a Nutshell 642
    Julian L. Simon

List of Figures   661
List of Tables   671
List of Contributors   673
Index   677

Acknowledgments

Greatest thanks go to the authors, of course. I thank them for being willing to go through many iterations, in many cases, to bring out the meaning of their material to the fullest. But even more, I thank them for having devoted their lives to producing the data which enable us to better understand the history of the human enterprise.

In the early stages, John Phelps rendered yeoman service in extensively editing some of the manuscripts. When illness made it impossible for him to continue, E. Calvin Beisner took on the load, and did a remarkable job in creating charts, clarifying meanings, shortening without causing offense, and everything else that a managing editor can do, until he had to leave the project. Karen Verde performed excellently as a copy-editor, as always. Chris Brest did a fine job with the remaining figures. After much despair, John Davey has helped the book because of his belief in it.

The book is indebted to the Grover Hermann Foundation for a helpful grant, and to the Cato Foundation for co-publishing the book.

Until his death, Herman Kahn was my partner in the predecessor to this volume, the 1984 *The Resourceful Earth*. His memory permeates this work. I miss the great fun of working with him.

This book may best be read in conjunction with the second edition of my *The Ultimate Resource*, due out in late 1995, which contains additional data as well as theoretical discussions of the mechanisms producing the trends discussed in this book.

For my friends Petr Beckmann, Herman Kahn, and Aaron Wildavsky, all dead too early

Searchers for truths
Finders of truths
Fighters for the truth

We miss them.

1

Introduction

Julian L. Simon

Executive Summary

If present trends continue, the world in 2000 will be more crowded, more polluted, less stable ecologically, and more vulnerable to disruption than the world we live in now. Serious stresses involving population, resources, and environment are clearly visible ahead. Despite greater material output, the world's people will be poorer in many ways than they are today.

For hundreds of millions of the desperately poor, the outlook for food and other necessities of life will be no better. For many it will be worse. Barring revolutionary advances in technology, life for most people on earth will be more precarious in 2000 than it is now – unless the nations of the world act decisively to alter current trends.

The above quotation from page one of the 1980 *Global 2000 Report to the President* began the Introduction to this volume's predecessor, *The Resourceful Earth*, which I edited with the late Herman Kahn. Then, in order "to highlight our differences as much as possible," we restated that paragraph with our substitutions in italics, as follows:

If present trends continue, the world in 2000 will be *less crowded* (though more populated), *less polluted, more stable ecologically, and less vulnerable to resource-supply disruption* than the world we live in now. Stresses involving population, resources, and environment *will be less in the future than now* . . . The World's people will be richer in most ways than they are today . . . The outlook for food and other necessities of life will be better . . . life for most people on earth will be *less precarious* economically than it is now.

The years have been kind to our forecasts – or more importantly, the years have been good for humanity. The benign trends we then observed have continued until the time of writing this volume. Our species is better off in just about every measurable material way. And there is stronger reason than ever to believe that these progressive trends will continue past the year 2000, past the year 2100, and indefinitely.

The outlook portrayed in this volume is even more happy than before for two reasons: (1) conditions have improved in the phenomena we discussed a decade ago, and (2) we now document a much wider range of phenomena pertaining to human welfare than in the previous volume, and almost all of these additional trends also point in a positive direction.

More specifically, the trends toward greater cleanliness and less pollution of our air and water are even sharper than before, and cover a longer historical period and more countries (though the environmental disaster in Eastern Europe has only recently become public knowledge). The increase in availability and the decrease in scarcity of raw materials have continued unabated, and have even speeded up. None of the catastrophes in food supply and famine that were forecast by the doomsayers has occurred; rather, the world's people are eating better than ever.

When we widen our scope beyond such physical matters as natural resources and the environment – to mortality, the standard of living, slavery and freedom, housing, and the like – we find that these trends pertaining to economic welfare are heartening, also. Most important, fewer people are dying young. And life expectancy in the rich countries has increased most sharply in the older age cohorts, among which many thought that there was no improvement. Perhaps most exciting, the quantities of education that people obtain all over the world are sharply increasing, which means less tragic waste of human talent and ambition.

The extrapolation of these trends into the future is discussed in chapter 58. The long-run prospects are so favorable and so certain that I am prepared to bet on them; the terms of the wager I offer are below.

Please notice that this benign assessment does not imply that there will not be increases in *some* troubles – AIDS at present, for example, and other diseases in the future, as well as social and political upheavals. New problems always will arise. Indeed, the solution to any existing problem usually creates new – albeit usually lesser – problems. But the assessment refers to broad aggregate measures of *effects upon people* rather than the bad phenomena themselves – life expectancy rather than AIDS, skin cancers (or even better, lifetime healthy days) rather than a hole in the ozone layer (if that is indeed a problem), and agriculture rather than global warming.

The complete failure of the dire forecasts of the doomsayers, starting in the 1960s and continuing through the 1980 *Global 2000 Report* until now, should confer credibility on the assessment given here. Regrettably, however, the doomsayers' failure has not reduced the frequency of forecasts of doom, or sapped the reputations or influence of the forecasters. This disregard of contrary evidence is one more sign of the absence of true science – whose essence is the comparison of theories and forecasts against the data – in the conventional literature of impending doom.

INTRODUCTION TO THE INTRODUCTION

The humour of blaming the present, and admiring the past, is strongly rooted in human nature, and has an influence even on persons endued with the profoundest judgment and most extensive learning. (David Hume, 'Of the Populousness of Ancient Nations,' in 1777/1987, 464)

This introduction charts humanity's welfare in the past, and shows where we are now. The concluding chapter forecasts the future.

The following long quotation from William Petty tells in a seventeenth-century context the central message of this book:

I have therefore thought fit to examin the following Perswasions, which I find too currant in the World, and too much to have affected the Minds of some, to the prejudice of all, viz.

. . . the whole Kingdom grows every day poorer and poorer; that formerly it abounded with Gold, but now there is a great scarcity both of Gold and Silver; that there is no Trade nor Employment for the People . . . that Trade in general doth lamentably decay; that the Hollanders are at our heels, in the race of Naval Power; the French grow too fast upon both, and appear so rich and potent, that it is but their Clemency that they do not devour their Neighbors; and finally, that the Church and State of England, are in the same danger with the Trade of England; with many other dismal Suggestions, which I had rather stifle than repeat.

. . . But notwithstanding all this (the like whereof was always in all Places), the Buildings of London grow great and glorious; the American Plantations employ four Hundred Sail of Ships; Actions in the East-India Company are near double the principal Money; those who can give good Security, may have Money under the Statute-Interest; Materials for building (even Oaken-Timber) are little the dearer, some cheaper for the rebuilding of London; the Exchange seems as full of Merchants as formerly; no more Beggars in the Streets, nor executed for Thieves, than heretofore; the Number of Coaches, and Splendor of Equipage exceeding former Times; the publique Theatres very magnificent; the King has a greater Navy, and stronger Guards than before our Calamites; the Clergy rich, and the Cathedrals in repair; much Land has been improved, and the Price of Food so reasonable, as that Men refuse to have it cheaper by admitting of Irish Cattle; And in brief, no Man needs to want that will take moderate pains. That some are poorer than others, ever was and ever will be: And that many are naturally querulous and envious, is an Evil as old as the World.

These general Observations, and that Men eat, and drink, and laugh as they use to do, have encouraged me to try if I could also comfort others, being satisfied my self, that the Interest and Affairs of England are in no deplorable Condition.

The Method I take to do this, is not yet very usual; for instead of using only comparative and superlative Words and intellectual Arguments, I have taken the course (as a Specimen of the Political Arithmetick I have

long aimed at) to express my self in Terms of Number, Weight, or
Measure; . . .
   . . . I hope all ingenious and candid Persons will rectifie the Errors,
Defects, and Imperfections, which probably may be found in any of the
Positions, upon which these Ratiocinations were grounded. (Petty, Hull
ed., 1986, pp. 241–5)

This volume portrays extraordinary progress for the human enterprise,
especially in the past two centuries. Yet many people believe that conditions of life are generally worse than in the past, rather than better. We
must therefore begin by discussing this perception, because it affects a
reader's reaction to the factual material presented in this book.

Please ask yourself: Is a big wheat harvest a good thing? It would seem
so; more wheat brought in implies more and cheaper food for consumers. Yet you'll often see such headlines as: "Good harvest, bad
news."

Yes, a big harvest is unwelcome to farmers because they receive low
prices; that is the "bad news." But is a small harvest better *on balance*
than a big harvest? Not many people would say so. Still, *The Washington
Post* headline just quoted is negative. And that "news" contributes to a
widespread perception that the overall course of events is bad.

Certainly the public view (or at least the public view of the public
view) is not wildly optimistic. At the time of writing, a major *Newsweek*
story of November 4, 1991, is headlined "That Sinking Feeling," and
goes on to say, "Americans feel gloomy, almost desperate, about the
future of the economy" (pp. 18, 28). And the front-page headline of *The
Washington Post* is "A Tide of Pessimism and Political Powerlessness
Rises" (November 3, 1991); the story cites a poll saying that "seven in
10 Americans think the country is off track." Pessimism about the
environment and resources is so universal that it needs no documentation.

The choice of comparison one makes always is crucial. A premise of
this book is that it usually makes sense to compare our present state of
affairs with *how it was before*. This is the comparison that is usually
relevant for policy purposes because it measures our progress. But many
private and public discussions instead compare a present state of *one
group* to the present state of *other groups*, as a supposed measure of
"equity," or as the basis for indignation and righteousness, or to support
their political positions. Others compare the actual situation to the *best*
possible, or to ideal purity, ostensibly to motivate improvement. A
typical front-page story from *The Washington Post* (July 5, 1991) does
both; it headlines a complaint of blacks that a nearby county "Isn't
Drawing Upscale Stores," and the caption under a picture says "Prince
George's resident Howard Stone is angered by the shortage of upscale
retail stores in his community." (Yes, that was on the front page.) This
issue is very different from the sorts of problems that most of humanity

has faced throughout most of its history, the sorts of problems that are addressed in this book.

Many events that, on first reaction, people tend to consider bad have more good than bad about them. And if we act on that first negative reaction instead of a balanced assessment, we risk making unsound social decisions.

Consider this example: Is the trend of black infant mortality encouraging? I've asked this question of many audiences, both laypeople and professionals – even demographers. Almost everyone's reaction in the United States is that black infant mortality is a bad situation. But look at figure 1.1, on infant mortality by race in the United States since 1915. White infant mortality in 1915 was almost 100 deaths per 1,000 births, and black infant mortality was fully 180 deaths per 1,000 births. Both rates are horrifying. And the rates were even worse in earlier years in some places – up to 300 or 400 deaths per thousand births. Nowadays, white infant mortality is about nine per thousand, and black infant mortality is about 18 per thousand. Of course it is bad that mortality is higher for blacks than for whites. But should we not be mainly impressed by the tremendous improvement for *both races* – rates falling to about ten percent of what they were – with the black rate coming ever closer to the white rate? Is not this extraordinary improvement for the entire population the most important story – and a most happy story? Yet the

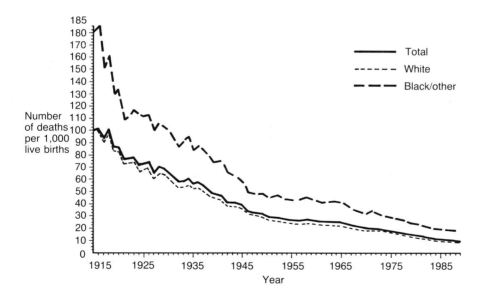

Figure 1.1 Infant mortality rate, total and by race, USA, 1915–1989
Sources: Historical Statistics of the US, Colonial Times to 1970, Series B 136–147; Statistical Abstract of the US, 1982–92 edn, Table 111; Statistical Abstract of the US, 1992 edn, Table 109.

press gives us the impression that we should be mainly distressed about the state of black infant mortality.

Someone said to Voltaire, "Life is hard." Voltaire replied, "Compared to what?" Every evaluation requires that we make some comparison. The comparisons one chooses are decisive in the judgments one makes about whether things are getting better or worse. (Here we have the old joke. Woman 1: "How is your husband?" Woman 2: "Compared to what?" Or in another version, "Compared to whom?")

Proof of the distorting effect of the negative slant in the press is seen in polls where people are asked about their own situations and about the situations of the public at large. If the poll is representative and people's assessment of others is accurate, the averages for assessments of self and of others should be equal. But figure 1.2 shows the typical result: that self-assessment is much more positive than assessment of others' situations; the gap between the two is a measure of the distortion. And the same result is found in poll after poll, of the state of the environment, economic welfare, crowdedness of neighborhood and nation, and so on; people have a more gloomy view of the situation at large than the objective facts warrant, by their own testimony.

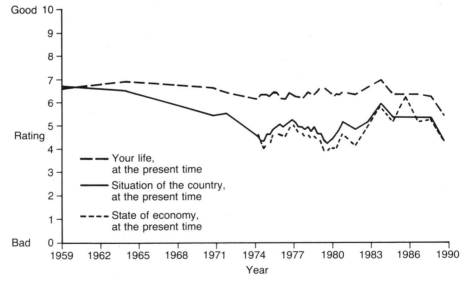

Figure 1.2 Ratings of own life, country, and economy, 1959–89
Note: No survey questions were asked after 1989.
Source: Seymour and Schneider, 1983, pp. 142, 143; data for 1980s courtesy of Cambridge Reports International.

THE PATH OF MATERIAL HUMAN WELFARE

Let us distinguish three types of economic change: (a) Change that is *mainly absolute rather than relative*. An example is a health improvement that benefits everyone worldwide. (b) Change that is *mainly relative*, but where there is also an important overall effect. An example is a productivity improvement in one country due to people working smarter, which allows that country to greatly increase its exports to the benefit of both exporters and importers, but causing problems for some other exporting countries. (c) Change that is *wholly relative*. An example is a change in the price charged by one trading partner to another, or the terms of trade between raw materials and consumer goods, or the dollar–yen exchange rate; in such zero-sum situations there is no on-balance change for bad or good. It is only the third category where one finds bad news, and indeed bad news is inevitable for one party or the other.

This is the central assertion of this book: Almost every absolute change, and the absolute component of almost every economic and social change or trend, points in a positive direction, as long as we view the matter over a reasonably long period of time. That is, all aspects of material human welfare are improving in the aggregate.

Most changes are like the infant mortality situation – some absolute or relative losers, but more mostly absolute gainers. However, many of the things that people worry about are purely relative, like the exchange rate of the dollar, or women's wages relative to men's. And this focus on the relative and the short run often misleads us into thinking that the overall course of things is bad, when in general just about every long-run trend in human welfare points in a positive direction.

For proper understanding of the important aspects of an economy we should look at the long-run movement. But the short-run comparisons – between the sexes, age groups, races, political groups, which are usually purely relative – make more news. And to repeat, just about every important long-run measure of human welfare shows improvement over the decades and centuries, in the United States as well as in the rest of the world. And there is no persuasive reason to believe that these trends will not continue indefinitely.

Would I bet on it? For sure. I'll bet a week's or month's pay (my winnings go to fund research) that just about any trend pertaining to human welfare will improve rather than get worse. First come, first served. But be warned that in economics, unlike the weather, it is easier to forecast the long run than the short run. (More about this in the epilogue, chapter 58.)

Amartya Sen has written that the standard of living is best measured, not by per capita income, but by a collection of measures of human welfare that he calls "functionings." This entire volume may be thought of as an operationalization of that notion. It contains essays on the long-run trends of all the important measures of human welfare for which there are sensible data and about which I could find a competent scholar to write. The volume contains essays on other relevant subjects as well, but long-run measures of human welfare comprise the core of the matter.

As soon as it is seen that the news is good even though the news stories are bad, the question arises: Why do we hear all these bad reports? Explaining why threats of doom are made, and why we choose to read and hear about them and then believe them, is a difficult subject about which there are many speculations but is beyond our scope here.

SOME KEY MATERIAL TRENDS

Let's start with the longest and deepest trends. Surprising though they may be, please be aware that these trends – documented in the chapters of the volume – represent the uncontroversial settled findings of the economists and other experts who work in these fields (except for the case of population growth). What you will read below on that subject was an unusual viewpoint until sometime in the 1980s, at which time the mainstream scientific opinion shifted almost all the way to the position set forth here.

Length of Life

Let's begin with the all-important issue, life itself. The most important and amazing demographic fact – the greatest human achievement in history, in my view – is the decrease in the world's death rate (chapters 2 and 3). The stylized graph in figure 1.3a shows that it took thousands of years to increase life expectancy at birth from just over 20 years to the high 20s. Then in just the past two centuries, the length of life one could

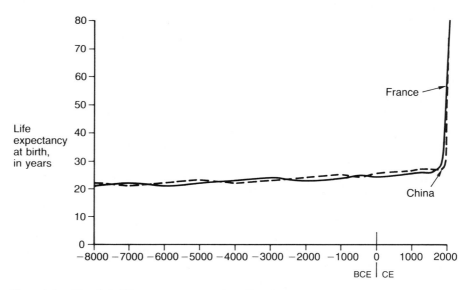

Figure 1.3a   Trends in life expectancy over the millennia (stylized)

expect for a newborn in the advanced countries jumped from less than 30 years to perhaps 75 years (figure 1.3b).

Starting well after World War II, length of life in the poor countries has leaped upward by perhaps 15 or even 20 years since the 1950s, caused by advances in agriculture, sanitation, and medicine. (China has excelled in this respect before developing her economy, which is exceptional.)

The extraordinary decline in child mortality shown in figure 1.4 is an important element in increased life expectancy, for which every parent must give fervent thanks. But contrary to common belief, in the rich countries such as the United States the gains in life expectancy among the *oldest* cohorts have been particularly large in recent years. For example, among US males aged 65–74, mortality fell 26 percent from 1970 to 1988, and among females of that age, mortality fell 29 percent and 21 percent from 1960 and 1970 to 1988, respectively (*US Statistical Abstract*, 1990, p. 75).

The decrease in the death rate is the root cause of there being a much larger world population nowadays than in former times. In the nineteenth century the planet Earth could sustain only one billion people. Ten thousand years ago, only four million could keep themselves alive. Now, more than five billion people are living longer and more healthily than ever before, on average. This increase in the world's population represents humanity's victory against death.

The trends in health are more complex. The decline in mortality is the most important overall indicator of health, of course. And the

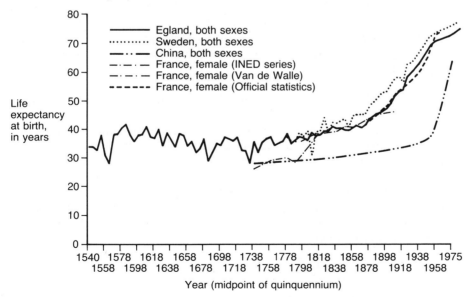

Figure 1.3b Life expectancy, England, Sweden, France, and China, 1541–1985
Sources: Preston in this volume for England and Sweden; Lee, 1979, p. 142 for France.

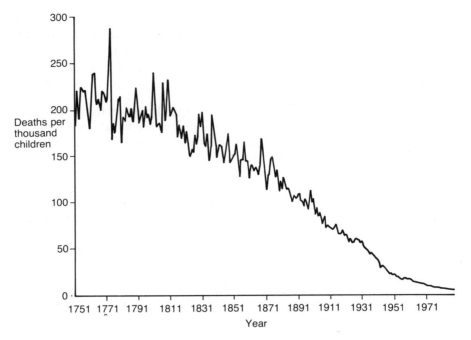

Figure 1.4 Child mortality, 1751 to present
Source: Hill in this volume.

increase in height in the Western countries in the past century is another strong index of health and nutrition (chapter 5). But whether the process of keeping more people alive into older age is accompanied by better or poorer health on average in those older years is in doubt (chapter 6).

Proportion of the Labor Force in Agriculture

The best simple measure of a country's standard of living is the proportion of the labor force devoted to agriculture. When everyone must work at farming, as was the case only two centuries ago, there can be little production of non-agricultural goods. Figure 1.5 shows the astonishing decline over the centuries in the advanced countries in the proportion of the population working in agriculture, now only about one person in fifty. This shift has enabled consumption per person to multiply by a factor of twenty or forty.

Raw Materials

People have since antiquity worried about running out of natural resources – flint, game animals, what-have-you. Yet, amazingly, all the histori-

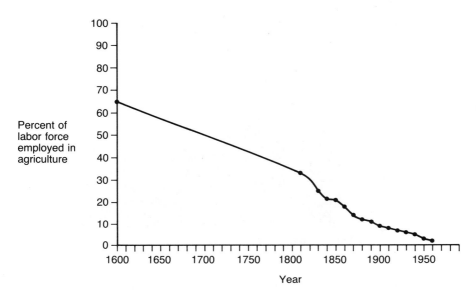

Figure 1.5 Percent of population employed in agriculture, Great Britain, 1600 to the present
Source: Mitchell and Deane.

cal evidence shows that raw materials – all of them – have become less scarce rather than more. Chapter 27 (Hausman) shows beyond any doubt that natural resource scarcity – as measured by the economically meaningful indicator of cost or price – has been decreasing rather than increasing in the long run for all raw materials, with only temporary and local exceptions. Copper gives typical evidence that the trend of falling prices has continued throughout all of history. In Babylonia under Hammurabi – almost 4,000 years ago – the price of copper was about a thousand times its price in the US now, relative to wages. At the time of the Roman Empire the price was a hundred times higher. And there is no reason why this trend should not continue forever.

The trend toward greater availability includes the most counter-intuitive case of all – oil. (See Adelman, chapter 28.) Concerning energy in general, there is no reason to believe that the supply of energy is finite, or that the price of energy will not continue its long-run decrease forever.

Food is an especially important resource. The evidence is particularly strong that the trend in nutrition is benign despite rising population. The long-run price of food is down sharply, even relative to consumer products, due to increased productivity (figures 1.6a–b). And per-person food consumption is up over the last 30 years. The increase of

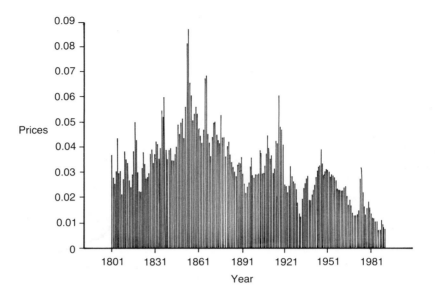

Figure 1.6a   Wheat prices indexed by consumer price index, USA, 1801–1990

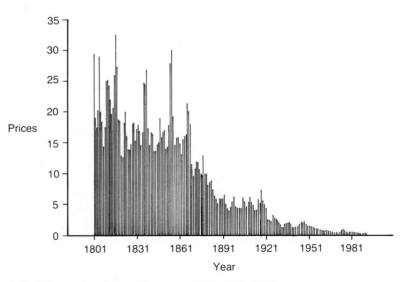

Figure 1.6b   Wheat prices indexed by wages, USA, 1801–1990

height in the West, documented in chapter 5 (Fogel), is another mark of improved nutrition.

(Africa's food production per person is down, but by the 1990s, few any longer claim that Africa's suffering has anything to do with a shortage of land or water or sun. Hunger in Africa clearly stems from civil wars and government interference with agriculture, which periodic droughts have made more murderous.)

Only one important resource has shown a trend of increasing scarcity rather than increasing abundance. It is the most important and valuable resource of all – human beings. Certainly there are more people on earth now than ever before. But if we measure the scarcity of people the same way that we measure the scarcity of other economic goods – by how much we must pay to obtain their services – we see that wages and salaries have been going up all over the world, in poor countries as well as in rich countries. The amount that one must pay to obtain the services of a barber or a professor has risen in India, just as the price of a barber or professor has risen in the United States over the decades. This increase in the price of people's services is a clear indication that people are becoming more scarce even though there are more of us.

THE STANDARD OF LIVING

The pure purchasing-power aspect of the standard of living is difficult to measure. Consider, for example, that before the collapse of communism, the conventional-data estimate of per-capita income in East Germany was 79 percent of that in West Germany, and the "purchasing power parity" estimate was fully 90 percent. It is now clear to all that these computations were misleading. And the clearest evidence comes from data on individual elements of consumption and wealth.

Chapters 13, 14 and 24 (by Burnett and Mokyr; by Lebergott; and by Rector) show unmistakably how the standard of living has increased in the world and in the United States through the recent centuries and decades, right up through the 1980s.

Aggregate data always bring forth the question: But are not the gains mainly by the rich classes, and at the expense of the poor? Chapter 18 (by Lindert and Williamson) shows that for a portion of US history, income distribution did widen (though this is hardly proof that the rich were exploiting the poor). But there has been little or no such tendency during, say, the twentieth century. And a widening gap does not negate the fact of a rising absolute standard of living for the poor. Nor is there evidence in chapter 23 (by Blank) that an increasing proportion of the population lives below some fixed absolute poverty line. And chapter 24 (by Rector) shows extraordinary gains by the US poor in consumption during this century, as well as a high standard of living by any historical and cross-national standards.

A related question concerns possible exploitation by the rich countries that might cause misery for the poor countries. But the distribution of the most important element of "real wealth" – life expectancy – has narrowed between rich and poor countries (as well as between the rich and poor segments of populations within countries) over previous decades – to wit, the extraordinary reduction in the gap between the mortality of China and the rich countries since World War II. The reduction in the gap between literacy rates and other measures of amount of education in rich and poor countries corroborates this convergence. Figures in chapter 13 (by Mokyr), showing convergence in economic productivity in the rich countries along with general growth, dovetail with the other measures of income distribution. Data on the *absolute* gap between yearly incomes of the rich and poor countries are beside the point; widening is inevitable if all get rich at the same proportional rate, and the absolute gap can increase even if the poor improve their incomes at a faster proportional rate than the rich. Here one should notice that increased life expectancy among the poor relative to the rich reduces the gap in lifetime income, which is a more meaningful measure than yearly income.

It is important that the convergence among nations be properly interpreted as a spreading of a better standard of living to the entire world, rather than as a leveling down of the rich.

Cleanliness of the Environment

Ask an average roomful of people if our environment is becoming dirtier or cleaner, and most will say "dirtier." Yet the air in the US and in other rich countries is irrefutably safer to breathe now than in decades past; the quantities of pollutants – especially particulates, which are the main threat to health – have been declining. And water quality has improved; the proportion of monitoring sites in the US with water of good drinkability has increased since the data began in 1961. More generally, the environment is increasingly healthy, with every prospect that this trend will continue. (See figures 1.7a–d)

When considering the state of the environment, we should think first of the terrible pollutions that were banished in the past century or so – the typhoid that polluted such rivers as the Hudson, smallpox that humanity has finally pursued to the ends of the earth and just about eradicated, the dysentery that distressed and killed people all over the world as it still does in India, the plagues and other epidemics that trouble us much less than in generations past, or not at all. Not only are we in the rich countries free of malaria (largely due to our intensive occupation of the land), but even the mosquitoes, that do no more than cause itches with their bites, are so absent from many urban areas that people no longer need screens for their homes and can have garden parties at dusk. It is a mark of our extraordinary success that these are no longer even thought of as pollutions.

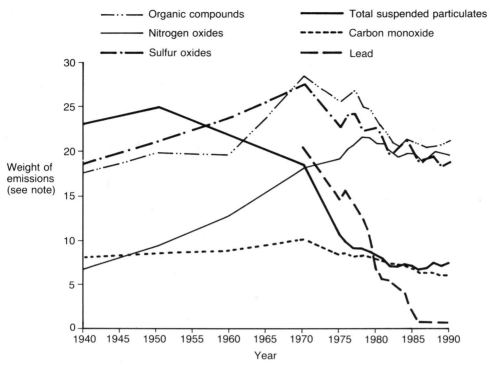

Figure 1.7a Emissions of major air pollutants in the USA, 1940–90
Note: In millions of metric tons per year, except lead in ten thousands of metric tons per year, and carbon monoxide in 10 million metric tons per year.
Source: Council on Environmental Quality, *Environmental Quality, 22nd Annual Report*, 1992, 273.

The root cause of these victorious campaigns against the harshest pollutions was the nexus of increased technical capacity and increased affluence – wealth being the capacity to deal effectively with one's surroundings (see Wildavsky, 1988).

Species Extinction

Fear is rampant about rapid rates of species extinction. The fear has little or no basis. The highest rate of observed extinctions is one species per year, in contrast to the 40,000 per year some ecologists have been forecasting for the year 2000. Species matter, and deserve thought. But the facts should matter, too, in deciding whether to spend tens of billions for research, "debt for nature" swaps, and other expensive programs. Furthermore, the new possibilities for genetic engineering, and for storage of seeds, reduce the dangers of extinctions that do occur. Chapter 34 (by Wildavsky and me) discusses how an issue of legitimate interest and concern has turned into an environmentalist scam.

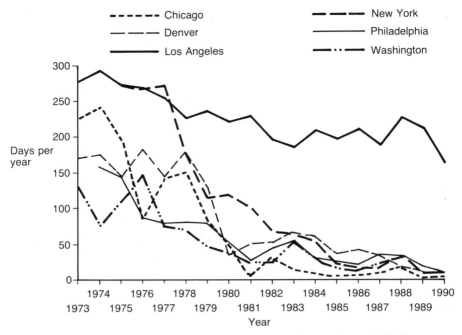

Figure 1.7b Air quality trends in major urban areas, USA (number of PSI days greater than 100)

Sources: Council on Environmental Quality, *Environmental Quality, 22nd Annual Report*, 1992, 277; id., *Environmental Quality, 12th Annual Report*, 1981, 244.

Population Growth

The predecessor volume did not discuss population growth because the topic was so controversial that we feared it would distract readers from the other topics. This volume follows the same policy, even though there have been major changes in the intellectual status of the topic. Only two comments will be made:

1. A score of competent statistical studies, starting in 1967 with an analysis by Nobel prizewinner Simon Kuznets, agree that there is no negative statistical relationship between economic growth and population growth for periods up to a century. And there is strong reason to believe that more people have a positive effect in the even longer run. That is, population growth does not lower the standard of living; it raises it in the long run.

2. There was a major turnaround in the 1980s in population economics. After decades of "Everyone knows . . ." that population growth hampers economic development, there has been a revolution in scientific thought on the matter. The consensus now is close to the position in the paragraph above, though it runs against intuitive "common

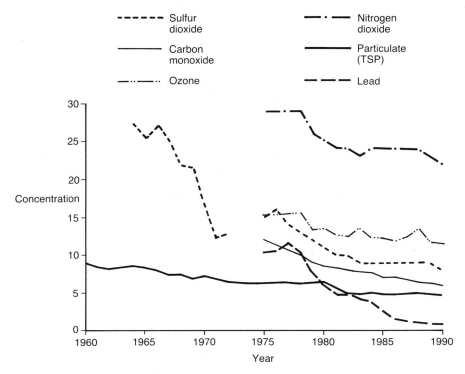

Figure 1.7c Pollutants in the air, USA, 1960–90
Sources: Council on Environmental Quality, *Environmental Quality, 12th Annual Report*, 1981, 243; Sulfur 1964 through 1972: EPA (1973): 32 stations.

sense." This turnaround was made "official" in 1986 by a report of the National Academy of Sciences, but the public and the media have not yet gotten the word on this.

The Greenhouse Effect, the Ozone Layer, and Acid Rain

What about the greenhouse effect? The ozone layer? Acid rain? The one certainty is that on all of these issues there is major scientific controversy and lack of consensus about what has happened until now, why it happened, and what might happen in the future. All of these scares are recent, and there has not yet been time for proper research to be done and for the intellectual dust to settle. There may be hard problems here, or there may not, as chapters 47–50 (Kulp, Singer, Michaels and Lave) discuss.

An important aspect of these atmospheric issues is that no threatening trend in *human welfare* has been connected to changes in these phenomena. There has been no increase in skin cancers from ozone, no

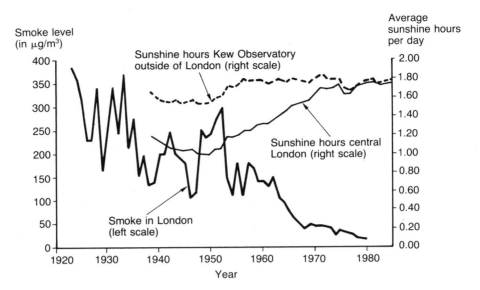

Figure 1.7d Smoke level and mean hours of winter sunshine in London, 1923–84
Sources: P. Brimblecombe and H. Rohde, "Air Pollution: Historical Trends," *Durability of Building Materials*, 5 (1988), 291–308; Derek Elsom, this volume.

damage to agriculture from a greenhouse effect, and slight or no damage to lakes from acid rain. It may even be that a greenhouse effect would benefit us on balance by warming some areas we'd like warmer and by increasing the carbon dioxide stimulus to agriculture.

Perhaps the most important aspect of these as well as the yet-unknown atmospheric scares that will surely be brought before the public is that we now have ever-increasing capacities to reverse such trends if necessary. And we can do so at costs that are manageable rather than being an insuperable constraint upon growth or an ultimate limit upon the increase of productive output or of population. So we can look these issues squarely in the eye and move on.

In summary of this section, I'll dramatize these trends toward a better life – which can be seen in most of our own families if we look – with an anecdote. I have mild asthma. Sometime ago I slept in a home where there was a dog, and in the middle of the night I woke with a bad cough and shortness of breath. When I realized that dog dander was the cause, I found my 12 dollar pocket inhaler, good for 3,000 puffs, and took one puff. Within ten minutes my lungs were clear. A small miracle. Forty years ago I would have been sleepless and miserable all night, and I would have had to give up the squash-playing that I love so much, because exercise causes my worst asthma in the absence of an inhaler. Or consider diabetes. If your child had diabetes a hundred years ago, you had to watch helplessly as the child went blind and died early. Now injections, or even pills, can give the child as long and healthy a life as other children. Or glasses. Centuries ago you had to give up reading when your eyes dimmed as you got to be 40 or 50. Now you can buy magnifying eyeglasses at the drugstore for ten dollars. And you can even wear contact lenses for eye problems and keep your vanity intact. Is there not some condition in your family that in earlier times would have been a lingering misery or a tragedy, that nowadays our increasing knowledge has rendered easily bearable? Does it not bear out the abstract trends presented above?

OTHER ASPECTS OF HUMAN WELFARE

People alive now are living in the midst of what may be seen as the most extraordinary three or four centuries in human history – the period before us and probably the period after us as well. The Industrial Revolution and the gross material aspects of life are only a tiny part of the change.

Some advances need only to be mentioned to be assented to. Consider the amount of physically-caused pain that people suffered in their lives without any help from scientific medicine. Childbirth was a nightmare for every woman; now anesthetics allow a woman to choose a level of pain that is tolerable to her. People whose limbs were injured in peace or war often had to have them amputated, with only liquor as painkiller. Those who lived long enough to die of cancer had no way to dull their agony. Nowadays, it is miraculously different.

So far only material aspects of life have been mentioned. Now let us consider some of the important non-material trends.

Education and Opportunity

Consider the astounding increase since World War II in the amount of education that the youth of the world are acquiring (chapter 21, by Boggs and me). This trend implies a vast increase in young people's opportunities to use their talents for their own and their families' benefits, and hence to the benefit of others in society as well. In my view, this trend is one of the most important, and one of the most happy, of all the trends experienced by the human enterprise. Already one can see results in the names on professors' doors in departments of computer science and chemistry, for example, in universities all over the United States, and in the Nobel Prize awards – Asian and African names that would not have been seen a decade or two ago. Less and less often will people of genius and strong character live out their lives in isolated villages unable to contribute to civilization at large.

The process of providing educational opportunity will not be complete for decades, at least. One can still see children sharing rickety desks and scarce books in a Colombian fishing village – just two miles from a busy modern international airport. But even this is vastly better than the way it was just a few years ago – no school at all. And we can be confident that a century from now, scenes like that will be quite scarce.

The spread of reading material is manifest – all in the period of time since Gutenberg, so short relative to the history of humanity that it seems only an eyeblink. The recent spread may be measured by the increased amount of newsprint used in the world from decade to decade, and in the prices of advertising – data that this volume regrettably could not collect. And though access to libraries is not yet universal, with the aid of computers and electronic communications, that day is not far away. To dramatize this progress, reflect on the childhood only two centuries ago of Abraham Lincoln, who could not afford paper and pencil. Among a world's population five times as large as it was then, there are almost no children so stricken with educational poverty.

Work, Leisure, and Boredom

Leisure is surely greater now than since our days as hunters and gatherers when people worked perhaps 10 or 20 hours a week. The marked increase in leisure time in modern countries is documented in chapter 22 (Robinson); compare this to the 80-hour work weeks in industrial jobs a century or two ago, backbreaking labor in mines and factories from before dawn to after dusk.

There is some evidence that work time has increased for some groups in the past decade or two. But for many people, work is a good in itself in the sense that its goal is not the production of additional income, which confuses the issue quantitatively; the opportunity to do this additional rewarding work is a great good in itself, rather than a bad thing (as is the case of the work that goes into this volume, for me and surely for many of the other authors).

Indeed, one of the greatest boons to humanity in recent decades is the diminution in the boring factory tasks that robotized and dehumanized workers. No one who has not experienced eight-hour shifts of such mind-numbing jobs as lifting cases of beer onto or off a conveyor belt, or stacking double, triple, or quadruple sets of empty cans eleven-high in a truck that requires petty arithmetic so that the worker cannot even daydream, can understand how bad the mind-torture is. (Postal workers who sort mail and move bags of it are some of the last of these sufferers in a modern society. And of course workers in poor countries still suffer this because their muscle power is cheaper than machines.)

Though I have no data to document the observation, we apparently do not suffer another of the old painful ills – leisure boredom with nothing to do – that people experienced in earlier times. Indeed, even people who work less than 40 hours a week complain that they have too many things that they want to do.

Boredom also is dispelled by electronic entertainment. This is an extraordinary gift to the old and shut-in; anyone who has been in a hospital bed suffering too much pain even to be able to read a newspaper knows the value of taped music and books, and even junk television. (I'll forever be grateful for the surcease that the two National Football League playoff games brought to me on my first postoperative day.)

Mobility and Travel

Mobility and travel speed constitute another aspect of life that shows extraordinary improvement (chapter 15, by Atack). Only two centuries ago, horse and sailing ship were the fastest modes of movement. Now there are bullet trains and supersonic flight. The 1992 price of much faster space travel is $6 million per trip, but that price is sure to fall rapidly.

These changes, along with the breathtaking gains in communications speed, are society-altering aids to the spread of human opportunity and liberty all over the world. They also ease the curse most of humanity has suffered – the boredom in isolated villages. Music was a rare pleasure only a century ago. Now there are few people on earth too poor to purchase music around the clock on a transistor radio.

QUALIFICATIONS TO THE ARGUMENT

I am not saying that all is well everywhere, and I do not predict that all will be rosy in the future. Children are hungry and sick; people live out lives of physical or intellectual poverty, and lack of opportunity; irrational war (not even for economic gain) or some new pollution may finish us off. What this book does show is that for most relevant economic matters, the aggregate trends are improving rather than deteriorating.

Please note, too, that a better future does not happen "automatically" and without effort. It will happen because women and men will struggle with problems with their muscles and minds, and will probably overcome most of them – as people always have eventually overcome economic problems in the past – if the social and economic system gives them opportunity to do so.

This volume deals with material and economic welfare, but not with emotional or sexual or spiritual welfare. Whether people are happier than in earlier times as a result of material progress is outside our scope. (To the extent that we can measure happiness, it does increase with the standard of living; see Simon, 1974; Easterlin, 1974.) Many of those non-economic matters are, however, connected to economic welfare, as political freedom is inevitably intertwined with an advanced economy; still, they are not the subject here.

Some supposedly worsening non-economic trends are negative only by one interpretation. Two related examples are: (a) decline in the proportion of the electorate who vote, and (b) decline in newspaper readership. Not surprisingly, print journalists consider both of these trends to be baneful signs of community and individual degeneracy. The interpretation that I favor, however, is that these are signs of an absence of really bad news. When there is no war or natural disaster, there is lots of space for political news. And if the political issues have relatively small bearing upon our personal welfare, they lose our attention to other entertainment, and we devote to other purposes the time and energy required to vote. (It is the church or the academic department that is in crisis that attracts all its members to the monthly meeting.) That is, these signs that the journalists fuss about I take to be the signs of a well-off and unthreatened public – to be greeted with smiles rather than with lament. But this is speculation beyond the scope of this volume.

And we now have a problem to which a solution is not visible: the lack of problems that will well challenge women and men. Anecdotal accounts suggest that in the 1950s, when it had the problem of rebuilding from the war and constructing a wholly modern society, Holland was a happier and more vibrant place to live in than now, when the problem people are most vocal about is the disposal of tens of millions of pig excrement each year.

THE DOOMSAYERS' FORECASTS

Every forecast of the doomsayers has turned out to be wholly wrong. Metals, foods, and other natural resources have become more available rather than more scarce throughout the centuries. The famous *Famine 1975* forecast by the Paddock brothers – that we would see millions of famine deaths in the US on television – was followed instead by gluts in agricultural markets. Paul Ehrlich's primal scream about "What will we do when the [gasoline] pumps run dry?" was followed by gasoline cheaper than since the 1930s. The Great Lakes are not dead; instead they offer better sport fishing than ever. The main pollutants, especially the particulates that have killed people for years, have lessened in our cities.

The false forecasts have not been harmless, however. They have caused economic disasters for (just a few examples) mining companies and for the poor countries which depend upon natural resources, by misleading them with unsound expectations of increased prices for metals; wasted expenditures for aircraft manufacturers and airlines who planned on high prices for gasolines; and chaos in the auto market for auto makers and for consumers, as a result of laws about fuel economy premised upon a growing scarcity. But neither their personal embarrassment nor their public damage has reduced the doomsayers' credibility with the press, or their command over the funding resources of the federal government.

Not much more than one century ago – after more than 50 centuries of recorded history and hundreds of centuries of unrecorded history – for the first time people had something better than a firelight or oil lamp to break the darkness after dusk. And the absence of electricity continued almost into the second half of the twentieth century for substantial portions of the population of the richest country in the world. Now all of us Americans take Edison's gift for granted.

Less than two centuries ago there appeared the first land transportation that did not depend entirely on animal muscles. Now we complain about the ubiquity of cars. In the last two centuries the first anesthetics other than alcohol became available to provide surcease from the hellish pains of hospital and dental operations, and of diseases such as cancer, as well as to give those women in childbirth who want it another option besides deep breathing. How can one *not* recognize that this is truly a new age of the liberation of humanity from the bonds in which nature has kept us shackled throughout all of our history? And what liberation in the future could possibly match this one?

CAN ALL THIS GOOD NEWS BE TRUE?

Some of the chapters reach documented conclusions that are unbelievable to most journalists and the public. For example, a Sunday *Washington Post* headline (November 10, 1991) reads "A House of Their Own is Out of Most Renters' Reach," and the article quotes young people saying that it is harder now to buy homes than it was for their parents. Richard Muth's chapter 20 shows that the contrary is true; the same quantity and quality house is much more easily affordable now for the representative young couple than it was for earlier generations. How many will believe that?

Hearers of the messages in this book often ask, "But what about the other side's data?" There are no other data. Test for yourself the assertion that the physical conditions of humanity have gotten better. Pick up the US Census Bureau's *Statistical Abstract of the United States* and *Historical Statistics of the United States* at the nearest library and consult the data on the measures of human welfare that depend upon physical resources, for the US or for the world as a whole. See the index for such topics as pollution, life expectancy, and the price indexes, plus the prices of the individual natural resources. While you're at it, check the amount of space per person in our homes, and the presence of such amenities as inside toilets and telephones. You will find "official" data showing that just about every single measure of the quality of life shows improvement rather than the deterioration that the doomsayers claim has occurred.

The long-run trends are mostly presented here in graphs rather than tables because graphs enable one to see trends especially well, and also because the volume reaches out beyond scholars in particular fields. But for scholars who wish to pursue these topics further, the underlying tables will (with luck) be published both in print and in computer files not long after this volume is out. And it is hoped that this collection of data will later be expanded to include material and subjects not included here, and that it will be updated regularly, to become an encyclopedia of how the human enterprise has fared throughout recent millennia. Inquiries from scholars who might participate in this project are warmly invited.

WHAT IS THE MECHANISM THAT PRODUCES PROGRESS RATHER THAN INCREASING MISERY?

How can it be that economic welfare grows in time along with population, rather than humanity being reduced to misery and poverty as population grows and we use more and more resources? We need some theory to explain this controversion of common sense.

The Malthusian theory of increasing scarcity, based on supposedly fixed resources, which is the theory that the doomsayers rely upon, runs exactly contrary to the evidence from the long sweep of history. This is because Malthusianism and contemporary doomsters omit from their accounts the positive long-run effects of the problems induced by additional people and economic activity. Maybe neighborhood kids running over your lawn do not benefit you by causing improvements in lawn care technology. But more people putting pollutants into the air eventually lead to agitation and the search for ways to prevent and clean up pollutions, a process that eventuates in our having a cleaner environment than before the pollution began to be bad. It is this crucial adjustment mechanism – the reason England's air and water are cleaner than they have been for centuries – that is too often left out of thinking on these matters.

More generally, the process operates as follows: More people and increased income cause problems in the short run – shortages and pollutions. Short-run scarcity raises prices and pollution causes outcries. These problems present opportunity and prompt the search for solutions. In a free society, solutions are eventually found, though many people seek and fail to find solutions at cost to themselves. In the long run the new developments leave us better off than if the problems had not arisen. This theory fits the facts of history.

Technology exists now to produce in virtually inexhaustible quantities just about all the products made by nature – foodstuffs, oil, even pearls and diamonds – and make them cheaper in most cases than the cost of gathering them in the wild natural state. And the standard of living of commoners is higher today than that of royalty only two centuries ago – especially their health and life expectancy, and their mobility to all parts of the world.

Consider this prototypical example of the process by which people wind up with increasing availability rather than decreasing availability of resources. England was full of alarm in the 1600s about an impending shortage of energy due to the deforestation of the country. People feared a scarcity of wood for both heating and the iron industry. This impending scarcity led to the development of coal.

Then in the mid-1800s the English came to worry about an impending coal crisis. The great English economist, Stanley Jevons, calculated that a shortage of coal would bring England's industry to a standstill by 1900; he carefully assessed that oil could never make a decisive difference. Triggered by the impending scarcity of coal (and of whale oil, whose story comes next), ingenious profit-minded people developed oil into a more desirable fuel than coal ever was. And in 1992 we find England exporting both coal and oil.

Another element in the story: Because of increased demand due to population growth and increased income, the price of whale oil for lamps jumped in the 1840s, and the US Civil War pushed it even higher, leading to a whale oil "crisis." This provided incentive for enterprising people to discover and produce substitutes. First came oil from rapeseed, olives and linseed, and camphene oil from pine trees. Then inventors learned how to get coal oil from coal. Other ingenious persons produced kerosene from the rock oil that seeped to the surface, a product so desirable that its price then rose from $0.75 a gallon to $2.00. This high price stimulated enterprisers to focus on the supply of oil, and finally Edwin L. Drake brought in his famous well in Titusville, Pennsylvania. Learning how to refine the oil took a while. But in a few years there were hundreds of small refiners in the US, and soon the bottom fell out of the whale oil market, the price falling from $2.50 or more at its peak around 1866 to well below a dollar.

Here we should note that it was not the English or American government that developed coal or oil, because governments are not effective developers of new technology. Rather, it was individual entrepreneurs and inventors who sensed the need, saw opportunity, used all kinds of available information and ideas, made lots of false starts that were very costly to many of those individuals but not to others, and eventually arrived at coal and oil as viable fuels – because there were enough independent individuals investigating the matter for at least some of them to arrive at sound ideas and methods. This happened in the context of a competitive enterprise system that worked to produce what was needed by the public. And the entire process of impending shortage and new solution left us better off than if the shortage problem had never arisen.

The extent to which the political-social-economic system provides personal freedom from government coercion is a crucial element in the economics of resources and population. Skilled persons require an appropriate social and economic framework that provides incentives for working hard and taking risks, enabling their talents to flower and come to fruition. The key elements of such a framework are economic liberty, respect for property, and fair and sensible rules of the market that are enforced equally for all.

What Happens if We Should Not Continue to Make New Discoveries?

We have in our hands now – actually, in our libraries – the technology to feed, clothe, and supply energy to an ever-growing population for the next seven billion years. Most amazing is that most of this specific body of knowledge was developed within just the past two centuries or so, though it rests on basic knowledge that had accumulated for millennia, of course.

Indeed, the last necessary additions to this body of technology – nuclear fission and space travel – occurred decades ago. Even if no new knowledge were ever invented after those advances, we would be able to go on increasing our population forever, while improving our standard of living and our control over our environment. The discovery of genetic manipulation certainly enhances our powers greatly, but even without it we could have continued our progress forever.

CONCLUSION

The decrease in the death rate, and the attendant increase in life expectancy – more than doubling – during the last two centuries in the richer countries, and in the twentieth century in the poorer countries, is the most stupendous feat in human history. The decline in mortality is the cause of the rapid increases in human population during these periods. This triumph against death would seem the occasion for great rejoicing. Instead we find gloom. One reason for this peculiar outcome is focusing on short-run inter-group comparisons rather than on long-run changes for the human group as a whole.

In the short run, all resources are limited. An example of such a finite resource is the amount of time you will devote to reading this introduction. The longer run, however, is a different story. The standard of living has risen along with the size of the world's population since the beginning of recorded time. There is no convincing economic reason why these trends toward a better life should not continue indefinitely.

The key theoretical idea again: The growth of population and of income create actual and expected shortages, and hence lead to price run-ups. A price increase represents an opportunity that attracts profit-minded entrepreneurs to seek new ways to satisfy the shortages. Some fail, at cost to themselves. A few succeed, and the final result is that we end up better off than if the original shortages had never arisen. That is, we need our problems, though this does not imply that we should purposely create additional problems for ourselves.

Progress toward a more abundant material life does not come like manna from heaven, however. My message certainly is not one of complacency. In this I agree with the doomsayers: our world needs the best efforts of all humanity to improve our lot. I part company with the doomsters in that they expect us to come to a bad end despite the efforts we make, whereas I expect a continuation of humanity's history of successful efforts. And I believe that their message is self-fulfilling, because if you expect your efforts to fail because of inexorable natural limits, then you are likely to feel resigned, and therefore to literally resign. But if you recognize the possibility – in fact the probability – of success, you can tap large reservoirs of energy and enthusiasm.

Adding more people causes problems. But people are also the means to solve these problems. The main fuel to speed the world's progress is our stock of knowledge; the brakes are our lack of imagination and unsound social regulations of these activities. The ultimate resource is people – especially skilled, spirited, and hopeful young people endowed with liberty – who will exert their wills and imaginations for their own benefit, and so inevitably they will benefit the rest of us as well.

RECOMMENDED READING

Other very useful references for more material on the subject of this book include Lebergott (1984), from which several of the chapters in this volume draw, Clark (1957 and 1967), and Clark and Haswell (1967).

REFERENCES

Clark, Colin (1957): *Conditions of economic progress.* 3d edn. New York: Macmillan.
—— (1967): *Population growth and land use.* New York: St Martin's.
—— and Margaret Haswell (1967): *The economics of subsistence agriculture.* New York: St Martin's.
Easterlin, Richard A. (1974): "Does Economic Growth Improve the Human Lot? Some Empirical Evidence," in P. A. David and W. R. Melvin (eds), *Nations and Households in Economic Growth.* Palo Alto: Stanford University Press.

Lebergott, Stanley (1984): *The Americans – An Economic Record*. New York: W. W. Norton & Company.

Simon, Julian L. (1974): "Interpersonal Comparisons Can Be Made – and Used for Redistribution Decisions," *Kyklos*, vol. XXVII, No. 1, pp. 63–98.

Wildavsky, Aaron (1988): *Searching for Safety*. New Brunswick: Transaction Press.

PART I

Life, Death, and Health

The most important indicators of the standard of living are the length of life, as discussed in Part I of this volume, and the proportion of the population that work in agriculture, as discussed by Richard Sullivan in Part II.

2

Human Mortality Throughout History and Prehistory

Samuel H. Preston

Accurate data for estimating life expectancy at birth in sizable populations do not become available until the sixteenth century. For earlier times, the analyst must rely on sources of questionable quality or representativeness: skeletal remains, burial inscriptions, and, after 1300 or so, records of unusual groups such as the European aristocracy or members of religious orders.

Most of these records suggest that life expectancy from prehistoric times until 1400 or so was in the range of 20–30 years. A detailed and comprehensive account of life expectancy estimates through classical antiquity is found in Acsadi and Nemeskeri (1970). The most satisfactory collection of skeletal remains is drawn from the Maghreb peninsula (North Africa, between Egypt and the Atlantic) during the Neolithic period. This population evidently had a life expectancy at birth of about 21 years. Its age pattern of mortality was remarkably similar to that of modern populations at similar levels of mortality (Acsadi and Nemeskeri, 1970, 173).

Burial inscriptions, mummies, and skeletons drawn from the Roman empire suggest that life expectancy was in the twenties for most of the geographic and occupational subgroups falling under its sovereignty. Few of these sources provided an adequate representation of infant deaths, so that some extrapolation to this age is required based on typical age patterns of mortality observed more recently.

Confidence in the range of 20–30 for life expectancy in the era before 1600 is enhanced by the use of demographic models. Since the world's population was growing very slowly during this period, life expectancy at birth was, to a very close approximation, the reciprocal of the birth rate. Given the age pattern of fecundity and the apparent absence of significant anti-natal practices, the birth rate was quite unlikely to have fallen outside the range of 0.033–0.050 births per capita per year, implying life expectancies in the range of 20–30 years.

Mortality rates among members of religious orders provide a very useful bridge between antiquity and the advent of modern death statistics.

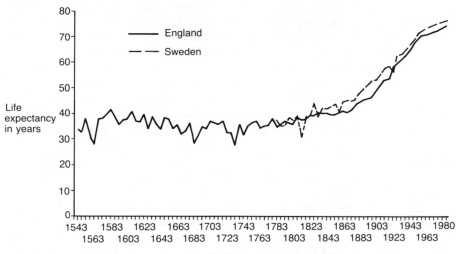

Figure 2.1 Life expectancy, England and Sweden, 1541–1985
Sources: For England and Wales: 1741–1875: Wrigley and Schofield (1981: tables 7.15);
1876–1970: Case et al. (1970); 1970–85 (individual years): Keyfitz and Flieger (1990). For
Sweden: 1778–1962: Keyfitz and Flieger (1968); for 1965–85: Keyfitz and Flieger 1990).

In an unusually well-documented study, Hatcher (1986) shows that
mortality among Benedictine monks in Canterbury, England from
1395–1505 corresponded to a life expectancy at birth of 22 years. He
notes that nutrition, clothing, sanitation, and shelter were much better
for this group than for the population as a whole, although its denser
living conditions were undoubtedly a negative factor in its survivorship.

The earliest satisfactory series achieving national coverage is based on
demographic reconstructions by a group of scholars at Cambridge.
Using a large sample of parish registers and adjusting their data to
achieve national representativeness, Wrigley and Schofield (1981) provide estimates of life expectancy at birth in England from 1541 to 1875.
Their quinquennial series of life expectancies is plotted in figure 2.1. It
is clear that, by the middle of the sixteenth century, life expectancy was
typically in the mid-thirties, with substantial fluctuation from period to
period. Some mild secular deterioration, possibly associated with increasing population density and repeated visitations of the plague, is
evident for a century and a half, followed by a slow advance. At the dawn
of the nineteenth century, life expectancy was about 37 years, much the
same level as it had been two centuries earlier. Life expectancy in France
in 1800 was about 30 years, a disparity that Fogel (1989) attributes to
poorer nutritional standards in France than in England.

The series of English life expectancies in figure 2.1 is completed by
using life tables computed from national vital statistics and censuses for
England and Wales. It is clear that a steady advance begins just after the

turn of the nineteenth century, and accelerates after about 1871–5. After the 1860s, there is no instance in which life expectancy declines from one period to the next. Not only were average conditions improving rapidly, but there was also less slippage from the gains that had been secured.

The first nation to produce reliable measures of mortality based on complete national counts of deaths and population is Sweden. A quinquennial series of Swedish life expectancies at birth beginning in 1778–82 is also plotted in figure 2.1. While its series begins at a level similar to that of England, Sweden gains an advantage in the course of the nineteenth century, probably because of much less rapid urbanization. (Urban areas exposed people more frequently to infectious diseases through direct personal contact and indirectly through contamination of water and food supplies.) As the urban health disadvantage is removed through public works during the twentieth century, the series for the two countries converge. As in England, life expectancy advances accelerate in the 1870s; the only instance of backsliding reflects the influenza epidemic after World War I. It has been argued that the acceleration in rates of mortality decline after the 1870s reflects primarily the implementation of personal and public health practices that took advantage of much clearer understandings of the nature of infectious diseases (Preston and Haines, 1991; Ewbank and Preston, 1990).

Thus far in the twentieth century, life expectancy in Sweden has increased by 24 years and in England and Wales by 27 years. These gains are typical for Western European countries and areas of overseas European settlement. Increases were larger in Southern and Eastern Europe, which began the century at lower levels. Italian life expectancy has increased by 32 years, from 43.0 to 74.7, and Czechoslovakian life expectancy by 31 years, from 40.3 to 71.1 (Preston and Haines, 1991, table 2.3; Keyfitz and Flieger, 1990). If we set prehistoric life expectancy at a midrange value of 25 years, it is clear that about half of the progress in European populations since prehistoric times has occurred during the short span of the twentieth century. The only notable setbacks have been the influenza epidemic after World War I and mild reversals in male mortality in Eastern Europe during the past two decades associated with alcohol consumption and other factors (Eberstadt, 1989; see also chapter 7 by Feshbach in this volume).

The United States completed its death registration system in 1933, the last industrialized country to do so. Nevertheless, it is evident from partial statistics that the course of mortality during this century was quite similar to that of England. In an innovative and convincing analysis, Lee and Carter (1992) show that the pace of decline in American age-specific death rates during the century has been virtually constant right through the 1980s. Projections by the US Census Bureau and Social Security Administration have repeatedly been too conservative about future gains in life expectancy, and it appears that these errors are being repeated in recent forecasts (Preston, 1993).

DEVELOPING COUNTRIES

Mortality improvements in developing countries during the twentieth century have been even more dramatic than in industrialized countries. The data base is less secure but there is little doubt that turn-of-the-century life expectancy for the aggregate of developing countries was less than 30 years, i.e., in the range that appears applicable to prehistoric populations. In China around 1930, a valuable demographic survey suggests, life expectancy was around 24 years (Barclay, et al., 1976). Intercensal analysis in India indicates a life expectancy of 24–25 years during 1901–11 (Bhat, 1987). Life tables for Taiwan in 1920 give a life expectancy of 27.9 years and for Chile in 1909 – one of the most advanced countries of the developing world – a life expectancy of 30.6 years (Preston, et al., 1972).

According to United Nations' (1991) estimates for 1985–90, China has a life expectancy of 69.4 years, India of 57.9, and Chile of 71.5. Taiwan's life expectancy, not available in UN sources, was 73.6 in 1985 (Keyfitz and Flieger, 1990). Thus, the mean increase from a level recorded earlier in this century for these four countries is about 42 years. Life expectancy more than doubled during the century for each of them. Since life expectancy for developing countries as a whole is estimated by the United Nations to be 61.4 years in 1985–90, it is apparent that this doubling also pertains to the aggregate of developing countries. Figure 2.2 presents a regional breakdown of the UN's life expectancy estimates. Even the poorest region, Africa, has a life expectancy (52.0

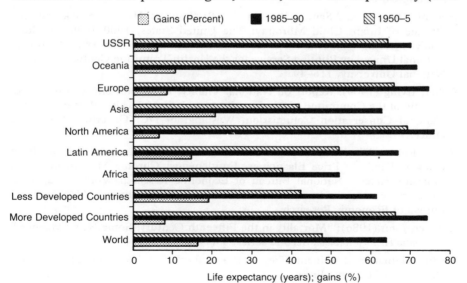

Figure 2.2 Life expectancy around the world, 1950–5, 1985–90, and gains
Source: United Nations (1991, 28).

years) that would have been the envy of Europe at the turn of the century.

In 1909, impressed by recent discoveries in bacteriology and their application to health practices, economist Irving Fisher declared that "the crowning achievement of science in the present century should be, and probably will be, the discovery of practical methods of making life healthier, longer, and happier than before" (p. 64). As the century closes, there is no doubt about the accuracy of Fisher's prophesy.

REFERENCES

Acsadi, George, and J. Nemeskeri (1970): *History of Human Life Span and Mortality*, Budapest: Akademiai Kiado.

Barclay, George, Ansley Coale, Michael Stoto, and James Trussell (1976): "A Reassessment of the Demography of Traditional Rural China." *Population Index*, 42(4), 606–35.

Bhat, Mari (1987): *Mortality in India: Levels, Trends, and Patterns*. Dissertation in Demography, University of Pennsylvania. Ann Arbor, Michigan. University Microfilms International.

Case, R. A. M., Christine Coghill, Joyce Harley, and Joan Pearson (1970): *The Chester Beatty Research Institute Serial Abridged Life Tables. England and Wales 1841–1960*. Supplemented sec. edn. London Chester Beatty Research Institute.

Eberstadt, Nicholas (1989): "Health and Mortality in Eastern Europe, 1965 to 1985." Joint Economic Committee, 101st Congress, First Session. Vol 1. *Pressures for Reform in the East European Economy*, Washington, DC: Government Printing Office.

Ewbank, Douglas, and Samuel Preston (1990): "Personal Health Behavior and the Decline of Infant Child Mortality: The United States, 1900–1930." In John Caldwell, (ed.) *What We Know About Health Transition*, Canberra. Australian National University Printing Service for the Health Transition Centre, Australian National University, 116–49.

Fisher, Irving (1909): Report on National Vitality, Its Wastes and Conservation. Bulletin of the Committee of One Hundred on National Health. Prepared for the National Conservation Commission, Washington, DC: Government Printing Office.

Fogel, Robert W. (1989): "Second Thoughts on the European Escape from Hunger: Famines, Price Elasticities, Entitlements, Chronic Malnutrition, and Mortality Rates." National Bureau of Economic Research Working Paper on Historical Factors in Long Term Growth. No. 1. Cambridge, MA: National Bureau of Economic Research.

Hatcher, John (1986): "Mortality in the Fifteenth Century: Some New Evidence." *Economic History Review*, Second series, 39(1), 19–38.

Keyfitz, Nathan, and Wilhelm Flieger (1990): *World Population Growth and Aging*, University of Chicago Press.

——— and Wilhelm Flieger (1968): *World Population: An Analysis of Vital Data*, University of Chicago Press.

Lee, Ronald, and Lawrence Carter (1992): "Modeling and Forecasting U.S. Mortality." *J. Amer. Statistical Association*, 87 (419), 659–71.

Preston, Samuel, Nathan Keyfitz, and Robert Schoen (1972): *Causes of Death: Life Tables for National Populations*, New York: Academic Press.

———— (1991): "Demographic Change in the United States, 1970–2050." In A. M. Rappaport and S. J. Scheiber, eds, *Demography and Retirement: The 21st Century*. Praeger.

———— and Michael Haines (1991): *Fatal Years: Child Mortality in Late Nineteenth Century America*. Princeton University Press.

United Nations (1991): *World Population Prospects: 1990*. Population Study No. 120. New York: United Nations.

Wrigley, E. A., and R. S. Schofield (1981): *The Population History of England, 1541–1871*. Cambridge, Mass: Harvard University Press.

EDITOR'S NOTE (JULIAN L. SIMON)

Here follow some estimates of world population through the ages. The rapid increase in population in recent centuries is mainly the result of the rapid decline in mortality which Samuel Preston documents in his article.

Because it runs contrary to the belief of many, it is worth considering here the number of human beings who have ever lived, according to two separate closely-agreeing estimates: (a) 12 billion up to 6000 BC, 42

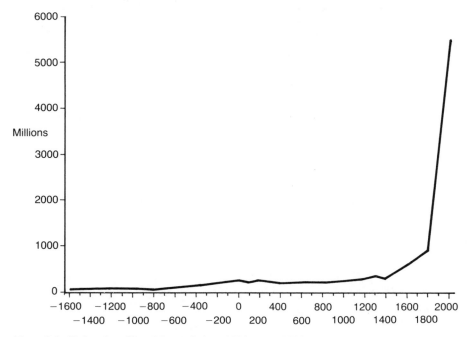

Figure 2.3 Estimation of world population, 1600 BC to 2000 AD

Source: Adapted from J. Bourgeois-Pichat, *From the 20th to the 21st Century: Europe and its population*.

billion from 6000 BC to 1650 AD, and 23 billion from 1650 AD to 1962 AD, making a total of 77 billion up until 1962 (Desmond, 1975), and (b) 3.8 billion before 40,000 BCE; 39 billion from 40,000 BCE to the start of the Common Era, and 22.6 billion from the year 1 up until the year 1750, plus another 10.4 billion from 1750 until 1950, and 4.3 billion from then until 1987, making a total of perhaps 80.3 billion (Bourgeois-Pichat, 1989, p. 90) a figure which is much larger than the number now alive.

REFERENCES

Bourgeois-Pichat, J., 'From the 20th to the 21st century: Europe and its population after the year 2000', *Population*, English Selection No. 1, Sept. 1989, 57–90.

Desmond, Anabelle, "How Many People Have Ever Lived on Earth?", *Population Bulletin* 18.1–19, reprinted in Kenneth C. W. Kammeyer, ed., *Population Studies*, second ed., Chicago, Rand McNally, 1975.

3

The Decline of Childhood Mortality

Kenneth Hill

INTRODUCTION

One of the most dramatic and significant changes in the human condition over the last two centuries has been the sharp and sustained decline in infant and child mortality. In the now-developed countries of Europe, North America, and Oceania, the probability of dying by the first birthday has declined from, in many cases, 200 per thousand live births to less than 10 in the span of 100 years. Child mortality has also declined sharply in developing countries, though the declines are more recent and have not progressed to such low levels. This change has had, and continues to have, profound implications for the family and society.

DATA AVAILABILITY AND DATA DEFICIENCY

Historically, reasonably good information about child mortality becomes available only with the emergence of sophisticated civil registration systems for recording births, deaths, and population. Such systems first developed in northwestern Europe in the late eighteenth century.[1] All world regions, with the major exception of sub-Saharan Africa, now include one or more countries with registration systems adequate for monitoring child mortality levels and trends. However, except in Europe, these countries are unlikely to be statistically representative of their regions; for those countries lacking such systems, and for sub-Saharan Africa as a whole, non-conventional methods[2] are required for measuring child mortality.

Estimates of child mortality for earlier periods also have to be derived from unconventional sources – for the sixteenth to nineteenth centuries, family reconstructions from parish registers or family histories, or records for especially well-documented population sub-groups such as the English aristocracy, and for earlier periods the analysis of ages at death based on excavation of burial grounds. Such data are both scanty and hard to interpret, so most hypotheses about child mortality prior to the sixteenth century have to be based either on general arguments about population dynamics or on extrapolations from recent conditions in some parts of the developing world.

We are on firm ground in tracing levels, age patterns, and cause structures of child mortality from the latter part of the nineteenth century in Europe, covering the major period of decline. However, for the more recent child mortality decline in the developing world, we can rarely trace changes in cause patterns of mortality, only in a minority of cases can we follow changing age patterns of mortality, and in some cases we cannot even track levels. Inevitably, the weaker the information base, the weaker the justification for explanations of observed changes.

CHILD MORTALITY IN PREHISTORIC AND EARLY HISTORIC TIMES

A number of studies have attempted to use skeletal remains, with estimated ages at death, to estimate average age at death and thus approximate expectation of life at birth (Acsadi and Nemeskeri, 1970; Johnston and Snow, 1961). The results of these studies are tenuous at best (Brothwell, 1975). Somewhat more promising are recent studies of hunter-gatherer populations with minimal exposure to modern life. One such study (Howell, 1979), with its results adjusted slightly in light of life table patterns,[3] gives some basis for hypothesizing a probability of death by age five among primitive hunter-gatherer peoples of between 400 and 500 per thousand. However, we cannot safely extrapolate from one modern primitive population to other modern primitive populations or to premodern populations.

We can, however, draw some conclusions about child mortality in the distant past from theoretical constraints of population dynamics. Over the long haul of pre-recorded history, the human population survived but grew very slowly, with an average annual growth rate (allowing for periodic ups and downs) of less than one per thousand. Births and deaths have to have been in very close balance, and the net reproduction rate (number of females surviving in the next generation to replace the mothers of one generation) must have averaged only very slightly over 1.0. For this to have occurred, the requirements of population dynamics indicate that, over the long haul of prehistory, the probability of dying before age five for females was probably not lower than 440 per thousand live births and not higher than 600. Risks for males would have been similar or higher.

The cause of death mix in childhood for prehistoric man would have been very different from that of a couple of centuries ago. The transformation from hunting and gathering to settled agriculture with higher population density, and from there to large cities in the classical world, probably had little effect on average mortality in childhood but much on the causes of mortality, with a shift away from starvation (as surpluses can be stored and later transported to protect against lean years) and violent deaths to infectious diseases supported by increased transmission resulting from greater crowding and larger susceptible populations. Firm evidence on the level of child mortality during this period is lacking (MacDonell, 1913; compare Hopkins, 1966). All we can do is conclude from the population dynamic argument that child mortality was on average high throughout early history.

CHILD MORTALITY IN EUROPE FROM 1500 TO 1800

Beginning around 1500, we start to have documentary evidence about child mortality levels in Europe, at least for some population sub-groups. The earliest observations are from family reconstructions for the aristocracy or high bourgeoisie, by no means representative of the population at large, but as we move forward into the seventeenth and eighteenth centuries, the number and variety of populations that can be reconstructed from religious records increases substantially. Table 3.1 summarizes child mortality estimates obtained by a number of these studies.

The first series in table 3.1, the British peerage (Hollingsworth, 1964), provides the most complete and detailed sequence. The probability of dying by age five in the mid-sixteenth century is around 250 per thousand live births, rising steadily to around 350 by the mid-seventeenth century, before starting a steady decline to below 200 by the mid-eighteenth and around 100 by the mid-nineteenth centuries. The rise from the sixteenth to the seventeenth centuries might be taken as evidence of data errors (omission of child deaths) in the earlier period but for the facts that both a less-detailed series for the English royal families (a group Henry [1965] argues would be less likely to suffer from omission) and adult mortality for the peerage itself show very similar general trends (though at a slightly higher level for the royal families). Thus child mortality probably did rise for this very small population sub-group from the sixteenth to the seventeenth centuries, perhaps reflecting the adoption of a more urban life style (which might also help to explain why the royal families had slightly higher child mortality than the peerage). With respect to the extent to which the peerage may be taken as representative of a broader population, it is instructive to compare the figures for the period after 1775 with those obtained from virtually complete registration for Sweden; around 1800, the British peerage have child mortality levels only half those of Sweden.

The second series, the Geneva bourgeoisie (Henry, 1956) show high child mortality, with a probability of dying by age five close to 400 until the end of the seventeenth century; thereafter, child mortality drops rapidly, following a trend very similar to that of the British peerage.

Geneva, being a city, would probably have experienced high overall child mortality, and being of bourgeois origins provided little protection, it seems, until the start of the eighteenth century.

Table 3.1 Child mortality (probability of dying by age five) in Europe, 1550 to 1849: selected historical studies

Period	British peerage[a] male	British peerage[a] female	Geneva bourgeoisie[b] male	Geneva bourgeoisie[b] female	Thirteen English parishes[c] male and female	Sweden (registration)[d] male	Sweden (registration)[d] female
1550–74	265	221	423	406	N/A	N/A	N/A
1575–99	256	240					
1600–24	268	247	388	312	237	N/A	N/A
1625–49	314	382					
1650–74	338	323	393	308	254	N/A	N/A
1675–99	291	296					
1700–24	276	279	283	225	281	N/A	N/A
1725–49	273	278					
1750–74	192	187	195	177	252	N/A	N/A
1775–99	147	154				332	308
1800–24	142	123	122	151	N/A	313	284
1825–49	109	105				258	229

Sources:
[a] Hollingsworth (1965).
[b] Henry (1956); reported probabilities of dying by age 20 have been converted to probabilities of dying by age five using the Coale–Demeny 'North' Model Life Tables.
[c] Wrigley and Schofield (1983).
[d] Keyfitz and Flieger (1968).

The third series shows child mortality in English parishes (Wrigley and Schofield, 1983), which were either rural or consisted of moderate-sized market towns. Among this population, which includes no major towns, child mortality apparently rose somewhat during the seventeenth and early eighteenth centuries before declining slightly in the late eighteenth century. The level of mortality is surprisingly low, however, starting at levels well below the British peerage, and only clearly exceeding the peerage's level in the late eighteenth century. There are also very marked differentials in child mortality among the 13 parishes studied; in the most rural parishes, the probability of dying by age five was only about 200 per thousand live births, whereas in the market towns the comparable figure exceeded 400 for some time periods. Overall, the series no doubt underestimates child mortality in England, since it excludes major towns, but the authors believe it probably captures trends. The fourth series, national-level data for Sweden, shows substantially higher mortality than others, with a sharp decline in the early nineteenth century.

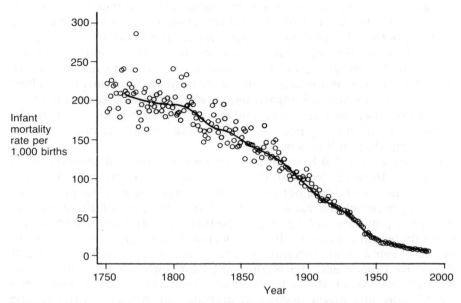

Figure 3.1 Infant mortality rate, Sweden, 1751–1988, with smoothed median trend
Source: B. R. Mitchell (1975) and United Nations Demographic Yearbooks.

These fragments of data provide a basis for some conclusions. First, even well-to-do families as recently as the seventeenth century suffered high child mortality levels not very different from some present primitive hunter-gatherer tribes (and presumably not very different from the contemporary population at large). Second, mortality decline for these privileged populations started early, apparently by 1700. Third, there were very large urban–rural differentials in child mortality, the risks in urban areas apparently being double those in rural areas. Fourth, overall child mortality was declining in Sweden from the late eighteenth century on.[4]

SUSTAINED DECLINE IN CHILD MORTALITY IN THE DEVELOPED WORLD, 1800–1990

Civil registration data on child mortality become increasingly available for European countries in the first half of the nineteenth century. The country with the longest sequence of infant mortality rates unbroken by war or other disturbance is Sweden. Figure 3.1 shows the decline in infant mortality in Sweden from 1751 to 1988 by single calendar years. Infant mortality appears to have fluctuated around an approximately constant level of about 200 per thousand live births until about 1810. It then declined fairly steadily to a level of about 100 around 1900, fell somewhat more rapidly to around 20 in 1950, and then continued to decline slowly toward its current level of about five. A dramatic feature of figure 3.1 is the way in which the variability of infant mortality has declined as the level itself has fallen; even in the early nineteenth century, before the level had changed much, the fluctuations are much reduced. Series for other European populations show similar patterns, though with different initial levels and mostly later starting points for sustained decline. For example, infant mortality in Germany was as high as 300 in the 1870s, did not fall below 200 until the first decade of the present century, was still over 100 in the 1920s, but fell to 10 in the 1980s; thus the starting point was higher and the decline both later and more rapid than in Sweden.

It is interesting also to examine how the age pattern of child mortality changed as the level declined. Figure 3.2 shows the relationship between the probability of dying between birth and age one and the probability of dying between the ages of one and five, by sex, for three western European populations (France, the Netherlands, and Sweden) from the mid-nineteenth century to the mid-twentieth century, thus covering the period of rapid child mortality decline. The general pattern, very clearly shown by France and the Netherlands, but only roughly followed by Sweden, was for mortality between ages one and five to decline somewhat before any major decline in mortality in infancy. In the second phase of decline, both infant and post-infant mortality decline proportionately, creating parallel straight lines in figure 3.2, with infant

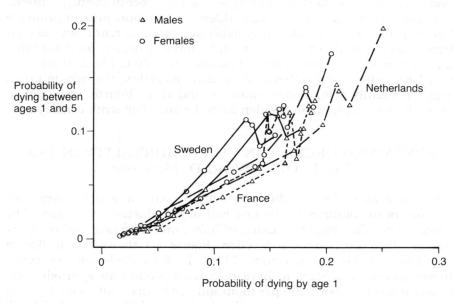

Figure 3.2 The relations between probabilities of dying in infancy and between ages one and five in the nineteenth and twentieth centuries by sex: the Netherlands, France and Sweden

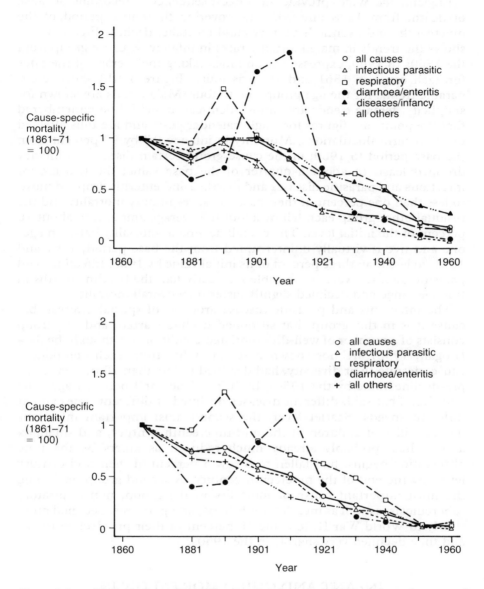

Figures 3.3a and 3.3b Indices of mortality rates in infancy and childhood by major cause of death group; males, England and Wales
Source: Preston et al. (1972).

mortality declining faster than post-infant mortality. The change from phase one to phase two occurs in the late nineteenth century and appears to be more a function of time period than of mortality level at the time.

England and Wales provide the longest sequence of recording of cause of death, from 1848 onward, thus covering the entire period of the nineteenth- and twentieth-century child mortality decline. Figure 3.3(a) shows the trends in male mortality rates in infancy by cause group from the 1860s to 1960, expressed as an index taking the average of the rates for each cause in 1861 and 1871 as unity. Figure 3.3(b) shows comparable trends for the age group one to four. Male trends are shown for simplicity; female trends are almost identical. It should be remembered that the points are figures for single calendar years and are thus affected by short-term fluctuations. Mortality in infancy fell by 90 percent from the base period to 1960; in the cause group certain diseases of infancy dropped least, to about a quarter of the base value; the two groups infectious and parasitic diseases and diarrhea and enteritis dropped most, to less than one percent of their base rates; respiratory mortality and the residual other group each fell by about the average amount, to about 10 percent of the initial level. The overall decline in mortality between ages one and five is virtually uninterrupted from the base period, 1861 and 1871, to less than three percent of its initial value by 1960. Infectious and parasitic diseases were responsible for nearly half the baseline deaths in this age range and declined slightly faster than overall mortality.

The infectious and parasitic disease group is of special interest, because it is in this group that sustained decline started, and the group consists of a number of well-differentiated conditions. Although the data (Logan, 1950) are not shown here, mortality from each component cause, despite their diversity, had declined to less than 10 percent of its pre-decline value by the 1950s, both in infancy and between ages one and five. That said, different diseases declined at different times and at different speeds. Scarlet fever, the second most important infectious disease killer of children in the mid-nineteenth century, and smallpox and typhus, probably already much reduced as killers by the time observation begins, had fallen to below 5 percent of their mid-century levels by the end of the nineteenth century. A second group, including the most important cause of mortality in the group, non-respiratory tuberculosis, as well as measles, diphtheria, and pertussis, declined more slowly to World War II, to some 10 percent of their pre-decline levels, and then declined very sharply in the 1950s.

INFANT AND CHILD MORTALITY IN
THE DEVELOPING WORLD

The number of countries with good records prior to 1950 is very small indeed. One of the few less-developed countries (LDCs) with an unbroken record of reasonably accurate infant mortality rates since 1900[5] is Chile. Figure 3.4 shows the decline in infant mortality in Chile since 1901. Early in the century, infant mortality was around 300 per thousand live births, with variations of close to 100 points between the highest and the lowest years. Around 1920, the variations become much less pronounced, and a pattern of sustained decline starts. However, the decline is interrupted twice, once in the 1930s, when infant mortality increases briefly, and again between 1955 and 1965, when the decline apparently stalls. There are thus both similarities (the trend beginning with a narrowing of fluctuations followed by fairly steady decline) and dissimilarities with the case of Sweden. In the case of Chile, the starting point is higher, the decline starts about 120 years later and proceeds twice as fast, and shows more pronounced fluctuations around the trend.

In the more distant past, the now less-developed countries experienced child mortality levels consistent with fairly high fertility and very slow population growth (i.e., high child mortality levels), but rather few observations of such pre-transition populations exist. Data for rural China in the 1920s indicate a probability of dying by age five of around 380 per thousand live births (Barclay, et al., 1978); the infant mortality rate for Chile prior to 1920 is consistent with a probability of death by age five of about 400; the probability observed for a rural village in the Gambia in the 1950s was close to 430 (McGregor et al. 1961); the value for the primitive hunter-gatherer !Kung tribe in the first half of the century was probably at least 400. Since then, all regions of the developing world have experienced rapid declines in child mortality.

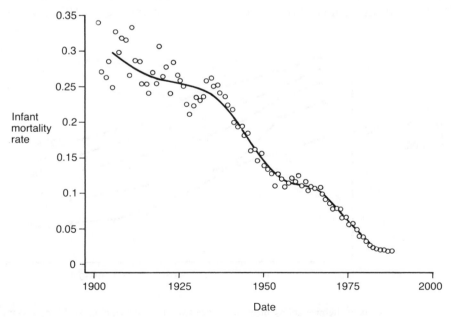

Figure 3.4 Infant mortality rate, Chile, 1901–87, with smoothed median trend
Source: B. R. Mitchell (1983) and United Nations Demographic Yearbooks.

Figure 3.5 shows the probability of death by age five estimated for each region of the developing world from the early 1950s to the early 1980s (United Nations, 1988). All regions have seen substantial declines. Africa, South Asia, and Latin America have all experienced declines of at least 100 points (roughly 50 percent) over the three decades. East Asia, dominated by China, showed an even more dramatic decline, falling by 80 percent in three decades.

The trends in figure 3.5 can be criticized on the grounds that, for many countries, neither levels nor trends of recent child mortality are known with any accuracy. However, an examination of trends in just those countries for which reasonably well substantiated estimates exist supports the broad trends in figure 3.5 (Hill and Pebley, 1989). The pace of child mortality decline in terms of percent change in probability of death by age five seems to have been roughly constant over the period from the early 1960s to the early 1980s. Africa has tended to underperform, but other regions show very similar rates of change. Age patterns of child mortality, and changes in such patterns, are rather similar to those found in Europe a century earlier; countries of the developing world have seen child mortality change in ways that would fit quite easily into figure 3.2.

Even less is known about trends and structures of cause of death in the developing world. Cause of death data from Chile for the early part of the century suggest a rather lower burden of infectious and parasitic diseases and higher levels of respiratory ailments than for England and Wales, but the quality of the data is suspect (Preston, et al., 1972). Early

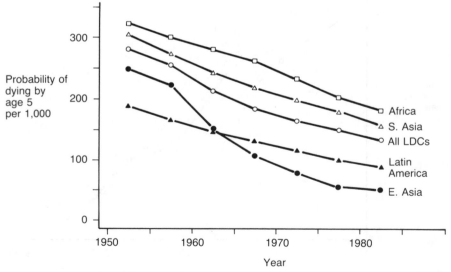

Figure 3.5 Trends in the probability of dying by age five in the developing world; by region, 1950–85
Source: United Nations (1988).

data from a tropical population, Taiwan, also show relatively high levels of respiratory mortality, but surprisingly low mortality from diarrhea, relative to European experience (Preston, et al., 1972). Though the data are fragmentary, there can be little doubt that malaria is, or has been, an important cause of child mortality in much of the tropics, particularly in Africa.

There are some important differences between developing country mortality declines and earlier declines in developed countries, such as the existence of lower child mortality in urban areas of LDCs and the existence of strong socio-economic differentials, particularly pronounced with female education. At the same time, there are very strong similarities, such as the apparent resistance to reversal in downward trends and the age patterns of improvement. A key difference is the link to economic development. Whereas in Europe child mortality fell during a period of unprecedented rise in per capita income, in LDCs it has fallen sharply in some countries, like Sri Lanka and China, that have experienced very little increase in per capita income. The strong link of child mortality to mother's education suggests a very important role for diffusion of health-related knowledge and behavioral change. However, there can be little doubt that medical interventions, often consisting early in the decline of public health style vector control programs (such as in Sri Lanka) and more recently of mass immunization programs have contributed substantially to the pace of decline. As a practical example, the village in the Gambia reported above as having high child mortality in the 1950s continued to have high mortality until the mid-1970s. At that time, an intensive program of medical intervention and surveillance was introduced, and the probability of dying by age five dropped from about 250 per thousand to about 50 (i.e., by 80 percent) in the period 1974–5 to 1982–3 (Lamb, et al., 1984). The authors attribute the pace of the decline largely to "regular, controlled surveillance of women [and children] by a physician and a qualified midwife. . . ." Clearly, such intensive services are not generalizable, but they show what medical services can potentially achieve without much alteration in underlying living conditions.

CONCLUSIONS

Over the last three or four centuries, child mortality has fallen dramatically. In developed countries, this transition from high to low mortality is essentially complete; whereas prior to the decline, a child had about a 40 percent risk of dying by age five, today that risk is less than 1 percent. The decline occurred through reduced exposure to pathogens, initially largely as a result of public health measures and later as a result of living in a healthier population, through improved household hygiene and nutrition and reduced crowding, and through specific medical preventive and therapeutic interventions. As the decline has progressed, the medical effects have become increasingly important. In developing countries, the transition is largely incomplete, though it has started everywhere. In a few countries in Africa, the risk of dying by age five is still over 300 per thousand; in parts of South Asia, it is still over 200 per thousand; in parts of Latin America, it is still over 150 per thousand. Practically everywhere, barring civil wars or natural disasters, it is declining at a fairly steady pace, and in some countries (Chile, Costa Rica) it has reached levels of which Europe was proud only three decades ago. This decline is much more strongly associated with medical knowledge and practice than was the earlier decline in Europe. Indeed, the development and implementation of cheap and effective preventive measures has probably been the decisive factor in child mortality decline in LDCs over the last two decades.

NOTES

1 The systematic collection of other useful information such as cause of death starts somewhat later still, in the late nineteenth century.
2 These methods use survey data for women concerning some or all of the children they have borne, and the survival of those children (see United Nations, 1983).
3 Model life tables (e.g., Coale and Demeny, 1966) express empirically-observed regularities in age patterns of mortality at different mortality levels.
4 It is worth noting that the relatively low mortality in rural England was associated with late and less-than-universal marriage, tending to hold fertility down, and that the sharp decline in child mortality among the Geneva bourgeoisie in the eighteenth century was associated with a rapid decline in marital fertility. Thus these relatively low levels of child mortality are not inconsistent with the requirements of population dynamics.
5 A recent United Nations (1988) exercise to describe levels and trends in child mortality worldwide since 1950 could use civil registration data alone for only 17, and civil registration in combination with other data for a further 16, developing countries, out of a total of 124.

REFERENCES

Acsadi, G., and J. Nemeskeri (1970): *History of human life span and mortality*. Budapest: Akademiai Kiado.

Banister, Judith (1987): *China's Changing Population*. Palo Alto: Stanford University Press, 26.

Barclay, G. W., A. J. Coale, M. A. Stoto, and T. J. Trussell (1976): "A Reassessment of the Demography of Traditional Rural China," *Population Index*, 44, 4.

Brothwell, D. (1975): "Paleodemography," in *Biological Aspects of Demography*, ed. W. Brass. London: Taylor and Francis.

Coale, A. J., and P. Demeny (1983): *Regional Model Life Tables and Stable Populations*, 2nd edn (with Barbara Vaughn). New York: Academic Press.

Daly, M., and M. Wilson (1984): "A Sociobiological Analysis of Human Infanticide." In *Infanticide: Comparative and Evolutionary Perspectives*, eds

G. Hausfater and S. B. Hardy (1984). Hawthorne, NY: Aldine Publishing Company.

Hausfater, G., and S. B. Hardy, eds (1984): *Infanticide: Comparative and Evolutionary Perspectives*. Hawthorne, NY: Aldine Publishing Company.

Henry, L. (1956): *Anciennes Familles Genevoises: Etude Demographique*. Paris: Institut National D'Etudes Demographiques.

—— (1965): "Demographie de la Noblesse Britannique," *Population*, vol. 20.

Hill, K., and A. R. Pebley (1989): "Child Mortality in the Developing World," *Population and Development Review*, 15, 4.

Hofsten, E., and H. Lundstrom (1976): *Swedish Population History: Main Trends from 1750 to 1970*. Stockholm: Statistika Centralbyvan.

Hollingsworth, T. H. (1965): "The Demography of the British Peerage," Supplement to *Population Studies*, 18, 2.

Hopkins, M. K. (1966): "On the Probable Age Structure of the Roman Population," *Population Studies*, 20, 2.

Howell, N. (1979): *The Demography of the Dobe !Kung*. New York: Academic Press.

Johnston, F. E., and C. E. Snow (1961): "The Reassessment of the Age and Sex of the Indian Knoll Skeletal Population: Demographic and Methodological Aspects," *American Journal of Physical Anthropology*, vol. 19.

Keyfitz, N., and W. Flieger (1968): *World Population: An Analysis of Vital Data*. University of Chicago Press.

Lamb, W. H., and F. A. Foord, C. M. B. Lamb, and R. G. Whitehead (1984): "Changes in Maternal and Child Mortality Rates in Three Isolated Gambian Villages Over Ten Years," *The Lancet*, October 20, 912–14.

Lancaster, H. O. (1989): *Expectations of Life: A Study in the Demography, Statistics and History of World Mortality*. New York: Springer-Verlag.

Logan, W. P. D. (1950): "Mortality in England and Wales from 1848 to 1947," *Population Studies*, 4, 1.

MacDonell, W. R. (1913): "On the Expectation of Life in Ancient Rome and in the Provinces of Hispania, and Lusitanis and Africa," *Biometrika*, vol. 9.

McGregor, I. A., W. Z. Billewicz and A. M. Thomson (1961): "Growth and Mortality in an African Village," *British Medical Journal*, 2, 1661–6.

McKeown, T. (1976): *The Modern Rise of Population*. New York: Academic Press.

Mitchell, B. R. (1975): *European Historical Statistics 1750–1970*. London: Macmillan.

Preston, S. H., N. Keyfitz and R. Schoen (1972): *Causes of Death: Life Tables for National Populations*. New York: Seminar Press.

—— and E. van de Walle (1978): "Urban French Mortality in the 19th Century," *Population Studies*, 28, 1.

United Nations (1983): *Indirect Techniques for Demographic Estimation*, Manual X. New York: United Nations.

—— (1988): *Mortality of Children Under Age 5; World Estimates and Projections 1950–2025*. New York: United Nations.

Woods, R. I., P. A. Watterson and J. H. Woodward (1988, 1989): "The Causes of Rapid Infant Mortality Decline in England and Wales, 1861 to 1921," *Population Studies*, Part I: 42, 3; Part II: 43, 1.

Wrigley, E. A. (1972): *Population and History*. London: Weidenfeld and Nicholson.

———— and R. S. Schofield (1983): "English Population History from Family Reconstitution: Summary Results 1600–1799," *Population Studies*, 37, 2.

4

Disease and Health through the Ages

Michael R. Haines

Perhaps mankind's greatest achievement in the modern era has been the increase in longevity and improvement in health. Although reductions in mortality and better health do not necessarily move together (Riley, 1989), there have been major improvements in human survival in the past several centuries. Much of the driving force in the modern population surge has been reduced mortality from infectious and parasitic diseases and, especially, diminished mortality peaks (i.e., years of severe excess mortality). Prior to the eighteenth century, epidemics and plagues were major killers. Since the eighteenth century, the incidence and importance of mortality peaks has sharply diminished and many other infectious and parasitic diseases have greatly decreased as well (Perrenoud, 1991; Schofield and Reher, 1991; Vallin, 1991; United Nations, 1973, pp. 110–11; 1982). As a consequence, most people now die of chronic and degenerative diseases such as cancer, cardiovascular diseases, and diabetes (Omran, 1971).

These developments have, in turn, been related to the modern agricultural revolution (beginning in the seventeenth century) and the industrial revolution (beginning in the late eighteenth century, Cipolla, 1965, chapter 4; McKeown, 1979). Compared to other animal species, human beings have suffered a disproportionately large share of mortality from infectious and parasitic diseases since the advent of sedentary agriculture about 8000 BCE (McKeown, 1979, p. 72). And it is precisely the reduction and control of the incidence of this class of diseases that have resulted in rapid increases in population growth and extended life.

History is replete with references to catastrophic epidemics and plagues: the destructive plague in 430 BCE in Athens during the Peloponnesian War (possibly smallpox or typhus); the repeated epidemics in the Roman Empire during the reign of Marcus Aurelius (ca. 165–180 CE), likely measles and smallpox; the great plague of Justinian (542–565 CE) and the Black Death of medieval Europe (1346–49 and subsequent waves), certainly bubonic plague; numerous references to periodic human mortality peaks in China (McNeill, 1976); and the tremendous decline in the indigenous population of the New World after contact with Old World inhabitants after 1492 from a variety of diseases, but particularly smallpox, influenza, and measles (Crosby, 1972).

Human viability is a balance with the environment that includes availability of food, adequate clothing, and shelter, plus the ecology of the parasitic microorganisms (microparasitism). The Neolithic revolution of ca. 8000 BCE increased food supplies but, at the same time, brought people together in situations favoring greater microparasitism and death through infectious disease. Changes in human mobility, densities, locations, and activities have often disturbed that balance and resulted in serious increases in infectious disease and mortality, and population stagnation or decline. McNeill (1976, chapter 3) has proposed that the incursion of measles and smallpox from Mesopotamia to the Roman Empire in the second century CE resulted in the longer-term decline of the demographic and economic basis of urban, commercial Roman civilization. It remained for these two highly contagious viral diseases to establish themselves as chronic endemic childhood diseases rather than periodic epidemic adult diseases to restore demographic balance. The effect of exposure to these infections on the indigenous populations of the New World after 1492 is better documented than earlier serious disturbances to the human–microparasite balance (Crosby, 1972; McNeill, 1976, chapter 5).

CAUSE OF DEATH

Some idea of the effects of changing causes of death over the course of mortality decline can be seen in figure 4.1. This shows the percentage distribution among major cause-of-death groups (infectious, parasitic, and respiratory diseases; cancer; diseases of the cardiovascular system; deaths by violence; and all other causes) for two different model populations. Population A has an expectation of life at birth of 50 years and a relatively young age structure, while population D has an expectation of life at birth of 70 years with an older age structure. (The information is reported in United Nations, 1973, table V.13 and is based on United Nations, 1963, pp. 106–12.) As is to be expected, the "epidemiologic transition" (Omran, 1971) from A to D is accompanied by a decline in the share of infectious and parasitic diseases and a relative increase in chronic and degenerative diseases (e.g., cancer and diseases of the cardiovascular system). The shift is dramatic. In population A, infectious and parasitic diseases account for 34 percent of all deaths, but they account for only 6.5 percent in population D. Similarly, cancer and cardiovascular diseases are responsible for 24.7 percent of all deaths in the high-mortality, younger population, but 62.9 percent in the low-mortality, older population.

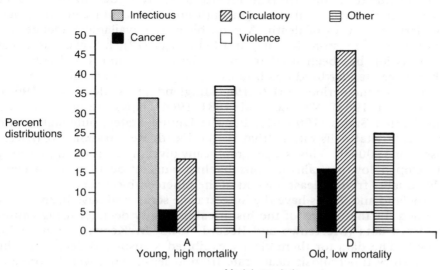

Figure 4.1 Causes of death, percent distribution, model populations

Before the modern era, the great killers came in epidemic forms: bubonic plague, smallpox, measles, typhus, scarlet fever, influenza. These diseases remained after the eighteenth century, but the mortality peaks began to subside (Perrenoud, 1991). Plague disappeared from most of Europe in the early eighteenth century (Biraben, 1975–6), and many diseases (e.g., smallpox, measles) were more commonly endemic. Periodic large-scale epidemics did recur even in the nineteenth century in Europe and other areas, notably when cholera moved from South Asia in the 1820s and created worldwide epidemics in the 1830s, late 1840s, 1860s, and early 1890s. By the nineteenth century, many endemic infectious diseases also began to diminish.

Table 4.1 Percentage distribution of deaths by cause in selected model populations

| Age structure | Young | Young | Old | Old |
Expectation of life at birth	50	70	50	70
Model Population	A	B	C	D
Cause of death				
All causes	100.0	100.0	100.0	100.0
Infectious, parasitic, and respiratory diseases	34.1	10.8	27.4	6.5
Cancer	5.6	15.2	7.9	16.4
Diseases of the circulatory system	18.7	32.2	26.0	46.5
Violence	4.3	6.8	4.0	5.2
All other causes	37.3	35.0	34.7	25.4

Source: United Nations (1973), 129.

Unfortunately, the historical record is relatively devoid of standard demographic data until the eighteenth and nineteenth centuries, particularly on causes of death. A good bit of the available evidence was compiled by Preston, Keyfitz, and Schoen (1972), however, and that information has been used to create figures 4.2 and 4.3. Figure 4.2 shows age-standardized death rates for respiratory tuberculosis for five nations for the period 1861 to 1964: England and Wales (1861–1964), Italy (1881–1964), New Zealand (1881–1964), Japan (1899–1964), and the United States (1900–64). For the United States, only states with adequate vital registration (that is, the Death Registration Area) were used for 1900–30. This is a geographically diversified group of presently developed, low-mortality nations that had cause-of-death data of acceptable quality from at least the end of the nineteenth century.

England and Wales have the longest time series and have been widely studied for the origins of the historical mortality decline, using cause-of-death data (McKeown and Record, 1962; McKeown, 1976, 1979, 1988). The data for them clearly exhibited a sustained decline in the respiratory tuberculosis death rate from at least 1861, well before any specific medical interventions. The tuberculosis bacillus was not identified by Koch until 1882, and specific chemotherapy and vaccination (BCG) were not available until the 1940s and 1950s, respectively (Lancaster, 1990, chapter 7). Yet mortality declined consistently. For England and Wales, 17.5 percent of the mortality decline from all causes over the period 1848–54 to 1971 was due to respiratory tuberculosis. The share for all air- and water-borne infectious diseases was 61.7 percent of all causes of death (McKeown, 1979, chapter 3). For the

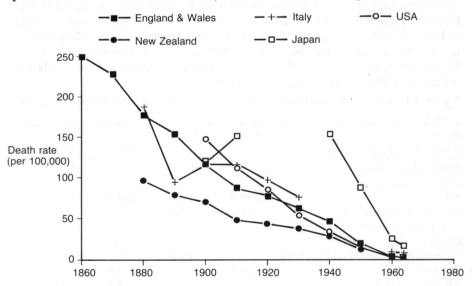

Figure 4.2 Respiratory tuberculosis, standardized death rates, 1861–1964

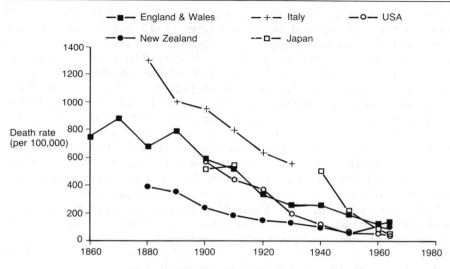

Figure 4.3 Other infectious disease, unstandardized death rates, 1861–1964

period 1851–60 to 1891–1900, the portion of the mortality decline in England and Wales due to respiratory tuberculosis was even larger – 43.9 percent – and the share of the decline over the same period attributable to infectious disease was 92 percent (McKeown, 1976, p. 65).

This English case is possibly atypical. Preston (1976, p. 20) notes that the average share of the decline in age and sex standardized death rates

Table 4.2 Death rates from respiratory tuberculosis. Selected nations 1861–1964. (Age-standardized. Both sexes combined.) Per 100,000 per year[a]

England/Wales		Italy		New Zealand		Japan		United States	
Year	Death rate	Year	Death rate	Year	Death rate	Year	Death rate	Year	Death rate
1861	249.29								
1871	227.66								
1881	177.22	1881	187.46	1881	95.51				
1891	153.41	1891	94.24	1891	77.88				
1901	115.98	1901	116.73	1901	69.41	1899	119.46	1900	146.56
1911	88.32	1910	115.37	1911	47.46	1908	150.76	1910	110.73
1921	77.68	1921	96.95	1921	43.49			1920	84.02
1931	62.68	1931	75.46	1926	38.05			1930	54.05
1940	45.61			1936	28.51	1940	153.34	1940	33.12
1951	19.61			1951	12.54	1951	86.78	1950	14.61
1960	3.54	1960	10.12			1960	25.17	1960	3.54
1964	2.54	1964	7.61	1964	2.02	1964	16.15	1964	2.02

Note:[a] Standardized to the age distribution of Coale and Demeny (1966). "West" model female stable population with a life expectancy of 45 years and an annual growth rate of 0.02.
Source: Preston, Keyfitz, and Schoen (1972).

from respiratory tuberculosis between 1851–60 and 1891–1900 was 44 percent for England and Wales but only 11–12 percent in the comparable stage of other populations studied. The cool climate, urban crowding, and polluted air probably all contributed to the high incidence of tuberculosis. The proportion of the decline due to other infectious and parasitic diseases was 46 percent in England and Wales over this period, as compared with the average for all nations studied of 14 percent. For Italy, the share of the early mortality decline (1881–1910) was only about 8–10 percent for respiratory tuberculosis, but it was 58–60 percent for all infectious diseases (Caselli, 1991, table 4.4).

Also, as figure 4.2 indicates, the timing of the decline in tuberculosis death rates clearly varied across nations. Italy did not begin its sustained decline in tuberculosis death rates until 1910 and Japan not until after 1940, whereas England and Wales, New Zealand (at much lower levels), and the United States showed consistent reductions. Overall, however, more than two thirds of the mortality decline over the past two centuries has been due to a reduction in infectious and parasitic diseases, of which tuberculosis was the most important (Puranen, 1991).

Tuberculosis may not have always been the major killer among infectious diseases. Evidence exists for Sweden and Finland that respiratory tuberculosis death rates increased in the eighteenth and into the nineteenth centuries as the effects of increased population density, crowding, urbanization, and (later) some industrialization took hold (Puranen, 1991, figure 5.1). In Italy, with its drier and warmer climate, gastroin-

Table 4.3 Death rates from infectious diseases, excluding tuberculosis. Selected nations, 1861–1964. (Not age-standardized. Both sexes combined.) Per 100,000 population per year[a]

England/Wales		Italy		New Zealand		Japan		United States	
Year	Death rate	Year	Death rate	Year	Death rate	Year	Death rate	Year	Death rate
1861	745.83								
1871	879.19								
1881	673.52	1881	1303.56	1881	388.15				
1891	785.34	1891	1001.94	1891	353.53				
1901	594.00	1901	950.68	1901	237.36	1899	512.28	1900	570.78
1911	515.08	1910	794.50	1911	184.30	1908	540.99	1910	439.06
1921	334.38	1921	635.14	1921	149.55			1920	373.40
1931	259.97	1931	554.79	1926	132.68			1930	195.54
1940	259.50			1936	102.07	1940	499.45	1940	118.42
1951	193.72			1951	60.11	1951	215.12	1950	52.55
1960	125.62	1960	102.55			1960	84.67	1960	51.69
1964	136.72	1964	85.17	1964	97.00	1964	53.62	1964	44.92

Note: [a] Death rates include deaths from the categories "Other Infectious and Parasitic Diseases," "Influenza, Pneumonia, Bronchitis," and "Diarrhea." "Respiratory Tuberculosis" was excluded.
Source: Preston, Keyfitz, and Schoen (1972).

testinal infections were relatively of more importance than respiratory infections as a cause of death (Caselli, 1991).

THE CONTROL OF DISEASE

The control of infectious diseases proceeded unevenly. For smallpox, inoculation existed in Europe from the early eighteenth century (and earlier in the Middle East) and vaccination from the 1790s. Inoculation actually gave the patient a case of smallpox under controlled conditions. Vaccination infected the patient with the related cowpox, a much milder and less dangerous infection in humans. Both conferred partial or total immunity. These measures, gradually adopted over the eighteenth and nineteenth centuries, brought this highly contagious killer under control. For England and Wales, for example, smallpox was responsible for only 1.2 percent of all deaths and 3.6 percent of deaths from airborne infections in the period 1848–54. It had become negligible as a cause of death by 1901 (McKeown, 1976, pp. 54–5). Much of the reduction in both epidemic and endemic incidence of smallpox had taken place before the mid-nineteenth century. Today, because of assiduous public health measures, smallpox is now extinct (or close to extinct) as a human disease.

Plague, as already mentioned, basically disappeared from Europe in the eighteenth century and only intermittently occurred in the rest of the world (usually Asia). This may have been due to ecological and biological changes in the transmission process, but government quarantines likely played some role. Scarlet fever and diphtheria (responsible for 4.6 percent of deaths in England and Wales in 1848–54) experienced significant declines in the late nineteenth century. Diphtheria was finally treatable with antitoxin (developed by Emil Behring in 1890 and first produced commercially in 1894), but scarlet fever appeared to diminish in virulence. Measles is a highly contagious viral infection whose effects are similar to smallpox: epidemic course with a high adult incidence and a high case fatality rate in populations without immunity acquired in childhood (e.g., the indigenous population of the New World in the sixteenth and seventeenth centuries, Pacific Island populations in the nineteenth century). Both measles and smallpox became largely endemic diseases of childhood with fewer catastrophic demographic consequences as populations became sufficiently dense (Lancaster, 1990, passim).

Progress in the control of infectious diseases may have been uneven, but much of the advance has been made since the second half of the nineteenth century. Table 4.3 demonstrates this for the same group of nations represented in the previous table. The unstandardized death rates are for a combination of the well-known air-, water-, and food-borne infectious diseases (excluding respiratory tuberculosis) as well as influenza, bronchitis, and pneumonia and diarrheal and gastrointestinal infections. (The data are from Preston, Keyfitz, and Schoen, 1972.) So, for example, the decline in infectious diseases in Italy (from 1881) was more rapid than in England and Wales and was concentrated in gastrointestinal diseases (as opposed to respiratory infections). In England and Wales, the course (and hence control) of these infections was uneven until after about 1890. New Zealand demonstrated a slow decline from much lower levels. Japan exhibited inconsistent progress until after World War II, whereas the United States had a steady and relatively rapid decline in infectious disease after 1900.

The causes of the decline are varied and may be grouped into four basic categories: improvements in the standard of living (including better nutrition, housing, and clothing); direct medical intervention; public health (including quarantine, improved water supplies and sewage disposal, hygienic milk supplies, compulsory immunizations, and health and hygiene education programs); and ecobiological factors (i.e., changes in the basic biological disease environment, organisms, method of disease transmission). A leading hypothesis on the origins of the modern mortality decline is that of Thomas McKeown (1976, 1979, 1983, 1988). He argues that, prior to the twentieth century, specific medical intervention was usually ineffective or nonexistent and that ecobiological changes were rare (affecting possibly plague, scarlet fever, and diphtheria). Thus, he gave most of the credit to rising standards of living, particularly better nutrition, although his later work accords a greater role to public health measures. In contrast, Samuel Preston attributes less than half of the mortality decline in the twentieth century to increases in income per capita, the remainder being due to medical, public health, and ecobiological factors (Preston, 1980).

CONCLUSIONS

The answers to the questions on the origins of the modern mortality transitions are not simple. Schofield and Reher (1991, pp. 16–17) recently noted, "One of the major conclusions . . . is that there was no simple or unilateral road to low mortality, but rather a combination of many different elements ranging from improved nutrition to improved education." As examples, public health measures succeeded in controlling cholera in a number of places before the microorganism was identified. Better water and sewage disposal was already having significant effects on urban mortality before the widespread acceptance of the germ theory of disease (Preston, Haines, and Pamuk, 1981; Preston and Haines, 1991). It is clear that medical science had contributed to, and had an impact on, public health in the nineteenth century (Woods, 1991). Improved standards of living were generated by the economic development process, which also created greater urbanization and, sometimes with a lag, industrialization. Cities had much higher mortality than rural areas. Urban crowding, housing problems, and poor sanitation created increased mortality risks, which were only controlled from the later nineteenth century onward with increased public health programs, administration, and expenditures. These measures were increasingly effective, and several decades into the twentieth century cities were healthier places than the countryside because of better public health, including water supplies and sewage disposal, more available medical care, and higher incomes. The industrialization process itself also produced new health and mortality risks (Haines, 1991).

Overall, the matter is far from settled. It is clear, however, that the "Age of Pestilence and Famine" (Omran, 1971) that characterized the premodern demographic regime has passed, even in most developing nations. We no longer face catastrophic epidemic infections, and even many pervasive endemic microparasitic infections have been banished, at least when civil order prevails. The virtual eradication of some diseases (e.g., smallpox) and control of others have freed humanity from a serious set of problems and are great events of the modern age. But mankind still faces the prospect of controlling chronic and degenerative diseases such as cancer, cardiovascular illness, and diabetes, as well as man-made environmental hazards. The quest to prolong and improve human life still poses many challenges.

REFERENCES

Biraben, J. N. (1975–76): *"Les hommes et la peste en France et dans les pays européens et méditerranéens."* Vol. 1. *La peste dans l'Histoire.* Vol. 2 *Les Hommes Face à la Peste.* Paris: Mouton.

Caselli, Graziella (1991): "Health Transition and Cause-Specific Mortality." In Roger Schofield, David Reher, and Alain Bideau, eds, *The Decline of Mortality in Europe,* 68–96. Oxford, UK: The Clarendon Press.

Coale, Ansley J., and Paul Demeny (1966): *Regional Model Life Tables and Stable Populations,* Princeton: Princeton University Press.

Cipolla, Carlo M. (1965): *The Economic History of World Population.* Baltimore, Md.: Penguin Books.

Crosby, Alfred W., Jr (1972): *The Columbian Exchange.* Westport, Conn.: Greenwood Press.

Haines, Michael R. (1991): "Conditions of Work and the Decline of Mortality." In Roger Schofield, David Reher, and Alain Bideau, eds, *The Decline of Mortality in Europe,* 177–95. Oxford, UK: The Clarendon Press.

Lancaster, H. O. (1990): *Expectations of Life: A Study in the Demography, Statistics, and History of World Mortality.* New York: Springer-Verlag.

McKeown, Thomas (1976): *The Modern Rise of Population.* New York: Academic Press.

——— (1979): *The Role of Medicine: Dream, Mirage, or Nemesis?* Princeton, NJ: Princeton University Press.

——— (1983): "Food, Infection, and Population." In Robert I. Rotberg and Theodore K. Rabb, eds, *Hunger and History; The Impact of Changing Food Production and Consumption Patterns on Society,* Cambridge, UK: Cambridge University Press, 29–49.

———— (1988): *The Origins of Human Disease.* Oxford, UK: Basil Blackwell.

———— and R. G. Record (1962): "Reasons for the Decline of Mortality in England and Wales during the Nineteenth Century." *Population Studies* 16, 2 (November): 94–122.

McNeill, William H. (1976): *Plagues and Peoples.* New York: Anchor Books.

Omran, Abdel R. (1971): "The Epidemiologic Transition." *Milbank Memorial Fund Quarterly* 49(1), 509–38.

Perrenoud, Alfred (1991): "The Attenuation of Mortality Crises and the Decline in Mortality." In Roger Schofield, David Reher, and Alain Bideau, eds, *The Decline of Mortality in Europe*, 18–37. Oxford, UK: Clarendon Press.

Preston, Samuel H. (1976): *Mortality Patterns in National Populations with Special Reference to Recorded Causes of Death.* New York: Academic Press.

———— (1980): "Causes and Consequences of Mortality Declines in Less Developed Countries during the Twentieth Century." In Richard A. Easterlin, ed., *Population and Economic Change in Developing Countries*, 289–360. Chicago: University of Chicago Press.

———— and Michael R. Haines (1991): *Fatal Years: Child Mortality in Late Nineteenth-Century America.* Princeton, NJ: Princeton University Press.

———— ———— and Elsie Pamuk (1981): "Effects of Industrialization and Urbanization on Mortality in Developed Countries." In International Union for the Scientific Study of Population. *International Population Conference: Manila, 1981*, "Solicited Papers," Vol. 2. Liège: IUSSP, 233–54.

———— Nathan Keyfitz, and Robert Schoen (1972): *Causes of Death: Life Tables for National Populations.* New York: Seminar Press.

Puranen, Bi. (1991): "Tuberculosis and the Decline of Mortality in Sweden." In Roger Schofield, David Reher, and Alain Bideau, eds, *The Decline of Mortality in Europe*, 97–117. Oxford, UK: Clarendon Press.

Riley, James C. (1989): *Sickness, Recovery, and Death: A History and Forecast of Ill Health.* Iowa City: University of Iowa Press.

Schofield, Roger, and David Reher (1991): "The Decline of Mortality in Europe." In Roger Schofield, David Reher, and Alain Bideau, eds, *The Decline of Mortality in Europe*, 1–17. Oxford, UK: Clarendon Press.

United Nations (1963): *Population Bulletin of the United Nations.* No. 6, 1962. "With Special Reference to the Situation and Recent Trends of Mortality in the World." New York: United Nations.

———— (1973): *The Determinants and Consequences of Population Trends. New Summary of Findings on Interaction of Demographic, Economic and Social Factors*, Vol. I. New York: United Nations.

———— (1982): *Levels and Trends of Mortality since 1950.* New York: United Nations.

Vallin, Jacques (1991): "Mortality in Europe from 1720 to 1914: Long-Term Trends and Changes in Patterns by Age and Sex." In Roger Schofield, David Reher, and Alain Bideau, eds, *The Decline of Mortality in Europe*, 38–67. Oxford, UK: Clarendon Press.

Robert Woods (1991): "Public Health and Public Hygiene: The Urban Environment in the Late Nineteenth and Early Twentieth Centuries." In Roger Schofield, David Reher, and Alain Bideau, eds, *The Decline of Mortality in Europe*, 233–47. Oxford, UK: Clarendon Press.

5

The Contribution of Improved Nutrition to the Decline in Mortality Rates in Europe and America

Robert William Fogel

The decline in mortality rates over the past three centuries is one of the greatest events in human history. Although improved nutrition, improved public and personal sanitation, decontamination of food and water, improved housing, and advances in medical technology all contributed to rising life expectancy, there are conflicting views about the relative importance of each factor. Resolution of the issue is essentially a particularly complicated accounting exercise that involves measuring not only the direct effect of particular factors but also their indirect effects and their interactions with other factors. The preliminary results indicate that the elimination of chronic malnutrition was an important factor.

Wrigley and Schofield (1981) and Weir (1984), relying on data developed by INED (1977) for 1740–1829, showed that both English and French crude death rates were high until the 1780s, when the English rate began to decline, bottoming out at about 22 per 1,000 in the 1830s and remaining stable until the beginning of the 1870s. The French rate began to decline about half a decade later, but much more rapidly, reducing the original gap of about 9 per 1,000 to about 2 per 1,000 by the late 1830s, after which the French rate stabilized at about 24 per 1,000 (see figure 5.1).

It is commonly thought that recurrent famines and other mortality crises (plagues, etc.) were the chief cause of high mortality rates in pre-industrial times. However, data developed by Wrigley and Schofield (1981) bear on the national impact of mortality crises on the annual crude death rate in early modern England. Out of the 331 years covered by their study, they found 45 crisis years. By combining the information from two of their tables, it is possible to assess the impact of crisis mortality on the average crude death rate. The result is summarized in figure 5.2. During the 210 years ending in 1750, crisis mortality

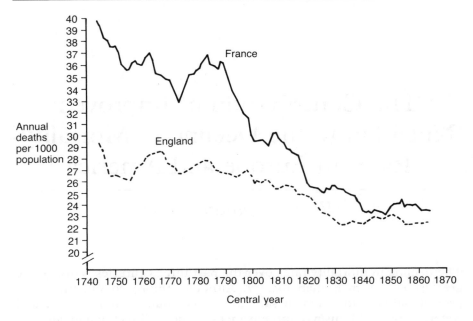

Figure 5.1   Crude death rates in France and England, 1740–1870
Sources: Weir (1984), corrected by Weir. Weir's sources were Wrigley and Schofield (1981),
531–5; INED (1977), 332–3; and Mitchell (1980), 116–19.

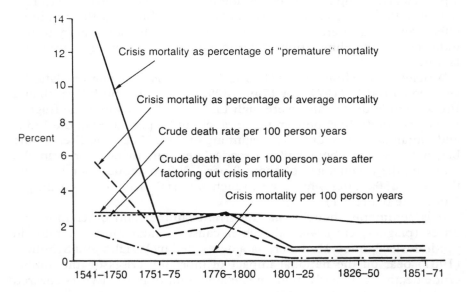

Figure 5.2   The impact of crisis mortality on the average crude death rate in England
Sources and methods of computation: see Fogel (1992), table 1. "Premature" mortality is defined
as the crude death rate of a given period minus the English death rate of 1980 standardized on
the English age structure of 1701–5.

accounted for less than 6 percent of total mortality and 7.6 percent of "premature" mortality. The data also indicate that just 6 percent of the decline in the annual crude death rate between the third quarter of the eighteenth century and the second quarter of the nineteenth could have been due to the elimination of crisis mortality ($0.27 \div 4.70 = 0.057$).

It follows that even if every mortality crisis identified by Wrigley and Schofield was due to famines, the elimination of famines would account for at most 6 percent of the first wave (1776–1850) of the secular decline in mortality. However, over 90 percent of crisis mortality was unrelated to famines (Wrigley and Schofield 1981, pp. 332–73; Lee 1981; Schofield 1983; Fogel 1986, pp. 494–5; Fogel 1987, appendix). Dramatic as famines were, then, their elimination could not have accounted for as much as 6 percent of the decline in average annual mortality between the third quarter of the eighteenth century and the second quarter of the nineteenth.

Demonstrating that famines and famine mortality are a secondary issue in the escape from the high aggregate mortality of the early modern era forces us to look more closely at the relationship between chronic malnutrition and the secular decline in mortality. In this connection, one must distinguish between diet (gross nutrition) and nutritional status (net nutrition: nutrients available to sustain physical development) (see Fogel 1986). In this essay, the term *diet* will always denote gross nutrition, while terms like *malnutrition, undernutrition, net nutrition,* and *nutritional status* will denote the balance between nutrient intake (diet) and claims on that intake.

Malnutrition can be caused either by an inadequate diet or by claims on that diet (including work and disease) so great as to produce malnutrition despite a nutrient intake that in other circumstances might be adequate. The high disease rates prevalent during the early modern era would have caused malnutrition even with extraordinary diets high in calories, proteins, and most other critical nutrients, but for many European nations prior to the middle of the nineteenth century, national food production was so low that the lower classes were bound to have been malnourished, and the high disease rates of the period were not merely a *cause* of malnutrition but also a *consequence* of exceedingly poor diets.

Recently improved economic analyses and biomedical estimates of survival levels of caloric consumption and of the caloric requirements of various types of labor; epidemiological studies of the connection between stature and the risk of both mortality and chronic diseases; and epidemiological studies of the connection between body mass indexes (BMI) and the risk of mortality make it possible to probe into the extent and consequences of chronic malnutrition in the eighteenth and nineteenth centuries.

Energy requirements beyond maintenance are normally divided between work and such discretionary activities as walking, community activities, games, optional household tasks, and athletics or other exercise. For a typical well-fed adult male engaged in heavy work, basal metabolic rate (BMR) and maintenance require about 60 percent of energy consumption, work 39 percent, and discretionary activity just 1 percent. For a well-fed adult male engaged in sedentary work (such as an office clerk), a typical distribution would be: BMR and maintenance 83 percent, work 5 percent, discretionary activity 13 percent. For a 25-year-old adult male engaged in subsistence farming in contemporary Asia, a typical distribution would be: BMR and maintenance 71 percent, work 21 percent, and discretionary activity 8 percent.

Toutain (1971) has estimated that the per capita consumption of calories in France was 1,753 during 1781–90 and 1,846 during 1803–12, or about 2,290 and 2,410 calories per consuming unit (equivalent adult male). English daily consumption during 1785–95 averaged about 2,700 calories per consuming unit (Fogel 1987; cf. Shammas 1984 and 1990). One way of assessing these two estimates is to consider their distributional implications. Table 5.1 displays the provisional caloric distributions for England and France implied by the available evidence. The estimates are consistent with the death rates of each nation. The crude death rate in France ca. 1785 was about 36.1 per 1,000; in England ca. 1790 it was about 26.7 (Weir 1984; Wrigley and Schofield 1981). It is plausible that much of the difference was due to the larger proportion of French than English who were literally starving (Scrimshaw 1987).

Two findings about caloric consumption at the end of the eighteenth century in France and England stand out. First is the exceedingly low

Table 5.1 A comparison of the provisional French and English distributions of the daily consumption of kcals per consuming unit toward the end of the eighteenth century

	A France ca. 1785 $\bar{X} = 2,290$ $(s/\bar{X}) = 0.3$		B England ca. 1790 $\bar{X} = 2,700$ $(s/\bar{X}) = 0.3$	
Decile *(1)*	Daily kcal consumption *(2)*	Cumulative % *(3)*	Daily kcal consumption *(4)*	Cumulative % *(5)*
1. Highest	3,672	100	4,329	100
2. Ninth	2,981	84	3,514	84
3. Eighth	2,676	71	3,155	71
4. Seventh	2,457	59	2,897	59
5. Sixth	2,276	48	2,684	48
6. Fifth	2,114	38	2,492	38
7. Fourth	1,958	29	2,309	29
8. Third	1,798	21	2,120	21
9. Second	1,614	13	1,903	13
10. First	1,310	6	1,545	6

Sources and procedures: See Fogel 1987, esp. tables 4 and 5 and note 6.

level of food production, especially in France, at the start of the Industrial Revolution. Second is the exceedingly low level of work capacity permitted by the food supply, even after allowing for reduced requirements for maintenance because of small stature and reduced body mass (cf. Freudenberger and Cummins 1976). In France, the bottom 10 percent of the labor force lacked sufficient energy for regular work, and the next 10 percent had only enough for less than three hours of light work daily (0.52 hour of heavy work). Although the English situation was somewhat better, the bottom 3 percent of England's labor force lacked the energy for any work, but the balance of the bottom 20 percent had enough energy for about six hours of light work (1.09 hours of heavy work) each day.

The available data on stature and body mass tend to confirm the basic results of the analysis based on energy cost accounting: chronic malnutrition was widespread in Europe during the eighteenth and nineteenth centuries.

Recent advances in biomedical knowledge make it possible to use anthropometric data for the eighteenth and nineteenth centuries to study secular trends in European nutrition, health, and risks of mortality. Extensive clinical and epidemiological studies show that height at given ages, weight at given ages, and weight-for-height (body mass index [BMI] = weight/height^2[kg/m^2]) are effective predictors of the risk of morbidity and mortality (Sommer and Lowenstein 1975; Chen, Chowdhury, and Huffman 1980; Billewicz and McGregor 1982; Kielmann et al. 1983; Martorell 1985; Marmot, Shipley, and Rose 1984; Waaler 1984; Fogel et al. 1986; Heywood 1983). Height and BMI measure different aspects of malnutrition and health. Height is a net rather than a gross measure of nutrition. Moreover, although changes in height during the growing years are sensitive to current levels of nutrition, mean final height reflects the accumulated past nutritional experience of individuals over all their growing years, including the fetal period. Thus, when final heights are used to explain differences in adult mortality rates, they reveal the effect not of adult nutrition on adult mortality rates but of infant, childhood, and adolescent nutrition on adult mortality rates. A weight-for-height index (BMI), in contrast, reflects primarily the current nutritional status. It is also a net measure in the sense that a BMI reflects the balance between energy intake and expenditure. Although *height* is determined by the *cumulative* nutritional status during an entire developmental age span, the *BMI* fluctuates with the *current* balance between nutrient intakes and energy demands. A person whose height is short relative to the modern US or West European standard is referred to as "stunted." One with low BMI is referred to as "wasted."

The predictive power of height and BMI with respect to morbidity and mortality is indicated by figures 5.3 and 5.4. Part A of figure 5.3 shows that short Norwegian men aged 40–59 between 1963 and 1979 were much more likely to die than tall men. Part B shows that height is also

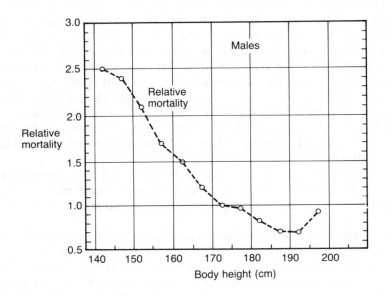

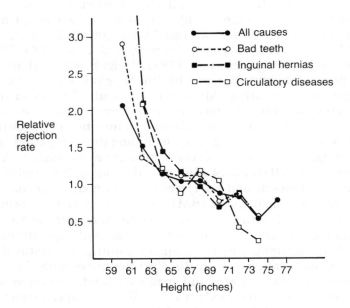

Figures 5.3a and 5.3b A comparison of the relationship between body height and relative risk in two populations. 5.3a: Relative mortality rates among Norwegian men aged 40–59, between 1963 and 1979. 5.3b: Relative rejection rates for chronic conditions in a sample of 4,245 men aged 23–49 examined for the Union Army
Sources: Waaler (1984); Fogel et al. (1986).

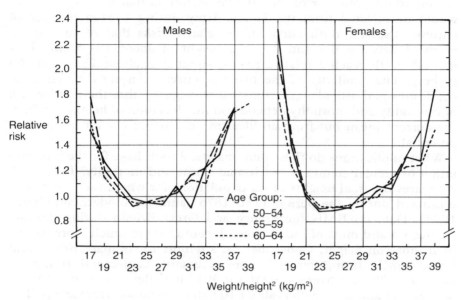

Figure 5.4 The relationship between BMI and prospective risk among Norwegian adults aged 50–64 at risk between 1963 and 1979
Source: Waaler (1984).

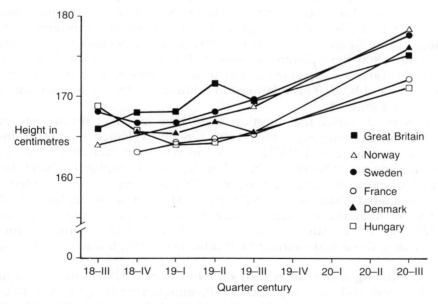

Figure 5.5 Estimated average final heights, European males
Source: Fogel (1992), table 9.

an important predictor of the relative likelihood that men aged 23–49 would be rejected from the Union Army during 1861–5 because of chronic diseases. Both curves have relative risks that reach a minimum of between 0.6 and 0.7 at a height of about 187.5 cm (73.9 inches). Both reach a relative risk of about 2 at about 152.5 cm (60 inches). The similarity of the two risk curves, despite differences in conditions and attendant circumstances, suggests that the relative risk of morbidity and mortality depends on the deviation of height not from the current mean but from an ideal mean associated with full genetic potential.

What implications do these new analytical tools have for the interpretation of secular trends in nutritional status and mortality? Figure 5.5 compares the final heights of six populations for which they have been estimated during the period 1750–1875 (and, for modern standards, 1975). Data on BMIs for France and Great Britain during the late eighteenth and most of the nineteenth centuries are much more patchy than those on height. Consequently, attempts to compare British and French BMIs during this period are necessarily conjectural. It appears that ca. 1790 the average BMI for British males about age 30 was between 21 and 22, which is about 10 percent below current levels. The corresponding figure for French males ca. 1785 may only have been about 19, which is about 25 percent below current levels (Fogel 1992). While the conjectural nature of these figures makes the attempt to go from the anthropometric data to differential mortality rates more illustrative than substantive, it appears that the French mortality rate predicted on this basis should have been about 35 percent higher than that of the English, which is quite close to the relative mortality rates indicated by figure 5.1. In other words, the available data suggest that in 1786–1800 the differences between France and England's average mortality rates are explained largely by differences in their distributions of height and BMI. This relationship is illustrated by the comparison of figures 5.1 and 5.5.

This result raises the question as to how much of the decline in European mortality rate since 1800 can be explained merely by increases in stature and BMIs, that is, merely by movements along an unchanging mortality risk surface. For the three European countries for which even patchy data are available – England, France, and Sweden – the estimated changes in height and BMI appear to explain about 90 percent of the decline in mortality during the century between 1775 and 1875. After 1875, increases in longevity involved factors other than those that exercise their influence through stature and body mass.

This finding is consistent with available data on US secular trends in both stature and mortality since 1720, summarized in figure 5.6. Both the series on stature and the series on life expectancy at age 10 rise during most of the eighteenth century, attaining both substantially greater heights and life expectations than prevailed in England during the same period (Floud 1985).

Figure 5.6 reveals not only that Americans achieved modern heights by the middle of the eighteenth century, but also that they reached levels of life expectancy not attained by the general population of England or even by the British peerage until the first quarter of the twentieth century (Fogel 1986, p. 467). The early attainment of modern stature and relatively long life expectancy is surprising. Yet it is by no means unreasonable. By the second quarter of the eighteenth century, Americans had achieved diets that were remarkably nutritious by European standards, and particularly rich in protein. Moreover, the low population density, marked by the low share of the population living in cities, reduced exposure to disease.

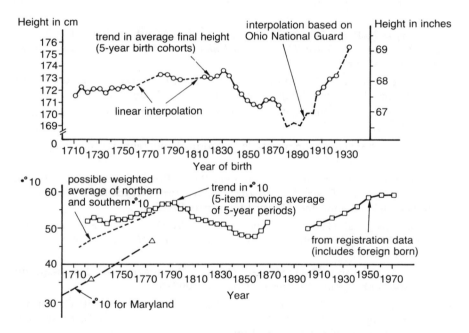

Figure 5.6 A comparison between the trend in the mean final height of native-born white males and the trend in their life expectation at age 10 a_{10}^0 (height by birth cohort; a_{10}^0 by period)
Source: Fogel 1986.

REFERENCES

Billewicz, W. Z., and I. A. MacGregor (1982): "A Birth to Maturity Longitudinal Study of Heights and Weights in Two West African (Gambian) Villages, 1951–1975." *Annals of Human Biology* 9(4), 309–20.

Chen, L., A. K. M. Chowdhury, and S. L. Huffman (1980): "Anthropometric Assessment of Energy-protein Malnutrition and Subsequent Risk of Mortality

Among Pre-school Aged Children." *American Journal of Clinical Nutrition* 33, 1836–45.

Floud, Roderick (1985): "Two Cultures? British and American Heights in the Nineteenth Century." Mimeographed. Birkbeck College, London.

Fogel, Robert W. (1986): "Nutrition and the Decline in Mortality Since 1700: Some Additional Preliminary Findings." In *Long Term Factors in American Economic Growth*, ed. Stanley L. Engerman and Robert E. Gallman. Conference in Research in Income and Wealth, vol. 41. University of Chicago Press (for National Bureau for Economic Research).

—— (1992): "Second Thoughts on the European Escape from Hunger: Famines, Chronic Malnutrition, and Mortality Rates." In *Nutrition and Poverty*, ed. S. R. Osmani. Oxford: Clarendon Press.

—— (1987): "Biomedical Approaches to the Estimation and Interpretation of Secular Trends in Labor Productivity, Equity, Morbidity, and Mortality in Western Europe and America, 1780–1980." Typescript, University of Chicago.

—— Clayne L. Pope, Samuel H. Preston, Nevin Scrimshaw, Peter Temin, and Larry T. Wimmer (1986): *The Aging of Union Army Men: A Longitudinal Study*. Cambridge, MA: photocopy.

Freudenberger, H., and G. Cummins (1976): "Health, Work, and Leisure Before the Industrial Revolution." *Explorations in Economic History* 13, 1–12.

Heywood, P. F. (1983): "Growth and Nutrition in Papua New Guinea." *Journal of Human Evolution* 12.

INED (1977): "Sixième rapport sur la situation démographique de la France." *Population* 32, 253–338.

Lee, R. (1981): "Short-term Variation: Vital Rates, Prices and Weather." In *The Population History of England, 1541–1871: A Reconstruction*, eds E. A. Wrigley and R. S. Schofield. Cambridge: Harvard University Press.

Kielmann, A. A., et al. (1983): *Child and Maternal Health Services in Rural India: The Narangwal Experiment*, vol. 1, *Integrated Nutrition and Health*. Baltimore: Johns Hopkins Press.

Marmot, M. G., and Rose Shipley (1984): "Inequalities of Death-specific Explanations of a General Pattern." *Lancet* 1, 1003–6.

Martorell, R. (1985): "Child Growth Retardation: A Discussion of its Causes and its Relationship to Health." In *Nutritional Adaptation in Man*, eds. Blaxter and Waterlow. London and Paris: John Libby.

Oddy, D. J. (1990): "Food, Drink, and Nutrition." In *The Cambridge Social History of Britain 1750–1950, Volume 2: People and Their Environment*, ed. F. M. L. Thompson. Cambridge: Cambridge University Press.

Schofield, R. (1983): "The Impact of Scarcity and Plenty on Population Change in England, 1541–1871." *Journal of Interdisciplinary History* 14, 265–91.

Scrimshaw, Nevin S. (1987): "The Phenomenon of Famine." *Annual Review of Nutrition* 7, 1–21.

Shammas, C. (1984): "The Eighteenth-century English Diet and Economic Change." *Explorations in Economic History* 21, 254–69.

Sommer, A., and M. S. Lowenstein (1975): "Nutritional Status and Mortality: A Prospective Validation of the QUAC Stick." *American Journal of Clinical Nutrition* 28, 287–92.

Toutain, J. (1971): "La consommation alimentaire en France de 1789 à 1964." *Economies et Societés, Cahiers de L'I.S.E.A.*, Tome, V, No. 11, 1909–2049.

Waaler, Hans Th. (1984): "Height, Weight and Mortality: The Norwegian Experience." *Acta Medica Scandinana* supplement no. 679. Stockholm.

Weir, D. R. (1984): "Life Under Pressure: France and England, 1670–1870." *Journal of Economic History* 44, 27–47.

Wrigley, E. A. (1990): *The Pre-industrial Consumer in England and America.* Oxford: Clarendon Press.

—— and R. S. Schofield (eds) (1981): *The Population History of England, 1541–1871: A Reconstruction.* Cambridge: Harvard University Press.

EDITOR'S NOTE (JULIAN L. SIMON)

Increasing heights and improving body mass indexes are closely correlated with increasing longevity over the past three centuries in Western countries (compare figures 5.1, 5.2, 5.3, 5.4, and 5.5). The increases in height and BMI are symptomatic of improving general health and longevity (figures 5.2 and 5.3) and are related to the conquest of chronic malnutrition by the improvement in water supplies and public sanitation, as well as by the provision in these countries of more adequate quantities and qualities of food not only to the wealthy and middle class but also to the poor of their populations. Indeed, differences in nutritional status as measured by height (figure 5.4) and BMI reasonably explain the differences in mortality (figure 5.1) between France and England prior to 1875.

6

Trends in Health of the US Population: 1957–89

Eileen M. Crimmins and Dominique G. Ingegneri

INTRODUCTION

The self-reported health of the US population has not improved over the past 30 years; in fact, there has been some trend toward a deterioration in self-reported health over the whole period from 1957 through 1989. Although change has not been consistent over this period, the evidence is that the 1970s were years when the reported health status of the population declined most markedly. Neither the 1960s nor the 1980s show such consistent change. Beginning in the early 1980s, some improvement in reported health occurred among those above age 44; the trend toward deteriorating health continued for those at younger ages.

What are the possible explanations for this pattern of change over time in self-reported health? One is that there has been a real change in health; health deteriorated during the 1970s and perhaps began to improve in some age groups during the 1980s. Another possible explanation is that there has been no change, or even improvement, in "real" health status, but that "reported" health has declined anyway. The decline in self-reported health could have occurred because people became more aware of their health, people's standards of good health and poor health changed, or people have changed their willingness to accommodate themselves to health problems. We review explanations for the trends in health of these two types after examining the data on the trends themselves.

DATA FOR TRENDS IN THE HEALTH OF THE US POPULATION

The National Health Interview Survey is the best source of data from which to construct information on trends in health over the longest period of time for the United States. This annual survey was begun in 1957 to monitor the health and health care usage of a representative sample of the noninstitutionalized population on a continuous basis. Each year over 100,000 individuals in approximately 40,000 households are interviewed for this purpose.

Only sketchy data are available before the early 1960s,[1] and individual records exist only from 1969 onward; thus, our data on trends are limited by what exists for the earlier years. However, over the more than 30 years the survey has been carried out, data have been collected annually on two separate indicators of health – "Limitation of Activity" and "Restricted Activity Days."

Both indicators are based on the idea that health is unimpaired functioning or the ability to carry out normal activities. Both measure the prevalence of ill health; that is, they indicate how much ill health there is in the population in a given period of time. This means that both will be influenced by how many episodes of ill health there are and how long each lasts. "Limitation of Activity" is an indicator of long-term disability, disability that is due to chronic conditions and diseases and for the most part has lasted at least three months. A person is limited in activity when he or she has difficulty performing his or her usual activity, or the activity that is normal for his or her age group.

"Restricted Activity Days" is designed to measure short-term disability. The respondent is asked how many days during the two weeks before the survey he or she had to cut down on normal activity because of health. Because restricted activity can be due to either acute conditions, like colds and sore throats, or chronic conditions, like heart disease, it is an indicator of the level of both acute and chronic illness. Reporting of restricted activity days is not limited to short-term disability by respondents. Many members of the population, especially older people, report that all their days are days of restricted activity.

While these two concepts have been applied fairly consistently throughout the period of the survey, changes in survey procedures and questionnaire format make data most comparable within the periods 1957 through 1968, 1969 through 1981, and 1982 to the present (Kovar and Poe, 1985; US DHEW, 1975). Some caution in examining change across the survey periods is warranted.

To present comparable series for as long a period as possible, in figures 6.1–6.4 we have adjusted several of the published series for obvious disturbances to the trend caused by procedural changes.[2] We feel confident that the 30-year trend is better portrayed with the adjusted data.

TRENDS IN HEALTH

The percentage limited in activity by age is shown in figure 6.1 from 1957 through 1989 for all ages and from 1962 through 1989 for broad

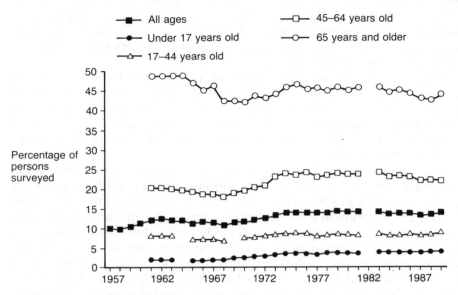

Figure 6.1 Persons reporting limited activity due to health, by age; USA, 1957–89
Sources: US Department of Health and Human Services, *Vital Health Statistics*, Series 10: Data from the national Health Interview Survey, nos. 5, 9, 13, 17, 25, 37, 43, 45, 51, 52, 60, 63, 72, 79, 85, 95, 100, 111, 115, 119, 126, 130, 136, 139, 141, 150, 154, 156, 160, 164, 166, 173, 176.

age groups. There is some increase in activity limitation for the entire population and for each age group under 65 over the whole period. For the whole population, the increase between 1957 and 1989 is 43 percent. This figure varies widely by age, ranging from an increase of 100 percent for youngsters less than 17 to a decrease of 10 percent for those 65 and over.

The change is not consistent over the whole time period, however. There is some decline in activity limitation or improvement in health during the 1960s, especially for the oldest age group. Beginning at the end of the 1960s there is a consistent and marked rise in activity limitation for all age groups through the late 1970s. For most age groups, the 1980s appear to be years of little change in health, as indicated by activity limitation, although there is some improvement in health above age 44 beginning in the early 1980s.

Activity limitation consists of difficulty in performing both major and secondary activities. Major activities are working, keeping house, and going to school, while secondary activities include recreational, civic, and cultural activities. Limitation in major activity might be considered more important in that it represents difficulty in fulfilling one's major social role. Trends in both types of limitation are shown in figure 6.2. The time trends in each type of activity are roughly the same as

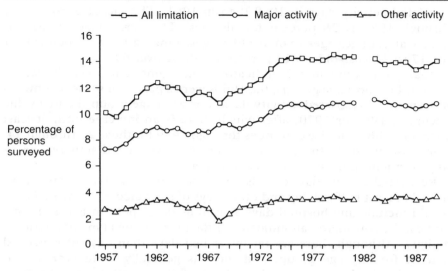

Figure 6.2 Persons reporting limited activity due to health, by degree of limitation
Source: see source for figure 6.1.

described earlier, except that there is a longer upward trend in limitation in major activity spanning both the 1960s and the 1970s.

The annual number of restricted activity days per year is shown for all ages and for broad age groups in figure 6.3. All age groups have more

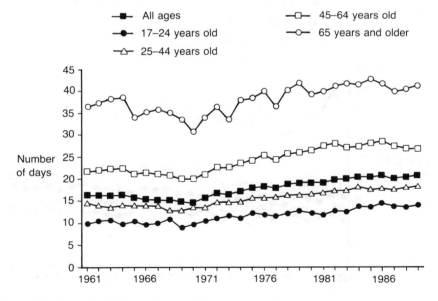

Figure 6.3 Annual days of restricted activity, per person, by age
Source: see source for figure 6.1.

restricted activity days in 1989 than in 1961. The increase from 1961 through 1989 is 28 percent for all ages. The percentage increase is higher at younger ages than at older ages, ranging from 45 percent for the 17 to 24-year-olds to 13 percent for those over 64.

Again, the trend in this indicator is not consistent over the whole period. For most age groups, the 1960s were a period of some downward movement in restricted activity days. For each age group the low value occurs in 1969 or 1970, after which there is an increase until at least 1985 or 1986. The rise continues through 1989 for those 25 through 44. For those above 44 there is some decrease in restricted activity days or improvement in health in the last few years.

Restricted activity days can also be divided into two types: days spent in bed at least half the day and other days of restricted activity. Bed days would include any hospital days as well as days spent in bed at home, and can be considered an indicator of the most serious type of restricted activity. Although there are far fewer bed days than days of restricted activity for any age group, and there is generally more year-to-year fluctuation in this measure, the pattern of change over time, shown in figure 6.4, is somewhat similar for bed days and all restricted activity days. For most age groups a low point in bed days is reached sometime in the late 1960s, followed by a very slow and gradual rise that continues to 1989 in those less than 45 years of age and ends in the early 1980s for those over 44.

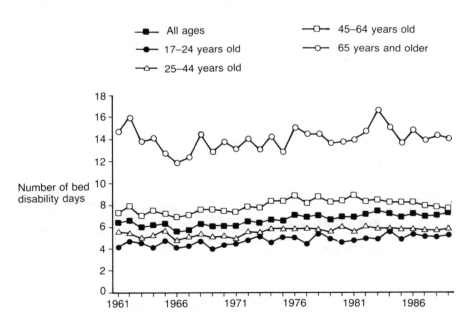

Figure 6.4 Annual bed disability days per person, by age; USA, 1961–89
Source: see source for figure 6.1.

EMPIRICAL EVIDENCE ON TRENDS IN
HEALTH FROM OTHER SOURCES

Whereas the data presented previously cover more age groups and a longer time period than most other studies that have investigated changes in health over time, other empirical work has tended to confirm the idea that the health of the population has deteriorated in the United States in recent years. Findings of this nature have been reported in a large number of studies based on National Health Interview Survey (NHIS) data like those presented here. These include studies of health change at all ages (Riley, 1990; Crimmins et al., 1989), as well as studies concentrating on segments of the population including children (Newacheck, et al., 1984, 1986), the working-age population (Verbrugge, 1984, 1989; Chirikos, 1986; Feldman, 1983; Colvez and Blanchet, 1981), and parts of the older population (Crimmins, 1987, 1990). Examination of health or disability change using other data, such as the decennial census and the Current Population Survey, have reached similar conclusions for the working age population (Bound and Waidmann, 1990; Wolfe and Haveman, 1990).

Most of this research has examined change during the 1970s. Riley (1990) and Ycas (1987) have incorporated data from the NHIS for several years of the 1980s into their work, and the conclusions from these two studies differ. Riley, using only the restricted activity days indicator, continued to see deteriorating health in the United States through 1986, while Ycas questions whether the rise in ill health ended with the 1980s. Both Bound and Waidmann (1990) and Wolfe and Haveman (1990) note that disability rates among the working age population improved markedly during the 1980s.

Health change has also been investigated in a variety of other countries where mortality is both low and continuing to decline. Surveys have shown deteriorating health in Canada during the 1970s (Wilkins and Adams, 1983), Australia during the 1980s (Mathers, 1990), Great Britain from the 1970s through the mid-1980s (Bebbington, 1988; Riley, 1990), and Japan from the 1950s through the 1980s (Riley, 1990).

POSSIBLE EXPLANATIONS OF THE OBSERVED
TREND IN HEALTH

As we mentioned in the introduction, explanations of the observed trends in perceived health fall into two categories: those that assume the change in reported health reflects change in "real" health, and those suggesting that what has changed is "reported," not "real" health. We will discuss explanations of each type. The idea that average real health has been or is deteriorating seems to be increasingly accepted by researchers in the field (Verbrugge, 1984; Riley, 1990). Some observers might question how it is possible that health has not improved during a period in which mortality declines have been dramatic. The answer is twofold: first, the causes of ill health differ markedly from the causes of death, meaning that change in mortality and change in morbidity do not have to be closely linked; second, the factors leading to the recent mortality decline itself may have increased the number of unhealthy people surviving in the population. We discuss each of these reasons in more detail.

First, recently we have recognized that a significant amount of ill health or morbidity is caused by a set of diseases that are not causes of mortality. Among the most important of these conditions are arthritis, musculoskeletal conditions, vision and hearing loss, and mental conditions (Verbrugge, 1984). Almost no one dies of these problems, but many people suffer from them for large portions of their lives. The proportion of people suffering from these conditions and their relative importance in causing ill health increases with age. About half of the long-term disability suffered by those 65 years of age and over is due to these nonfatal diseases (Crimmins, 1990). As people survive from fatal diseases and live to older ages, they are more likely to acquire these nonfatal diseases, leading to more health problems among the surviving population because of the higher proportion of people with multiple diseases or "co-morbid" conditions.

Now to turn to how the characteristics of the recent mortality decline itself could be linked to increasing ill health. While mortality decline has been dramatic over the course of the twentieth century, its causes have changed in the last two decades. From 1900 through the 1950s, mortality decline was due primarily to reduction in deaths from infectious disease. From the mid-1950s through the mid-1960s there was a period of relative stability in mortality, which ended in about 1968 when another period of mortality decline began that has continued up to 1990. In contrast to earlier years, however, the post-1967 mortality decline has been due to a reduction in deaths from chronic diseases, especially heart disease (Crimmins, 1981).

Death rate reduction in age groups that do not die in significant numbers from heart disease could also be related to increasing rates of reported disability. Infants and children have experienced significant mortality declines from some neonatal and chronic conditions during the period from the mid-1960s to the present. For instance, survival rates of low birthweight babies have soared upward. These children are often born with significant long-term health problems that would result in an increasing level of disability in the younger population with survival of increasingly frail babies. Gruenberg's (1977) pessimistic outlook on the results of lowering mortality on the health of the surviving population is related to his interest in Down's Syndrome. He notes that until recently children with this and other conditions would have died fairly young; now they have been saved to live a lifespan not much shorter than the rest of the population. Also, people in the early adult years have benefited from improved survival rates from trauma, which may leave people alive but with permanent or long-term disabling conditions.

The switch in causes of mortality decline from infectious to chronic conditions may have resulted in a new relationship between mortality change and health change. The decline in infectious diseases came about either because people no longer got diseases like polio, measles, and smallpox, or because they were treated and cured of diseases like tuberculosis and scarlet fever. There is little evidence that death rates today have been reduced because people are not developing chronic diseases. In fact, one study of US morbidity in the 1970 to 1980 period showed that chronic conditions among the elderly were developing at slightly younger ages (Crimmins, 1987).

In contrast to the situation for infectious diseases, there are no known "cures" for most chronic conditions. Once an individual develops a chronic disease or condition, he or she is likely to have it for the remainder of his or her life. The result of people living longer and diseases developing at the same age or earlier is that people will live longer with diseases. When mortality decline occurs because people who have chronic diseases survive death and continue to live longer with disease, the proportion of the surviving population suffering from disease will increase.

It is possible that while the number of people with a disease will increase, the average level of disease severity could be reduced. If the treatment of the disease prevents its progression as well as death, it is possible that the disease severity among the surviving population could be reduced (Manton, 1982). This would be an offsetting factor, resulting in more sick people but with less severe sickness and, perhaps, less incapacitation.

The previous discussion is based on the assumption that the population is homogeneous in its likelihood of acquisition and progression of disease. However, within a population there is a distribution of frailty, or the likelihood that one will be affected by diseases. When mortality rates decline, the group that is "saved" from death tends to come from the frailest end of the distribution. This means that the new survivors are at greater risk of developing other conditions and diseases than the rest of the population (Alter and Riley, 1989; Verbrugge, 1989).

Most recent theoretical work on the relationship of health change to mortality change (Manton, 1982; Verbrugge, 1990; Riley, 1990) has come to accept the idea that mortality change of the nature that we have experienced is likely to be accompanied by a rising prevalence of morbidity as people live longer with diseases. This rise in morbidity should be viewed as an effect of our success in preventing death from a wide variety of diseases and a necessary stage in an "epidemiologic transition." Only when prevention and delay of the onset of a large number of chronic conditions becomes possible is the trend likely to change to one of decreasing morbidity.

If one accepts this explanation for rising morbidity during the 1970s, the logical explanation for declining morbidity in the 1980s among those over middle age is that the era of disease prevention has begun. To document this, further investigation into changes in age at onset and incidence of diseases is required. If age at onset of disease rose and incidence was reduced in the 1980s, this is a plausible explanation of the trend in ill health from the 1970s through the 1980s.

Riley (1990) has noted that the inconsistent pattern of health change in the United States in the 1960s occurred during a period of relative stability in mortality and thus is not inconsistent with this explanation of the relationship of health change to mortality change.

Some analysts find it hard to accept the conclusion that real health has deteriorated and emphasize the fact that the trend in self-reported health can be affected by factors other than changes in health. For instance, National Center for Health Statistics researchers have cautioned against accepting the idea that health has deteriorated and have stressed that survey procedures and perhaps completeness of reporting have improved over the period under investigation here (Wilson and Drury, 1984). However, we are led to discount this as an important explanation of the time trend by the following evidence: the pervasiveness of similar change in health in the 1970s across a number of data sets in the United States and the evidence of similar change in other countries, the change in the trend in the 1980s, and estimates of the likely small size of the effect of changing survey procedures (Bound and Waidmann, 1990).

Others who believe that there was no real change in health think that reported health changed because standards of good health changed among Americans. This syndrome has been called "Worried Sick" by Barsky (1988), who feels that while real health has improved, Americans have become so fixated on good health that they now worry increasingly over less significant health problems and report themselves in increasingly worse health as time goes on. Although there is some intuitive appeal to this argument as an explanation for trends in the 1970s, it is difficult to see how people above age 44 would have changed their views in recent years to reverse the direction of the trend in reported health.

Some researchers believe that accommodation to health problems may have changed among Americans (Verbrugge, 1989, 1990; Crimmins and Pramaggiore, 1988). People may be more willing to change their lives at an earlier phase of illness in order to prevent the progression of disease. These arguments have been put forth in order to reconcile the findings that trends in reported health and ease of obtaining disability benefits seem to go together for the working age population (Bound and Waidmann, 1990; Wolfe and Haveman, 1990; Bailey, 1987). This explanation certainly fits the time trend for the older working age and retirement age population of deteriorating health during the 1970s, a period when benefits became increasingly available, and improving health during the 1980s, a time when benefits became more difficult to obtain. This explanation makes sense for the working age population, but it is difficult to reconcile this with the pervasiveness of the trend toward deteriorating health during the 1970s for all other ages.

Of course, it is possible that not one but some combination of these explanations is responsible for the observed trend. Unfortunately, we cannot choose among them or assign relative weighting to their import-ance on either theoretical or empirical grounds.

SUMMARY AND CONCLUSIONS

Evidence on trends in health for the United States indicates that self-reported health deteriorated over the 1957 through 1989 period with much of the deterioration concentrated in the 1970s. There is some evidence that the trend toward deterioration in self-reported health among the middle-aged and older population was arrested and perhaps even reversed in the 1980s.

One hypothesis offered as an explanation is that the deterioration in health during the 1970s was due to the remarkable progress during that decade in reducing mortality from a number of chronic diseases by improving the diagnosis and treatment of existing conditions and dis-eases. The pattern of change in health in the 1980s may indicate that a new phase of disease prevention and delay at the older ages has begun, but further research is necessary to confirm this. We cannot rule out change in attitudes and behavior accompanying ill health as an explana-tion of the trend, but we believe that this is a less plausible explanation given the empirical evidence. We should note that the measures of health presented do not control for any change in the severity of health problems affecting normal activity over the period. It is possible that the average level of impairment has changed along with the number of people affected; thus, perhaps a trend controlled for severity would tell a different story.

These trends in health do not mean that even in the 1970s the expected years of healthy life decreased for the average individual, but rather that increases in life expectancy were likely to be in years with some diminished functioning from chronic disease. It is possible to link measures of mortality and measures of morbidity using life table tech-niques to divide life expectancy into years with a disability and years free of disability (Crimmins et al., 1989). For the period of increasing ill health in the 1970s, combining the measures of restricted activity days and limitation of activity described earlier with mortality change results in stability in expected years of life with disability but increases in total life expectancy. Even the small declines in limitation of activity and restricted activity days that appear to have taken place in the late 1980s may mean that we have begun to add healthy years to life expectancy.

NOTES

1 Riley (1989), however, makes the case that even in the latter part of the nineteenth century, the length of sickness was increasing for people who fell ill. He emphasizes that this increasing duration of sickness may have resulted in adverse trends in health even before the 1970s.

2 While the effect of the change from the first set of procedures to the second on our indicators is not obvious, the changes instituted in 1982 had striking effects on the level of ill health reported for some age groups on both limitation of activity and restricted activity days. In addition, even a cursory glance at the time trend data for limitation of activity indicates that 1977 was an unusual year. A special supplement on disability was added to the main survey during this year to be administered to those with some limitation in activity in order to reduce the number of supplements they would have to administer.

The method applied to determine the adjustment for the series of limitation of activity was to regress the annual proportion reporting limitation of activity for all ages and each age group separately on year, year squared, a dummy variable for 1977, and a dummy variable indicating the period 1983 through 1989. Data for the years 1968 through 1989 were used in the regression. No data are available for limitation of activity in 1982 because of the procedures employed in the original data collection, thus the proportion of the population limited in 1981 was substituted for 1982 in the regression.

For each series the equation resulted in an R^2 greater than 0.90. Adjustments were made when the probability level of the coefficient was 0.1 or smaller.

For adjusting restricted activity days the procedure was similar except that no indicator for 1977 was included in the regression, and the dummy variable represented the period 1982 through 1989.

REFERENCES

Alter, G., and J. C. Riley (1989): "Frailty, Sickness and Death: Models of Morbidity and Mortality in Historical Populations." *Population Studies*, 43, 25–46.

Bailey, M. N. (1987): "Aging and the Ability to Work: Policy Issues and Recent Trends." In: Gary Burtless, ed., *Work, Health and Income Among the Elderly*, Washington, DC: Brookings, 59–97.

Barsky, Arthur J. III (1988): *Worried Sick: Our Troubled Quest for Wellness*. Boston, Little, Brown.

Bebbington, A. C. (1988): "The Expectation of Life Without Disability in England and Wales." *Social Science and Medicine*, 27, 321–6.

Bound, J., and T. Waidman. (1990): "Disability Transfers and the Labor Force Attachment of Older Men: Evidence from the Historical Record." NBER Working Paper, No. 3437.

Chirikos, T. N. (1986): "Accounting for the Historical Rise in Work-Disability Prevalence." *The Milbank Quarterly*, 64, 271–301.

Colvez, A., and M. Blanchet (1981): "Disability Trends in the United States Population 1966–76: Analysis of Reported Causes." *Amer. J. Public Health*, 71, 464–71.

Crimmins, E. M. (1981): "The Changing Pattern of American Mortality Decline, 1940–1977, and Its Implications for the Future," *Population and Development Review*, 7, 229–54.

——— (1987): "Evidence on the Compression of Morbidity." *Gerontologica Perspecta*, 1, 45–9.

——— (1990): "Are Americans Healthier as Well as Longer-Lived?" *J. Insurance Medicine*, 22, 89–92.

——— and M. Prammaggiore (1988): "Changing Health of the Older, Population and Retirement Patterns Over Time." In: R. Ricardo-Campbell and E. Lazear (eds), *Issues in Contemporary Retirement*, Stanford, CA: Hoover Institution Press.

——— Y. Saito, and D. G. Ingegneri (1989): "Changes in Life Expectancy and Disability-Free Life Expectancy in the United States." *Population and Development Review*, 15, 235–67.

Feldman, J. J. (1983): "Work Ability of the Aged under Conditions of Improving Mortality." *Milbank Memorial Fund Quarterly/Health and Society*, 61, 430–44.

Fries, J. F. (1980): "Aging, Natural Death and the Compression of Morbidity." *New England J. Medicine*, 303, 130–5.

Gruenberg, E. M. (1977): "The Failures of Success." *Milbank Memorial Fund Quarterly*, 55, 3–34.

Kovar, M. G., and G. S. Poe (1985): *National Center for Health Statistics: The National Health Interview Survey Design, 1973–84, and Procedures, 1975–83. Vital and Health Statistics*, Series 1, No. 18, DHHS Pub. No. (PHS) 85–1320. Washington, DC: US Government Printing Office.

Manton, K. G. (1982): "Changing Concepts of Morbidity and Mortality in the Elderly Population." *Milbank Memorial Fund Quarterly/Health and Society*, 60, 183–244.

Mathers, C. D. (1990): "Disability-Free and Handicap-Free Life Expectancy in Australia." *Australia Institute of Health: Health Differentials Series No. 1.*

Newacheck, P. W., P. P. Budetti, and P. McManus (1984): "Trends in Childhood Disability." *Amer. J. Public Health*, 74, 232–6.

———, ——— and N. Halfon (1986): "Trends in Activity-Limiting Chronic Conditions Among Children." *Amer. J. Public Health*, 76, 178–84.

Palmore, E. B. (1986): "Trends in the Health of the Aged." *Gerontologist*, 26, 298–302.

Riley, J. C. (1989): *Sickness, Recovery, and Death: A History and Forecast of Ill Health*. Iowa City: University of Iowa Press.

——— (1990): "The Risk of Being Sick: Morbidity Trends in Four Countries." *Population and Development Review*, 16, 403–32.

United States Department of Health, Education, and Welfare (1975): *Health Interview Survey Procedure 1957–1974, Vital and Health Statistics*. Series 1. No. 11. DHEW Pub. No. (HRA) 75–1311. Rockville, Md.: National Center for Health Statistics.

Verbrugge, L. M. (1984): "Longer Life but Worsening Health?: Trends In: Health and Mortality of Middle-Aged and Older Persons." *Milbank Memorial Fund Quarterly/Health and Society*, 62, 475–519.

——— (1989): "Recent, Present, and Future Health of American Adults." In: L. Breslow, J. E. Fielding, and L. B. Lave (eds), *Annual Review of Public Health*. Vol. 10. Palo Alto, CA: Annual Reviews Inc.

——— (1990): "Pathways of Health and Death." In: Rima D. Apple (ed.), *Women, Health, and Medicine in America: A Historical Handbook*, New York: Garland, 41–79.

Wilkins, R., and O. B. Adams (1983): "Health Expectancy in Canada, Late 1970's: Demographic, Regional, and Social Dimensions." *Amer. J. Public Health*, 73, 1073–80.

Wilson, R., and T. Drury (1984): "Interpreting Trends in Illness and Disability: Health Statistics and Health Status." *Annual Review of Public Health*, 5.

Wolfe, B., and R. Haveman (1990): "Trends in the Prevalence of Work Disability from 1962 to 1984, and Their Correlates." *The Milbank Quarterly*, 68, 53–80.

Ycas, M. A. (1987): "Are the Eighties Different?: Continuity and Change in the Health of Older Persons." *Proceedings of the 1987 Public Health Conference on Records and Statistics*. DHHS Pub. No. (PHS) 88–1214. Hyattsville, Md: National Center for Health Statistics.

7

Mortality and Health in the former Soviet Union

Murray Feshbach

The former Soviet Union is *sui generis* – in demography no less than in other facets of Soviet society and economy. It follows no discernible pattern of other developed nations in terms of health, morbidity, and mortality. Life expectancy at birth has not yet recovered its low level from the mid-1960s and even declined in the last two years of the union, and in Russia and Ukraine since then. Maternal mortality persists at a rate 6 to 7 times that of the United States, while infant mortality rates are only 15 percent higher. The rate of typhoid in 1989 was 15 times the rate in the United States, while the rate of diphtheria was nearly 190 times higher and a major epidemic occurred in Moscow in 1990 and 1992 (close to 700 and 2,000 cases, respectively, compared with 94 in 1989 – and 0 to 4 cases annually in the United States). New cases of tuberculosis per year are 6 to 10 times higher than in the United States; only rising US rates due to AIDS have brought the comparative levels as low as 6 in recent years. The survival ratio of all Soviet males aged 25 to 50 in 1986–7 was only slightly above the terribly low rate for US blacks (about 85 percent) in 1987 (both were about 9 percentage points below the rate for US whites).

MORTALITY RATES AND LIFE EXPECTANCY

The overall situation can be seen from Soviet-prepared data comparing Soviet cause-of-death rates, by sex, with those of the West (defined as the United States, West Germany, France, Great Britain, and Japan). Only the rates for cancer among women are close to the same. Most others, as shown in figure 7.1, are about 1.4 to 2 times as high; the disparity for males dying from respiratory illnesses is 2.83 times.

Crude death rates, crude as they are, give some indication of broad trends over decades (figure 7.2). From a low of 6.9 deaths per 1,000 in 1964, the rate increased by some 57 percent to a postwar recorded high

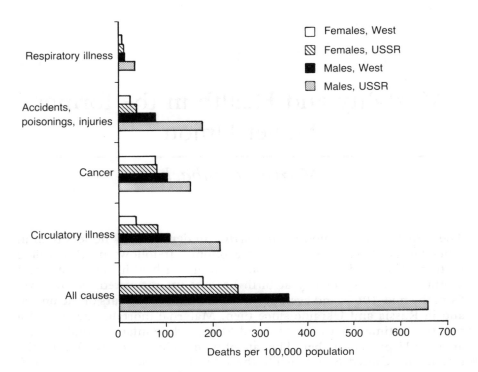

Figure 7.1 Causes of death, USSR and the West, able-bodied ages, by sex, 1987
Note: "West" is defined as USA, West Germany, France, Great Britain, and Japan. Population of able-bodied ages is defined as 16–59 years for men and 16–54 years for women, inclusive, in Soviet official statistics.
Source: Base data from Goskomstat SSSR, Press Release no. 191, May 5, 1989, 2.

of 10.8 in 1984. Most countries of western Europe showed an increase of only 10 to 12 percent over this period, the United States a small but steady decline. Aging population only partly explains the Soviet increase. The increase in infant mortality and premature mortality of males 20 to 44 years old accounts for more. After a decline between 1984 and 1986, the crude death rate rose again, to 10.4. The trends for men and women are similar.

Figure 7.3 depicts recent trends in life expectancy at birth, with isolated data also for 1896–7, and 1926–7. There was a significant decline between 1964 and 1985. Afterward, life expectancy rose significantly for both men and women but fell slightly again after 1987 for men. Despite *glasnost*, and the collapse of the Soviet State, the Russians had not, as of the time of writing, published life expectancy data (as age-specific death rates) for 1981/1982. I cannot prove it, but I firmly believe that this was the low year for men at least – closer to 60 than to 63 years. For women, however, the rates show an interesting pattern of

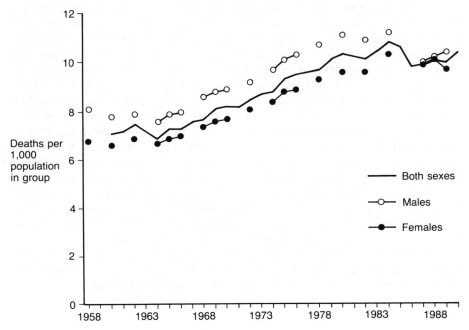

Figure 7.2 Crude death rates, by sex, USSR, 1958–90
Sources: Naseleniye SSSR 1988, 414–15; *Naseleniye SSSR 1987,* 320–1; and Goskomstat SSSR,
Demograficheskiy yezhegodnik SSSR 1990 (Moscow: Finansy i statisticka, 1990), 362.

increasing levels in urban areas and a fairly continuing decline in rural
areas until 1984–5, some increase, and then a decrease in 1989. Other
evidence indicates that rural environmental conditions, as well as poor
medical services, led to the premature deaths of many women, especially
in the rural south.

Recent data on mortality by basic cause of death (figure 7.4) show a
remarkably high rate for infectious and parasitic diseases that, while it
declined slightly after 1985, is still about 20 times higher than in the
United States. Cardiovascular disease rates are about 1.5 to 2 times
higher. Given Soviet problems with air pollution, the respiratory disease
rates are not surprising. A former Russian minister of health has written
that if someone wants to live longer, he should breathe less!

Until perhaps five years ago, infant mortality data (figure 7.5) were
largely unavailable. Even when first published in the postwar period,
they included only benchmark years (e.g., every fifth). Now, however,
we can observe a significant decline from the early twentieth century to
the 1960s, with a low in 1971. The increase from 22.9 infant deaths per
1,000 live births in 1971 to 31. Tragically, however, the rate rose by 37
percent to 31.4 per 1,000 by 1976, after which it declined only very
gradually until, in 1989, at 22.6 per 1,000, it surpassed the previous low
for the first time.

The former Soviet Union's infant mortality data are made suspect, however, by significant lying, with such errors of ommission ranging from 19 to 86 percent among rural communities. Combined with some errors of omission and adjusted for rural and urban settings, the net result could be a national infant mortality rate some 50 percent higher than reported. Moreover, Soviet methodology has excluded children born live who die before their seventh day of life and meet three other critieria: weight of less than 1,000 grams at birth, under 28 weeks' gestation, and less than 35 centimeters in length. Those who die before the seventh day – some 15 percent or more – are considered stillbirths. Adding these deaths to the earlier adjusted data indicates a national infant mortality rate of about 38 per 1,000 – 68 percent above the reported rate and over four times the US rate. The Soviet criteria have been even more restrictive than the United Nations' minimum of 500 grams and 22 weeks' gestation. Adjusting for this differential would raise the rate by 25–30 percent according to Russian officials.

Also shown in figure 7.5 are recent – and recently released – data on maternal mortality rates. The decline of about 25 percent through the 1980s, the rate in 1988 was still over four times the US rate.

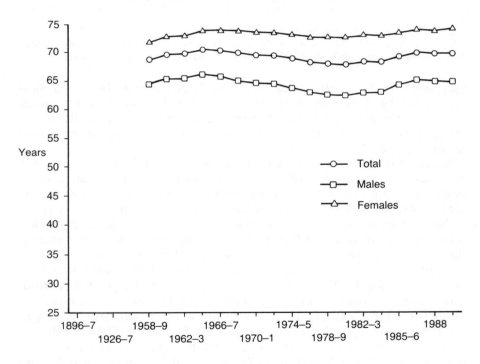

Figure 7.3 Life expectancy at birth, by sex, USSR, 1897–1989
Source: *Naseleniye SSSR 1987*, 351–2; *Narkhozy 1989*, 43; and *Press-vypusk No. 358*, Sept. 20, 1990, 14.

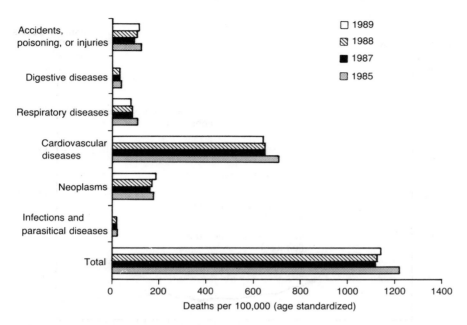

Figure 7.4 Mortality by cause of death (rate), USSR, 1985–9
Sources: Goskomstat SSSR, *Press-vypusk No. 42*, Feb. 10, 1989; and *Press-vypusk No. 62*, Feb. 13, 1990.

MORBIDITY RATES

The former Soviet Union has also suffered from disease rates that tend to exceed those in other developed countries. Particularly alarming have been recent trends in tuberculosis. The number of new cases registered annually is presently about 500 percent higher than in the United States, while the population is only about 15 percent higher. Although there was a sharp decrease between 1970 and 1989, most of that occurred by 1980 (down by 69 percent), not from 1980 to 1989 (down by 10 percent). During the two decades, the share of tuberculosis of the respiratory system rose from 85.8 percent to 89.7 percent.

The rate of infectious hepatitis reached its low in 1969, a high in 1983, and declined to a level still 78 percent higher than 1969 in 1988, and increased to 1989. Information in secondary sources indicates that this rate increased again in 1990. My rough estimate of diphtheria in the USSR for 1990 is not less than 1,400 cases (higher than at any time since 1968), with about 700 cases in Moscow alone – an incredible increase from 840 cases for the whole country and 94 in Moscow in 1989. Since some 21 percent of Soviet children did not receive DPT (diphtheria–pertussis–tetanus) shots in 1989, the potential for major increases in diphtheria is clear.

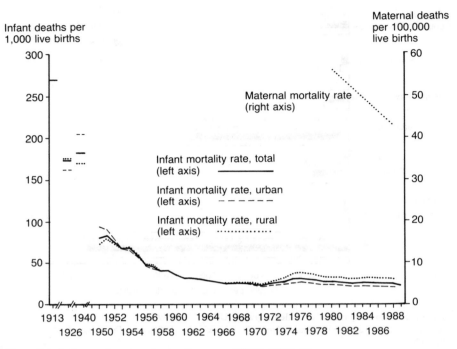

Figure 7.5 Infant and maternal mortality rates, USSR, 1913–89
Sources: Goskomstat SSSR, *Naseleniye SSSR 1987*, 344; *Narkhozy 1989*, 40.

The rate for polio is nearly 40 times that in the US, though we may hope a recent decline will continue and not repeat an increase from 1984 to 1987. Measles has finally dropped below 100,000 cases per year (compared with five to 10,000 cases per year in the US). Typhoid has dropped below 10,000 cases per year, but I expect it to increase because of major deterioration in water quality and the pipeline system. The general trend for influenza and upper respiratory infections is upward, but it is not possible to be certain of the pattern given problems with medical delivery, the availability of heating fuel, food supply, and air pollution. Overall, high levels of infectious diseases prevail in the former Soviet Union.

8

Worldwide Historical Trends in Murder and Suicide

Jean-Claude Chesnais

The rate of suicide tends to increase with stages of social and economic development, while the rate of homicide tends to decrease. In literate societies, suicide rates are high, while murder rates are low, and vice versa. The cross-sectional evidence suggests that in more highly structured societies, where duties are heavily stressed, the suicide/murder ratio is high, while in relatively unstructured societies the situation is the reverse.

HOMICIDE TRENDS SINCE 1860

In contemporary developed nations, death at the hands of a second party is very rare. Excluding the United States, annual mortality from homicide in the West today is around one per 100,000 population (figure 8.1). A clear pattern of convergence has emerged during the last two centuries between the early modernized countries of northwest Europe, which had a low rate of mortal crime, and the agricultural nations to the south and east of the continent. A century ago in Italy, for instance, the homicide rate was about five per 100,000 population. By 1930, this rate had been halved, and it has since declined to a low of one per 100,000 in the 1960s. There are only two exceptions to this declining historical trend in Italy: peaks at the end of each world war. In most European societies, the pattern is similar. Up to the mid-twentieth century, there is a downward trend decade by decade. By 1860–80, in Sweden as well as in England, the death rate from murder was about two per 100,000 inhabitants; in the following decades, it declined to less than one.

In many Western nations, however, this trend has reversed slightly since the mid-1960s. For the United States, the contrast is sharper between the last two decades and the previous ones. But even in the United States, reported property crimes (burglary, larceny, theft) have

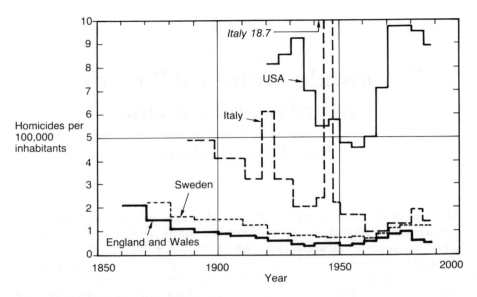

Figure 8.1 Homicide rates since 1850

increased much faster than crimes against persons, and the country is more lawful and safer than popular opinion imagines. International comparisons lend perspective. In El Salvador, for instance, the rate is above 30 per 100,000 inhabitants; by 1970, in Thailand and the Philippines, it was close to 20; in Mexico and Colombia, close to 15; in areas that produce and sell drugs, the incidence is much higher still.

Rates of recorded infanticide are clearly down; in more developed countries, they are commonly under 10 per 100,000 children under age one. France is interesting in this respect: during the second half of the nineteenth century, its infanticide rate was constantly in the range of 15–20; however, the proportion of unwanted pregnancies was surely lower than in any other society since France had the lowest fertility rate in the world (in 1850, 3.5 children per woman, instead of 5–6 in most nations). This infanticide rate fell progressively to one per 100,000 during the 1970s. The actual decrease is probably even greater, since in rural areas many infanticides were recorded as accidental deaths.

SUICIDE TRENDS SINCE 1750

Traditional societies have a low rate of suicide. In the middle decades of the last century, no country outside the German cultural area of central Europe had a rate above 10 per 100,000 inhabitants. Even Denmark, then known to have by far the highest rate in the world, had a lower rate

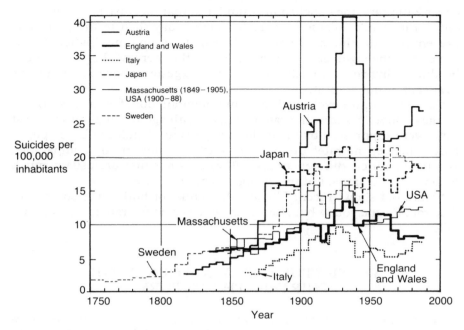

Figure 8.2 Suicide rates since 1750

than it does now (25 instead of 30 per 100,000). Currently, crude suicide rates for the two sexes combined tend to be well below 10 per 100,000 for the less-developed countries, while the only contemporary Western nations with rates below 10 are the Catholic nations of southern Europe (Italy, Spain, Portugal, and Greece) and the British Isles (figure 8.2). At the end of the 1980s, the recorded rates for the most populated Western countries were: United States, 12.2; Japan, 16.4; West Germany, 14.5; France, 19.7; England and Wales, 7; Italy, 7. In the Western world, most rates are between 10 and 20 (figure 8.3).

The one major exception is Hungary, which, from the Soviet-invasion in 1956 until the mid-1980s, had a steadily rising rate that culminated at 45 – unprecedented in the world – at the beginning of the 1980s. The previous world peak was in Austria in the 1930s (41). Neither Sweden nor Japan, both with reputations for high suicide rates, has reached such levels. Although the Swedish rate has climbed from very low levels over the last 240 years, only in 1965–75 did it go just over 20. In Japan, where suicide is still positively viewed and often considered an ultimate act of honor, the rate reached its maximum at about 24 in 1956–60. Central Europe, continuing its nineteenth-century pattern, sustains especially high rates – rates about 30 are not exceptional, but the rising trend seems to have stopped. In Austria, for example, in the 1930s, the rate was much higher than it is today.

Let us return to long-run trends. According to Swedish data, which cover 240 years, the conclusion is similar to that for Austria: a consistent pattern in which suicide is more frequent today than in the past. The present recorded incidence of suicide is six to seven times higher than at the end of the eighteenth century. (Part of the increase, however, could be due to improved registration, as we suggested above in discussing France, but not all of it.)

Suicide is almost exclusively an adult phenomenon: few suicides occur at ages under 15 years; the rate increases with age throughout life. Past rates were suppressed by the higher proportion of children in the population; hence, the rates quoted previously are not strictly comparable. But the historical differences are still valid, since they are usually much larger than the impact of the changing age structure.

Figure 8.4 shows recent US suicide trends for both sexes combined. The rate for the young has risen sharply, whereas the rate for the old has fallen.

A CASE STUDY: ENGLAND SINCE THE THIRTEENTH CENTURY

A systematic study of deaths from violence since the medieval period is possible only for England, where the data are sufficient.

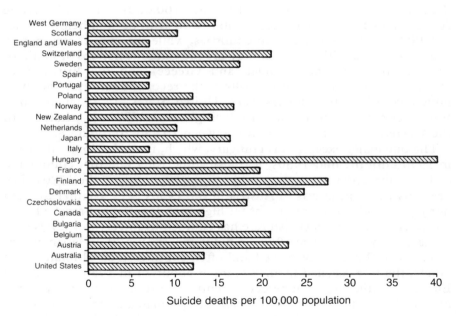

Figure 8.3 Age-standardized suicide rates, selected countries, late 1980s
Source: World Health Organization, *1990 World Health Statistics Annual* (Geneva: WHO).

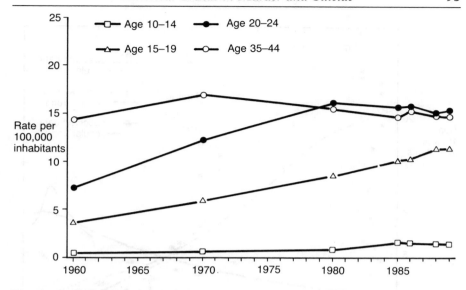

Figure 8.4a Suicide trends in the USA
Source: US Bureau of the Census, Statistical Abstract of the United States, various years.

Violence as a way of life

In past centuries, violent deaths were a prominent feature of peasant life.
The homicide rate varied from county to county, but it was universally

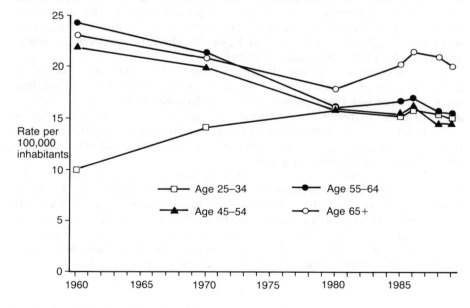

Figure 8.4b Suicide trends in the USA
Source: US Bureau of the Census, Statistical Abstract of the United States, various years.

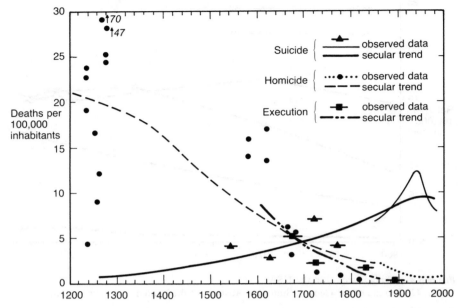

Figure 8.5 Deaths from suicide, homicide, and execution in England since the thirteenth century

very high compared with rates in modern industrialized societies (see figure 8.5). Annual rates above 20 per 100,000 inhabitants were quite common. In the rural area of Warwick, for instance, in the thirteenth century, the homicide rate was 47. In cities, like London or Bristol, it was not so high: during the period 1244–76, the rate was between eight and 15 in London, and four in Bristol.

This high rate of criminal violence declined from century to century. In the late seventeenth century, there were probably close to five homicide deaths per 100,000 inhabitants in Britain; four centuries earlier, according to scattered observations, the rate was much higher – probably about 20. Although the exact figure for Britain as a whole is only a rough estimate, there can be no doubt that the incidence was substantially higher than in later centuries.

A new phase evolved in the nineteenth century with the rise of centralized states with governmental and political structures on a larger-than-local scale, and the emergence of local police as a professional group devoted to capturing and arresting criminals. Victims of violence were taught to seek official help. The move to cities produced a more "civilized" generation than its predecessors; schooling was also an important factor influencing the long-term decline of violence.

Executions

According to the London Bills of Mortality from the mid-seventeenth to the mid-nineteenth century, the metropolitan rates of execution were higher than those of recorded homicide. This could be partly because prisoners were brought to London for execution, but also the death sentence was common even for minor crimes like burglary and the theft of cattle, horses, or sheep. The number of capital offenses exceeded 200. In the late eighteenth century, annual executions were about five per 100,000 inhabitants – including even some children. The number of offenders sentenced to death and subsequently executed decreased in both absolute and relative volume, decade by decade, from the eighteenth century to the present. In the early twentieth century, the rate of execution became negligible (figure 8.5). A similar trend was observed in France: average yearly executions fell from 72 in 1826–30 to 31 in 1851–5 to only two in 1901–5.

Suicides

There are few quantitative observations on suicide before the nineteenth century. Data collection improved only in the second half of the eighteenth century and was limited to the most advanced nations of Europe – Sweden and Finland, for example. Early rates in these countries were usually very low, well under five per 100,000.

But for England, local data are available for much earlier periods. They also show, for the sixteenth and seventeenth centuries, a low suicide incidence. In London, the incidence was three per 100,000 inhabitants in 1629–36; from the mid-sixteenth century until the end of the seventeenth, it fluctuated between four and seven per 100,000.

The Bills of Mortality surely missed many suicides, but the historical evidence consistently shows much higher rates in urban areas than in rural areas. The suicide rate in Nottinghamshire between 1530 and 1558 was four; suicides reported to the coroners in Shropshire in 1780–4 were at the same rate. The available data show that the early incidence of suicide was about half that of today. (But, as noted above, Britain has the lowest current suicide rate among the most advanced nations of the world.) In the Middle Ages, suicide was virtually not mentioned.

In sum, longer-term English data confirm the impression based on more recent evidence from other countries that the homicide rate has consistently declined, while the suicide rate has consistently increased.

REFERENCES

Chesnais, J. C. (1981): *Histoire de la Violence en Occident de 1800 à nos jours*. Paris: Laffont.

Given, J. B. (1977): *Society and Homicide in Thirteenth-Century England*. Stanford: Stanford University Press.

Hair, P. E. H. (1971): "Deaths from Violence in Britain: A Tentative Secular Survey." *Population Studies*, March.

9

The History of Accident Rates in the United States

Arlene Holen

Although danger makes headlines, the success that has been achieved in reducing major causes of accidents and injuries receives relatively little notice. The record shows that the death rate from accidents has fallen dramatically during the twentieth century.

HISTORY OF THE DECLINE

Fatal accidents are the fourth leading cause of death in the United States, exceeded by heart disease, cancer, and stroke. Accidents are the leading cause of death among persons aged one to 37. Overall accident fatality rates in the United States have fallen by well over half since the turn of the century, from 87.2 deaths per 100,000 population per year in 1903 to 38.1 in 1989 (figure 9.1).

Accident fatalities in virtually all categories have declined substantially, and the record of progress has been strongest in the 1980s. The total death rate from accidents fell by more than 20 percent over the past 10 years, and the rate of motor vehicle deaths exhibited a similar decline. The death rate from accidents at work fell by almost 30 percent, and the death rate from accidents at home fell by about 15 percent. The death rate from public non-motor vehicle accidents, including falls, drownings, and airplane crashes, dropped by about 30 percent. Accidental death rates associated with firearms also declined during the 1980s.

The age distribution of the population affects the overall death rate from accidents, and it has shifted substantially with increases in longevity. An increase in the aged population by itself tends to increase accidental death rates, since the elderly suffer a much higher than average incidence of accidents. Thus, accident fatality rates have fallen even faster when adjusted for age.

Although accident fatality rates have fallen dramatically for all age groups, the greatest improvement has been among children under five

Figure 9.1 US accidental death rate
Source: National Safety Council, *Accident Facts*, annual (Chicago: NSC, 1990), 26–7.

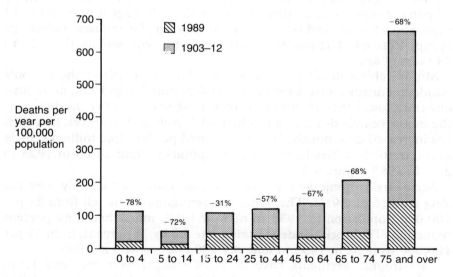

Figure 9.2 Accidental death rates by age, 1903–12 and 1989, with percent change
Source: National Safety Council, *Accident Facts: 1990 Preliminary Condensed Edition*, March 1990, 1.

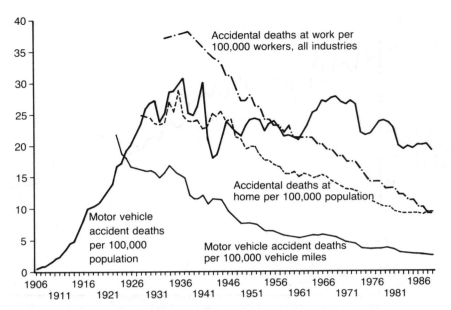

Figure 9.3   Accidental death rates by cause; motor vehicle, work, and home
Source: National Safety Council, *Accident Facts*, annual (Chicago: NSC, 1990), 26–7, 32–3, 37, 72–3.

(figure 9.2). Death rates per 100,000 population for that youngest group fell by 78 percent from the 1903–12 rates to 1989. Death rates fell by 68 percent over the same time period for the oldest age groups, 65 to 74 years and 75 years and over, and by 67 percent for the next oldest age group, 45 to 64. The smallest decrease, 31 percent, was for those 15 to 24 years of age.

Motor vehicle deaths, which account for about half of the nation's accidental deaths, were 18.9 per 100,000 population in 1989, more than one third lower than they were at their peak year, in 1937. In that year, the motor vehicle death rate reached 30.8. Automobile travel, however, has increased enormously. When measured per hundred million vehicle miles, traffic fatalities have fallen precipitously, from 21.65 in 1923 to 2.25 in 1989 (figure 9.3).

Accidents at home and at work have also declined sharply over the long term (figure 9.3). The home accident fatality rate fell from 24 per 100,000 population in 1933 to nine in 1989, more than a 60 percent reduction. The work accident fatality rate fell by 75 percent, from 37 per 100,000 workers in 1933 to nine per 100,000 workers in 1989.

Catastrophic accidents have declined sharply over the long term. Total deaths due to accidents in which five or more persons were killed declined from 13,250 during the 1940s to 10,090 during the 1970s. The death rate per 100,000 population due to such catastrophic accidents

Table 9.1 Largest US disasters by category

	Type and Location	No. of Deaths	Date of Disaster
Floods	Galveston tidal wave	6,000	Sept. 8, 1900
	Johnstown, Pa	2,209	May 31, 1889
	Ohio and Indiana	732	Mar. 28, 1913
	St. Francis, Calif., dam burst	450	Mar. 13, 1928
	Ohio and Mississippi River valleys	380	Jan. 22, 1937
Hurricanes	Florida	1,833	Sept. 16–17, 1928
	New England	657	Sept. 21, 1938
	Louisiana	500	Sept. 29, 1915
	Florida	409	Sept. 1–2, 1935
	Louisiana and Texas	395	June 27–28, 1957
Tornadoes	Illinois	606	Mar. 18, 1925
	Mississippi, Alabama, Georgia	402	Apr. 2–7, 1936
	Southern and Midwestern states	307	Apr. 3, 1974
	Ind., Ohio, Mich., Ill. and Wis.	272	Apr. 11, 1965
	Ark., Tenn., Mo., Miss. and Ala.	229	Mar. 21–22, 1952
Earthquakes	San Francisco earthquake and fire	452	Apr. 18, 1906
	Alaskan earthquake–tsunami hit Hawaii, Calif.	173	Apr. 1, 1946
	Long Beach, Calif., earthquake	120	Mar. 10, 1933
	Alaskan earthquake and tsunami	117	Mar. 27, 1964
	San Fernando–Los Angeles Calif., earthquake	64	Feb. 9, 1971
Marine	"Sultana" exploded – Mississippi River	1,547	Apr. 27, 1865
	"General Slocum" burned – East River	1,030	June 15, 1904
	"Empress of Ireland" ship collision – St. Lawrence River	1,024	May 29, 1914
	"Eastland" capsized – Chicago River	812	July 24, 1915
	"Morro-Castle" burned – off New Jersey coast	135	Sept. 8, 1934
Aircraft	Crash of scheduled plane near O'Hare Airport, Chicago	273	May 25, 1979
	Crash of scheduled plane, Detroit Michigan	156	Aug. 16, 1987
	Crash of scheduled plane in Kenner, La.	154	July 9, 1982
	Two-plane collision over San Diego, Calif.	144	Sept. 25, 1978
	Crash of scheduled plane, Ft. Worth/Dallas Airport	135	Aug. 2, 1985

Table 9.1 *(Continued)*

Type and Location	No. of Deaths	Date of Disaster
Railroad Two-train collision near Nashville, Tenn.	101	July 9, 1918
Two-train collision, Eden, Colo.	96	Aug. 7, 1904
Avalanche hit two trains near Wellington, Wash.	96	Mar. 1, 1910
Bridge collapse under train, Ashtabula, Ohio	92	Dec. 29, 1876
Rapid transit train derailment, Brooklyn, N.Y.	92	Nov. 1, 1918
Fires Peshtigo, Wis., and surrounding area forest fire	1,152	Oct. 9, 1871
Iroquois Theatre, Chicago	603	Dec. 30, 1903
Northeastern Minnesota, forest fire	559	Oct. 12, 1918
Cocoanut Grove nightclub, Boston	492	Nov. 28, 1942
North German Lloyd Steamships, Hoboken, N.J.	326	June 30, 1900
Explosions Texas City, Texas, ship explosion	552	Apr. 16, 1947
Port Chicago, Calif., ship explosion	322	July 18, 1944
New London, Texas, school explosion	294	Mar. 18, 1937
Oakdale, Pa., munitions plant explosion	158	May 18, 1918
Eddystone, Pa., munitions plant explosion	133	Apr. 10, 1917
Mines Monongha, West Va., coal mine explosion	361	Dec. 6, 1907
Dawson, New Mexico, coal mine fire	263	Oct. 22, 1913
Cherry, Ill., coal mine fire	259	Nov. 13, 1909
Jacobs Creek, Pa., coal mine explosion	239	Dec. 19, 1907
Scofield, Utah, coal mine explosion	200	May 1, 1900

Source: National Safety Council, *Accident Facts*, annual (Chicago: NSC, 1990), 15.

fell from 0.97, on an annual basis, during the decade 1941–50 to 0.21 in 1983 (figure 9.4). Around the turn of the century, major disasters in the US took thousands of lives in floods, shipping mishaps, fires, and hurricanes. However, the last recorded disaster to claim more than 500 lives – a ship explosion – occurred in 1947 (table 9.1).

Over the past several decades, accident fatality rates in other industrial countries have followed trends generally similar to those of the US. Many countries have had strong declines in accident rates during the 1970s and 1980s, following some increase during the 1960s (figure 9.5). Accident fatality rates in Japan fell by 44 percent from 1967 to 1987. In Sweden they fell by 20 percent during the same period. From 1967 to 1988, accident fatality rates dropped by 48 percent in West Germany.

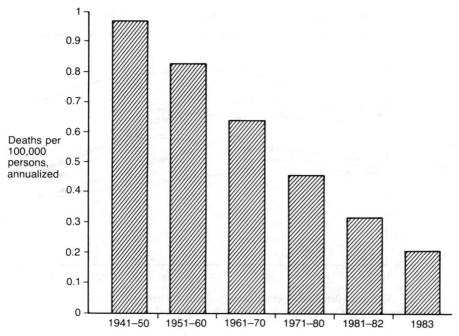

Figure 9.4 Death rates from catastrophic accidents

Note: Catastrophic accidents are accidents in which five or more persons are killed.

Sources: Metropolitan Life Insurance Company and Statistical Abstract of the United States.

In Australia, Canada, and Iceland, rates fell by 36 percent from 1968 to 1988. Over these same years, accident fatality rates fell by 34 percent in England and Wales, and by 19 percent in France. In comparison, the decline in the United States during the same two decades was 32 percent.

REASONS FOR THE DECLINE

Analysts have not yet quantified the factors that explain long-term declines in accident rates. Many point to changes in public attitudes and to more stringent government regulation to account for the progress that has been made. Campaigns against unsafe behavior, such as drunk driving, have been credited for reducing automobile accidents. Some analysts focus on psychological factors, such as the desire to control one's environment, or on demographic factors, such as a population that is reproducing more slowly and later in life. Some emphasize political developments, such as the consumer movement of the 1970s or the efforts of labor unions to enhance safety regulation in the workplace.

These explanations are plausible, but empirical evidence is generally lacking. Some studies of the effects of government regulation, moreover,

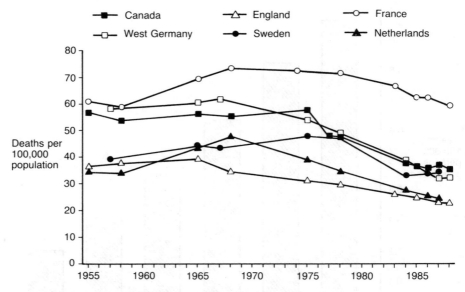

Figure 9.5a Accidental death rates, other industrialized countries
Source: National Safety Council, *Accident Facts*, selected annual issues.

have found that, rather than reducing risk, regulation sometimes re-
places one kind of risk with another. Some regulations have been
counterproductive, requiring the replacement of existing consumer pro-

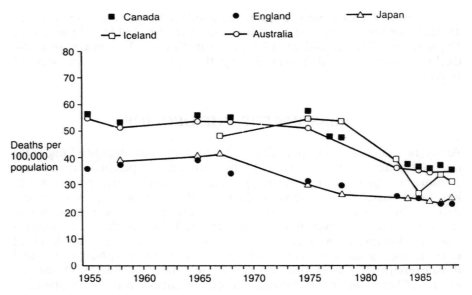

Figure 9.5b Accidental death rates, other industrialized countries
Source: National Safety Council, *Accident Facts*, selected annual issues.

ducts with products that turned out to be equally or even more dangerous. One flame retardant chemical, for example, intended to improve the
safety of children's sleepwear, later was discovered to be carcinogenic.

An underlying cause of reductions in accident fatalities has been a
rising standard of living. Higher incomes enable consumers to choose
safer products, homes that are more safely designed and constructed,
and safer forms of transportation. Communities with more ample resources are able to build safer roads, airports, swimming pools, and
ice-skating rinks. Voters with higher incomes are also more likely to
demand more stringent and more effective regulations affecting safety
and health.

Workers with more education and technical skills can get safer jobs.
An important cause of the reduced rate of accidental deaths at work has
been the marked shift in employment patterns that occurred during this
century, away from agriculture and other relatively dangerous goods-
producing industries toward the relatively safer service industries.

Along with rising standards of living, technological progress has enabled safer products to be produced and safer environments to be constructed. Advances in medicine have improved the treatment of accident
victims and have mitigated the effects of potentially disabling injuries.

The rate of reduction in accident fatalities has been especially strong
in the last decade and does not show signs of slowing in the near future.
With continuing economic and technological progress, the nation may
look forward to further declines in accident fatality rates.

NOTE

The author is a Commissioner of the Federal Mine Safety and Health Review
Commission in Washington, D.C. The views expressed are those of the author
and do not necessarily represent those of the Commission.

10

World Trends in Smoking

Allan M. Brandt

As a major form of tobacco consumption, the cigarette is a twentieth-century phenomenon. Although cigarettes were produced and smoked before 1900, only in this century has their use become widespread. Annual consumption per capita in the United States rose from 54 in 1900, to 1,365 in 1930, to 4,318 in 1965 (figure 10.1). This shift was

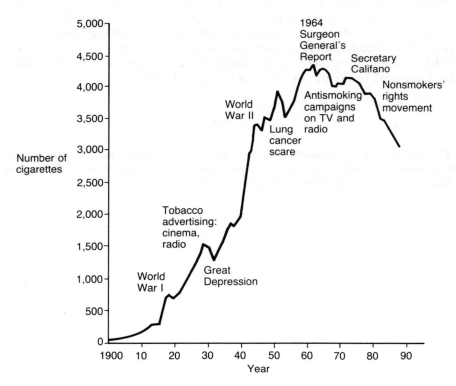

Figure 10.1 Number of cigarettes consumed annually per adult (age 18 and older): USA, 1900–88
Source: Economic Research Service, US Department of Agriculture.

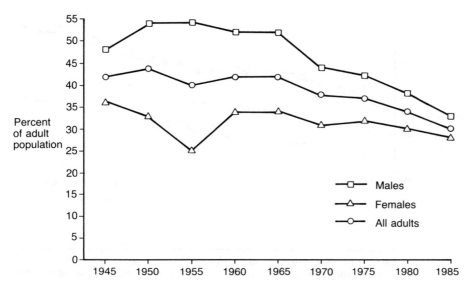

Figure 10.2 Percent of adults smoking regularly, by sex, USA, 1945–85
Source: From survey data collected by the Gallup Poll, National Center for Health Statistics, and Centers for Disease Control.

the result of a number of factors. Developments in agricultural technique, production technology, and industrial organization, as well as such factors as the introduction of the portable match, all contributed to the growth of the tobacco industry (Bennett, 1980).

As consumption of cigarettes rose, so too did concern about their impact on health. In the last twenty-five years, American society has witnessed a revolution in attitudes and behaviors relating to cigarette smoking. When Surgeon General Luther Terry announced in January 1964 that his commission on smoking and health had concluded that smoking causes lung cancer and other diseases, the 70 million adult smokers in the United States consumed some 400 billion cigarettes each year (Terry, 1964). By 1985, in response to spreading knowledge of the connection between smoking and disease (e.g., Doll and Hill 1956; Hammond and Horn, 1958), the proportion of Americans who smoked had fallen from 42 percent to 30 percent (figure 10.2). A large percentage of all living Americans who have ever smoked have now quit (figure 10.3). According to the 1989 Surgeon General's Report, approximately 750,000 smoking-related deaths have been avoided since 1964 because people have quit or not started smoking (Koop, 1989).

The decline in cigarette smoking in the United States has been the result of a complex combination of particular scientific, social, and political forces that led, over time, not only to a fundamental shift in how the risks of smoking were perceived but also to a social environment that

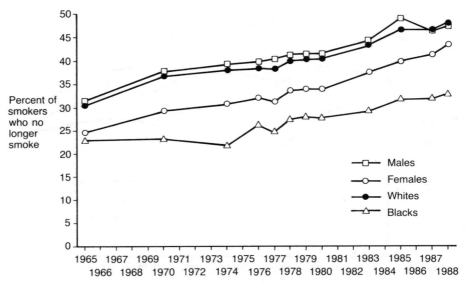

Figure 10.3 Cigarette smoking quit ratio, by sex and race; USA, 1965–88
Source: US Statistical Abstract, 1992 and 1990 editions.

discredited smoking (Brandt, 1990). Despite the decline, however, cigarettes continue to exact a significant toll on health in the United States.
The United States still ties with Japan for second in per capita consumption among developed nations, behind only Greece (table 10.1). And
cigarettes continue to exact a significant toll on health in the United
States. According to recently revised figures, 390,000 deaths each year
are attributed to cigarette smoking (*Washington Post*, 1989). Smoking is
estimated to cause 30 percent of all cancer deaths, 21 percent of all
deaths from coronary artery disease, and 82 percent of all deaths from
chronic obstructive pulmonary disease. Since 1986, lung cancer has
become the leading cause of cancer deaths among American women,
surpassing breast cancer – the epidemiological result of the rise in
smoking among women from the 1940s to the 1960s. Smoking remains
the "single most preventable cause of death" in the United States
(Koop, 1989). It is currently estimated that health care costs associated
with smoking total more than $16 billion each year; indirect costs for
lost productivity, disability, and premature death cost an additional
estimated $40 billion annually.

While consumption of tobacco in the USA and other Western nations
such as Canada and Great Britain has declined moderately since the
1970s, it has dramatically increased in the developing world. According
to the World Health Organization, there has been a 77 percent increase
in consumption on the African continent in the last 25 years. In Asia
and Latin America, consumption outpaced population growth by 30

Table 10.1 Adult per capita consumption of manufactured cigarettes, by country, 1985

Country		Country		Country		Country	
Cyprus	4,050	Saudi Arabia	2,110	Korea, People's Republic	1,180	Laos	490
Cuba	3,920	Romania	2,110	Guadeloupe	1,080	Togo	460
Greece	3,640	Syria	2,050	Morocco	1,070	Madagascar	450
Poland	3,300	Belgium	1,990	Indonesia	1,050	Liberia	450
United States	3,270	Turkey	1,970	Honduras	1,010	Mozambique	430
Japan	3,270	Norway	1,920	Chile	1,000	Zambia	400
Hungary	3,260	Colombia	1,920	Paraguay	1,000	Malawi	390
Canada	3,180	Philippines	1,910	Guyana	1,000	Ghana	380
Iceland	3,100	Venezuela	1,890	Iraq	980	Nigeria	370
Yugoslavia	3,000	Egypt	1,860	Dominican Republic	980	Peru	350
Switzerland	2,960	Malaysia	1,840	Reunion	940	Bolivia	330
Lebanon	2,880	Argentina	1,780	Congo	920	Tanzania	330
Libya	2,850	Uruguay	1,760	Thailand	900	Central African Republic	280
Kuwait	2,760	Portugal	1,730	Ecuador	880	Bangladesh	270
Spain	2,740	Finland	1,720	Panama	850	Uganda	260
Australia	2,720	Jordan	1,700	Sierra Leone	830	Haiti	240
Korea, Republic of	2,660	Brazil	1,700	Jamaica	820	Cape Verde	210
Austria	2,560	Mauritius	1,700	El Salvador	750	Zaire	210
Ireland	2,560	Netherlands	1,690	Benin	740	India	160
Czechoslovakia	2,550	Sweden	1,660	Ivory Coast	710	Chad	150
New Zealand	2,510	Suriname	1,660	Vietnam	670	Burma	150
Italy	2,460	Trinidad and Tobago	1,600	Pakistan	660	Nepal	150
Bulgaria	2,410	Algeria	1,590	Iran	620	Sudan	130
France	2,400	China	1,590	Senegal	610	Niger	100
Germany, Federal Republic	2,380	Hong Kong	1,580	Cameroon	610	Ethiopia	60
Germany, Democratic Republic	2,340	South Africa	1,550	Guatemala	550	Afghanistan	50
Israel	2,310	Tunisia	1,470	Kenya	550	Papua New Guinea	30
Singapore	2,280	Barbados	1,380	Angola	530	Guinea	30
Soviet Union	2,120	Nicaragua	1,380	Zimbabwe	500	Burkina Faso	30
United Kingdom	2,120	Costa Rica	1,340	Sri Lanka	500		
Denmark	2,110	Fiji	1,320				
		Mexico	1,190				

Note: "Adult" defined as 15 years of age and over.
Source: World Health Organization (1985).

percent from 1970 to 1985 (WHO, 1987). These increases are related, according to a recent study by the World Bank, to higher levels of national incomes in developing countries.

This increase in worldwide tobacco consumption has significant implications for patterns of disease. According to WHO, 600,000 new cases of lung cancer now occur worldwide every year, most resulting from smoking. By 2000, the annual number of lung cancer cases may be as high as 2 million, with 900,000 in China alone (Chandler, 1986). British epidemiologist Richard Peto has projected that an estimated 2 million Chinese men will die annually from tobacco-related health problems by the year 2025; cumulatively, 50 million deaths among the 500 million Chinese under age 20 are anticipated (Peto, 1987). Rates of lung cancer have been rising in Japan and Singapore as well. In India, where consumption has risen by 400 percent since 1960, an estimated 630,000 people die each year of tobacco-related disease (Gupta, 1988). Current surveys suggest that worldwide tobacco consumption is now responsible for 2.5 million excess or premature deaths each year – almost 5 percent of all deaths.

Correlating with rising trends in cigarette consumption worldwide is a rise in antismoking activity (table 10.2). By 1986, 55 countries had enacted legislation to limit or ban cigarette advertising (20 had total bans, 35 partial or moderate bans). Fifty-two countries have mandated health warnings on cigarette packages; only six, however, have required that such warnings be periodically rotated. Forty-seven countries have

Table 10.2 Worldwide antitobacco activities*

Country	Advertising restrictions†	Health warnings‡	Package informations§	Smoking in public places¶	Smoking in workplaces#
Argentina	Strong partial ban	Required	–	–	–
Brazil	–	Required	–	Moderate	–
Bolivia	Strong partial ban	Required	–	Moderate	–
Canada	Total ban (pending)	Voluntary	NIC/TAR‖	Moderate	–
Chile	Moderate ban	–	–	–	–
Costa Rica	–	Required	–	–	–
Colombia	Moderate ban	Required	–	–	–
Ecuador	Moderate ban	Required	–	–	–
Guatemala	–	Required	–	–	–
Mexico	Moderate ban	Required	–	–	–
Panama	Moderate ban	Required	–	–	–
Paraguay	Moderate ban	–	–	–	–
Peru	Moderate ban	Required	–	–	–
United States	Moderate ban	Rotating	–	Moderate	Legislation
Uruguay	Moderate ban	Required	–	Moderate	–
Venezuela	Moderate ban	Required	–	Moderate	–
Algeria	Total ban	Required	–	Moderate	–

Gambia	Strong partial ban	–	–	–	–
Ivory Coast	–	–	–	Moderate	–
Kenya	–	Required	–	Moderate	–
Mozambique	Total ban	–	–	Moderate	–
Nigeria	–	–	–	Moderate	–
Senegal	Strong partial ban	Required	NIC/TAR	Moderate	–
Cyprus	Strong partial ban	Required	–	Moderate	–
Egypt	Strong partial ban	Required	NIC/TAR	Moderate	–
Jordan	Total ban	Required	NIC/TAR	Stringent	–
Kuwait	–	Required	NIC/TAR	–	–
Lebanon	Moderate ban	Required	–	Moderate	–
Pakistan	–	Required	–	–	–
Saudi Arabia	–	–	–	Moderate	–
Sudan	Total ban	Required	–	Moderate	–
Hong Kong	Moderate ban	Required	NIC/TAR	Moderate	–
India	–	Required	–	–	–
Macao	Moderate ban	Required	–	Moderate	–
Sri Lanka	Moderate ban	Required	–	Moderate	–
Thailand	Moderate ban	Required	–	Moderate	–
Austria	Moderate ban	–	NIC/TAR	Moderate	Legislation
Belgium	Strong partial ban	Required	NIC/TAR/ CO	Stringent	Legislation
Bulgaria	Total ban	Required	–	Stringent	–
Czechoslovakia	Total ban	Required	–	Moderate	–
Denmark	Strong partial ban	Required	NIC/TAR‖	Moderate	–
Finland	Total ban	Rotating	NIC/TAR/ CO	Stringent	–
France	Strong partial ban	Required	NIC/TAR	Stringent	Legislation
German Democratic Republic	Total ban	–	–	Moderate	–
German Federal Republic	Strong partial ban	Voluntary	NIC/TAR	–	Legislation
Greece	–	–	–	Moderate	–
Hungary	Total ban	Required	–	Stringent	–
Iceland	Total ban	Required	–	Moderate	Legislation
Ireland	Strong partial ban	Voluntary/ rotating	–	–	–
Israel	–	–	–	Moderate	–
Italy	Total ban	–	–	Stringent	–
Malta	Moderate ban	–	–	–	–
The Netherlands	Moderate ban	Required	NIC/TAR	–	Legislation

Table 10.2 *(Continued)*

Country	Advertising restrictions†	Health warnings‡	Package informations§	Smoking in public places¶	Smoking in workplaces#
Norway	Total ban	Rotating	NIC/TAR	–	–
Poland	Total ban	–	–	Stringent	Legislation
Portugal	Total ban	Required	NIC/TAR	Stringent	–
Romania	Total ban	–	–	Stringent	–
Spain	Total ban†	Required	NIC/TAR	Moderate	Legislation
Sweden	Strong partial ban	Rotating	NIC/TAR/ CO	Moderate	Legislation
Switzerland	Moderate ban	Required	NIC/TAR	–	–
Turkey	–	–	–	Moderate	–
Soviet Union	Total ban	Required	–	Stringent	Legislation
United Kingdom	Strong partial ban	Voluntary/ rotating	TAR‖	–	–
Yugoslavia	Total ban	–	–	Moderate	–
Australia	Strong partial ban	Required	–	–	–
French Polynesia	Total ban	Required	–	–	–
Japan	–	–	–	Moderate	–
Malaysia	Moderate ban	Required	–	Moderate	–
New Zealand	Strong partial ban	Voluntary	–	Moderate	–
Singapore	Total ban	Required	–	Moderate	–

Source: Council on Scientific Affairs, American Medical Association, 1990.
Notes: ★Data from Roemer.
† Total ban indicates ban on advertising of tobacco products in all media; strong partial ban, restrictions in several media; and moderate ban, only minor restrictions in several media, or major restrictions in several media.
‡ Required indicates required by legislation; voluntary, required by voluntary agreement with industry; and rotating, rotating warnings required by legislation.
§ NIC indicates nicotine level required by legislation; TAR, tar level required by legislation; and CO, carbon monoxide level required by legislation.
‖ Required by voluntary agreement with industry.
¶ Stringent indicates restrictions to widely control smoking in public places for health reasons; and moderate, limited restrictions to control smoking in public places.
# Legislation indicates legislation exists to control smoking in workplaces.

enacted legislation to limit smoking in public places, and 11 have passed legislation limiting smoking in the workplace (Roemer, 1986). Enforcement of restrictions on smoking in public places, however, remains highly variable. Since cigarette smoking is highly addictive, patterns of consumption are difficult to reduce once established.

Tobacco is deeply entrenched within national and local economic structures, as well as through powerful social and cultural conventions. Even significant public health and policy initiatives have not radically altered patterns of production, marketing, and use. It will, no doubt, require a special resourcefulness to alter these patterns of consumption and disease in the decades ahead.

REFERENCES

Bennett, William (1980): "The cigarette century." *Science* 80 (September/October), 37–43.

Brandt, Allan M. (1990): "The cigarette, risk, and American culture." *Daedalus* 119, 155–76.

Chandler, W. U. (1986): "Banishing tobacco." *World Watch Paper* 68.

Doll, R., and A. B. Hill (1952): "A study of the aetiology of Carcinoma of the lung." *British Medical Journal* (December 13) 2, 1271–86.

————— ————— (1954): "The mortality of doctors in relation to their smoking habits: a preliminary report." *British Medical Journal* (June 26) 1 (4877), 1451–5.

————— ————— (1956): "Lung cancer and other causes of death in relation to smoking. A second report on the mortality of British doctors." *British Medical Journal* (November 1) 2, 1071–81.

Gupta, P. C. (1988): "Health consequences of tobacco use in India," *World Smoking and Health* 13, 5–10.

Hammond, E. C., and D. Horn (1958a): "Smoking and death rates – report on forty-four months of follow-up on 187,783 men. I. Total Mortality." *JAMA* (March 8), 166 (10), 1159–72.

————— ————— (1958b): "Smoking and death rates – report on forty-four months of follow-up on 187,783 men. II. Death rates by cause." *JAMA* (March 15), 166 (11), 1294–308.

Terry, L. (1964): *Smoking and Health Report of the Advisory Committee to the Surgeon General of the Public Health Service.* US Department of Health, Education, and Welfare, Public Health Service. Surgeon General's Report. PHS Publication No. 1103, 1964.

Koop, C. Everett (1989): *Reducing the Health Consequences of Smoking. Twenty-five Years of Progress.* US Department of Health and Human Services, Public Health Service. Surgeon General's Report. DHHS Publication No. (CDC) 90-8411.

Peto, R. (1987): "Tobacco related deaths in China." *Lancet* 2, 221.

Ravenholt, R. T. (1990): "Tobacco's Global Death March." *Population and Development Review* 16, 213–40.

Roemer, Ruth (1983): *Legislative Action to Combat the World Smoking Epidemic.* Geneva. World Health Organization.

————— (1986): *Recent Development in Legislation to Combat the World Smoking Epidemic.* Geneva. World Health Organization.

Washington Post (1989): "U.S. report raises estimate of smoking toll." (January 11), A20.

WHO (World Health Organization) (1987): *Tobacco Alert.* 20 (April–June), 3.

11

Long-Term Trends in the Consumption of Alcoholic Beverages

James S. Roberts

This chapter presents data on long-term trends in the consumption of alcoholic beverages in Western societies since the beginning of the nineteenth century. The statistical series reflect significant shifts in both production and consumption practices over the last two centuries. Although these data are usually presented as consumption statistics,

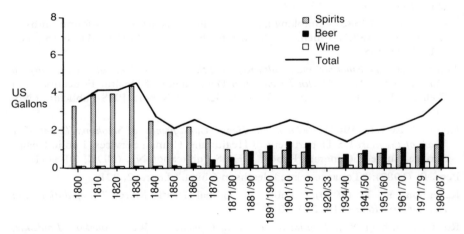

Figure 11.1 Apparent absolute alcohol consumption, USA, US gallons per capita aged 15 and over

Note: Based on a drinking age population of 15 and over through 1973 and 14 and over thereafter.

Sources: Compiled from Dan E. Beauchamp, *Beyond Alcoholism: Alcohol and Public Health Policy* (Philadelphia: Temple University Press, 1980), 100–1; Merton M. Hyman et al., *Drinker, Drinking and Alcohol Related Morality and Hospitalizations* (New Brunswick: Rutgers Center of Alcohol Studies Publications, 1980), 3; W. J. Rorabaugh, *The Alcoholic Republic: An American Tradition* (New York: Oxford University Press, 1979), 232–3; and editions of the US Statistical Abstract.

they generally derive from the taxation of production or retail sales. Such data, of course, are often blind to the not insignificant volume of home production and to production and sale of alcohol outside legal channels. Nevertheless, these figures provide the best evidence available of long-term movements in the consumption of alcoholic beverages.

CONSUMPTION TRENDS

Figures 11.1–11.4 describe consumption trends in the United States, Germany, France, and the United Kingdom. All data are expressed in terms of US gallons of "absolute" alcohol, or the total ethanol consumed in the various beverage categories. The data are derived in large part from major monographic studies of each country which used varying population bases. In order to facilitate comparison across national boundaries, consumption data have been normalized over a "drinking age" population aged 15 and over.[1] This is the norm used in most recent comparative studies.

Information on long-term consumption trends in the United States[2] is presented in figure 11.1. The early years of this time series, based on estimates developed in W. J. Rorabaugh's *The Alcoholic Republic*,[3] indicate the dramatic effects of the spread of domestic distilling. According to Rorabaugh's estimates, per capita consumption of both distilled beverages and absolute alcohol in the United States peaked in the 1830s. Thereafter, spirits consumption entered a long-term decline that

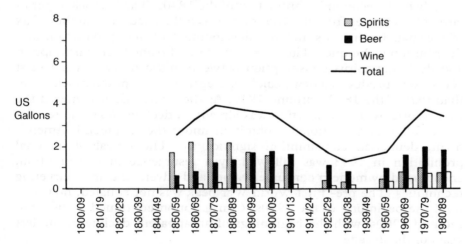

Figure 11.2 Apparent absolute alcohol consumption, Germany, US gallons per capita aged 15 and over

Note: West Germany after 1945.

Sources: Compiled from W. G. Hoffmann, *Das Wachstum der deutschen Wirtschaft* (Berlin, 1965), 170–4, 650–2; and editions of the *Statistisches Jahrbuch fuer das Bundesrepublik Deutschland*.

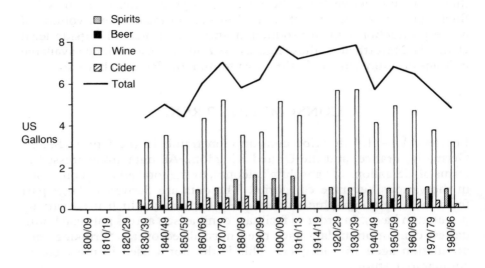

Figure 11.3 Apparent absolute alcohol consumption, France, US gallons per capita aged 15 and over
Sources: Compiled from Sully Ledermann, *Alcool, alcoolisme, alcoolisation* (Paris, 1956), 60–7; and editions of the *Statistique annuaire de la France.*

stretched through the Great Depression. The break in the trend line during the 1840s was due in some part to the influence of the early temperance movement, still in its moral suasionist phase, but more directly to the economic contraction of the 1840s. The increasing popularity of beer dates from the 1850s and was influenced by both the influx of German immigrants and the development of new production and distribution techniques. The scissors pattern of rising beer consumption and declining spirits consumption between mid-century and the First World War produced a new peak in aggregate consumption, much lower than that of the 1830s, around 1910. As the United States entered the First World War consumption was already in decline. Wartime restrictions, followed by national prohibition under the Eighteenth Amendment, depressed consumption significantly. The repeal of national prohibition in 1933 was followed by a resurgence in consumption, encouraged by modern consumer values and advertising and marketing techniques. The pre- First World War peak was not equaled, however, until the beginning of the 1970s. Consumption continued to rise into the mid-1980s, with the beginnings of an apparent decline in the last years of the decade.

In the German case (figure 11.2), systematic data are not available for the first half of the nineteenth century. The available evidence suggests, however, that spirits consumption was increasing from the 1790s, with rapid increases after 1815. According to the contemporary economist and statistician Ernst Engel, the consumption of spirits in Prussia nearly

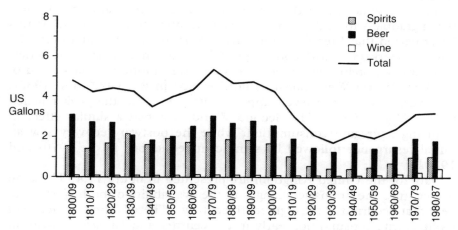

Figure 11.4 Apparent absolute alcohol consumption, United Kingdom, US gallons per capita aged 15 and over
Sources: Compiled from G. B. Wilson, *Alcohol and the Nation* (London, 1940); G. P. Williams and G. T. Brake, *Drink in Great Britain 1900 to 1979* (London, 1980); and editions of the *Annual Abstract of Statistics*.

tripled between 1806 and 1831, increasing from three to eight Prussian quarts per capita.[4] At the same time, brewing was commercially moribund and wine was of only local significance. As in the United States, there was a significant consumption peak in the 1830s, with a marked decline in the 1840s, attributable in Germany as in the United States both to the influence of the early temperance movement and, more importantly, to the depressed economic conditions that culminated in the Revolution of 1848. Consumption began to increase again from the late 1850s, with beer and spirits consumption increasing in parallel to produce what was probably the peak of German alcohol consumption in the early 1870s. Thereafter, the two beverages follow a scissors pattern, with spirits consumption decreasing and beer consumption increasing into the first years of the twentieth century. These movements produced a subsidiary peak just before the turn of the century. As in the United States, consumption had begun to decline even before wartime restrictions limited consumptions. Aggregate consumption remained below pre-First World War levels until the mid-1970s, with per capita figures eventually exceeding those of a century earlier. Figures for the 1980s suggest the beginnings of a moderate decline.

Data for France, presented in figure 11.3, date from the 1830s and are based on the pioneering work of Sully Ledermann.[5] French consumption statistics are dominated by wine, whose movements were highly dependent on quantity and quality of annual harvests. As in Germany, however, there is a significant peak in per capita consumption of alcohol in the 1870s. A second, in this case higher, peak occurred in the first decade of the twentieth century. Beer remained relatively unimportant in France, while spirits consumption assumed significant proportions, rising gradually from the 1830s and rapidly after the 1870s. It was in this latter period that French temperance sentiment crystallized around the issue of workers' spirits consumption.[6] As in other Western societies, there was a notable reduction of alcohol consumption in the period of the First World War. Consumption revived in the 1920s and 1930s, however, and reached levels approximately equal to the peak reached before World War I. The Second World War had a similar effect in reducing aggregate consumption, while the postwar recovery saw an initial resurgence followed by significant reductions in the 1970s and 1980s.

Data for the United Kingdom are presented in figure 11.4. These do not capture the dramatic changes in the availability of both spirits and beer that occurred in the eighteenth century.[7] There is some evidence of rising spirits consumption early in the century and an early plateau in aggregate consumption in the 1820s and 1830s. The subsequent reduction can again be attributed both to the influence of the early temperance movement and to the significant economic contraction of the 1840s. The all-time peak consumption was reached in the late 1870s, with a secondary peak at the turn of the twentieth century. The low point in aggregate consumption was reached during the depression of the 1930s. Consumption rebounded in the 1940s and rose gradually through the 1960s and 1970s, with continuing but more moderate growth in the 1980s.

CONCLUSIONS

What general conclusions can be made about these data on the history of alcohol consumption? The evident diversity of the four cases presented here suggests the extent to which alcohol consumption is embedded in each society's social, cultural, economic, and political fabric. The French case in particular, so heavily influenced by the dietary importance of wine, defies virtually every generalization that could be made about the American, German, or British data. Yet the French case is probably fairly typical of other wine-producing regions like Italy, Spain, and Portugal.

Despite this diversity, however, some more far-reaching generalizations can be offered, perhaps less derivations than interpretations of the long-term consumption trends presented here. It can be argued, first, that alcohol consumption is generally a product of growing prosperity, not an index of social misery. Compared to earlier centuries, the nineteenth- and twentieth-century levels of alcohol consumption, notwithstanding significant variations within and among societies, represent significant increases. Moreover, periods of rising aggregate consumption (e.g., 1850s to 1870s, 1950s to 1970s) are invariably periods of increasing prosperity; the converse, however, does not hold since not all periods of rising prosperity are accompanied by increasing consumption (e.g., 1880s to 1900s). There are obviously limits to the extent to which drinkers want more (or better) alcoholic beverages.[8] More detailed studies of alcohol consumption and business cycles generally support this link between prosperity and increasing alcohol consumption, as do cross-sectional studies across income groups.[9] Despite well-organized, often well-meaning, and sometimes highly successful temperance and prohibition movements in all these countries, then, most people in Western societies have consistently viewed alcohol as part of the good life, rejecting efforts to establish an inevitable link between drinking and degradation.

Second, the dynamic relationship between production and consumption seems to have been driven first on the supply side and then on the demand side. The eighteenth- and nineteenth-century revolution in distilling, the industrialization of brewing, and the extension of viticulture, and the commercialization of distribution first in regional and then national and international markets increased availability and lowered prices at a time of increasing purchasing power. This supply side dynamic was displaced after the period of the world wars as modern marketing and merchandising helped stimulate demand, further strengthening the hold of alcohol in modern Western societies and reinforcing the dominant notion that alcohol is part of the good life, not its nemesis. Each of these endeavors has had a commercial or business logic requiring capital and technical ingenuity, each success helping to legitimate the widespread use of beverage alcohol. Whatever the costs and benefits of contemporary drinking practices, they are deeply entrenched in economic, social, and political relationships.[10]

Finally, drinking behavior has been to a large extent self-regulating in responding to the surge in availability that began in the late eighteenth and early nineteenth centuries. This point is not meant to deny the importance of formal regulatory mechanisms, but to underscore the underlying social dynamic of drinking behavior, which can be shaped by regulation but never totally controlled. Just as important as formal regulations, which are often circumvented (prohibition, licensing regulations, drinking ages, "blue laws"), are the complex and historically changing social conventions that permit greater or lesser use according to a variety of situational factors (work, leisure, home, tavern, competing demands for time and money) and personal characteristics (age, sex, class, ethnic and religious identification). The twentieth-century history of alcohol consumption in the United States, the United Kingdom, and Germany (and even France, if one focuses on spirits and beer consumption) are strikingly similar. The fundamental similarities in national consumption trends despite very different regulatory histories and temperance reform traditions, including an extended period of national prohibition in the United States, suggest the importance of the underlying social dynamics of countless individuals making personal choices about their drinking behavior in the context of competing claims for their time and resources. These choices seem to be strongly influenced by similarities in the business cycles affecting each society.

As a human activity, drinking behavior thus has a natural history. That history is shaped by regulation and fiscal policy; by technologies and business practices; by medical, religious, and civic teachings; but it depends ultimately on individual choices made on the basis of a more or less clear calculus of the personal costs and benefits of alcohol consumption. That there are costs as well as benefits remains the enduring source of the ambivalent relationship between man and alcohol.

NOTES

1 Information on historical age structures was derived from Nathan Keyfitz and Wilhelm Flieger, *World Populations: An Analysis of Vital Data* (Chicago: University of Chicago Press, 1968); and B. R. Mitchell, *European Historical Statistics 1750–1975*, 2nd edn (New York, 1981).

2 For an overview, see Mark Edward Lender and James Kirby Martin, *Drinking in America* (New York: Free Press, 1982).

3 (New York: Oxford University Press, 1979), 225–34. I have excluded from my figures Rorabaugh's impressionistic estimates of cider consumption.

4 James S. Roberts, *Drink, Temperance and the Working Class*. London and Boston: George Allen and Unwin, 1982 16.

5 *Alcool – Alcoolisme – Alcoolisation: Donnees scientifiques de caractere physiologique, economique et social* [Institut national d'etudes demographiques, Travaux et Documents, Cahier n. 29] (Paris: Presses Universitaires de France, 1956).

6 Patricia E. Prestwich, *Drink and the Politics of Social Reform: Antialcoholism in France Since 1870*. Palo Alto: SPSS, 1988, 75–107.

7 For a fuller discussion, see Brian Harrison's classic *Drink and the Victorians: The Temperance Question in England*. Pittsburgh: University of Pittsburgh Press, 1971. See also A. E. Dingle, "Drink and Working Class Living Standards in Britain, 1870–1914," *Economic History Review* 25 (1972), 608–22.

8 Cf. Dingle, "Drink," passim; Roberts, *Drink, Temperance and the Working Class*, 109–27.

9 Cf. Harrison's *Drink and the Victorians*, 313–14; Dorothy Swain Thomas, *Social Aspects of the Business Cycle*. New York: Knopf, 1927, 127–32. For examples of cross-sectional data, see James S. Roberts, "Drink and Working Class Living Standards in Late 19th Century Germany," in U. Engelhardt, ed., *Arbeiterexistenz in 19. Jahrhundert*. Stuttgart, 1981, 74–91; and Vera Efron, et al., *Statistics on the Consumption of Alcohol and Alcoholism*. New Brunswick: Rutgers Center of Alcohol Studies Publications, 1974, 9.

10 For a cogent public health perspective on contemporary alcohol issues, see Dan E. Beauchamp, *Beyond Alcoholism: Alcohol and Public Health Policy*. Philadelphia: Temple University Press, 1980.

EDITOR'S NOTE (JULIAN L. SIMON)

Perhaps a paper on the consumption of alcohol should be included in the Standard of Living section of the book, rather than here. Perhaps the same is true of smoking and the use of other mild drugs such as caffeine and perhaps marijuana. The decline in the consumption of alcohol as people become affluent, probably after a rise in consumption, however, makes it seem less of a "superior good," in the technical term. Placing it here just seems less problematic, and a touch of easygoingness may not be amiss.

PART II
Standard of Living, Productivity, and Poverty

12

Trends in the Agricultural
Labor Force

Richard J. Sullivan

Because the nonagricultural sector relies on the agricultural sector for sustenance, the size of the nonagricultural sector will be constrained by the surplus produced in the agricultural sector. The share of the total labor force devoted to agriculture, therefore, has long been used as an indicator of the extent and pace not only of agricultural development, but also of economic development generally. The absolute size of the agricultural labor force is also of considerable interest because of its implications regarding the agricultural land–labor ratio. This chapter reviews the trends and patterns of the share and size of the labor force devoted to agriculture for a select group of countries.[1]

THE DEVELOPED WORLD

Figure 12.1 shows the share and size of the labor force in agriculture for the period from 1500 to the present in England, The Netherlands, and France.[2] The Netherlands, with an approximate 60 percent share for agriculture as early as 1550, and an approximate 40 percent share in 1650, was clearly exceptional. It was able to attain this low proportion of its labor force devoted to agriculture by a combination of specializing in overseas trade and importing food. The fact that the share was essentially stable from 1650 to 1850 demonstrates that an economic structure not dominated by agriculture is no guarantee of continued structural change. The stability of the share for The Netherlands, and the very modest decline in the share for France from 1500 to 1800, was likely representative of continental Europe. Since urban residents represent a major consumer of agricultural surplus, urbanization will be closely (though inversely) tied to the share of agriculture in the labor force. For the period 1600–1800, urbanization "advanced rather modestly on the [European] continent as a whole" (Wrigley, 1986, p. 148). Continental Europe's experience is a sharp contrast with the steady and

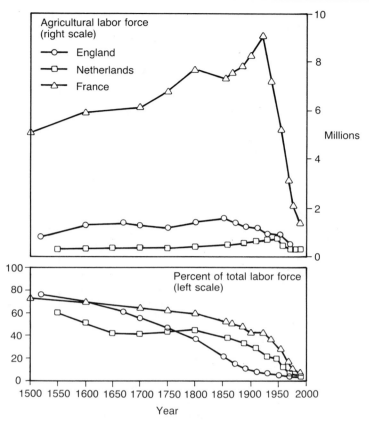

Figure 12.1 Agricultural labor force, England, the Netherlands, and France

large decline in the share of the agricultural workforce in England (Wrigley, 1986, pp. 147–50).

The declining share of agricultural workers in England for the period 1500–1750 shows that significant structural change can occur in a pre-modern context. The proximate source of the change is no mystery. Since imports of food were minor, and significant job growth outside of agriculture had to be supported by larger agricultural surpluses, the downward trend is evidence that productivity of English agriculture was improving.[3]

The most outstanding feature of figure 12.1 is the clear influence of modern economic growth on economic structure. France and The Netherlands are typical of today's industrially advanced economies: the share and size of agricultural workers in the total workforce can drop significantly, often in only a few decades. The reason is again tied to agricultural productivity – modern economic growth benefited farming dramatically, initially through invention of farm machinery, later through invention of artificial fertilizers, and lately through development of hybrid seeds and methods to control weed and insect infestation.

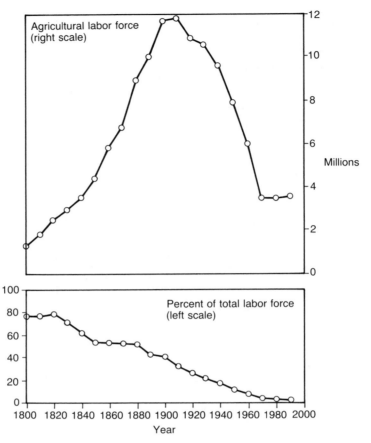

Figure 12.2 Agricultural labor force, USA

In a well-known article, Dovring (1959) noted that a country's share of the agricultural labor force may decline for a considerable period, while at the same time the country may see an increase in the size of the agriculture labor force. The experience of the United States provides a clear example; figure 12.2 shows that the size of the US agricultural labor force peaked in 1910, though the share had declined for the previous century.[4] Figure 12.1 show that the agricultural labor force peaked in England in the mid-nineteenth century, after centuries of a slow fall in the share of the agricultural labor force. A similar pattern occurred in The Netherlands and France, though the decline in the size of the agricultural workforce occurred in the twentieth century. Figure 12.3 shows the same results in Belgium and Finland; similar graphs would show the same pattern for Germany, Switzerland, Denmark, Norway, Sweden, Canada, and Australia. Invariably, with sufficient economic development, the absolute level of the agricultural labor force declines.

In the mid-1950s, Dovring (1959, p. 86) could note that France had experienced only modest declines in its agricultural labor force, a fact that figure 12.1 shows has changed dramatically. Figure 12.3 indicates that Finland experienced rapid reductions in its agricultural workforce in the course of a few recent decades. Similarly, in the past 30 years Spain and Italy have seen large absolute declines in their agricultural workforce, as revealed in figure 12.4. Dovring (1959, pp. 91–2) wrote also that Japan's agricultural labor force had not declined, but was confident that "the pull from urban occupations may prove strong enough to cause a decisive reduction in the . . . high density of [Japanese] on the land." Figure 12.4 indeed shows the large reduction in the Japanese agricultural workforce that Dovring predicted. The speed of the decline in the size of the agricultural workforce experienced by France, Finland, Italy, Spain, and Japan, is striking.

THE DEVELOPING WORLD

The circumstances facing the developing world are considerably different from the those experienced by the developed world.[5] The major

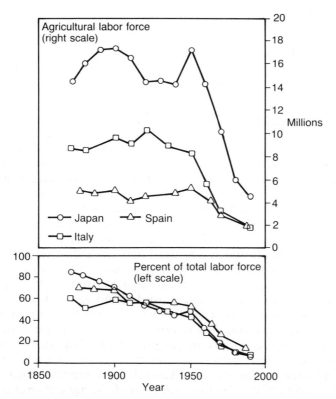

Figure 12.3 Agricultural labor force, Belgium and Finland

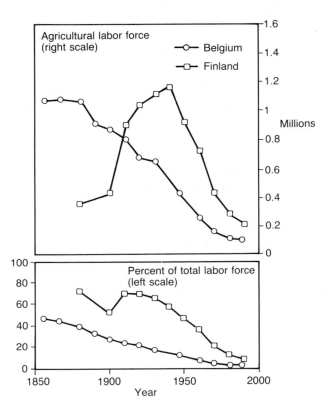

Figure 12.4 Agricultural labor force, Japan, Italy, and Spain

differences are the extensive international trade in food and the high levels of productivity of the farming sectors of the developed countries. Today, imports of inexpensive food rid the developing world of the constraint on nonagricultural population imposed by agricultural productivity.[6] But this does not necessarily mean that the historical experiences of the more developed countries are irrelevant to the less developed countries. The difficulty that developing countries face in competing in international agricultural markets enhances the attractiveness of creating nonagricultural jobs. Moreover, techniques to increase the productivity of the agricultural sectors of the less developed countries are available, and as strategies to develop agriculture are pursued these techniques will be adopted, thus leading to the same reduction in the agricultural workforce experienced by the more developed world.

Figures 12.5 and 12.6 present data for Brazil, Mexico, Turkey, Egypt, and Colombia.[7]

Currently, these five countries all have an agricultural share of the labor force in the range of 20 to 45 percent, and each has experienced several decades of decline in the share. Similar graphs (not shown) for Argentina,

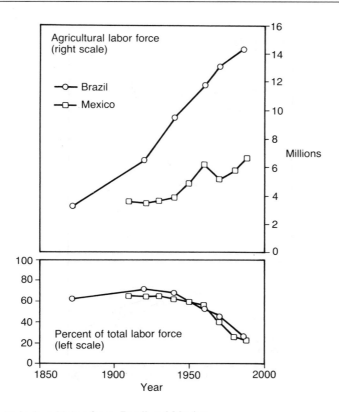

Figure 12.5 Agricultural labor force, Brazil and Mexico

Venezuela, Chile, Peru, and South Africa would also reveal several decades with declining shares.

The most recent labor force surveys show small declines for recent years in the size of the agricultural labor force in Egypt, Brazil, and Chile; given the uncertainty of survey data, however, time must pass before we know if the sizes of the agricultural labor force in these countries have reached an apex. Among the five countries shown in figures 12.5 and 12.6, only Turkey has experienced a significant decline in the size of her agricultural labor force, and it was rising until 1980. It is impressive that the size of Turkey's agricultural labor force declined even when its share was as high as 58 percent, because it requires particularly rapid growth of nonagricultural employment.[8] A larger non-agricultural sector has a greater capacity for creating nonagricultural employment; it is therefore likely that the other four countries will also see significant declines in the size of their agricultural workforce in the not-too-distant future.[9]

Figures 12.7 and 12.8 present data for the Philippines, Indonesia, Thailand, and India. Currently, these four countries all have an agricultural

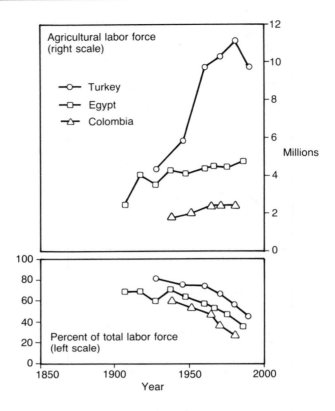

Figure 12.6 Agricultural labor force, Turkey, Egypt, and Colombia

share of the labor force at or above 41 percent and show trends indicative of the early stages of structural change. In recent decades, the Philippines, Indonesia, and Thailand, have had significant reductions in the share of their labor force devoted to agriculture. And even India, whose share had remained at about 72 percent for the period from 1950–70, had a share of 61 percent in 1991. All four countries continue to show positive growth of the size of their agricultural labor forces, and we may expect growth to continue into the near future.

CONCLUSION

From the perspective of the 1950s, Dovring (1959, p. 94) could write that "[e]ven in phases of advanced industrialization, declines in absolute numbers [of agricultural workers] has on the whole remained moderate." Clearly, this is no longer the case: in recent decades, the agricultural workforce has declined in large numbers in all of the developed countries. And in developing countries we are seeing similar patterns.

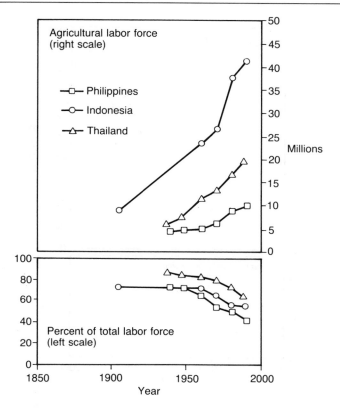

Figure 12.7 Agricultural labor force, Philippines, Indonesia, and Thailand

The great majority of developing countries have a declining share of the agricultural labor force. Many have or will soon have a declining absolute size of the agricultural workforce. In some countries, for the near future, growth of the agricultural labor force will result in increasing pressure on land. If previous experience is a useful guide, however, the most likely pattern the developing world will see is a continued downward trend in the share of the agricultural labor force, followed by a fall in the absolute size of the agricultural labor force. Thus, we will most likely see a long-run upward trend in the world's land–labor ratio, and continued decreasing pressure on the world's agricultural land.

NOTES

The opinions expressed in this article are not necessarily those of the Federal Reserve Bank of Kansas City or of the Federal Reserve System.

1   The data I present include forestry and fishing with agricultural workers. To simplify exposition, I use agriculture as a term that includes forestry and fishing.

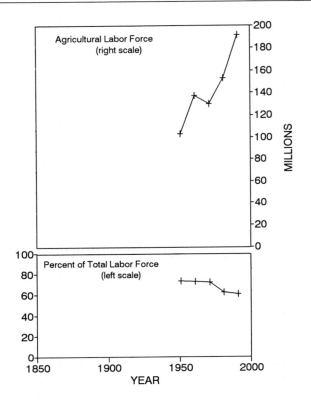

Figure 12.8 Agricultural labor force, India

The countries in this survey are not a random sample. Enough countries are included, however, to reveal general experience among market-oriented economies. Selection criteria include the extent of available data, geographic representation, and graphical presentability. Except where noted, data come from Mitchell (1981, 1982, 1983), with updates taken from various issues of the *Yearbook of Labour Statistics*.

In order to obtain as long a perspective as possible, data are used that may not be strictly comparable due to changes in definition and compilation. The main problem is in the short-run fluctuations in the data, which often can be meaningless. The long-run trends, however, are unlikely to change even if comparable data were available.

For the purposes of presentation, the vertical scale for the size of the agricultural labor force varies from figure to figure.

2 Data for the period from 1840 to the present are from Mitchell (1980) and the *Yearbook of Labour Statistics* for The Netherlands and France, and from Lee (1979) for England. For the period 1500–1800, the data are from Wrigley (1986), with the following adjustments. Agriculture's share of the total labor force are Wrigley's share of the agricultural population. The absolute size of the agricultural labor force was calculated as 45% of Wrigley's agricultural population. The 45% is a rough estimate, based on

the agricultural labor force participation rate for early nineteenth-century England.

3   Wrigley (1986, table 10, p. 160) calculates that English agricultural productivity rose by 88% in the period 1520–1801. Where the growth of productivity came from is less clear, although it is likely that England's relatively market-oriented farming industry rewarded invention and overcame obstacles to innovation in a manner not repeated elsewhere (Wrigley, 1986, pp. 166–9). Empirical research has shown that England's population growth stimulated agricultural invention (Simon and Sullivan, 1989), so that a pool of inventive ideas for farming was increasingly available as population grew. Theoretically, rising income per person should exert downward pressure on the growth of the agricultural labor force because individuals would devote a smaller proportion of their budget to food (H. Simon, 1957), a prediction that has been confirmed empirically (Simon, Reisler, and Gobin, 1983). Thus, rising income per person would have led to a fall in the growth rate of the English agricultural labor force.

The decline in the share in The Netherlands for the period 1550–1650 was almost certainly influenced by productivity growth. Dutch agriculture was at least as advanced as English agriculture in this period, so it is likely that English agricultural productivity growth was paralleled in The Netherlands.

4   Data for the United States are taken from Weiss (1986) for 1800–1860, from US Bureau of the Census (1976, p. 139, series D167 and D170), for 1870–1970, from US Bureau of the Census (1990, 378) for 1980, and from the *Yearbook of Labour Statistics* (1991) for 1988.

5   I use The World Bank's (*World Development Report 1992*, pp. 218–19) designation of high-income countries to define developed countries, and define developing countries as the low-, lower-middle-, and upper-middle-income countries.

6   Bairoch (1988, pp. 459–63) has documented the results of food imports to the Third World. Urbanization for their stage of development is far higher compared to urbanization in industrially developed countries at similar stages of development.

7   The developing countries in figures 12.4 to 12.7 have been grouped roughly according to per capita income; countries with the highest per capita income are presented first. For the sake of presentability, Egypt was included in figure 12.6 (with Turkey and Colombia), although Egypt's per capita income is closer to that of Indonesia and India (figures 12.7 and 12.8). Similarly, Thailand's data are presented in figure 12.7 (with Indonesia and the Philippines), although Thailand's per capita income is closer to that of Turkey and Colombia (figure 12.6).

8   Let L = total labor force, A = agricultural labor force, and N = nonagricultural labor force. Then $dL = dA + dN$, where d indicates a change in the variable. If we wish to hold the agricultural labor force constant, $dA = 0$, so that $dL = dN$. Now, the growth of the total labor force is given by $g_L = dL/L$; then $g_L = dN/L = (N/L)(dN/N) = (N/L)g_N$, where g_N = growth of nonagricultural employment. Rearranging, we get $g_N = (L/N)g_L$.

If the agricultural employment were 50% then L/N is 2, and growth of nonagricultural employment would need to be 2 times the growth of the labor force to keep the number of agricultural jobs constant. If the agricultural employment were 25%, the multiplier falls to 1.33. Since Brazil, Mexico, Egypt, and Colombia have shares of less than 40%, the multiplier would be at most 1.67; it is therefore likely that we should soon see declines in the size of the agricultural work force in some or all of these countries.

9   The list of developing countries with a declining size of the agricultural labor force also includes Argentina and South Africa.

REFERENCES

Bairoch, Paul (1988): *Cities and Economic Development*. Chicago: University of Chicago Press.

Dovring, Folke (1959): "The Share of Agriculture in a Growing Population." *Monthly Bulletin of Agricultural Economics and Statistics*, vol. 8 (September), 1–11. Reprinted in Carl Eicher and Lawrence Witt, eds, *Agriculture in Economic Development*. New York: McGraw-Hill, 1964.

Lee, C. H. (1979): *British Regional Employment Statistics 1841–1971*. Cambridge: Cambridge University Press.

Mitchell, Brian R. (1981): *European Historical Statistics*, 2nd edn. New York: Facts on File, 159–73.

—— (1982): *International Historical Statistics, Africa and Asia*. New York and London: New York University Press, 83–93.

—— (1983): *International Historical Statistics, The Americas and Australasia*. Detroit: Gale Research Company, 149–60.

Simon, Herbert A. (1957): "Productivity and the Urban-Rural Population Balance." Chapter 12 in his *Models for Man*. New York: Wiley. Originally in *Econometrica*, 15 (January 1947).

Simon, Julian, William Reisler, and Roy Gobin (1983): " 'Population Pressure' on the Land: Analysis of Trends Past and Future." *World Development*, 11(9), 825–34.

—— and Richard Sullivan (1989): "Population Size, Knowledge Stock, and Other Determinants of Agricultural Publication and Patenting: England, 1541–1850." *Explorations in Economic History*, 26(1), 21–44.

US Bureau of the Census (1976): *Historical Statistics of the United States: Colonial Times to 1970*. Washington, DC: US Government Printing Office.

—— (1990): *Statistical Abstract of the United States*. Washington, DC: US Government Printing Office.

Weiss, Thomas (1986): "Revised Estimates of the United States Workforce, 1800–1860." In Stanley L. Engerman and Robert E. Gallman, eds, *Long-Term Factors in American Economic Growth*. Chicago: University of Chicago Press for the NBER, 641–71.

Wrigley, E. Anthony (1986): "Urban Growth and Agricultural Change: England and the Continent in the Early Modern Period." In Robert I. Rotberg and Theodore K. Rabb, eds (guest edited by Roger S. Schofield and E. Anthony Wrigley), *Population and Economy*. Cambridge: Cambridge University Press, 123–68.

World Bank (1992): *World Development Report 1992*. Oxford: Oxford University Press.

Yearbook of Labour Statistics. 1979, 1980, 1983, 1985, 1988, 1991, and 1994. Geneva: International Labour Office, Table 2A.

13

The Standard of Living Through the Ages

Joyce Burnette and Joel Mokyr

The standard of living in the West today is higher than it has ever been. A careful study of the standard of living, however, requires careful definition to allow cross-cultural comparisons and quantification (for which see Sen, 1987). National income is quantifiable, but we must use it carefully. It does not measure exactly what we want, especially in societies farther away from the modern Western world. It does not capture non-market transactions, availability of new products, changes in product quality, or other aspects of a quality life such as leisure, lack of pollution, etc. It may even rise when the standard of living declines (as when rising crime leads people to buy locks). Life expectancy, height, infant mortality, death rates, and similar examples of Sen's "functionings" measure aspects of the standard of living directly. Health, longevity, and nutrition are valuable in all societies, so we can compare different societies and cultures. Consumption statistics of individual commodities are also a useful measure, though here we must also be careful. Whatever commodities we pick, they will not be as inherently valuable as health and longevity. Measuring fuel consumption in tropics is a biased indicator of standards of living. Such pitfalls notwithstanding, this chapter will examine evidence of changes in the standard of living over the long run. Given this evidence, there is no doubt that the standard of living in the Western world has increased considerably.

ECONOMIC GROWTH BEFORE THE INDUSTRIAL REVOLUTION

Our knowledge of the standard of living before the eighteenth century is limited by the lack of aggregate data, but there is evidence suggesting that economic growth is not a novel phenomenon. Indicators such as technological innovation, structural change, or life expectancy can be used to identify periods of growth in per capita incomes. It is clear that we should abandon any Eurocentric notion that growth is exclusively a Western phenomenon. Eric Jones points to Sung China and Tokugawa Japan as examples of intensive growth outside Europe (Jones, 1988). In China under the Sung empire, markets for labor and land grew, transactions became more monetized, trade expanded, and consumer goods became more abundant. Tea and pepper were widely consumed, iron production expanded, and productivity increased due to technological progress in energy use and agriculture. The Chinese introduced sophisticated machinery for textile production, built seaworthy ships, printed books, and practiced advanced medicine (Mokyr, 1990, chapter 9). For a couple of centuries the Chinese economy prospered, so that in all likelihood the Chinese bettered their standard of living. In Tokugawa Japan, agricultural output nearly doubled between 1600 and 1850, while the population increased by 45 per cent (Jones, 1988, p. 155). Susan Hanley presents evidence that the standard of living in Japan around 1850 was well above subsistence, and even similar to that of Britain. The common people had money to spend on festivals and funerals, as well as on clothing and tools. Housing in Japan was less capital-intensive than that in Europe, but of good quality. Life expectancy was 35 or 40 years for 1800–35, which compares well to Britain's 35.9 in 1799–1809 and 40.4 in 1854–8 (Hanley, 1986 and 1979; Yasuba, 1986).

Jones concludes that growth of per capita income occurred in Asia before 1750, though on the whole it was rare. The real difficulty seems to have been that growth, even when it took place, could not be sustained in the long run. Before the Industrial Revolution, growth was too often stifled by rent-seekers, corruption, wars, and inept government. In that regard much of Europe was not different. Europe before the Industrial Revolution seems to have experienced some per capita growth, but there is disagreement about how much. Growth was often reversed by cataclysmic effects such as the German Thirty Years War or the invasions of Northern Italy. In much of Europe, living standards before 1800 were not high, even by absolute standards. It has been suggested that Europeans worked a shorter work-week before the Industrial Revolution because they simply did not have enough nourishment to work full-time (Freudenberger and Cummins, 1976; Fogel, 1989; Bekaert, 1991; and chapter 5 by Fogel in this volume). Periodic famines remained a serious threat until deep into the eighteenth century and in some regions until much later. There was growth, but it was sporadic, and certainly slower before 1700 than after (Landes, 1969).

If growth rates were lower than after the Industrial Revolution, how low were they? Some scholars claim that growth was nearly nonexistent. Maddison suggests per capita income did not grow between 500 and 1500, and that the average annual growth rate of per capita GDP was 0.1 percent between 1500 and 1700 (Maddison, 1982). Although this is probably a good approximation for much of the continent over the long run, there were important exceptions. In a recent study comparing the

Domesday Book (1086) with Gregory King's statistics (1688), Graeme Snooks finds much more growth between the eleventh and seventeenth centuries. He calculates that real per capita income in 1688 was 5.8 times that of 1086. This corresponds to an average annual rate of growth of 0.29 percent over this period (Snooks, 1990). This finding, though, uses only two data points, so the result is sensitive to the representativeness of the two dates. The estimate also will not reveal exactly when the growth occurred, and there is no reason to expect that growth was steady. Indeed, it seems likely that most of it occurred between 1100 and 1300, and that after that growth slowed down.

Of course, if Snooks is right about the extent of growth, the standard of living in Western Europe must have been very low in 1086. If the average person in 1086 had about one-sixth the income of the average person in 1688, he or she did not have much. Snooks claims that English peasants in 1086 had little more than enough food to keep them alive, and sometimes not even that. Houses were crude, temporary structures. A peasant owned one set of clothes, best described as rags, and little else. As late as the fifteenth century expenditures of the masses on non-food items such as clothing, heat, light, and rent were probably only 13 percent of all expenditures (Snooks, 1990, p. 18). In comparison, a mason in Berlin in 1800 spent about 27 percent of his budget on non-food items (Braudel, 1967). This suggests a low standard of living indeed.

Jones points to many factors that indicate living standards in Europe increased between the Middle Ages and the eighteenth century (Jones, 1988, chapter 10). The capital stock, especially buildings, harbors, and roads grew, and the quality of housing greatly increased with the introduction of chimneys and bricks. Ships and clocks were perfected, wind and water power were widely diffused, and coal came widely into use. Europeans introduced new crops from the New World (potatoes, maize, tobacco), ate more and better-processed fish, and devoted more resources to producing products other than food. Furniture, flat- and tableware, clothing, bedding, books, jewelry, musical instruments, and guns were owned widely if not universally, as probate records indicate. An unusual and shining example is the Dutch Republic, where rising living standards spread to wide population groups in the seventeenth century, creating a substantial urban middle class who enjoyed an "Embarrassment of Riches" (Schama, 1988). Britain and The Netherlands were exceptional in the large size of their "middle class," but high levels of consumption can also be found in parts of Germany, Scandinavia, and Northern Italy.

ECONOMIC GROWTH FROM THE INDUSTRIAL REVOLUTION INTO THE TWENTIETH CENTURY

During and after the Industrial Revolution we have more measures of growth. The British Industrial Revolution itself has often been identified as a process of economic growth. Yet despite the dramatic technological changes and the rise of the factory system, macroeconomic measures are surprisingly ambiguous and the exact effect of the Industrial Revolution on the standard of living of the British between 1750 and 1850 has been, and still is, ground for heated controversy (Mokyr, 1985). In part, the problem is that the Industrial Revolution coincided with rapid population growth. A large increase in production was required just to keep income per capita from falling. In part, the innovations, dramatic as they were, affected at first only a few sectors and limited regions, so that the effects on the country as a whole were diluted. In any event, modern scholarship has concluded that per capita income growth in Britain was slow during the Industrial Revolution (Harley, 1993). The path of per capita national income in constant prices is depicted in figure 13.1, which clearly demonstrates that the revised figures by Crafts raised serious doubts whether the Industrial Revolution represented a rapid take-off into sustained growth. Yet even these figures present an average rate of growth of 0.3 percent per annum between 1700 and 1830, which is higher than Maddison's estimate for pre-industrial Europe. It may

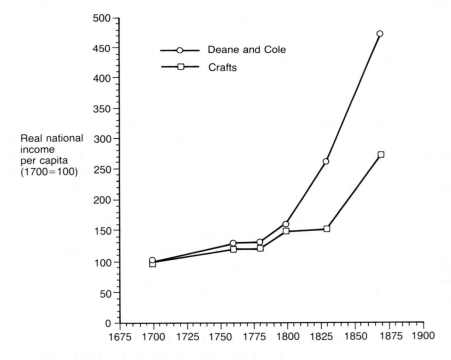

Figure 13.1   Two estimates of economic growth in Britain 1675–1900
Source: N. F. R. Crafts, *British Economic Growth During the Industrial Revolution*, Oxford: Clarendon Press, 1985. Phyllis Deane and W. A. Cole, *British Economic Growth 1688–1959*, 2nd edn, Cambridge: Cambridge University Press, 1959.

seem that the figure is consistent with Snooks's estimate for the period 1086–1688, but it should be kept in mind that *total output* grew much faster in 1700–1830 (on average 1.08 percent per annum by Crafts's figures).

For purpose of the standard of living (traditionally defined), it is *per capita* variables that count. In the per capita statistics, though, there seems to be little evidence for a general improvement in the standard of living for the preponderant majority of Britons before the middle of the nineteenth century (Mokyr, 1988; Brown, 1990; Huck, 1992). The evidence on real wages for employed workers seems clearly to show improvement (Lindert and Williamson, 1985), but per capita consumption of desirable goods shows no significant improvement before the mid-1840s. It is striking, indeed, that for most European economies, whether they industrialized like Britain or Belgium, or did not, like Ireland or The Netherlands, the evidence shows little or no overall improvement in living standards in the traditional sense before 1850 (Mokyr, 1991). Mortality figures, though, do show improvement by 1850. Life expectancy rose and death rates fell (see figure 13.2 for evidence from Britain). Much of the improvement, however, predated the Industrial Revolution and seems hardly affected by it.

There was thus a long lag between the onset of the modernization of the European economy and the rise in the standard of living. In the

Table 13.1 Annual growth rates of per capita income

	1 GDP 1700–1820	2 GDP 1820–1979	3 GNP 1830–1910	4 GNP 1830–1973	5 GNP 1850–1950
Austria		1.5		1.53	
Belgium		1.7	1.34	1.55	1.59
Denmark		1.6	1.60	1.81	1.26
France	0.3	1.6	1.18	1.72	1.27
Germany		1.8	1.33	1.74	0.99
Italy			0.40	1.31	
Netherlands	– 0.1	1.5	0.89	1.34	
Norway		1.8	1.10	1.78	
Sweden		1.8	1.40	2.03	
Switzerland		1.6		1.60	
UK	0.4	1.4	1.21	1.30	1.03
Europe			0.92	1.52	
Japan		1.8			
USA		1.8			1.90

Sources: Columns 1 and 2: Angus Maddison, *Phases of Capitalist Development*, Oxford: Oxford University Press, 1982, p. 44. Columns 3 and 4: Paul Bairoch, "Europe's Gross National Product: 1800–1975," *Journal of European Economic History*, Fall 1976, 5:273–340, pp. 283 and 309. Column 5: B. R. Mitchell, "Conseguenze della rivoluzione industriale," in Melograni and Ricossa, eds, *Le Rivoluzioni del Benessere*, Roma-Bari: Gius. Laterza and Figli Spa, 1988.

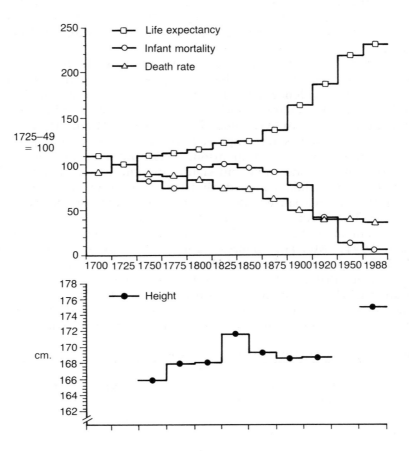

Figure 13.2   Longevity, mortality, and height in Britain, 1700–1988

Sources: Life expectancy to 1871: E. A. Wrigley and R. S. Schofield, *The Population History of England 1541–1871*, Cambridge, Mass.: Harvard University Press, 1981, 529. 1875–1931: J. Whitaker, *Whitaker's Almanac*, London: Whitaker, 1900, 1924, 1930, 1938, 1965, and 1970. 1988: The World Bank, *World Development Report*, 1990, Oxford: Oxford University Press, 1990.

Death rates to 1871: Wrigley and Schofield. 1875–1914: Neil Tranter, *Population since the Industrial Revolution*, London: Croom Helm, 1973, 53. 1920–38: B. R. Mitchell, *European Historical Statistics 1750–1970*, New York: Columbia University Press, 1975 (statistics for England and Wales). 1960–5 and 1988: The World Bank, *World Development Report*, 1978 and 1990.

Infant mortality: 1725–1824: Michael W. Flinn, *The European Demographic System 1500–1820*, Baltimore: Johns Hopkins University Press, 1981. Estimates are averages of studies whose years include the given dates. 1838–54: J. Whitaker, *Whitaker's Almanac*, 1900, 684. 1850–1938: B. R. Mitchell, *European Historical Statistics 1750–1970*. 1960–5 and 1988: The World Bank, *World Development Report*, 1978 and 1990.

Height: Robert Fogel, "Second Thoughts on the European Escape from Hunger," NBER Working Paper on Historical Factors in Long Run Growth, 1989; and Floud, Wachter, and Gregory, *Height, Health, and History*, Cambridge: Cambridge University Press, 1990.

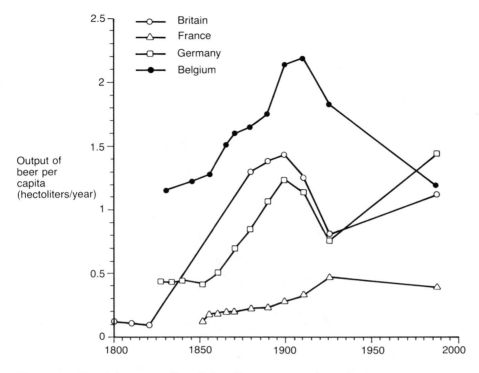

Figure 13.3 Trends in output of beer in four European countries
Sources: B. R. Mitchell, *European Historical Statistics 1750–1970*, New York: Columbia University Press, 1975. "Beer Per Caps Grow in Europe," *Beverage Industry*, March 1990, 46–7.

longer run, however, the effects of economic growth on the standard of living were inevitable. For Europe as a whole, Maddison and Bairoch have estimated GNPs for Europe after 1800 (table 13.1). Growth in per capita income was far higher after the Industrial Revolution than before it. Maddison estimates that per capita GDP for 16 developed countries was 13 times larger in 1979 than in 1820. This implies an annual average growth rate of 1.6 percent over the same period. Bairoch estimates average annual growth in per capita GNP to be 0.92 percent between 1830 and 1910 for Europe as a whole, and 1.21 percent for the UK. These rates of growth are much higher than anything experienced anywhere before 1800.

During this period of growth in Europe, the gap between Europe and the Third World grew. According to Bairoch's estimates of per capita GNP in the Third World, the average annual growth rate of per capita GNP was 0.37 percent between 1830 and 1970, which pales in comparison to Europe's 1.52 percent (Bairoch, 1981). Bairoch estimates that the ratio of per capita GNP in developed countries to that in the Third World grew from 1.1 in 1800 to 7.2 in 1970. Note, however, that GNP

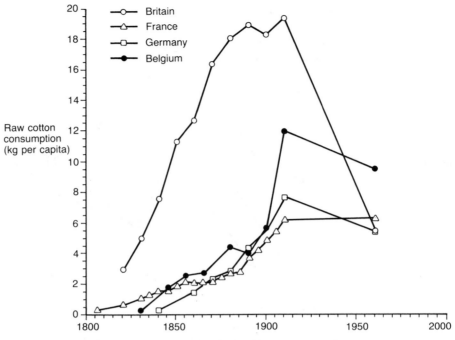

Figure 13.4 Growth in cotton consumption in four European countries
Source: B. R. Mitchell, *European Historical Statistics 1750–1970*, New York: Columbia University Press, 1975.

is a less accurate measure in Third World countries because more activity takes place outside the money economy. Other measures, though, tell the same story. Some Third World countries have daily per capita caloric intakes that are not above what England and France had before 1800 (see table 13.2), and Third World countries still suffer famines. Most of the Third World countries in table 13.2 experienced

Table 13.2 Daily per capita calories

	ca. 1790	*ca. 1810*	*1965*	*1986*
England, North	2540–2880		3353[*]	3256[*]
England, South	1900–2250		3353[*]	3256[*]
France	1753	1846	3217	3336
Belgium		2039		3679[#]
USA			3324	3645
Japan			2687	2864
Bangladesh			1971	1927
Brazil			2404	2656
China			1926	2630
El Salvador			1859	2160
Ethiopia			1824	1749
India			2111	2238

Mexico	2644	3132
Nigeria	2185	2146
Pakistan	1761	2315
Rwanda	1665	1830

*United Kingdom
#1985
Sources: England 1790: Carole Shammas, "The Eighteenth-Century English Diet and Economic Change," *Explorations in Economic History*, July 1984, 21:254–69. Estimates are from household budgets, and the upper bound adds an estimate of home-produced food not included in the budgets. The numbers given have been converted from calories per adult equivalent to calories per capita. France, 1790 and 1810: Robert Fogel, "Second Thoughts on the European Escape From Hunger," NBER Working Paper on Historical Factors in Long Run Growth, 1989. Belgium: Geert Bekaert, "Caloric Consumption in Industrializing Belgium," *Journal of Economic History*, September 1991. 1965 and 1986: The World Bank, *World Development Report 1990*, Oxford: Oxford University Press, 1990.

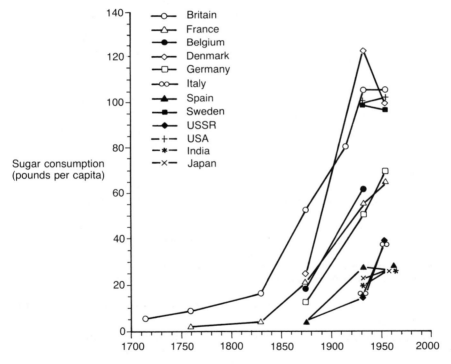

Figure 13.5 Growth in sugar consumption around the world
Sources: France 1730–89: Robert Stein, *The French Sugar Business in the Eighteenth Century*, Baton Rouge, Louisiana State University Press, 1988, 164. Belgium 1812 and 1846: Appendix to Geert Bekaert, "Caloric Consumption in Industrializing Belgium," *Journal of Economic History*, Sept. 1991. France 1790–1849 and all but Britain in 1850–99: Michael Mulhall, *Dictionary of Statistics*, London: George Routledge and Sons, 1899, 550. 1955: E. W. Mayo, ed., *Sugar Reference Book*, vol. xxv, New York: Mona Palmer, 1957, 112–120. All other: Noel Deer, *The History of Sugar*, London: Chapman and Hall Ltd., 1950, 532.

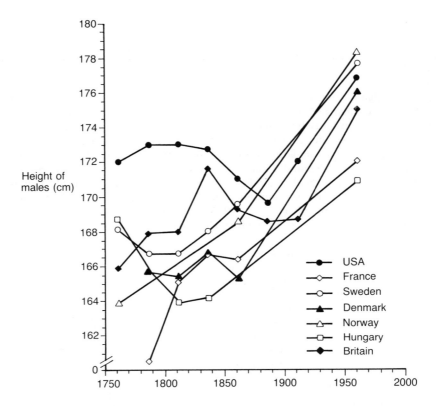

Figure 13.6 Trends in height in selected developed countries
Sources: US, 1750–1924: R. W. Fogel, "Nutrition and the Decline in Mortality since 1700: Some Findings." In Engerman and Gallman, eds, *Long-term Factors in American Economic Growth*, Chicago: University of Chicago Press, 1986. US 1980s: Phyllis Eveleth and James Tanner, *Worldwide Variation in Human Growth*, 2nd edn, Cambridge: Cambridge University Press, 1990. UK, 1874–1907: R. Floud, K. Wachter, and A. Gregory, *Height, Health, and History: Nutritional Status in the United Kingdom, 1750–1980*, Cambridge: Cambridge University Press, 1990; table 4.1. All others: R. W. Fogel, "Second Thoughts on the European Escape from Hunger," NBER Working Paper on Historical Factors in Long Run Growth, 1989.

gains in per capita calories between 1965 and 1986, but some experienced declines.

In spite of variation in growth rates, there is no doubt that living standards increased substantially after 1850. With the steady rise in incomes, consumption increased, even when a large portion of the increment was absorbed by higher gross investment and two major wars. Consumption, of course, is not a perfect measure for the standard of living, but it can serve as a proxy. Table 13.2 shows the increment in per capita food consumption. Counting calories *under*states the increase in living standards because of the improvement in the quality of the diet. Consumption of other goods increased equally rapidly after 1850. New goods, such as bicycles, electrical appliances, and personal entertainment, unimaginable in 1830, were widely diffused by 1914. Figures 13.3–5 depict the increases in consumption of beer, cotton, and sugar. These figures also underline the fact that consumption of one commodity is not a fool-proof measure of the standard of living. The consumption of cotton in Britain dropped dramatically between 1910 and 1960, but this is not very surprising, since synthetic materials became important over this period. Evidence on consumption, though, does not stand alone, and it confirms the conclusion drawn from other evidence that the standard of living in the Western world has grown rapidly since 1850. Evidence on heights also suggests that living standards have increased dramatically (see figure 13.6 and chapter 5 by Fogel in this volume).

CONCLUSION

All these measures tell the same story. As a result of economic growth, the standard of living has increased dramatically in the past century and a half. It does not matter which of the measures one chooses: a believer in aggregate statistics can consult the national income accounts, whereas a skeptic can choose among infant mortality, literacy rates, mean height, or the ownership of refrigerators. The difference between modern economic growth and economic growth before 1750 is that today's prosperity in the Western world is sufficiently strong and resilient that it is unlikely to be undone by the political and social disruptions that destroyed the fruits of economic growth in the more remote past. It is possible, however, that the historical enemies of economic progress – rent seeking, corruption, violence, and intolerance – could play a role in keeping the rest of the world from enjoying similar improvements.

REFERENCES

Bairoch, Paul (1976): "Europe's Gross National Product: 1800–1975." *Journal of European Economic History*, 5, 273–340.

——— (1975): *The Economic Development of the Third World Since 1900*. London: Methuen.

——— (1981): "The Main Trends in National Economic Disparities since the Industrial Revolution." In Bairoch and Levy-Leboyer, eds, *Disparities in Economic Development since the Industrial Revolution*. New York: St Martin's Press.

Bekaert, Geert (1991): "Caloric Consumption in Industrializing Belgium," *Journal of Economic History*, 51, 633–55.

Braudel, Fernand (1967): *Capitalism and Material Life 1400–1800*. New York: Harper and Row.

Brown, John (1990): "The Condition of England and the Standard of Living." *Journal of Economic History*, 50, 591–614.

Floud, Roderick, Kenneth Wachter, and Annabel Gregory (1990): *Height, Health, and History: Nutritional Status in the United Kingdom, 1750–1980.* Cambridge: Cambridge University Press.

Fogel, Robert (1989): "Second Thoughts on the European Escape from Hunger." National Bureau for Economic Research Working Paper on Historical Factors in Long Run Growth.

—— (1991): "New Sources and New Techniques for the Study of Secular Trends in Nutritional Status, Health, Mortality, and the Process of Aging." National Bureau for Economic Research Working Paper on Historical Factors on Long Run Growth.

Freudenberger, Herman, and Gaylord Cummins (1976): "Health, Work and Leisure before the Industrial Revolution." *Explorations in Economic History*, 13, 1–12.

Hanley, Susan (1986): "A High Standard of Living in Nineteenth Century Japan: Fact or Fantasy?" *Journal of Economic History*, 46, 217–24.

—— (1979): "The Japanese Fertility Decline in Historical Perspective." In Lee-Jay Cho and Kazumasa Kobayaski, eds. *Fertility Transition of the East Asian Populations.* Honolulu: University Press of Hawaii, 24–48.

Harley, C. Knick (1993): "The State of the British Industrial Revolution: a Survey of Recent Macroeconomic Reassessment." In Joel Mokyr, ed., *The British Industrial Revolution: an Economic Assessment.* Boulder: Westview Press.

Huck, Paul (1992): "The Standard of Living of British Workers." Ph.D. Dissertation, Northwestern University.

Jones, E. L. (1988): *Growth Recurring: Economic Change in World History.* Oxford: Clarendon Press.

Landes, David S. (1969): *Unbound Prometheus.* Cambridge: Cambridge University Press.

Lindert, Peter, and Jeffrey G. Williamson (1985): "English Workers' Living Standards During the Industrial Revolution." In Joel Mokyr, ed., *The Economics of the Industrial Revolution.*

Maddison, Angus (1982): *Phases of Capitalist Development.* Oxford: Oxford University Press.

Mokyr, Joel (1980): "Industrialization and Poverty in Ireland and the Netherlands." *Journal of Interdisciplinary History*, 429–58.

—— ed. (1985): *The Economics of the Industrial Revolution.* Totowa NJ: Rowman and Littlefield.

—— (1985): "Is There Still Life in the Pessimist Case? Consumption during the Industrial Revolution, 1790–1850." *Journal of Economic History*, 48, 69–92.

—— (1990): *The Lever of Riches: Technological Creativity and Economic Progress.* New York: Oxford University Press.

—— (1991): "Dear Labor, Cheap Labor, and the Industrial Revolution." In Henry Rosovsky and Patrice Higonnet, eds, *Favorites of Fortune: Technology, Growth, and Economic Development since the Industrial Revolution.* Cambridge: Harvard University Press.

Nordhaus, William, and James Tobin (1972): "Is Growth Obsolete?" In *Economic Growth.* New York: National Bureau of Economic Research.

Schama, Simon (1988): *The Embarrassment of Riches*. Berkeley: University of California Press.

Sen, Amartya (1987): "The Standard of Living: Lecture I, Concepts and Critiques" and "The Standard of Living: Lecture II, Lives and Capabilities." In Geoffrey Hawthorn, ed., *The Standard of Living*. Cambridge: Cambridge University Press.

Snooks, G. D. (1990): "Economic Growth During the Last Millennium: A Quantitative Perspective for the British Industrial Revolution." Working Paper from the Australian National University.

Steckel, Richard (1983): "Height and Per Capita Income." *Historical Methods*, 16, 1–7.

Usher, Dan (1980): *The Measurement of Economic Growth*. New York: Columbia University Press.

De Vries, Jan (1975): "Peasant Demand Patterns and Economic Development: Friesland, 1550–1750." In William N. Parker and Eric L. Jones, eds, *European Peasants and Their Markets*. Princeton: Princeton University Press.

Yasuba, Yasukichi (1986): "Standard of Living in Japan Before Industrialization: From What Level Did Japan Begin? A Comment." *Journal of Economic History*, 46, 217–24.

EDITOR'S NOTE (E. CALVIN BEISNER)

Some Americans wonder how their standard of living compares with those of other nations around the world. Figure 13.7 depicts the real

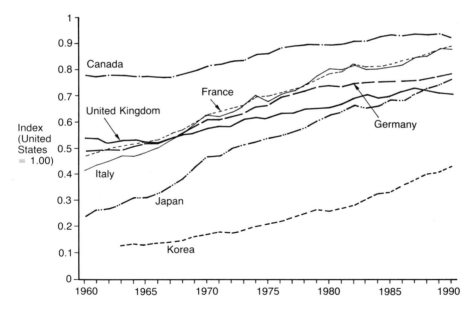

Figure 13.7 Indices of real GDP per employed person, purchasing power parity exchange rates
Source: Unpublished data from the US Department of Labor reported in "Is the US 'Uncompetitive'?" *Jobs & Capital*, vol. 2, no. 1 (Winter 1993), 10–11.

purchasing power of gross domestic product per employed person in the seven member nations of the Organization for Economic Cooperation and Development – the seven leading economies in the world – relative to the purchasing power of GDP per employed person in the United States over a 30 period. The purchasing power parity comparison is more realistic than a monetary exchange rate comparison because what can be bought for a given amount of money in one country may significantly exceed what can be bought for the same amount in another country.

14

Long-Term Trends in the US Standard of Living

Stanley Lebergott

The goal of every economy is to provide consumption. So economists of all persuasions have agreed, from Smith and Mill to Keynes, Tobin, and Becker. How much has the American economy provided? We review changes in the American standard of living in the century just ending. The subject is not how happiness changed, or welfare. There is wide disagreement on what these words mean. But what of the standard of living, measured simply by what consumers bought?[1]

Changes in economic well-being are measured better by that standard than by the usual empirical measure – workers' incomes. For the story told by income must be heavily qualified: (1) Consumers never see "total income": the government taxes away 15 percent of that total. (2) Most people save, thus consuming less than their total income. (Any claims to consumption they leave at death will be used up by their survivors, the inheritance tax authorities, and the lawyers.) (3) Money income estimates for low-income consumers are particularly biased, for the value of free consumption – food stamps, medicare, subsidized housing – is typically ignored.

Remembering Lincoln's remark – "the Lord must have loved the common man since he made so many of them" – we focus on the average American. This chapter first summarizes the essentials, then goes on to discuss housing, household operation, medical care, and food. Major trends are summarized in the concluding section.

THE ESSENTIALS

Survival in North America depends on three items: clean water, food, and heat.

Water

By 1982 Americans were spending six million times as much for household water as in 1800.[2] Gallons consumed per person rose tenfold for washing clothes and cars, watering lawns, sprinkling the children.[3] The average family doubled the share of its income spent for water between 1900 and 1990 – from one to two days' worth of annual income.

Two facts largely explain that rise:

1. From 1800 to 1920 the American washing machine was a housewife. In 1800 she carried nearly 9,000 gallons of water a year into the house from river, brook, or well (with some help from the children). Water was free. So was the work.

2. Spending for water increased massively as Americans sought to cut the death rate from typhoid. That rate fell by 75 percent in Albany, and 79 percent in Lawrence, Massachusetts, when they began filtering their water (about 1900). A few years earlier, Zurich and Hamburg achieved similar declines.[4] If drinking water had not been cleaned up, and typhoid death rates had persisted, 79,000 more Americans would have died in 1990 – five times as many as then died from AIDS.[5] Instead, clean drinking water had become available throughout the country, as in certain other nations.

Food

What about food? Starvation is possible in any society, however rich the land, however good the climate. Food supply is determined by the way society uses its resources. But resources are not, in any decisive sense, its stock of minerals or land – as the history of Martha's Vineyard and Hong Kong, of Venice and Amsterdam, reminds us. It is not changes in resources but effort and organization that bring revolutionary changes in levels of living. (Few Chinese starved to death in the 1940s and early 1950s, but 30 million did so in three years of Chairman Mao's "Great Leap Forward."[6])

How did US society organize food production and distribution? In 1800, some 60 percent of the American labor force was busy producing raw food. Today 3 percent do the job (97 percent produce *other* goods and services).

Americans have eaten better foods since 1900, and more of them. Yet the share of family income spent for food fell substantially. By 1990, American farmers used the soil so much more efficiently that they fed three times as many Americans as in 1900 (plus several million Russians and Chinese), and did so with only one third as many farm workers. They had adopted much more efficient farm machinery, more fertilizer and pesticides, better farm organization. They thereby provided more pounds of food, and more variety, yet took less of the consumer's dollar.

Heat

Space heating in 1990, unlike 1900, was largely provided by cartels and monopolies of oil, gas, and electricity. Yet they took a smaller share of

family income than in 1900. The efficiency with which petroleum and natural gas were manufactured and distributed had persistently increased, and most of the gain was passed along to the
consumer.

Changing "Poverty" Levels

What of the "minimum" essentials? "Poverty" in America may have
meant bread and water for some in 1900. By 1990 the typical family
classed by the Census as "in poverty" had its own automobile.[7] Meanwhile, the millionaires who consumed the most expensive yachts,
homes, watches, automobiles, and clothing in history were neither Mor-
gans nor Rockefellers. They were increasingly singers, dancers, actors,
and athletes, few from *Social Register* backgrounds.

HOUSING

Americans in 1900 did not live as their grandparents had lived across the
ocean. Few shared one or two rooms with the family cow or pig,
removing the manure each spring to fertilize the fields. But one third of
the foreign born, as of blacks, did share their space with boarders or
lodgers, according to a major study by the Immigration Commission in
1910. (So did 14 percent of native whites.) Some 39 percent of the
foreign born squeezed three (or more) persons into each sleeping room –
compared to 17 percent of native whites.

   But rising incomes changed what Bessie Pierce delicately called the
"disgusting" housing conditions in which "large undigested groups of
foreigners" lived.[8] It turned out that the foreign born too reached for
privacy when their income permitted. By 1980 the percentage of American families with three or more persons per sleeping room had fallen to
under 5 percent from 39 percent (immigrant), and 17 percent (non-
white) in 1910.

   The second most visible change was in facilities. Few Americans had
running water in 1900, or bathtubs, hot water, or flush toilets. Few had
central heating. A mere handful had electricity. Public baths were rare,
and baths in homes were so infrequent that the New York Tenement
House Committee didn't even bother to count them in 1894. After all,
far less than 1 percent of the tenement house apartments investigated
even had flush toilets.[9] By 1980, however, 97 percent of American
families had indoor flush toilets, plus hot and cold piped water, plus a
bathtub or shower.[10] Electricity and central heating had become universal (99 percent). About a fifth of the rent for American housing went for
plumbing, heating, and electrical facilities – all almost completely absent
in 1900. Such items cost about $100 billion – or about as much as was
spent for new automobiles.[11]

HOUSEHOLD OPERATION

From 1900 to 1990 the proportion of US households with domestic servants fell to about 1 percent, from about 10 percent.[12] The supply of willing immigrant and black women had dwindled in the face of better job opportunities.

Did American women therefore spend more hours in housework? On the contrary. They spent 42 fewer hours a week, buying mechanical devices to cut their work (table 14.1). They opted for what a disgusted English economist called "gimmickry, motorized implements, novelties,"[13] and what two of America's best known sociologists called "a gorged stream of new things to buy . . . radios, automatic refrigerators and all but automatic ways to live."[14]

Although only 30 percent of British and Italian households had "automatic" refrigerators by 1960, over 96 percent of American households did. The hours of shopping time they saved, their ability to keep leftovers without waste or bacterial contamination, all reduced the time and effort they had to put into meals. Though the washboard was universally used in 1900 (98 percent), its charms were celebrated by few. It gave way to "automatic" washing machines (owned by 73 percent of American families in 1990). Apartment residents and singles relied on laundromats or laundries. Homes universally lacked cold running water in 1900. But "the gorged stream of new things to buy" added running hot and cold water to 98 percent of all households by 1990.

Finally, one must note the switch away from wood and coal for heating (about 50 percent of American families relied on each in 1900). The steady distribution of coal ash, oil smuts, and fuel dust on floors, rugs, furniture, and into curtains, clothing, and bed clothes was largely over by 1960. The shift to natural gas (60 percent of families) or oil and electricity fairly well ended the six hours a week spent on dusting and

Table 14.1 Percentage of households with appliances

	1900	1987	
	All	All	Family Income $50,000 or more
Refrigerator	0	100	100
Radio	0	100	100
TV	0	93	99
Washing machine	0	73	92
Microwave oven	0	61	81
Air conditioner	0	62	
Dishwasher	0	43	79
Freezer	0	35	41

Source: Data for 1987 from Department of Energy, *Housing Characteristics, 1987*, table 30.

cleaning. (It also ended the chore of shoveling six tons of coal into the furnace each heating season, and removing barrels of ash and clinkers.) The provision of the newer fuels by petroleum cartels and electricity monopolies naturally changed the share of the consumers' dollar going to fuel producers. It fell – from 3 percent to 2 percent. (Persistent productivity advance in mining and electric production was chiefly responsible.)

MEDICAL CARE

The twentieth century may seem the century of death, with two overwhelming world wars, the Holocaust, a nearly endless slaughter of ex-colonial tribes killing one another. But if due proportion be kept, it is even more truly the century of life. Its immense reduction in death rates has been without precedent. Many more children were saved than adults slaughtered.

US consumer spending to improve health, and defer death, increased with unusual speed after 1900. Health had long been a superior good. But the elasticity of demand for health expenditure never before reached such levels. Between 1900 and 1929, spending for physicians, dentists, and the like nearly doubled (in constant dollars). Drug spending quadrupled. Spending for hospitals rose tenfold.

But these gains were dwarfed after 1945. In 1905 Sir William Osler had observed that "we may safely say (reversing the proportion of fifty years ago) that for one damaged by dosing, one hundred were saved."[15]

Patent medicines then yielded increasingly to more reliable products. But not until the late 1930s did pharmaceuticals demonstrate their ability to save millions of lives. Neither the sulfanilimides nor penicillin, however, reached civilian markets until World War II ended. A staggering possibility then became a reality: ancient, widespread and overwhelming menaces – influenza, tuberculosis, diarrhea, measles – could now be blocked. The US death rate from infectious disease fell to less than a tenth that in 1900.[16] For the first time lower income families could, in some real sense, buy the gift of life.

What of the other components of medical spending? The number of active physicians per capita actually fell for decades after 1900. It then rose after World War II and soared as tax deductible medical insurance for employees expanded. (By 1976 the US had 30 percent more physicians per capita than the UK.)

The increase in spending for physicians, however, as for hospitals, had no perceptible impact on overall death rates.[17] But together they provided quicker diagnosis, less pain, shorter hospital stays, more assurance to those caught in the turmoil of sickness. And they helped patients in their gory preference for dying later rather than now, and from some cause other than the instant one.

FOOD

The common human encounter with famine is present reality in Ethiopia, Bangladesh, and India, and appeared through the millennia in Britain, France, Poland, Russia, across to China and through Africa. But the past two centuries reveal no famine in the United States, Europe, or Canada. Table 14.2 shows that even in the worst depression year, 1932, American consumption of key nutrients was virtually identical with that in the prosperity of 1929 or the 1970s.[18]

Nor did food production, and distribution, break or falter in our twentieth century depressions. (Since the former USSR has wasted 40 percent or more of its vegetable, potato, and fruit crops in recent years, and 20 percent of its wheat, the importance of distribution as well as production is clear.[19])

Food in the United States has cost less and less in real terms. Taking about a third of the consumer's dollar in 1900, it required only one quarter by 1929, and 18 percent by 1987. Thus, in periodic economic recessions Americans could, and did, leave their nutrition unchanged, but cut spending for less urgent items – autos, appliances, home purchase, recreation, clothing.

Food trends

The major trend in food consumption in the twentieth century is clear: spending for food (in constant dollars) doubled between 1900 and 1987,[20] and not because Americans ate more pounds. (They actually ate less. They consumed 100 fewer pounds of potatoes, despite officials and scientific nutritionists who declared they needed 786 pounds a year.[21]) Instead of filling up on potatoes and grain products, they added other vegetables, fresh and frozen fruit, restaurant meals.

What were the major elements in that change?

Table 14.2 Nutrients available

	1909–11	1929	1932	1949	1969	1988
Calories	3,530	3,460	3,320	3,200	3,300	3,600
Fat (g)	125	137	133	140	157	168
Protein (g)	104	94	91	94	98	105
Calcium (g)	0.83	0.88	0.86	0.98	0.84	0.89
Vitamin A (int'l units)	7,800	8,300	8,400	8,500	7,800	10,600
Thiamin (mg)	1.68	1.57	1.53	1.89	2.00	2.20
Ascorbic (mg)	105	111	107	109	100	118

Sources: Data for 1909–49 from USDA Agricultural Economic Report 138, *Food Consumption Prices Expenditures*. Data for 1969–88 from USDA Statistical Bulletin 825, *Food Consumption, Prices and Expenditures, 1968–89*.

Table 14.3 Food consumption (pounds per person)

	1909–11	1929	1949	1969	1989
Red meat	130	115	125	143	123
Beef	56	39	50	82	69
Pork	61	65	63	55	49
Poultry	18	16	23	33	61
Fish	11	12	11	11	16
Dairy products[a]	759	811	735	572	568
Fats and oils[a]	(38)	45	42	52	61
Fruits: fresh	128	128	121	77	94
processed[c]	8	18	42	77	(109)
Vegetables					
fresh: home and commercial[b]	187	194	175		
fresh: commercial			98	92	(119)
processed (farm weight)[c]	15	26	41	61	(54)*
Potatoes	190	168	113	(88)	(92)
Flour and cereal	295	236	169	140	169
Sugar and syrup[d]	90	117	113	121	134
Aspartame, Saccharin[e]	0	0	0	5	20*
Coffee	8	10	16	11	7
Ice cream	2	11	20	29	29

Notes:
Estimates in parentheses are approximations comparable with earlier data.
* 1988.
[a] Fat content basis.
[b] Not available for later years because home production not estimated.
[c] USDA figure for later years are adjusted to be comparable with 1909–49.
[d] Includes honey, excludes artificial sweeteners.
[e] Includes use as food additive.
Sources: Data for 1909–49 from USDA Agricultural Economic Report 138, *Food Consumption, Prices Expenditures*. Data for 1969–89 from USDA Statistical Bulletin 825, *Food Consumption, Prices and Expenditures, 1968–89*.

1. A switch from work by the housewife to work in the factory and grocery store. The 1900 housewife spent hour after hour preparing fruits and vegetables, and then washing, peeling, coring, cutting, boiling, blanching. The housewifely tasks customary in 1900 were all reduced. Only 7 percent of all vegetables had been processed (e.g., canned) in 1900, but that figure increased to 25 percent by 1982. For fruits, the rise was from 6 percent to 40 percent. Since 100 percent of families had refrigerators by 1990, frozen vegetables, juices, and other prepared foods further reduced kitchen work.

2. More and more food was prepared in restaurants, factory cafeterias, school lunchrooms, and take-out establishments. Such spending doubled between 1929 and 1990. Nutritionists and social workers had

urged "scientifically planned" hot meals at school. Unions and benevolent employers had pressed for clean, convenient eating places at work. Few factory workers carried lunch pails by 1990. And few students carried lunch boxes.

3. The diet changed.

Calories did not change, but significant nutrients rose (table 14.2). Central components of the European diet for two thousand years declined persistently (table 14.3). A steady flow of grain exports showed that the decline had nothing to do with availability. Americans consumed 100 fewer pounds of potatoes per capita, 120 pounds less flour and cereals, and 200 pounds less of dairy product fat. Given the choice, they had switched to a more varied diet.

Fewer Americans consumed "inferior foods" – salt pork, salt beef, lard, molasses. In 1900, 94 percent consumed lard, 84 percent salt pork, and 69 percent molasses; by 1988 fewer than 5 percent of Americans ate any of these. Nor did the clean, safe, carefully packaged foods consumed in 1990 much resemble the residues in the sugar barrel or the strong salt meat (with or without maggots) once staples of the American diet on land and sea.

The most spectacular quantity gains over the century included: +300 percent for poultry, +50 percent for sugar, and +1,350 percent for ice cream. The most unexpected changes in the last few decades were (health related?) declines for red meat and coffee, increases for orange juice, yogurt, and sugar substitutes. Poultry increasingly substituted for red meat, vegetable oils for animal fats and oils.

SUMMARY

What, finally, of the major categories of consumer spending? Figures 14.1a and 14.1b report on those long-term changes. Three observations stand out. (1) Every sector increased. Whatever the problems of estimate and dollar deflation, the 1990 consumer clearly consumed more goods and services in every sector. (2) Nor did the wildest increases reflect merely useless novelties, "automatic" appliances, etc. Medical care led the list. Recreation was not far behind. (3) Gains for individual items ranged from enormous (electricity and telephone) to very modest (alcohol and tobacco). No single theory of human behavior, capitalism, the free market, or the twentieth century explains these divergent trends. Multiple facets of personality, plus multiple opportunities opened out in free markets, together led to the rich complexity of actual change.

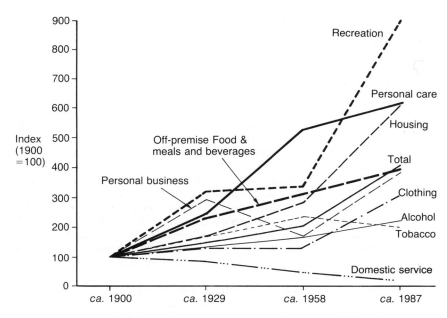

Figure 14.1a Personal consumption expenditures, per capita, constant dollars, 1900–87

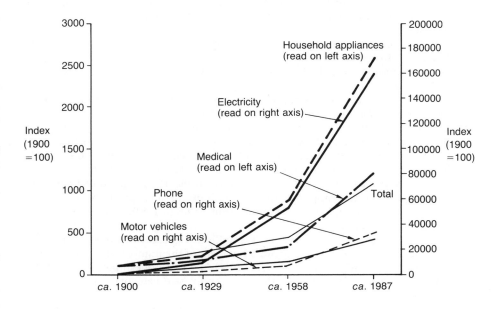

Figure 14.1b Personal consumption expenditures, per capita, constant dollars, 1900–87

NOTES

1   Data not elsewhere footnoted come from the writer's forthcoming study, *Consumer Expenditure Since 1900: New Estimates and Old Motives* (Princeton), and his *Pursuing Happiness: American Consumers in the Twentieth Century* (Princeton: Princeton University Press, 1993).

2   The 1982 Census of Governments, *Finances of Special Districts, Governmental Finances*, Vol. 4, p. 2, T1 reports expenditure for water supply approximately equaled general expenditure for sewerage. We therefore take half the Bureau of Economic Affairs, personal consumption expenditure estimate of $12 billion for both water and sanitation charges.

3   Household family consumption of water in Boston and middle-sized New England towns in 1800 was about 25 gallons a day. (Only 16 water works existed then.) F. E. Turneaure and H. L. Russell, *Public Water Supplies* (1911), 11, 22, and G. E. Waring, Jr, *Journal of Social Science* (December 1879), 183. With 4.76 persons per household, per capita consumption was thus about five gallons. In 1990 the average person consumed about 100 gallons each day. U.S. Geological Survey Circular 1081, *Estimated Use of Water in the United States in 1990* (1993), 28, Water Withdrawals and Deliveries. With 2.71 persons per household, about 279 gallons were consumed per household.

4   Turneaure and Russell, 460.

5   Center for Disease Control, *HIV/AIDS Surveillance* (December 1990). The national death rate from typhoid in 1900 was 31.3 per 100,000. *Historical Statistics of the United States from Colonial Times to 1970*, 1, 58.

6   Basil Ashton et al. "Famine in China, 1958–61," *Population and Development Review* (December 1984), 10, 619.

7   U.S. Bureau of Labor Statistics, *Consumer Expenditure Survey 1984–86*, Bulletin 2333, table 2. Consumer units with incomes below $5,000 averaged 1.0 automobiles.

For a discussion of trends in "poverty" see chapters 23 and 24 by Rebecca M. Blank and Robert Rector in this volume.

8   Bessie Louise Pierce, *A History of Chicago* (Chicago: University of Chicago, 1937) 3, 55.

9   51 of 33,385. New York State, Assembly No. 37, January 17, 1895, *Report of the Tenement House Committee of 1894*, 114–15.

10   U.S. Census, 1980, *Detailed Housing Characteristics, U.S. Summary*, 1–66. The figure for the New York City–New Jersey standard metropolitan statistical area, was 95 percent and 94 percent for all black households. Ibid., pp. 1–69, 1–204.

11   The 20 percent estimate, for 1968, is from an FHA study cited in Lebergott, *The American Economy*, p. 260. It is applied to the 1988 Bureau of Economic Affairs' personal consumption expenditures housing total. Given the increase in second bathrooms and electrical outlets from 1968 to 1990, 20 percent may be a slight underestimate.

12   The number of domestic servants, 1.8 million (Lebergott, *Manpower in Economic Growth*, p. 513) came to about 12 percent of the number of families (1900 Census of Population, *Supplementary Analysis*, 736). We estimate 15.8 million ordinary families. We exclude the 11+ family size category, as it included lodging houses, etc.

13   E. J. Mishan in Mancur Olson and Hans Landsberg, *The No-Growth Society* (New York: Norton, 1974), 85.

14 Robert and Helen Lynd, *Middletown in Transition* (1937; New York: Harcourt Brace Jovanovich, reprint 1982), 46.

15 William Osler, *Aequanimitas* (1905, reprinted 1932) 125.

16 American Chemical Society *Chemistry in the Economy* (Washington: American Chemical Society, 1973) 171.

17 Cf. Lester Breslow, et al., "Relationship of Health Practices and Mortality," *Preventive Medicine* (March 1973), and J. P. Newhouse and L. J. Friedlander, "The Relationship Between Medical Resources and Health," *Journal of Human Resources* (Winter 1980).

18 Nutrient data from: *1990 Statistical Abstract*, table 205; USDA Statistical Bulletin 702, *Food Consumption, Prices and Expenditures*; and U.S. Census, *Historical Statistics . . . to 1970*, 328. "Nutrient available in the food supply." Changes in the diet at higher income levels – chiefly reductions in potatoes, grains, and red meat – accounted for the 1909–29 decline in calories and protein.

19 The World Bank, *The Economy of the USSR* (1990), 39; the *Economist*, Dec. 1, 1990, US losses are well under 5 percent.

20 All data on food expenditures in this section are per capita and in 1982 dollars.

21 Cf. Bureau of Labor Statistics, *How American Buying Habits Change* (Washington: BLS, 1947), 238, and the Bureau of Applied Economics Report cited there.

EDITOR'S NOTE (JULIAN L. SIMON)

Figure 14.2 shows consumption per capita for selected countries in 1985. US consumption is relatively higher than US published data on

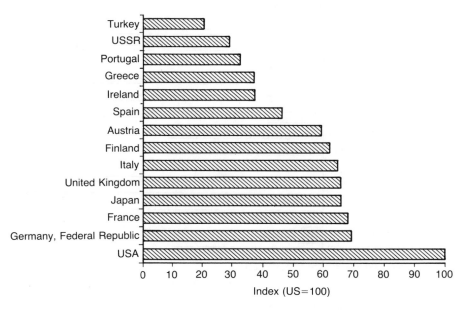

Figure 14.2 Comparative consumption per capita, selected countries, 1985
Source: A. Bergson, "The USSR Before the Fall: How Poor and Why," *Journal of Economic Perspectives*, 5:4 (Fall 1991), 29–44, 31.

income per capita in part because of the structure of consumer prices. (This figure should be given particular attention because its source, Abram Bergson, has been one of the most respected analysts of the Soviet economy for half a century.) The USSR consumption figures are considerably below their published income per person data in part because of a tilt toward (wasteful) investment. Additionally, the quality of consumer goods was so poor that the stated consumption figure surely was too high, low as it was.

A few specific consumption data (which are the best way to compare the standard of living and the income among countries) for Japan:

	Japan	US
Food as a percent of total consumption expenditure	20	11
Per person electricity use	5,418	10,658
Autos per person	1/4.3	1/1.9
Living space per household in square feet	874	1,456
Percent of homes with central heating	10	
85 percent of homes with flush toilets	45	99.8
Television sets per 100 persons	27	83

Figure 14.3 shows additional data for the US since 1900. Despite these changes, the front-page right-hand-column headline of the *Wall Street Journal* reads, "U.S. Living Standards are Slipping and Were Even Before Recession" (June 17, 1991, A11; see also *Newsweek*, Feb. 27, 1989, p. 20, and *Washington Post*, Nov. 9, 1991, A22).

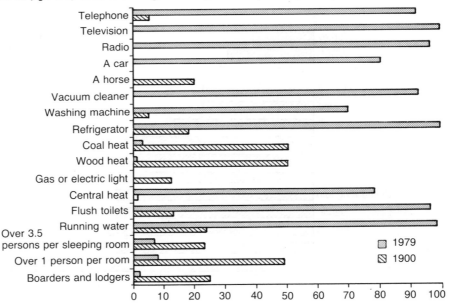

Figure 14.3 Percent of families with consumption changes, 1900–79, USA
Source: S. Lebergott, *The Americans: An Economic Record* (New York: Norton, 1984), 492.

15

Long-Term Trends in Productivity

Jeremy Atack

Since the early nineteenth century, productivity – defined as output per unit of input – in the economy as a whole has doubled about every 75 years, while labor productivity growth has been somewhat more rapid (figure 15.1). These gains have been crucial for our aggregate economic standing and well-being but they pale by comparison with productivity growth in narrowly defined activities experiencing rapid technological change.

In some areas of human endeavor a doubling every four or five years would even appear slow. New ideas, particularly those resulting in new mechanical or electrical devices, have resulted in some extraordinary changes in productivity.

Figure 15.1 Total labour and factor productivity, USA, 1840–1980

Sources: E. F. Denison, *The Sources of Economic Growth in the United States* (Washington, D.C.: Brookings Institution, 1962); Lana Davis et al., *American Economic Growth* (New York: Harper and Row, 1972); E. F. Denison, *Accounting for Slower Economic Growth* (Washington, D.C.: Brookings Institution, 1979): US Dept of Commerce, *Survey of Current Business* (Washington, D.C.: Government Printing Office), various issues.

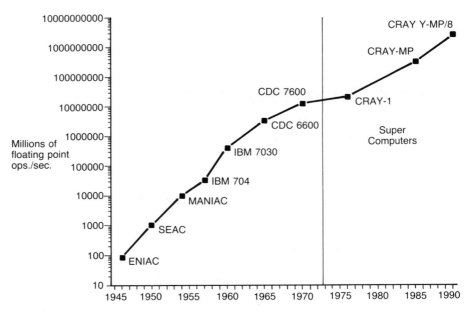

Figure 15.2   Increasing speed of digital computers
Sources: Science (Feb. 8, 1985), p. 595; *Business Week* (Dec. 10, 1990), p. 216.

Forty years ago, a cutting-edge mainframe computer, such as a UNIVAC, had the computing power of a second-generation personal computer, such as a now obsolete IBM PC/AT. The performance of a 33 MHz Intel 80486SX-based machine – now considered an entry-level machine and available in three- to four-pound notebook computers – is computationally speaking more than a match for an IBM-370-168 mainframe circa 1975. The IBM-370-168 sold for $3.4 million, excluding the operating system and applications programs; the cheapest 80486-computers currently retail for about $1,000 and often include a wide selection of software products. The increasingly common Pentium-based computers are four to six times faster, and even faster chips are now on the drawing boards.

Early mainframe computers performed perhaps a million floating point operations per second (figure 15.2); the so-called super-computers perform two to four *billion* floating point operations per second, and up to 20 billions are claimed for NEC's largest SX-3 machine with four parallel processors.

Unfortunately, productivity in the activities using these increasingly fast computers (particularly if measured by *useful* output) has not grown as rapidly. Rather, these machines are asked to address new tasks, not merely to perform the old ones more quickly. Thus, for example, researchers at the University of Illinois once expended 1,200 hours of computer time to solve the famous four-color problem.

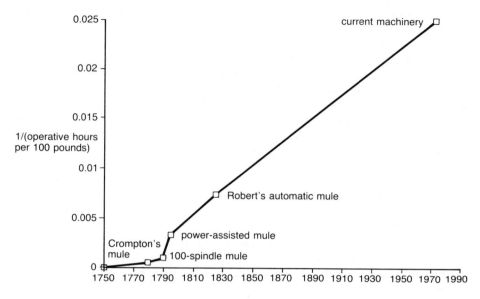

Figure 15.3 Productivity growth in spinning cotton thread
Source: H. Catling, *The Spinning Mule* (Newton Abbott: David & Charles, 1970).

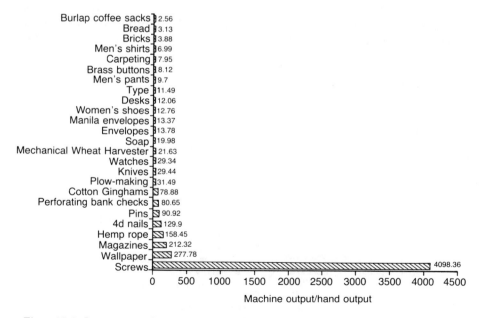

Figure 15.4 Improvement in output per man-hour from mechanization
Source: US Bureau of Labor, *Thirteenth Annual Report of the Commissioner of Labor, 1898: Hand and Machine Labor* (Washington, D.C.: Government Printing Office, 1899).

Dramatic productivity gains are not just an artifact of recent techno-logical change. Prior to the Industrial Revolution, it took 50,000 hours or more to convert 100 pounds of raw cotton to thread by hand. The earliest spinning machines at the end of the eighteenth century accom-plished the same task in 2,000 hours. Within 15 years, though, this had been cut to 300 hours, while today machines take less than 40 (figure 15.3). A late nineteenth-century study of 672 industrial processes recently converted from hand to machine production found similar dramatic gains in productivity from mechanization. For example, that the switch from hand to machine weaving of 36" wide gingham cloth reduced man-hours to produce 500 yards from 5,039 to 64, a 79-fold increase in labor productivity. Nor was this gain extraordinary. Machine printing and binding was 212 times more efficient than hand produc-tion, while machine production of screws was 4,032 times more efficient (figure 15.4).

During the initial phases of invention and innovation, productivity seems to grow exponentially as successive generations refine and im-prove the original idea (figure 15.5). Such growth, however, is not sustainable indefinitely. Eventually, incremental gains come only at increasing cost and complexity. The law of diminishing marginal returns eventually begins to operate, and there is some limit to performance. This is most readily apparent within a particular technology such as incandescent or fluorescent lighting (figure 15.6). In some cases there is a physical limit. For example, in computers the speed of electrons imposes a finite upper bound to the speed of computation. More and more of the increased speed in mainframes has come from reducing the

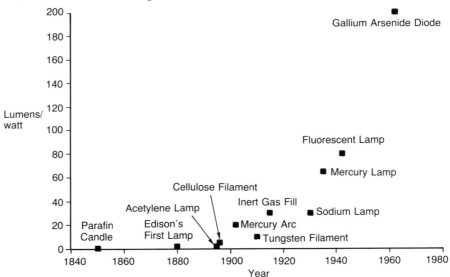

Figure 15.5   Light output per watt for different devices
Source: *Science* (Oct. 26, 1973), p. 362.

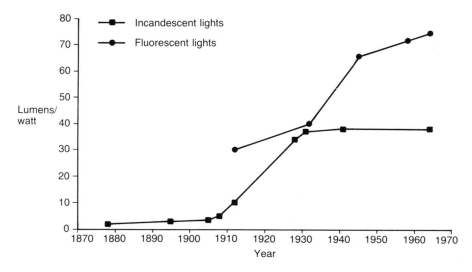

Figure 15.6 Illumination per watt by different technologies
Source: Science (Oct. 26, 1973), p. 360.

distance electrons must travel through the development of very large integrated circuits and increasing speed by reducing resistance through super-cooling and the like. Some of these same solutions are now showing up in personal computers too. Intel, for example, recently adopted a 0.6 micron technology for its Pentium processors over the earlier 0.8 micron technology to increase speed and reduce heat build-up. One solution to these constraints has been the development of parallel processing – the subdivision of tasks into fragments that can be worked on simultaneously by separate central processors. The gains, however, are not quite exponential. Two processors do not process the information twice as fast as one. Time is lost in dividing up the problem and allocating its parts between processors.

While productivity growth from a single invention is finite, there is no evidence of a secular slowdown in the rate of invention that might indicate a slowing in the overall rate of productivity growth. This is fortunate because productivity growth – not growth in the labor force or capital stock – is now the leading source of economic growth in America (see table 15.1), and it was the temporary dip in productivity growth in the 1970s and early 1980s that was largely responsible for the slowdown in America's rate of economic growth.

EDITOR'S NOTE (JULIAN L. SIMON)

A sign of productivity in the transportation industry is the purchase for $12 million of an eight-day round trip into space arriving at a Soviet

Table 15.1: Growth of factor inputs and output and decomposition of growth rate by epoch, 1840–1980

	Rate of Growth (percent per year)			
Period	Labor	Land	Capital	Net National Product
1840–60	3.42	3.73	6.57	4.75
1860–1930	2.24	2.55	4.35	3.75
1940–80	1.59	0.34	3.14	3.22

	Percent of Growth of Net National Product Attributable to:			
	Labor	Land	Capital	Productivity
1840–60	49.0	10.2	26.3	14.5
1860–1930	42.5	4.1	26.6	26.8
1940–80	40.6	0.4	13.7	45.4

Sources: See figure 15.1.

spacecraft (*Washington Post*, December 6, 1990, p. D1). Of course the price will fall sharply in coming decades. But even now the price per mile compares favorably with past costs in land travel. And of course the speed of space travel dwarfs the speed of the fastest land travel and even air travel, for which see figures 15.7 and 15.8.

Figures 15.9 and 15.10 together show long-term trends in comparative levels of labor productivity in fifteen developed countries from 1870 to 1987 relative to the US. Figure 15.11 shows worker productivity in the United States, France, Germany, and Japan for a somewhat

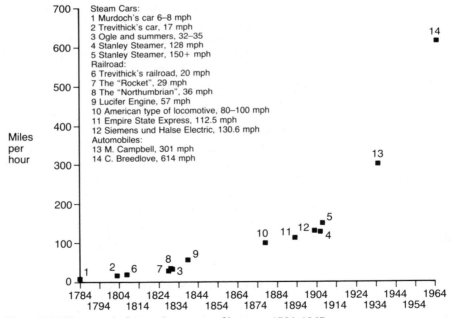

Figure 15.7 Top speed of ground transport of humans, 1784–1967
Source: Atack, forthcoming.

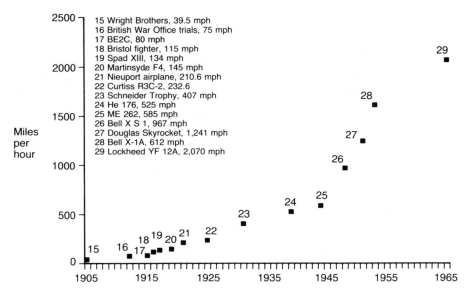

Figure 15.8 Top speed of air transport of humans, 1905–1965 (excluding space travel)
Source: Atack, forthcoming.

earlier period from 1950 to 1990 relative to worker productivity in the US. Figure 15.12 shows overall relative labor productivity for eight developed countries in 1989.

Genetics shows increasing promise of contributing to human health and well-being. Figure 15.13 shows progress in human gene mapping.

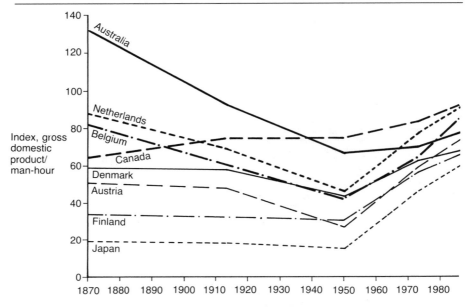

Figure 15.9 Comparative levels of productivity, 1870–1987 (US GDP per man-hour = 100)
Source: A. Maddison, *Dynamic Forces in Capitalist Development* (New York: Oxford University Press, 1991), 53.

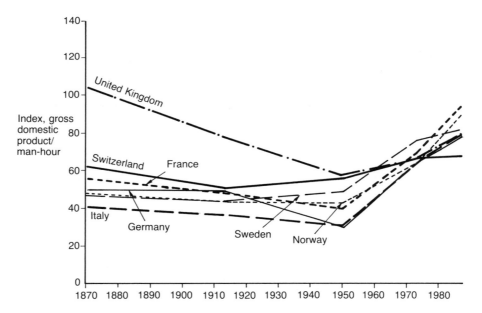

Figure 15.10 Comparative levels of productivity, 1870–1987 (US GDP per man-hour = 100)
Source: A. Maddison, *Dynamic Forces in Capitalist Development* (New York: Oxford University Press, 1991), 53.

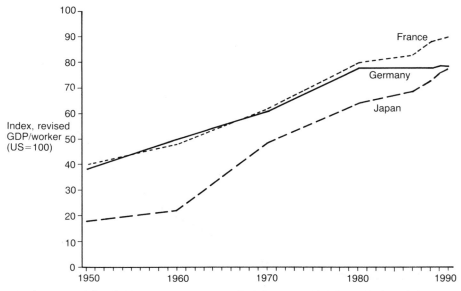

Figure 15.11 Productivity levels as percent of USA: France, Germany, and Japan, 1950–90
Source: W. J. Baumol and E. N. Wolff, "Comparative US Productivity Performance and the State of Manufacturing: The Latest Data," *CVStarr Newsletter* (New York University), vol. 10 (1992), 1, 4.

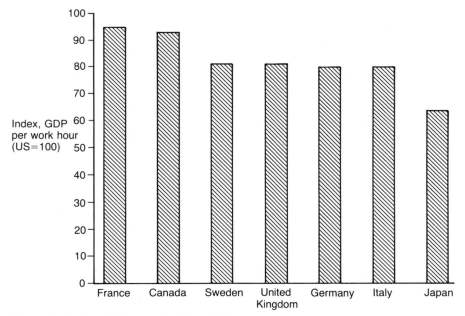

Figure 15.12 Overall labor productivity, 1989
Source: W. J. Baumol and E. N. Wolff, "Comparative US Productivity Performance and the State of Manufacturing: The Latest Data," *CVStarr Newsletter* (New York University), vol. 10 (1992), 1, 4.

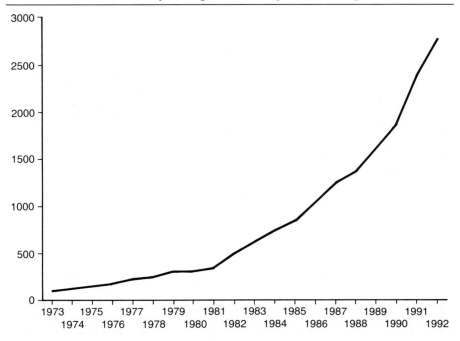

Figure 15.13 Number of human genes mapped, 1973–92
Source: *The Washington Post.*

16

The Extent of Slavery and Freedom Throughout the Ages, in the World as a Whole and in Major Subareas

Stanley L. Engerman

Precise estimates of the share of a population living in slavery are often difficult to obtain. Although slavery, as it existed in the European settlements in the Americas, was rather precisely defined in legal codes to include ownership by another individual, the prospects of being bought and sold, and inheritance of status through the slave condition of the mother, in other societies the provisions were often less clear and specific agreement as to the meaning of various rights of masters and of enslaved was often lacking. Serfdom, wherein sale of the individual apart from land was presumably restricted, causes similar difficulty for issues of measurement. Because of the changing magnitudes and importance of an international slave trade, as well as the particular mortality and fertility experience of the slave (and free) population, the basic measures of the share of population enslaved will vary with particular years. And, since the most profitable use of slave labor varies by climatic and geographic area, the choice of region or nation to be described will also affect the measured shares.

Moreover, to the extent that slavery has become an all-purpose term to describe unfavorable laboring and political conditions, it has been applied to "wage slavery," to international movements of contract labor (which preceded and succeeded slavery in several parts of the world – the later migration described by the nineteenth-century British as a "new system of slavery"), to debt bondage and other forms of local contractual labor, convict labor, apartheid and other forms of racial domination, and even the persistence of poverty – all of which have been described as variants of slavery. A corollary of this, the existence, at times, of more favorable material and social conditions for legally enslaved than for free, has correspondingly led some to question the applicability of the term *slave* to individuals in different societies and conditions. While slavery has long existed in almost all societies over long periods of time, continuing into the nineteenth and twentieth centuries, and discussions of the nature of slavery persist for many societies, estimates of the magnitude of slavery and its full economic impact are known for considerably fewer. Similarly, while the nature of legal codes in modern societies makes it possible to date the legal endings of slavery, in many earlier periods the gradual economic and social transition from slavery to other social forms makes any attempt at a precise dating rather uncertain.

The distinguished classicist, Moses I. Finley, has claimed that throughout recorded history there have been only five societies that are to be considered major slave societies.[1] Slave societies are those in which the social and economic impact of slavery was extensive, and even though the share of slaves in the total population may have been below one half, their presence had a pervasive influence. The major slave societies were also large in terms of area and population, with a heavy influence on world political and economic conditions. Those considered slave societies by Finley are: Ancient Greece, the Roman Empire, the United States, the Caribbean, and Brazil – two from the ancient European world, three from the Americas between the sixteenth and nineteenth centuries. Yet slavery also existed in the Americas prior to European settlement, and throughout Africa and Asia, and, indeed, the ending of legal slavery on the latter two continents generally postdated that in the Americas, frequently not coming until into the twentieth century.

Two wide-ranging political economists of the mid-eighteenth century have provided some perceptions of the then-present extent of political liberty and of what they considered slavery. Arthur Young, writing in his *Political Essays* in 1772, argues that of the estimated 775,300,000 souls inhabiting the world, only 33,500,000 lived under free governments – over one third of those subjects of the British Empire. The bulk of mankind, over 95 percent, were "miserable slaves of despotic tyrants." Adam Smith, in his Report of 1762–3 published in the *Lectures on Jurisprudence*, comments that "slavery is entirely abolished at this time . . . in only a small part of Europe." More generally, he argued that: "Slavery therefore has been universal in the beginnings of society, and the love of domination and authority over others will probably make it perpetual." This, it should be noted, was just prior to the first successes of the British attack on slavery in the Americas, a move that successfully spread to the colonies of other European countries in America, Asia, and Africa, over the next 150 years.

THE SUBAREAS

Europe

The estimated share of slaves in the population of the Greek states was about one third.[2] In Roman Italy, ca. 225 BC, the ratio was about 10 to

15 percent, rising to about one third in 28 BC.[3] With the decline and disappearance of slavery in Europe, the basic form of labor control shifted to serfdom. The dating and nature of this transition varied, and still remains somewhat controversial in the historical writing, but it is possible that slavery remained important in parts of Europe into the eleventh and twelfth centuries. According to the Domesday Book, 9 percent of the population of England in 1086 was legally enslaved, and slavery continued in areas such as Scandinavia for several centuries past this. Black slavery emerged in Iberian Europe with the trans-Saharan trade, and then increased with the expansion into Africa after the fifteenth century. In the mid-sixteenth century, Portugal had a black slave population equal to about 2.5–3.0 percent of its overall population, while for Spain the estimated share was about 1.25–1.50 percent.[4]

Serfdom persisted in most parts of continental Europe (having "withered away" in England by the close of the sixteenth century and disappeared centuries earlier in the Dutch areas) until the third quarter of the nineteenth century.[5] Of the 38 continental areas surveyed by Jerome Blum, five had initial decrees of serf emancipation between 1771 and 1798, 11 between 1804 and 1820, and 22 between 1831 and 1864, the last being Romania (the Danubian Principalities). The years between 1848 and 1853 had the most frequent incidence (14 cases) of emancipation. In the largest of serf nations, Russia, in the eighteenth and nineteenth centuries, the ratio of (private) male serfs to total male population generally ranged between 45 and 55 percent, being about 45 percent in 1858 (a share then below that of state peasants) just prior to the legislated emancipation of 1861.[6] The estimated proportions of serfs in eastern Europe were higher than in western Europe in the eighteenth and nineteenth centuries, with some estimates being of serf populations of over 80 or 90 percent for parts of the former region. France, just prior to the Revolution of 1789, apparently had somewhere between 140,000 and one million serfs in a population of about 27 million, with lower shares apparent in central Germany.[7] In central and eastern Europe, serfdom was both more important and lasted longer. Prior to emancipation in 1848, estimates place the share of serfs in the population of Austria at about 72 percent, while in Hungary (where emancipation occurred in 1853) the ratio was about 50 percent.[8]

Americas

Slavery did long persist among native Americans in both North and South America, but there are no reliable quantitative estimates of the population shares. The data for the post-European-settlement Americas are rather more clear-cut. After European contact, slavery developed throughout the two continents, but tended to be relatively more important in the Caribbean than in areas north and south of that area. Most northern states of the USA passed legislation ending slavery in the period between the start of the Revolutionary War and the start of the nineteenth century, and many South and Central American countries ended slavery after their successful revolutions in the first half of the nineteenth century (the share of slaves in total population in both these areas generally being below 10 percent). The large-scale endings of slavery are those in Haiti (by slave uprising, 1804), and, as the result of political legislation (in the case of the USA due to the outcome of a major civil war) in: the British Empire (1834), the French and Danish colonies (1848), the Dutch colonies (1863), the United States (1865), Puerto Rico (1873), Cuba (1886), and Brazil (1888). By the end of the nineteenth century, slavery no longer existed in the Americas.[9] The transatlantic slave trade to the Americas, accounting for some 10 to 15 million Africans, was ended by the United States and the English by 1808, but the last African-born slaves did not land until the 1850s in Brazil and the 1860s in Cuba.[10]

To provide some context by area, for two time periods before the ending of slavery, the following data are presented for the percentages of slaves in the total population for 1750 and 1830:[11]

Percentage of slaves in selected populations in the Americas, 1750 and 1830

	1750	*1830*
United States	20.2 (Negro)	15.6
(southern states)	38.0 (Negro)	34.7
British Caribbean	85.4	81.2
French Caribbean	87.1	80.4
(St Domingue)	90.4 (included in French)	0 (excluded from French)
Danish Caribbean	88.2	65.1
Spanish Caribbean	15.0	30.7
Dutch Caribbean	86.5	73.4
Brazil	48.7 (1798)	31.2 (1850)

The varying proportions of the enslaved in overall populations reflect, in part, the changing racial composition of the population, since whites were not legally enslaved. These were influenced by the different ratios of migrants and the differences in demographic performance of blacks and of whites, as well as the different importance of manumissions of individual slaves, by grant, self-purchase, or other methods. The manumissions, however, lack the dramatic effect on the overall numbers of slaves that resulted from the legal ending of the slave systems.

Asia and Africa

Much less can be said, even approximately, about the share of enslaved in Asia and Africa, particularly since less is also known about the overall levels of population. It is probable that in ninth and tenth century Iraq – the site of the great Zanj slave rebellion – over 50 percent of the population was enslaved.[12] For the Islamic empire as a whole around this time, it is estimated that the share enslaved was likely to have been below 10 percent.[13] Elsewhere, in the nineteenth century, in Thailand, between one quarter to one third of the population was enslaved, full emancipation there not occurring until the start of the twentieth century.[14]

The estimates for Africa are difficult to evaluate, particularly since there was great variation by political groupings, and because the extent of enslavement varied over time depending on the volume of the transatlantic and trans-Saharen slave trades. In West Africa, in the early nineteenth century, estimates are that somewhere between one quarter and one half of the populations were enslaved. In South Africa, in 1820, the proportion was about two fifths, and in Zanzibar, at mid-century, the share was over three quarters. It is claimed that by 1850 there were more slaves in Africa than in the Americas, and that by the end of the nineteenth century overall possibly between one half and two thirds of the population in many African societies was enslaved.[15] Portugal passed a decree in 1858 to bring about the end of slavery in its colonies, which were mainly in Africa, but those then enslaved did not achieve freedom until 1875.[16] The ending of slavery in Portuguese Africa, as in many other parts of the world, was followed by the expansion of a system of contract labor migration. Slave emancipation elsewhere in Africa was introduced in the early twentieth century by the European colonial powers.[17]

CONCLUSIONS

It was not until about 1970 that the world was considered to be freed of legal slavery, with the final emancipations on the Arabian peninsula. Nevertheless, slavery is still considered widespread, whether due to illegal practices, or, more generally as the result of debt bondage, child labor, contract labor, and other varieties of coerced work for limited periods of time, with limited opportunities for mobility, and with limited political and economic power. But, clearly, the legal characteristics of slavery as defined in Rome and the Americas are no longer present nor are those practices anywhere near as widespread as earlier. What may be noted, for those examining the developments of legal freedom, are the simultaneous restrictions upon and the ultimate legal endings of slavery by Europeans in their overseas offshoots and the endings of serfdom throughout Europe in the late eighteenth and nineteenth centuries, and the impact of these upon the ending of legal slavery in European colonies and other overseas dependencies. From institutions that were of great importance in almost all parts of the world as late as the end of the eighteenth century, as argued by Young and Smith, legal slavery and serfdom, involving private ownership of other individuals, declined sharply in magnitude over the course of the nineteenth century and has effectively disappeared during the twentieth century.

NOTES

1   M. I. Finley, *Ancient Slavery and Modern Ideology*. New York: Viking Press, 1980, 9.

2   Many of the estimates of the share of slavery in different societies presented in the next sections are drawn from the extremely useful compendium in Appendix C of Orlando Patterson, *Slavery and Social Death*. Cambridge: Harvard University Press, 1982.

3   Keith Hopkins, *Conquerors and Slaves*. Cambridge: Cambridge University Press, 1978, 68, 101.

4   A. C. de C. M. Saunders, *A Social History of Black Slaves and Freedmen in Portugal, 1441–1555*. Cambridge: Cambridge University Press, 1982, 59–60.

5   Most of the information on serfdom and its endings is taken from Jerome Blum, *The End of the Old Order in Rural Europe*. Princeton: Princeton University Press, 1978, particularly 29–49, 356.

6   Jerome Blum, *Lord and Peasant in Russia from the Ninth to the Nineteenth Century*. Princeton: Princeton University Press, 1961, 420.

7   See the estimates for France in Seymour Drescher, *Capitalism and Antislavery*. London: Oxford University Press, 1986, 201–2.

8   This information was provided by John Komlos.

9   The dating of these emancipations can be found in Robert William Fogel and Stanley L. Engerman, *Time on the Cross*. Boston: Little, Brown, 1974, vol. 1, 33–4.

10 See Philip D. Curtin, *The Atlantic Slave Trade: A Census*. Madison: University of Wisconsin in Press, 1969, and David Eltis, *Economic Growth and the Ending of the Transatlantic Slave Trade*. New York: Oxford University Press, 1987.

11 Sources include Patterson, *Slavery and Social Death*; Herbert S. Klein, *African Slavery in Latin America and the Caribbean*. Oxford: Oxford University Press, 1986; US Bureau of the Census, *Historical Statistics of the United States*. Washington, D.C., G.P.O., 1975, vol. 1, 15–18, vol. 2, 1168; and Stanley L. Engerman and B. W. Higman, "The Demographic Structure of the Caribbean Slave Societies in the Eighteenth and Nineteenth Centuries," in *UNESCO General History of the Caribbean, vol. III, Slave Societies of the Caribbean*, ed. Franklin W. Knight. Forthcoming.

12 See Patterson, *Slavery and Social Death*, 357.

13 As estimated in Raymond W. Goldsmith, *Premodern Financial Systems*. Cambridge: Cambridge University Press, 1987, 62.

14 See Patterson, *Slavery and Social Death*, 357.

15 Sources for these comments include Patterson, *Slavery and Social Death*; Suzanne Miers and Igor Kopytoff, eds, *Slavery in Africa*. Madison: University of Wisconsin Press, 1977; Paul Lovejoy, *Transformations in Slavery*. Cambridge: Cambridge University Press, 1983, 195–6; Patrick Manning, *Slavery and African Life*. Cambridge: Cambridge University Press, 1990, 23; Fred Cooper, *Plantation Slavery on the East Coast of Africa*. New Haven: Yale University Press, 1977, 56; and Martin A. Klein, "The Impact of the Atlantic Slave Trade on the Societies of the Western Sudan," in Joseph E. Inikori and Stanley L. Engevman, eds, *The Atlantic Slave Trade*. Durham: Duke University Press, 1992, 25–47. The extent to which estimates apply to all Africa or only to west Africa is sometimes unclear. There remain major debates, recently centered on the

writings of Walter Rodney, concerning the relation between European contact and the development and expansion of slavery in Africa, and on the social nature of what has been called slavery in Africa, raising the question of the application of the same term to black slavery in Africa as to black slavery in the Americas.

16 James Duffy, *A Question of Slavery*. Cambridge: Harvard University Press, 1967, 9, 60–2.

17 Suzanne Miers and Richard Roberts, eds, *The End of Slavery in Africa*. Madison: University of Wisconsin Press, 1988.

17

Black Americans: Income and Standard of Living from the Days of Slavery to the Present

Robert Higgs and Robert A. Margo

During the centuries when the great majority of American blacks were slaves, the blacks' average income and standard of living did not change very much over the long run in absolute terms. Relative to the whites, who enjoyed some economic growth, blacks lost ground. Since emancipation, average black income and standard of living have increased greatly both absolutely and relative to whites, although the improvement has not occurred during every subperiod of the past 125 years.

UNDER SLAVERY

Abolitionists frequently charged that the slave diet was "inadequate," and much recent research has focused on the specific composition of an "adequate" diet. Early studies examined per capita consumption of foods and caloric intake. These studies indicated that, by European standards, slaves had a relatively high per capita consumption of meat. Caloric intake also was high – for adult males an average of 4,000 calories per day. It soon became clear, however, that such studies might be misleading. The average adult male slave might have been well fed while other groups (e.g., children) were poorly fed. Also, nutritionists have shown that an adequate diet cannot be defined independent of the disease environment and work intensity. Slaves are known to have worked harder than free laborers.

To deal with these complications, scholars have turned their attention to other measures of nutritional status, especially to anthropometric indicators such as height or weight by age. (See chapter 5, "The Contribution of Improved Nutrition to the Decline in Mortality Rates in Europe and America," in this volume.) Adult height is a cumulative result of the nutritional status of individuals throughout their childhood;

heights of young children capture contemporaneous influences more closely. Adult heights of male slaves were well below modern levels but compared favorably with heights of northern urban whites before the Civil War. The shortfall in adult height resulted from low birth weight and poor childhood nutrition. Once a slave child reached working age, the diet improved enough to permit some "catch-up" growth. Low birth weight, a consequence of inadequate maternal nutrition and overwork of pregnant women, caused high rates of infant mortality, about 250–300 per 1,000 live births (Steckel, 1986). Over time slave heights cycled rather than trending up or down. Heights declined for cohorts born just after the Revolution; heights of later cohorts increased. Heights of cohorts born in the late 1830s and early 1840s declined, suggesting nutritional deterioration in the late antebellum period (Margo and Steckel, 1982).

Various estimates of slave mortality in the 1850s agree that the expectation of life at birth was only about 30–33 years, mainly because of high infant and child mortality. At that time US whites had an expectation of life at birth of about 40–43.

FROM THE CIVIL WAR TO WORLD WAR II

Black health deteriorated during the mid-1860s as the socioeconomic disruptions associated with the war and emancipation facilitated the spread of infectious diseases and diminished blacks' resistance, but these abysmal conditions did not persist for more than a few years. How black mortality changed during the late nineteenth century remains in doubt, as demographic historians' estimates disagree widely, some indicating virtually no change, others showing major improvement. The urbanization that was occurring worked against favorable changes in average life expectancy, because urban mortality rates exceeded rural rates considerably at that time. After 1900, mortality undoubtedly dropped more or less steadily (figure 17.1). By 1940 black life expectancy at birth had reached about 53 years (versus about 64 for whites), that is, perhaps twice its level during the mid-1860s and almost certainly two-thirds greater than its level in the 1850s (Higgs, 1989). It is extremely unlikely that such a large gain in life expectancy could have been achieved without substantial improvements in living standards, especially in diet and housing.

Direct estimates of black income per capita are available for selected years between the Civil War and World War II, though none of the estimates can be regarded as very precise. According to the estimates of Higgs (1977, 1989), black income per capita relative to white income per capita was about 24 percent ca. 1867–8, about 35 percent ca. 1900, and about 34 percent ca. 1940. The implication is that blacks increased their incomes a good deal faster than whites during the first three

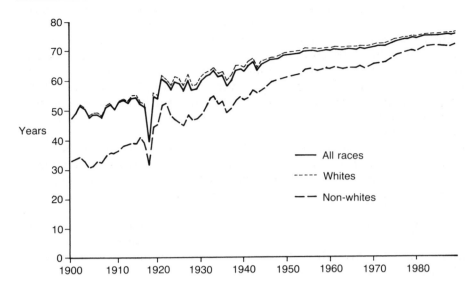

Figure 17.1 Life expectancy at birth, USA, by race, 1900–89
Note: Data for pre-1993 are for reporting areas only and are not precisely comparable with post-1933 data.
Source: US Census Bureau.

decades and about as fast as whites during the next four decades. Of course major swings may have occurred, and probably did occur, within these subperiods.

Evidence on specific forms of consumption (e.g., food, clothing, and housing) and property ownership, including owner-occupancy of homes, confirms that blacks made substantial economic gains during the first 75 years after emancipation, both absolutely and relative to whites (Higgs, 1977, chapter 5; 1989). Higgs (1982, 1984) and Margo (1984) analyzed tax assessment data for six Southern states and found that blacks generally increased their wealth holdings much faster than whites during the period between the Civil War and World War I. In Georgia, the state with the longest and most reliable time series, the real assessed value of black property per capita increased more than threefold between 1880 and 1910; the per capita black/white wealth ratio rose from 1/36 to 1/16. Despite these gains, it is obvious that the blacks' *relative* socioeconomic condition remained very low on the eve of World War II.

FROM 1940 TO THE PRESENT

The years since 1940 have witnessed both absolute and relative improvement of the economic status of American blacks. In 1940 the average annual earnings of black men relative to white men equaled 42 percent. By 1980 the ratio had reached 69 percent (US Commission on Civil Rights, 1986). An index of relative occupational status shows an increase of 13 percentage points during the same period (Smith, 1984). The blacks' faster growing incomes afforded improving housing, food, clothing, and health. On the eve of World War II, 58 percent of black families lived in homes without indoor plumbing, and only 24 percent owned their homes. In 1980, far fewer lived in substandard housing, and 45 percent of black families were homeowners. Black life expectancy at birth increased from 53 years in 1940 to nearly 72 (versus 76 for whites) in 1989; over the same period infant mortality (figure 17.2) fell from 74 to 18 per 1,000 births (versus eight per 1,000 for whites).

Economists usually measure living conditions in terms of income and tangible consumption, but a broader conception would include the enjoyment of personal and political rights. In 1940 blacks in the South were subject to de jure segregation in many aspects of daily life; virtually none could vote. The Civil Rights revolution of the 1950s and 1960s, along with associated legislative and federal court actions, led to the abolition of de jure segregation and gave rise to a great increase in black voting and office holding.

Not all indexes of relative black status show continuous improvement since 1940. Black incomes tend to decline during recessions. In some dimensions the improvement of relative black economic status stalled after 1975. Since the 1950s, black unemployment rates have remained high relative to white rates. Labor force participation by adult black

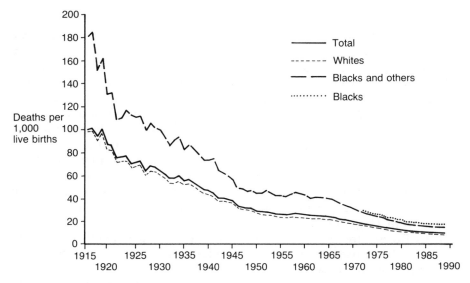

Figure 17.2 Infant mortality rates, USA, 1915–89, by race
Note: Prior to 1933, data reflect only areas reporting.
Source: US Census Bureau.

men, even in prime working ages (35–54), has fallen relative to that of white men. The racial difference of family income has remained large, partly because of a substantial increase in the proportion of black families headed by poor, single women. Especially notable has been the rise of joblessness among black youths, accompanied by an increase in criminal activity. De facto segregation, especially of housing, continues in many cities.

Scholarly debate is polarized with respect to explaining the post-World War II improvement of relative black economic status. The most common explanation emphasizes increases in demand for black labor associated with the Civil Rights movement and federal anti-discrimination laws. Another view is that long-term changes in human capital account for the improvement. Between 1940 and 1970, millions of blacks moved out of the rural South, where earnings were far below the national average. Successive generations entered the labor force with more years (and better quality) of schooling than their parents. Eventually the racial gap in effective schooling declined, leading to a narrowing of racial income differences. Proponents of the human capital argument also note that relative black economic status improved substantially in the 25 years before the Civil Rights laws of the mid-1960s took effect. Although the debate continues, recent research suggests that a synthesis of both explanations does the best job of accounting for the facts (Heckman, 1990; Margo, 1990).

THE LONG VIEW

Discussions of black economic conditions tend to focus on the ups and downs of relatively recent times. People lose sight of what has happened over the longer term. When American blacks gained their freedom in the mid-1860s, they were overwhelmingly destitute, illiterate, ignorant, unskilled, concentrated in the rural South, and subject to appallingly bad sanitary and health conditions and horrendous morbidity and mortality. Only 40 years ago the Jim Crow system remained solidly entrenched, imposing racial segregation by law on people throughout the South.

Although not every aspect of the long-term improvement in black economic conditions can be quantified, it is possible – and instructive – to calculate crudely comparable measures of black income per capita for a few widely separated benchmark years. The estimates in figure 17.3 derive from Higgs (1977, 1989) for the first three dates and from the Census Bureau for the most recent dates. The standard consumer price index has been used to express the underlying estimates in a common unit of purchasing power.

Crude as the estimates no doubt are, they warrant an incontrovertible conclusion: real black income per capita has grown enormously. Between the immediate post-emancipation period and 1900 it increased by

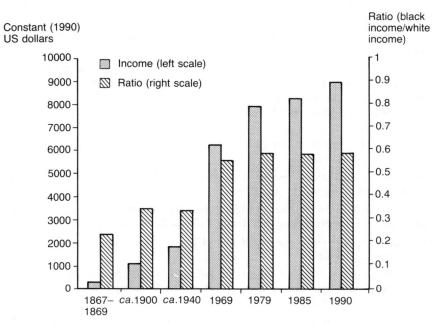

Figure 17.3   Black income per capita, 1867–1990
Sources: Higgs, 1977, 1989; US Census Bureau, Current Population Reports.

261 percent; between 1900 and 1940 by 61 percent; between 1940 and 1985 by 342 percent. The cumulative result: real black income per capita is now more than 26 times greater than it was in the late 1860s. Few if any of the world's large population groups enjoyed comparable economic growth during the same period. Certainly US whites did not. As figure 17.3 shows, over the long run, average black income increased much faster than average white income: black income rose from 24 percent to nearly 60 percent of white income.

Although segments of the black population still live in relatively poor economic and social conditions, for the average black the most striking long-term development is immense improvement. In addition, all de jure racial segregation has been swept away. Overall, the socioeconomic condition of today's black Americans bears no similarity whatever to that of their forebears of 125 years ago. In many respects the conditions of even 40 years ago are no longer recognizable.

REFERENCES

Heckman, James J. (1990): "The Central Role of the South in Accounting for the Economic Progress of Black Americans." *American Economic Review* 80 (May), 242–6.

Higgs, Robert (1977): *Competition and Coercion: Blacks in the American Economy, 1865–1914.* New York: Cambridge University Press.

—— (1982): "Accumulation of Property by Southern Blacks before World War I." *American Economic Review* 72 (September), 725–37.

—— (1984): "Accumulation of Property by Southern Blacks before World War I: Reply." *American Economic Review* 74 (September), 777–81.

—— (1989): "Black Progress and the Persistence of Racial Economic Inequalities, 1865–1940." In Steven Shulman and William Darity, Jr, eds, *The Question of Discrimination: Racial Inequality in the U.S. Labor Market*. Middletown, Conn.: Wesleyan University Press.

Margo, Robert (1984): "Accumulation of Property by Southern Blacks before World War I: Comment and Further Evidence." *American Economic Review* 74 (September), 768–76.

—— (1990): *Race and Schooling in the South, 1880–1950: An Economic History*. Chicago: University of Chicago Press.

—— and Richard H. Steckel (1982): "The Heights of American Slaves: New Evidence on Slave Nutrition and Health." *Social Science History* 6 (Fall), 516–38.

Smith, James P. (1984): "Race and Human Capital." *American Economic Review* 74 (September), 685–98.

Steckel, Richard H. (1986): "A Peculiar Population: The Nutrition, Health, and Mortality of American Slaves from Childhood to Maturity." *Journal of Economic History* 46 (September), 721–42.

US Bureau of the Census (1975): *Historical Statistics of the United States, Colonial Times to 1970*. Washington: US Government Printing Office.

—— (1986): *Statistical Abstract of the United States 1987*. Washington: US Government Printing Office.

—— (1989): *Current Population Reports, Series P–60, No. 162*. Washington: US Government Printing Office.

—— (1992): *Statistical Abstract of the United States 1992*. Washington: US Government Printing Office.

US Commission on Civil Rights (1986): *The Economic Progress of Black Men in America*. Washington: US Commission on Civil Rights.

EDITOR'S NOTE (JULIAN L. SIMON)

Some data showing both black and white educational rates are shown in chapter 21. Additional data are shown in figures 17.4 to 17.6. Shown in figure 17.7 are data regarding lynchings, for whites and blacks, and lynchings prevented, from 1889 through 1932.

The following anecdote may be relevant to the data in this chapter and appendix – and indeed, to the entire book:

> I ask, "But is it better?"
> "Is it better?" Dempsey asks, scornfully. "White boy, I don't know what's wrong with you. What the hell are you sayin'?"
> "Is it better or worse for blacks today?"
> Suddenly the contentious Dempsey Travis gets calm, even solicitous,

Figure 17.4 Unemployment rates in the USA, by race, 1890–1990
Source: R. Vedder and L. Gallaway, *Out of Work: Unemployment and Government in Twentieth-Century America* (New York: Holmes & Meier, 1993), 272.

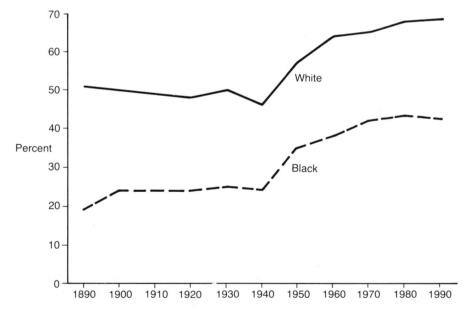

Figure 17.5 Percent owner-occupied housing units, USA, by race, 1890–1990
Source: For 1890–1970: B. Wattenberg, *The Real America* (New York: Doubleday, 1974); for 1980–1990, *Statistical Abstract of the United States*, 1992, 716, table 1225.

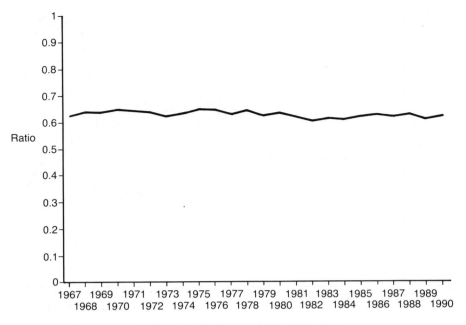

Figure 17.6 Ratio of black-to-white family income, USA, 1967–90
Source: US Census Bureau, *Current Population Survey*, annual.

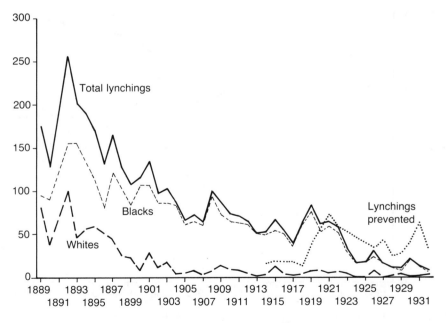

Figure 17.7 Lynchings and prevented lynchings, by race, USA, 1889–1932
Source: A. F. Raper, *The Tragedy of Lynching* (New York: Dover, 1970), 480–1, 484.

like he is taking me to school. He says whites are always asking if life isn't better for blacks today. He refuses to answer that question because it carries an implicit question white's won't ask: If it's better, why are you still complaining? Well, if somebody is beating you three times a day and they switch to twice a day, is life better? If you discover that your wife hasn't been killed but only paralyzed in a car crash, is life better? Is life in South Africa better for blacks today than it was ten years ago? The question is dumb. "The point," Dempsey says, still solicitous, "is that racism is pervasive."

[Walt Harrington, *Crossings* (New York: Harper Collins, 1993, p. 272]

Nothing here is intended to suggest that racism is not pervasive. But just as surely, I believe that publishing the truth about these trends is good for both blacks and whites in the long run.

18

The Long-Term Course of American Inequality: 1647–1969

Peter H. Lindert and Jeffrey G. Williamson

There have been important movements in inequality in America over the past three centuries, even after all possible technical adjustments to raw data. (See figure 18.1.) The ratio of income or wealth of the richest 10 percent to that of the poorest 10 percent is not constant. Rather, certain epochs of American history experienced rising or falling inequality in

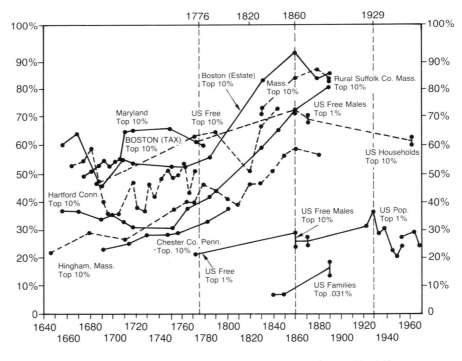

Figure 18.1 Shares of wealth held by top wealth holders in America, 1647–1969
Source: Lindert and Williamson, 1976, 81.

income and wealth. Furthermore, these inequality epochs have about the same dates whether they are documented by the behavior of earnings, income, or wealth.

Inequality among *free* Americans before the Revolution was not too different from that today. Yet, inequality was hardly stable for the long period in between. The main epoch of increasing inequality appears to have been the last four decades before the Civil War, which itself reduced inequalities within regions but also increased inequality between regions by opening a severe income and wealth gap between North and South. Although we find no evidence of rising inequality in the remainder of the nineteenth century, trending inequality emerges once more between 1900 and World War I. The war administered a brief, strong dose of equality, but the effects had worn off by 1929. Considerable equalization in income and wealth occurred between 1929 and mid-century. After World War II, there was no sharp reversion to earlier high degrees of inequality, as happened after World War I. Instead, postwar distributions appear to exhibit a curious stability.

We shall now separately discuss the major periods in American history. Although we shall use wealth data where only that is available, the pictures obtained from combined wealth and income data are much the same. Hence the wealth data provide a satisfactory broad picture for the long term.

TRENDS IN COLONIAL WEALTH INEQUALITY

Probate wealth inequality trends are revealing sources for the colonial period. If one were to take 1690 or 1700 as a base, the wealth inequality series would suggest mixed trends, but, on average, a drift toward greater wealth concentration for the seven or eight decades prior to the Revolution. This characterization holds for rural Connecticut (but *not* for Hartford County), for rural Massachusetts (but *not* for rural Suffolk County), for Boston as well as Portsmouth (New Hampshire), and for Philadelphia as well as nearby Chester County. It does *not* hold for Maryland, however, which exhibits stability from the 1690s onward. New York City is another exception, with stable wealth distribution between 1695 and 1789.

Among those probate wealth inequality series that extend backward before the 1690s, Worcester County (Massachusetts) and Philadelphia reveal the minority position: a clear secular drift toward inequality for the entire colonial era. Connecticut, Boston, rural Suffolk County (Massachusetts), and Maryland represent the majority: they do *not* reveal inequality trends. If, instead, one is content to start the analysis with 1700, then a modest drift toward inequality seems to characterize these colonial "local histories" best. The 1700 benchmark may impart a spurious upward trend to wealth concentration indices.

WEALTH CONCENTRATION IN THE
FIRST CENTURY OF INDEPENDENCE

The top 1 percent of free wealth holders in 1774 held 12.6 percent of American total assets, while the richest 10 percent held a little less than half. In 1860, the top 1 percent held 29 percent of total assets, and the richest decile 73 percent. Thus, the top-percentile share more than doubled and the top-decile share increased by half again its previous level. Equally dramatic surges are implied for the South and non-South separately.

The rise in wealth inequality is still evident if one includes slaves as part of the population. Counting slaves both as potential wealth holders and as wealth effectively raises estimated inequality before the Civil War. This follows from the reasonable assumption that slaves had zero assets and net worth. Adding extra "wealth holders" with zero wealth is equivalent to shrinking the share of the population represented by the same number of top wealth holders. This adjustment should be greater for 1770 than for 1860, since the slave-population share peaked at about 21.4 percent in 1770 and declined to about 11 percent by 1860. Thus, counting slaves as both people and property, a defensible procedure, should have raised the inequality measure more for 1774 than for 1860. Nevertheless, this adjustment has little effect on the net rise in inequality between these two dates.

The 1774 colonial wealth distribution bears a close resemblance to the (revised) distribution implied by the Federal Reserve survey for 1962. The share held by the richest 1 percent was apparently a little lower in 1774, both among the free and among the free plus slaves. On the other hand, the top-decile share appears to have been somewhat higher on the eve of the Revolution than it was nearly two centuries later. Then there occurred an epochal rise in wealth concentration between 1774 and 1860.

THE UNEVEN HIGH PLATEAU: CIVIL WAR
TO GREAT DEPRESSION

During the seven decades following the Civil War, wealth inequality remained very high and exhibited no significant long-term trend. This judgment is based on slim evidence, since the period is illuminated statistically only near its start and finish.

Concerning *income* inequality between 1860 and 1929, our indicators seem to mark out the entire period from Civil War to Wall Street crash as one of far greater income inequality than today. This plateau contains three episodes that may have seen the highest income inequalities in American history: (a) the end of the Civil War decade and the early

1870s; (b) the eve of the First World War, especially 1913 and 1916; and (c) the eve of the Great Crash, or 1928 and the first three quarters of 1929.

While the six decades following the Civil War clearly registered persistent inequality levels far in excess of mid-twentieth-century standards, and although no dramatic long-term trends like the antebellum surge or the post-1929 collapse can be identified, we might be a bit more precise in dating America's peak inequality watershed. For the "Income Tax Age," the years of greatest income inequality were 1916 and 1929, with size distributions more skewed in the former.

TWENTIETH-CENTURY LEVELING

Post-World War I Estimates

Since World War I, the top-quintile shares reveal unambiguous and well-known trends. Top wealth holders increased their share markedly between 1922 and 1929, apparently recovering their pre-World War I position. Their share then dropped over the next 20 years, hitting a trough around 1949. This leveling in wealth distributions also parallels the "revolutionary" income leveling over the same period, about which more will be said later. Furthermore, as with incomes, the wealth leveling is not solely a wartime phenomenon, since an equally dramatic leveling took place early in the Great Depression. Whereas this revolutionary change in the distribution of wealth has become a permanent feature of the mid-twentieth century, the postwar period has recorded no further trend toward wealth leveling.

Income Data, 1929–1951

There apparently was a dramatic and pervasive shift toward more equal incomes between the Wall Street crash and the Korean War. The entire income spectrum seemed to converge, since every series tells the same tale of pronounced leveling in income. The change is impressive. The only factual question requiring a closer look is whether the Great Depression saw more or less leveling than World War II. The classes gaining in relative shares differed between the two decades. During the Depression, higher-paid employees, such as skilled and white-collar workers and professionals, suffered less than others simply by keeping their jobs at negotiated nominal wage rates that were less sensitive than others to the cycle. Meanwhile, the urban unskilled, farmers, and profit recipients all suffered an erosion of incomes. While the top 5 percent of all income recipients suffered a drop in their relative share across the 1930s, the shares of the top 5 percent of *employees* and the top income *regions* actually peaked at the bottom of the Depression and were still no worse in 1940 than in 1929. The 1940s, by contrast, saw a clear contraction of the entire income spectrum. The shares of top-percentile individuals dropped again, but this time the biggest gainers were those at the bottom – farm workers, blacks, southern states, women, and unskilled urban white males.

The greatest changes over the two decades as a whole were the rise in the share received by the poorest fifth and the decline in the share received by the richest fifth (especially the top 5 percent). In 1929, the average income of the richest fifth was 15.5 times that of the poorest fifth. By 1951 this ratio had dropped to 9.0. An impressive leveling also occurred in regional inequality as revealed by estimates of personal income per capita derived from state production data. The North–South gap in average incomes dropped dramatically, in part owing to the heavy migration of low-income workers from the South to northern urban centers. In no other extended period of American history did the available indicators swing so sharply toward equality.

This leveling was remarkable in two respects. First, it spanned a 22-year period that was far from uniform. Between these two full employment dates, America sank into its greatest depression, surged back with the help of World War II, had a postwar boom, and then entered the Korean War. Such turbulent times might be expected to have brought reversals in inequality trends, but the leveling appears to have continued unabated throughout, although it seems to have accelerated during World War II. Second, the trends are all the more remarkable, since they document a leveling of incomes *before* the effects of government are included. Furthermore, this decrease in pre-fisc inequality appears to have been as great as the entire equalization achieved by all government programs in 1950 and almost as great as the total equalization by government programs in 1970. This last statement bears repeating: the leveling in pre-fisc incomes between 1929 and 1951 was as great as the difference between the distribution of pre-fisc and post-fisc incomes in 1950, the latter including *all* state, local, and federal tax-transfer expenditure policies.

Information from Income Since World War II

The leveling ceased by 1950. By almost any yardstick, inequality has changed little since the late 1940s. If there has been any trend, it is toward slightly more inequality in pre-fisc income and slightly less inequality in post-fisc income. This stability has been extraordinary even by twentieth-century standards.

OVERVIEW

Colonial wealth inequality was stable and low from the mid-seventeenth century to the Revolution. Between 1774 and the outbreak of the Civil

War, there occurred a near tripling in the ratios of the average wealth of the top 1 percent or 10 percent of wealth holders to the average wealth of all other groups. Regional estimates suggest that most of the antebellum shift to wealth concentration occurred from the 1820s to the late 1840s, though the supply of such shorter-run data is still inadequate. In addition, the apparent rise in wealth inequality before the Civil War cannot be explained by mere shifts in age distribution, by the increasing share of foreign born, or by urbanization, though this last factor does contribute noticeably to it.

We still know little about wealth inequality trends from the Civil War to World War I. Slave emancipation unambiguously leveled wealth inequality within the South and for the nation as a whole in the 1860s. For the half century after 1870 we are in the dark, so that we cannot with confidence identify peak wealth inequality with 1929, 1914, or 1860. Nevertheless, it appears that no significant long-term leveling occurred during the period and that inequality persisted at very high levels.

The twentieth-century figures suggest a clear pattern. Wealth inequality, like income inequality, dipped across World War I and rose across the 1920s, though it is hard to say whether 1929 wealth was more or less equally distributed than that of 1912 or some nearby year. From 1929 until mid-century, wealth inequality seems to have undergone a permanent reduction, again paralleling the movement in income inequality. After mid-century, neither wealth nor income inequality has

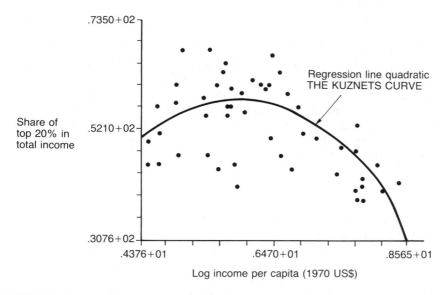

Figure 18.2 The Kuznets curve: international 60-country cross-section from the 1960s and 1970s
Source: Ahluwalia, 1976, 340, 341, table 8.

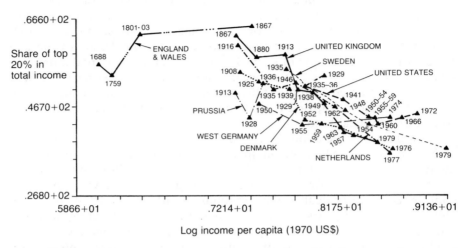

Figure 18.3 The Kuznets curve: historical time series from five European countries and America
Source: Lindert and Williamson, 1985.

shown a trend that can be judged significant based on existing data. The American record thus documents a "Kuznets inverted-U" for wealth inequality, and it appears that significant wealth inequality did not become a part of the American scene until the onset of modern economic growth in the early nineteenth century.

The evidence from 150 years of American income inequality trends – particularly for the twentieth century – strongly suggests that movements in the size distribution of income are paralleled by movements in the basic pay structure. When measures of overall inequality were on the rise, so too were measures of dispersion in the rates of pay for occupational groups. Trends in skill premia, occupational wage differentials, earnings dispersion, and thus in rates of return to human capital seem to correlate well with size-distribution trends. In addition, however, we observe a remarkably close correspondence between trends in income inequality and wealth inequality.

Corroboration

Two kinds of data corroborate the general pattern that we have portrayed in this paper. Figure 18.2 is the famous Kuznets curve that relates income level to level of inequality. As in the history of the United States, inequality is low when the country is poor and mostly agricultural, rises as income rises for a while, then falls as income rises even farther to modern levels of a developed country. Figure 18.3 shows data for several countries arrayed in a similar fashion, with the dates showing the same progression as in the United States.

NOTE

This chapter is drawn from Peter H. Lindert and Jeffrey G. Williamson, "Essays in Exploration: Growth, Equality, and History," *Explorations in Economic History 22* (1985), 341–77.

19

Trends in Unemployment in the United States

Alexander Keyssar

Although American society has always contained men and women who wanted to work but could not find jobs, unemployment became a recognized phenomenon in the United States only in the nineteenth century – particularly in the second half of the nineteenth century. The word "unemployed" did not acquire its modern definition (being out of work and needing or wanting to work) until after the Civil War; the more encompassing and abstract term unemployment appeared only in the 1880s. Before the Civil War (or, to be more precise, before the industrial revolution), the nature of the economy and the social structure (the presence of slavery in the South, coupled with widespread self-employment and the interlacing of agriculture and manufacturing in the North) delayed the appearance of unemployment as a significant phenomenon or as a social problem. The unemployed were counted for the first time (in Massachusetts) in the 1870s; the earliest (and notably unsuccessful) attempt at a national survey was in 1880.[1]

Therefore, there is no way to gauge, with any reasonable accuracy, the extent of unemployment in the United States before the 1870s. And even thereafter, the statistical evidence is uneven and must be used with caution. This is so, in part, because unemployment is intrinsically difficult to measure with precision: there is a subjective component to the very definition of the term ("needing" or "wanting" to work), and enumerations of the unemployed demand other judgments as well (e.g., is a part-time worker employed or unemployed?). In addition, the methods that have been used to count the unemployed in the United States have changed considerably over the last 100 years, significantly complicating the task of comparing statistics from different time periods with one another. Before 1935, the most abundant data were collected in conjunction with census surveys; since 1940, the monthly Current Population Survey (CPS, along with its annual work experience survey) has been the primary source of statistics dealing with joblessness.[2]

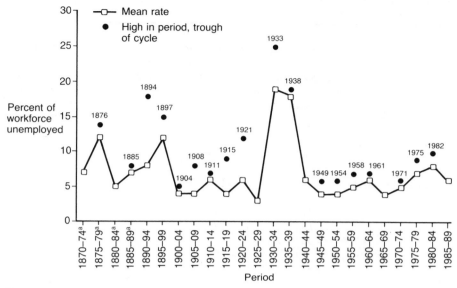

Figure 19.1   Unemployment rates in the USA, 5-year averages, 1870–1989
Note: [a] Author's estimates for these years.
Sources: *Historical Statistics of the United States*, p. 135; US Bureau of Labor Statistics, Bulletin 2340, p. 129; Bulletin 2361, p. 3.

Still, the available data do reveal several notable long-term trends. The first is that unemployment levels, on the whole, have not changed much since the 1870s. To be sure, the unemployment rate has fluctuated continually, but average unemployment rates are not much different now than they were during the final third of the nineteenth century or the first third of the twentieth century (figure 19.1). Peak unemployment levels (during the troughs of cyclical downturns) appear to have been somewhat lower after World War II than they were between 1870 and 1930, but the deep recession of the early 1980s (when unemployment rates were higher than they had been since the Great Depression) suggests that this trend may prove to have been short-lived.[3] The decade of the 1930s stands out clearly as a departure from prevailing patterns, but the overall volume of unemployment since the 1930s closely resembles the pre-1930 norm.[4]

There does, however, appear to have been a long-term shift in the distribution of unemployment within the labor force. As table 19.1 suggests – with imperfect and rather unwieldy data – the annual incidence or frequency of unemployment (the percentage of labor force members unemployed at some point in the course of a year) in column 1 seems to have declined in the long run: from a cyclical range of roughly 20–35 percent in the late nineteenth century to a range of 13–22 percent between 1957 and 1987. At the same time, the mean annual duration of

Table 19.1 Annual incidence of unemployment and unemployment durations, 1885–1989

	1 *% Experiencing some unemployment in the course of the year*	*2* *Mean weeks of unemployment per year*	*3* *Mean weeks of unemployment per spell*	*4* *% of unemployed with unemployment >15 weeks*
1885 (Massachusetts)	30	18		57[a]
1890	16	15		42[a]
1895 (Massachusetts)	28	14		
1900	22	15		52[a]
1901	27	11		
1909	19	17		
1933	>35			
1931–9				60–80[b]
1938	>30			
1948			17[c]	27
1952			17	24
1957	15		21	40 (1958)
1962	18		29	36
1967	13		17	23
1969		16		
1972	15		24	34
1974	18	20–32		
1975	20			
1977	18		29	38
1977–9	14(17)[d]	22		
1982	22		31	52
1987	14		29	45

[a]   Duration of 4 months or more for these years.
[b]   Author's estimate.
[c]   Double the CPS duration figure. See note 5 and Summers (1990), p. 300.
[d]   17 percent is the figure from the annual work experience survey. The 14 percent incidence and 22 week duration are from Margo (1990).
Sources: Keyssar (1986), pp. 51, 146, 302–05; unpublished tabulations of 1910 Census; Margo (1990), pp. 217–20; Bolino (1966), pp. 31–47; USBLS, *Bulletin 2340*, pp. 146, 220–6; *Eleventh U.S. Census, Population*, Part II, pp. 302, 448; *Twelfth U.S. Census, Occupations*, pp. lxxxvii–xciii, 214; Summers (1990), pp. 21–2. *18th Annual Report of the U.S. Commission of Labor*, pp. 290–7.

unemployment (the number of weeks lost to unemployment in the course of a year) in column 2, as well as the mean duration of unemployment spells in column 3, seems to have increased, particularly since the late 1940s.[5] Stated simply, these figures mean that proportionately fewer people are now becoming unemployed, but those who do experience unemployment are remaining jobless for longer periods. In all likelihood, this change was the consequence of institutional shifts that took place in the 1930s and 1940s, particularly the widespread formal adoption of seniority as the principle governing layoffs. Such developments rationalized the distribution of joblessness and diminished the breadth of its threat; however, as a consequence, the burden of unemployment came to be borne by a smaller proportion of the labor force.[6]

Two other (and perhaps related) shifts in the distribution of unemployment have also occurred. The first, as table 19.2 indicates, is that the gap between white and African-American unemployment rates seems to have widened over time. This was, in part, the consequence of dramatic shifts in the occupational structure of the African-American labor force, as African-Americans left the rural South and entered the industries of the North. But the salience of persistent discrimination is suggested by the fact that the gaps between whites and African-Americans in recent decades have been far larger than the gaps between immigrants and natives at the turn of the century. Similarly, the disparity between teenage and adult unemployment rates has also become more pronounced (table 19.3). Teenagers have always been jobless more frequently than adults, but the gap was far greater after World War II than

Table 19.2 Unemployment by race and by place of birth

Year	*Unemployment frequency*		
	Native-born whites	Foreign-born whites	African-Americans
1885 (Massachusetts)	26	36	
1890	15	16	16
1900[a]	22	20	29

	Unemployment rates		
	White	Hispanic	African-American
1931 (Philadelphia)	24		35
1934 (Massachusetts)	19		32
1937 (Massachusetts)	16		23
1957	4		8
1962	5		11
1967	3		7
1972	5		10
1977	6	10	14
1982	9	14	19
1987	5	9	13

[a] In 1900, 52 percent of the native-white unemployed were jobless for 4 months or more; among foreign-born whites, the figure was 51 percent, and among African-Americans, 42 percent. Samples of Massachusetts data suggested similar durations among natives and immigrants.

Sources: *Eleventh U.S. Census, Population*, Part II, pp. 744–50; *Twelfth, U.S. Census, Occupations*, pp. ccxxvi, ccxxxiv; Keyssar (1986), pp. 80–6; Chandler (1970), pp. 40–7; USBLS, *Bulletin 2217*, p. 64; *Bulletin 2340*, pp. 136–41.

Table 19.3 Unemployment among different age groups

Year	Unemployment rate		Unemployment frequency	
	Teenagers	*≥ 20 Years*	*Teenagers*	*≥ 20 Years*
1885 (Massachussetts)				
M	14	10		
F	13	8		
1890[a]				
M			19	16
F			15	13
1934 (Massachusetts)	48	23		
1937	37	16[b]		
1948	9	3		
1952	9	3		
1957	12	4		
1962	15	5		
1967	13	3		
1972	16	5		
1977	18	6		
1982	23	9		
1987	17	5		

[a] 1890 data are "uncorrected"; "corrected" data were never developed in different age groups. See *Twelfth Census, Occupations,* ccxxv–vi.
[b] Rate for all labor force members, including teenagers.
Sources: *Eleventh U.S. Census, Population*, Part II, pp. 744–50; Keyssar (1986), pp. 92, 105; Chandler (1970), pp. 36–8; USBLS, *Bulletin 2340*, p. 136.

it had been at the end of the nineteenth century.[7] To a considerable degree, then, the concentration of unemployment among the young and among African-Americans is, in historical perspective, a new phenomenon.

Several dimensions of the distribution of unemployment have, however, remained fairly constant. One is that unemployment rates for men and women have been roughly similar to one another for the past century and a quarter.[8] The second is that unemployment remains primarily a working-class and blue-collar experience. Although white-collar employees and even professionals are jobless from time to time (layoffs in the upper echelons of the occupational hierarchy are generally regarded as a sign of serious recession), the burden of unemployment has, for more than a century, been shouldered largely, and routinely, by manual workers in manufacturing, construction, and extractive industries.[9] Third, it must be mentioned that national unemployment rates have always masked very significant, and constantly changing, geographic variations in unemployment. At any given moment, different states and cities generally have widely varied unemployment rates, and, over time, high and low unemployment areas have changed dramatically and sometimes with remarkable speed.[10]

Finally, it should be noted that the unemployment experience of the United States, in comparison with the industrial nations of Europe, has improved somewhat in recent decades. Although the available data prohibit precise comparisons, it appears that, prior to the Great Depression (as well as during much of it), the United States tended to have higher unemployment rates than did most European nations. Although this pattern continued into the 1950s and 1960s, it no longer obtains. Since 1973, some, but not all, European countries have experienced far more unemployment than the United States; by the 1980s, indeed, the United States seemed to have roughly average unemployment levels for an industrialized nation.[11]

NOTES

1 Keyssar (1986), 1–38.
2 For a discussion of these statistics and a comparison of the census data with the CPS figures, see Keyssar (1986), 342–58.
3 In recent years, there has been some debate about whether the standard historical statistics on national unemployment, assembled by Stanley Lebergott and published in the *Historical Statistics of the United States*, exaggerate the volatility of unemployment prior to 1930, particularly in comparison to the post-World War II data. Christina Romer (1986a, 1986b) has argued this case; Lebergott, however, has published a (to this author) persuasive rebuttal (Lebergott, 1986). Two substantive points warrant mention here. First, accepting Romer's statistics would buttress the argument that there has been little long-run change in unemployment in the United States. Second, although Romer's own series displays less volatility than that of Lebergott and peak unemployment levels are lower, her figures, on average, are little different than those of Lebergott (which are used in figure 19.1 here). Five-year average unemployment rates, in Romer's series, are, with only exception (1925–9), within one percentage point of those presented in figure 19.1.
4 See Keyssar (1987).
5 This trend is almost certainly much more pronounced than the data in table 19.2 can show. Recent studies by Summers and others have demonstrated that official government duration statistics significantly understate the duration of unemployment – largely because the CPS survey inappropriately and erratically categorizes workers as having left the labor force when they remain between jobs and are really unemployed. The post-World War II figures in table 19.2 should therefore be adjusted upward to accurately reflect actual unemployment durations. Just how great that adjustment should be remains unclear (which is why it has not been done in table 19.2), but, in all likelihood, it would be at least 25 percent. The figures presented in table 19.2 regarding the durations of unemployment spells are derived from the CPS data indicating the duration of spells up to the moment of the survey. The CPS figures have been doubled to offer a rough (and minimal) gauge of durations for entire spells. See Summers (1990), 3–108, 286–316.
6 See Keyssar (1987).
7 This is discussed in more detail in Keyssar (1987).
8 See the sources listed for tables 19.1–19.4 and Keyssar (1986), 96–108.
9 For a discussion of these issues, see Keyssar (1986), 39–76 and Keyssar (1987).

10 Keyssar (1986), 111–29, 299–307; for examples with modern data, see U.S. Bureau of Labor Statistics (USBLS), *Bulletin 2340*, pp. 205–210.
11 Keyssar (1986), 447–8; Summers (1990), 227–85; USBLS, *Bulletin 2217*, 417–19; Therborn (1986).

REFERENCES

Bolino, A. C. (1966): "The Duration of Unemployment; Some Tentative Historical Comparisons." *Quarterly Rev. Econ. and Bus.* 6 (Summer), 31–47.

Chandler, L. V. (1970): *America's Greatest Depression, 1929–41.* New York: Harper and Row.

Keyssar, A. (1986): *Out of Work: The First Century of Unemployment in Massachusetts.* New York: Cambridge University Press.

——— (1987): "Unemployment Before and After the Great Depression." *Social Research,* (Summer), 201–22.

Lebergott, S. (1964): *Manpower in Economic Growth: The American Record Since 1800.* New York: McGraw-Hill.

——— (1986): "Discussion," *J. Econ. Hist.* (June), 367–71.

Margo, R. A. (1990): "The Incidence and Duration of Unemployment: Some Long-Term Comparisons." *Econ. Lett.,* 32, 217–20.

Romer, C. (1986a): "Spurious Volatility in Historical Unemployment Data." *J. Polit. Econ.,* 1–37.

——— (1986): "New Estimates of Prewar Gross National Product and Unemployment." *J. Econ. Hist.* (June), 341–52.

Summers, L. H. (1990): *Understanding Unemployment.* Cambridge: MIT Press.

Therborn, G. (1986): *Why Some Peoples Are More Unemployed Than Others.* London: Verso.

US Bureau of Labor Statistics (1985): *Handbook of Labor Statistics, Bulletin 2217.* Washington, DC, June.

——— (1989): *Handbook of Labor Statistics, Bulletin 2340.* Washington, DC, August.

——— (1990): *Geographic Profile of Employment and Unemployment, 1989, Bulletin 2361.* Washington, DC, May.

U.S. Department of Commerce, Bureau of the Census (1975): *Historical Statistics of the United States, Colonial Times to 1970.* Washington, DC: GPO, part I.

Unpublished unemployment statistics from the Thirteenth Census of the United States.

U.S Department of Commerce and Labor (1904): *Eighteenth Annual Report of the Commissioner of Labor, 1903.* Washington, DC: GPO.

U.S Department of the Interior, Census Office (1897): *Eleventh Census of the United States: 1890, Population,* part II, Washington, DC: GPO.

U.S Bureau of the Census, Department of Commerce and Labor (1904): *Occupations at the Twelfth Census.* Washington, DC: GPO.

EDITOR'S NOTE (JULIAN L. SIMON)

Data comparing black and white unemployment trends may be found in the editor's note following chapter 17 by Higgs and Margo.

20

Trends in Costs and Quality of Housing

Richard F. Muth

Housing is the largest single item of expenditure in many persons' budgets. It is also frequently the most important repository of accumulated savings. Trends in the cost and quality of housing are thus of considerable importance to individuals and society.

The terms *cost* and *price* applied to housing have at least two distinct meanings: (1) rental and (2) structural value. The first refers to the expenditure necessary to cover the costs of occupying a given dwelling for a certain time period, most commonly a month or a year. The second refers to the expenditure necessary to acquire the occupancy rights to a given dwelling in perpetuity – the sales price. To be meaningful, such

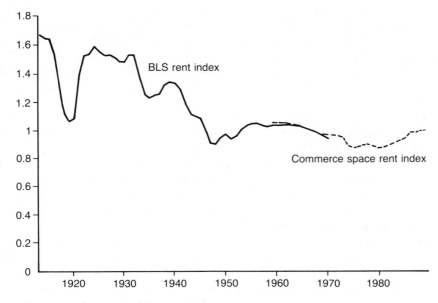

Figure 20.1 Index of real housing prices (1967 = 1.0)
Source: Bureau of Labor Statistics.

measures should refer to identical dwelling units when making comparisons over time or among different places. Many frequently cited measures of sales prices, however, do not meet this requirement.

Measures of housing rents are shown in figure 20.1 for the period 1913 to 1989. These measures are "real" in that they are adjusted for changes in the general level of prices, or, conversely, the value of money. The solid line refers to the rent index of the Bureau of Labor Statistics (BLS) and the broken line refers to the space rent index (strictly the implicit deflator) from the National Income Division of the Department of Commerce. The latter is available only for 1959 and subsequent years. The BLS rent index or its successor in the consumer price index was technically flawed for the period following 1970 – especially for the late 1970s and early 1980s – and hence is not shown.

Figure 20.1 suggests that the occupancy cost of a given dwelling has fallen rather sharply from the 1920s to the present. Both currently and in the middle 1960s, this occupancy cost is only about two thirds its level in the 1920s. Contrary to popular perception, housing occupancy costs declined even further in the late 1970s and early 1980s; they were then about 10 percent below their current level. The principal reason for this decline in occupancy costs, paradoxically, was the rapid increase in structure prices in that period. Increases in structure prices serve as an offset to interest, tax, and other costs associated with occupying a dwelling, so greater increases in structure prices reduce occupancy costs.

Figure 20.2 shows measures of structure purchase prices. The solid line is the Boeckh index of construction costs for residential structures, which is available from 1915 to the present. (The Boeckh index is a fixed-weight index of the prices of construction labor and building materials. Of the many construction cost indexes available, it is the only one that measures the costs of building residential structures.) The broken line is the residential structure price deflator of the National Income Division of the Department of Commerce; it is available only since 1947.

The two structure price measures diverge considerably during the first 20 years following World War II, making an assessment of their long-term trend difficult. From 1947 to 1967 the Boeckh index increased by somewhat over 20 percent; other construction cost indexes behaved in much the same way. The Commerce index, however, declined on the order of 10 percent from the 1950s to the late 1960s. The divergence may reflect the fact that the Commerce index does a better job of capturing the effects of technological improvements than fixed-weight averages of labor and material costs such as the Boeckh index.

The two measures are in rough agreement since the late 1960s, however. They indicate that the cost of acquiring structures has risen about 10 to 15 percentage points more rapidly than the average of all prices. And, indeed, at their peak in 1978–9, structure prices were about 20 to 25 percent higher than they had been a decade earlier. The Boeckh index also suggests that real structure prices increased by perhaps 40 percent in the period between the two world wars.

One might think the rent and sales-price trends noted above to be contradictory. After all, the cost of acquiring ownership of a structure is a principal determinant of the cost of occupying it. One of the main reasons for the divergence is the income tax treatment of (implicit) income from owner-occupied housing. Over most of the post-World War II period, certainly, both home ownership and income tax rates rose. By increasing the magnitude of the tax saving, these trends reduced the average occupancy cost of owner-occupied housing in the United States. Rental housing also became progressively more favorably taxed until the tax revision of 1986. It is well to remember, moreover, that there is considerable uncertainty in the trend of structure prices prior to the late 1960s.

Concerning consumption of housing, the available data strongly suggest that in the United States it has increased greatly in the postwar period. Figure 20.3 shows two measures of housing consumption per capita, both in dollars of 1967 purchasing power. The dashed line represents expenditures on housing for the period since the end of World War II. The solid line represents the stock of housing structures in thousands of dollars, available since 1925. Both are estimates by the Department of Commerce. These suggest that the consumption of housing per person has risen two to three times over the past 45 years. The stock measure also suggests that – perhaps because of the Great

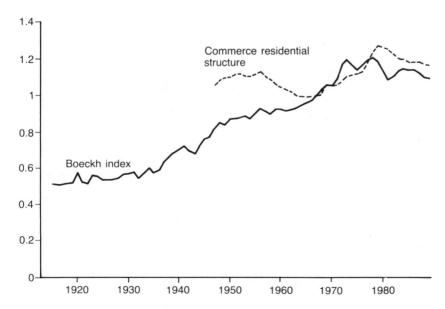

Figure 20.2 Index of real structure prices (1967 = 1.0)
Source: Dept of Commerce.

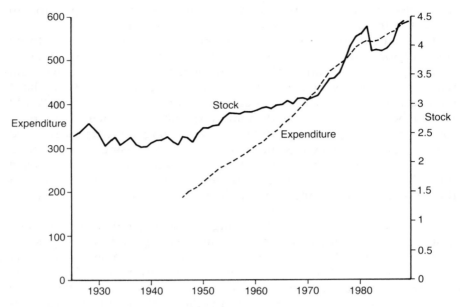

Figure 20.3 Per capita real housing consumption
Source: Dept of Commerce.

Depression and World War II – there was little overall increase in
housing consumption from the mid-1920s to the mid-1940s.

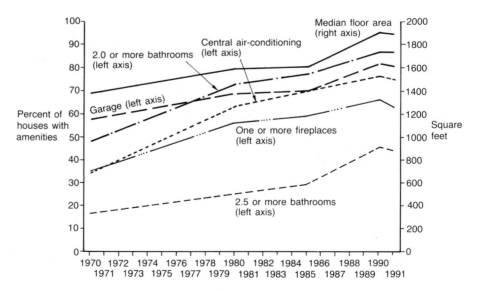

Figure 20.4 New privately owned one-family houses completed 1970–91
Source: *Statistical Abstract of the United States, 1992*, 711, table 1213.

In conclusion, this century – especially its second half – has seen great improvement in the housing conditions of the US population. Housing has become considerably cheaper to inhabit, though more expensive to build, relative to other kinds of consumption. Since the end of World War II housing consumption per person has increased by a factor of two to three times. Though further improvement is possible, and will no doubt occur in the next half century, the record of the past half century is impressive indeed.

EDITOR'S NOTE (JULIAN L. SIMON)

Data comparing black and white home-ownership trends may be found in the Editor's Note following chapter 17 by Higgs and Margo.

Data for other countries of percentages of households without flush toilets or bathing facilities are shown in chapter 24 by Rector. Fewer than 20 percent of Japanese households have central heating (*Washington Post,* March 4, 1991, A9).

Data on rooms per person, crowding, and amenities (running water, flush toilet, electricity) since 1900 in the United States are shown in chapter 24 by Rector and in figure 20.4.

21

Trends in the Quantities of Education[1]–USA and Elsewhere

Julian L. Simon and Rebecca Boggs

> . . . Perhaps in this neglected spot is laid
> Some heart once pregnant with celestial fire;
> Hands, that the rod of empire might have sway'd,
> Or waked to ecstacy the living lyre:
>
> But Knowledge to their eyes her ample page
> Rich with the spoils of time, did ne'er unroll;
> Chill Penury repress'd their noble rage,
> And froze the genial current of the soul.
>
> Full many a gem of purest ray serene
> The dark unfathom'd caves of ocean bear:
> Full many a flower is born to blush unseen,
> And waste its sweetness on the desert air.
>
> . . . Th'applause of list'ning senates to command,
> The threats of pain and ruin to despise,
> To scatter plenty o'er a smiling land,
> And read their history in a nation's eyes.
>
> *Thomas Gray[2]*

Perhaps the most impressive, exciting, and heartening trend – starting in the nineteenth century in the advanced countries, and accelerating in the twentieth century in all the world – is the increase in the amounts of education youths receive. This trend is crucial for broad economic development, of course. But in human terms it represents the opportunity for fulfillment rather than frustration for the many talented people of the world, as well as benefit to all humanity now and forever from the contributions to knowledge that these talented people will make.

Despite the solid good news about education in almost all respects, however, the press and the educational establishment give the impression that the news is bad. Consider for example how a story on high school dropouts – considered important enough to be on the front page of *The Washington Post* – suggests that bad things are happening. The

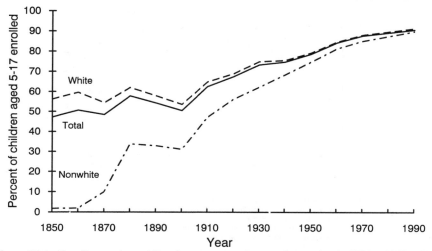

Figure 21.1 Enrollment in public elementary and secondary school, USA, 1850–1990, proportion of children ages 5–17, by race

Sources: 1850–1980: U.S. Department of Commerce, Bureau of the Census, Historical Statistics of the United States: Colonial Times to 1970, Series H 433–441. 1990: U.S. Department of Education, National Center for Education Statistics, Statistics of Public Elementary and Secondary Day Schools 1990, 376–7. Education Information Branch, 1992.

headline was "D.C. Dropout Rate Among Worst in U.S.," and a university president was quoted as saying, "It is probably the most dangerous national security risk we're facing today." The president of the Carnegie Foundation for the Advancement of Teaching "explained" what he called the "high dropout rate nationally." The two "reasons he gave were "anonymity and irrelevance."

We will return to the special topic of dropouts with data in hand, and you will then be able to evaluate for yourself the relationship of the news to the facts.

The graphs in this chapter document the trends, which are so obvious that they need little commentary.

TRENDS IN THE UNITED STATES

Figure 21.1 shows the long-run trend in the proportion of children of school age enrolled in school from 1850 to the present. (The proportion among non-whites – mostly blacks – also is shown here because it is so eye-catching; more data on comparative educational trends for blacks and whites may be found in the editor's appendix to chapter 17, Higgs and Margo, in this volume).

The data in figure 21.1 understate the growth in education because the proportion of enrolled students who actually attended school on any given day was much lower in past years than now; that increase in

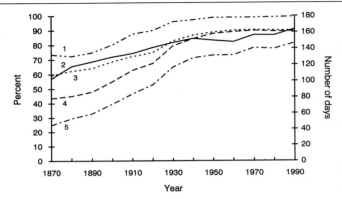

1. Average length of school term (right scale)
2. Percent aged 5-17 enrolled (left scale)
3. Percent of enrollees attending daily (left scale)
4. Average number of days attended per pupil (right scale)
5. Average number of days attended per person of school age (right scale)

Figure 21.2 Amount of schooling per child, USA, 1870–1990, public elementary and secondary school
Sources: U.S. Department of Education, National Center for Education Statistics, Statistics of Public Elementary and Secondary Day Schools 1990, 47. Education Information Branch, 1992.

attendance is shown in figure 21.2. The length of the school term also has increased, also seen in figure 21.2. The composite result of those two elements is the number of days in school per enrolled student. The figure also combines the data for enrollment and for number of days attending school into a display that shows the number of days in school per year per person of school age. This is the best overall measure of the amount of education received up to the end of secondary school.

Figure 21.3 shows the trends in the enrolled pupil–teacher ratio, and also shows that ratio adjusted for the number of students actually in school each day. If these data were combined with the composite data in figure 21.2, we would have a measure of the amount of teacher time per child of school age per day or year.

Data on attendance at and completion of secondary school are shown in figure 21.4 – enrollment among youth aged 14–17, proportion of 17-year-olds who graduate, and the proportion of the 25–29-year-old cohort who have not finished high school. Figures 21.5 and 21.6 provide data on tertiary schooling.

Figure 21.7 shows percentages with five years of school or fewer, four or more years of high school, and four or more years of college. The summary statistic – median years of schooling achieved by those aged 25–29 – is displayed in figure 21.8, separately for whites and blacks (combined data not available); data on mean years are shown in the editors' appendix to chapter 17, Higgs and Margo, in this volume.

The data on amount of education measured in years undoubtedly are misleading – biased toward showing convergence – because there are

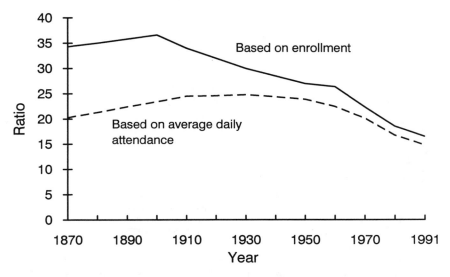

Figure 21.3 Pupil/teacher ratios, USA, 1870–1991, public and private, elementary and secondary school
Sources: U.S. Department of Commerce, Bureau of the Census, Historical Statistics of the United States, Colonial Times to 1970, and U.S. Department of Education, National Center for Education Statistics, Digest of Educational Statistics, various issues.

important differences in the quality of the educations received by students of different races. But it is also true that the apparent trends toward convergence are not *entirely* misleading, as shown by the earnings of blacks and whites with equal amounts of education measured in years.

Before moving from the USA to the world situation, let us return to investigate how the newspaper story mentioned in the introduction could paint a bleak negative picture of the "dropout rate" when the trends in figure 21.4 (and 21.5 and 21.6) are positive. Indeed, the diagram accompanying the newspaper story shows a positive trend toward less dropping out from 1972 to 1991.

One cause of a negative impression is focusing on the District of Columbia, which one may argue is a legitimate local negative story. But the story offers a negative outlook for the nation as a whole, too; this is accomplished by quotes from supposed authorities. The factual positive trend is noted only in the last two paragraphs of the news story, for "balance." The facts presented would have been even more positive if the newspaper chart had been carried backward even just a bit before 1972, as seen in figure 21.4.

Before leaving the case of the United States, a word about SAT and other standardized test scores: Comparison of scores over a long period of years is fraught with many difficulties, most especially the change in the composition of the groups taking the tests, as well as the tests themselves. We have yet to see any analysis that is sufficiently complete

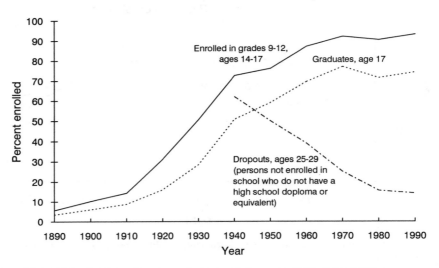

Figure 21.4 High school enrollment, graduates, and dropouts, USA, 1890–1990
Sources: dropouts, Ben J. Wattenberg, *The First Universal Nation* (New York: Free Press, 1991), 91; others, U.S. Department of Education, National Center for Educational Statistics, Digest of Educational Statistics, 1992, 67, 107.

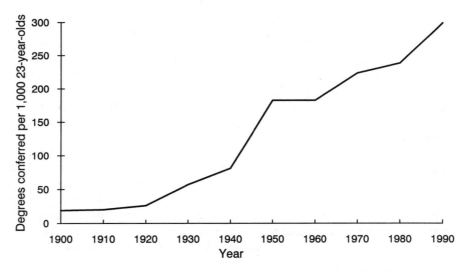

Figure 21.5 Bachelor's and first professional degrees conferred, USA, 1900–1990
Sources: 1900–70: Historical Statistics of the United States, Colonial Times to 1970, Series H 751–765. 1980–90: U.S. Bureau of the Census, 1980 Census of Population, PC 80–01–B1, 1–26, and 1990 Census of Population, CP–1–1, p. 17, Table 13, and U.S. Department of Education, National Center for Educational Statistics, Digest of Educational Statistics, 1992, p. 241, Table 229.

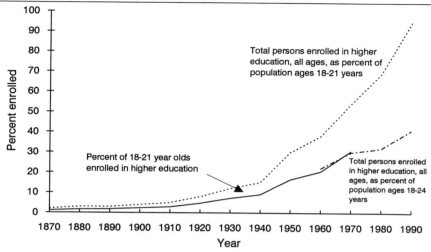

Figure 21.6   Higher education, selected enrollment rates, 1870–1990
Sources: U.S. Department of Education, National Center for Educational Statistics, Digest of Educational Statistics 1992, 171, Table 159; U.S. Department of Commerce, Bureau of the Census, Historical Statistics of the United States, Colonial Times to 1970, Series H 700–715, and Current Population Reports, Series P-20, School Enrollment: Social and Economic Characteristics, various issues.

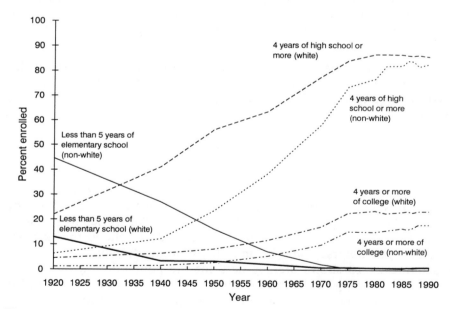

Figure 21.7   School years completed, by race, 25–29 years old, 1920–1990
Sources: U.S. Department of Commerce, Bureau of the Census, Census of Population, 1960, vol. 1, part 1; Current Population Reports, Series P-20; Series P-19, no. 4; 1960 census monograph, "Education of the American Population", by John K. Folger and Charles B. Nam; and U.S. Department of Labor, Bureau of Labor Statistics, Office of Employment and Unemployment.

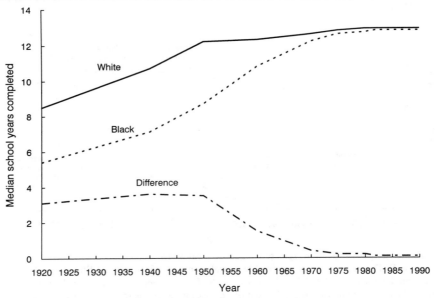

Figure 21.8 Median school years completed, 25- to 29-year-olds, by race, 1920–90
Sources: U.S. Department of Commerce, Bureau of the Census, Census of Population, 1960, vol. 1, part 1; Current Population Reports, Series P-20; Series P-19, no. 4; 1960 census monograph, "Education of the American Population", by John K. Folger and Charles B. Nam; and U.S. Department of Labor, Bureau of Labor Statistics, Office of Employment and Unemployment.

in both data and analysis to convince us that its conclusions about the trends are reliable, and we ourselves are not now in position to conduct such an inquiry even if this were an appropriate place to publish its outcome.

TRENDS IN THE WORLD

Long-run series for the poorer countries of the world are hard to come by. Figures 21.9–11 provide aggregate attendance data for the world, and for the developed and developing parts of the world, for persons 6–11, 12–17, and 18–23 respectively. Figure 21.12 shows the skimpy, rough data on illiteracy rates by birth cohort in the developing countries, and figure 21.13 shows literacy data for the population at large (less precise or meaningful than age cohorts) of India. Figure 21.14a and b provide some long-run enrollment rates for selected countries.

From the point of view of the creation of new knowledge, the *total number* of well-educated persons is more relevant than the *proportions* of the populations that are well-educated. Therefore, figure 21.15 shows the total number of persons aged 18–23 who were enrolled in schools above the high-school level.

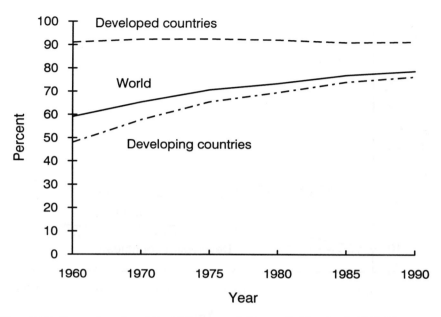

Figure 21.9 Proportion of world's children, ages 6–11, enrolled in school, 1960–90
Source: UNESCO Statistical Yearbook, World (Paris: UNESCO, 1991), Table 2.11, p. 2–31.

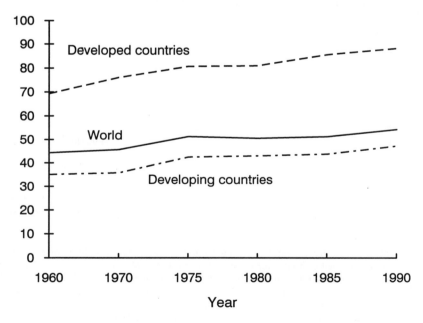

Figure 21.10 Proportion of world's children, ages 12–17, enrolled in school, 1960–90
Source: UNESCO Statistical Yearbook, World (Paris: UNESCO, 1991), Table 2.11, p. 2–31.

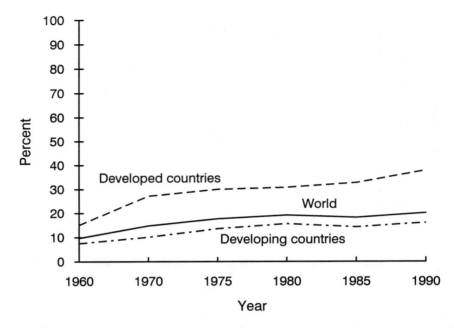

Figure 21.11 Proportion of world's children, ages 18–23, enrolled in school, 1960–90
Source: UNESCO Statistical Yearbook, World (Paris: UNESCO, 1991), Table 2.11, p. 2–31.

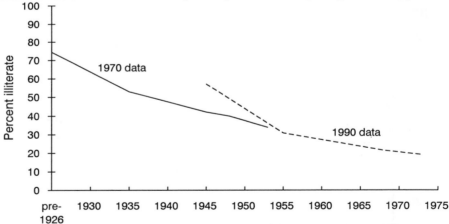

Figure 21.12 Adult illiteracy, developing countries, by birth year, pre-1925 to 1973
Note 1: The two lines represent estimates made from two surveys taken in different years, 1970 and 1990. Both provided estimates for the groups born from 1945 to 1953. *Note 2*: UNESCO provides data on illiteracy rates among different age groups for the years 1970 and 1990. In this graph, the middle of each age range is represented as the birth year for the purposes of comparing 1970 and 1990 data and showing the overall trend. Example: the illiteracy rate among persons age 20–4 in 1970 (born 1946–50) was 52.9% for 1948.
Source: UNESCO, Compendium of Statistics on Illiteracy (Paris: UNESCO Office of Statistics, 1990).

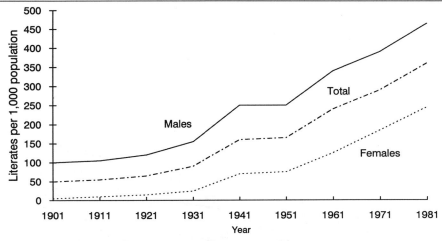

Figure 21.13 Literacy rates per thousand population, India, 1901–81
Source: India: A Statistical Outline (New Delhi: Oxford abd IBH Pub., 1987, p. 9).

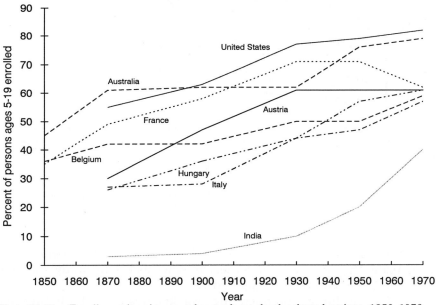

Figure 21.14a Enrollment in primary and secondary school, selected nations, 1850–1970
Source: Piero Melograni and Sergio Ricossa, eds., *Le Revoluzioni Del Benessere* (Roma-Bari:
Gius, Laterz & Figli Spa, 1988), 32.

COMPARATIVE US EDUCATIONAL
EXPENDITURES AND ATTAINMENTS

It is certainly difficult, and perhaps impossible, to find meaningful
indices of how well a country is educating its people. Therefore, it is
customary to examine crude measures of output such as years of

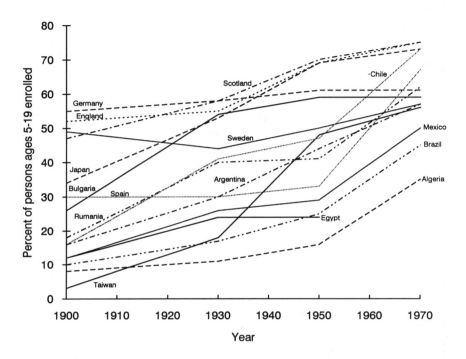

Figure 21.14b Enrollment in primary and secondary school, selected nations, 1900–70
Source: Piero Melograni and Sergio Ricossa, eds., *Le Revoluzioni Del Benessere* (Roma-Bari: Gius, Laterz & Figli Spa, 1988), 32.

education, and measures of input such as amounts of expenditures. Recognizing the crudity of such measures, let us briefly study a case in which the US is said to be doing badly.

"U.S. Lags in Education Spending" was the headline of a *Washington Post* story (September 24, 1992, p. A9). The tone in the text of this article is negative in every sentence in terms of the supposed data being reported. And it leaves no doubt that the supposed low spending by the USA is a bad thing.

Figures 21.16 and 21.17 show that in absolute money terms, the United States spends more per student at the primary and secondary levels (in purchasing power parity measures, which is a bit tricky) than does almost any other country – exactly the opposite of the impression the story gives by its focus on percentages rather than absolute amounts.

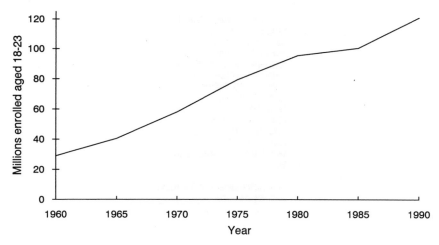

Figure 21.15 Number of persons ages 18–23 enrolled in school worldwide, 1960–90
Sources: UNESCO, Statistical Yearbook, 1991, page 2–31; United Nations, Population Studies, No. 122, The Sex and Age Distributions of Population: The 1990 Revision of the United States Global Population Estimates and Projections, p. 226.

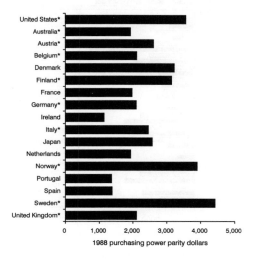

Figure 21.16 Expenditures per student for primary education from public and private sources
Source: Center for Educational Research and Innovation, Education at a Glance: OECD Indicators, 1992, 57.

Japan and Germany – which certainly have not done badly economically – spend much less than the US not only in dollar terms but in percentage terms. If one wants to connect economic success to this measure, the cases of Japan and Germany would hardly support the idea.

At the tertiary level, the United States has a larger proportion of students attending than do most other countries. Combined with the expenditure

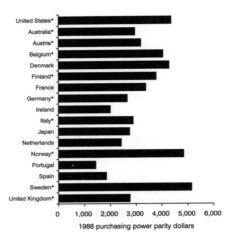

Figure 21.17 Expenditures per student for secondary education from public and private sources
Source: Center for Educational Research and Innovation, Education at a Glance: OECD Indicators, 1992, 57.

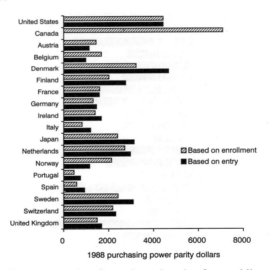

Figure 21.18 Expenditures per student for tertiary education from public and private sources
Sources: Data calculated from data in Center for Educational Research and Innovation, Education at a Glance: OECD Indicators, 1992, 57, Table 6; pp. 79–80, Tables P14 and P15.

per student shown in Figure 21.18, the United States clearly spends more per tertiary student, also, than does almost any other country.

The news story discusses only *percentages* of income spent for education, rather than absolute dollars. By the most tortured interpretation, this measure might not be absolutely false in saying that the United States "lags." The justification given is that this magnitude represents "effort." But this is not valid, in our view.

Even if it were true that the United States were at the bottom of the list in percentage terms – which is most certainly not so – this would not necessarily be bad evidence. As people get richer, they tend to spend a smaller proportion of their income for certain important goods – for example, food. On average, people in the richest countries – just like the richest families – spend a smaller proportion of their income on necessities such as food than do poorer persons. Indeed, a small proportion of one's income spent on such things is an exceedingly good indicator of how well off one is in general. In aggregate, a poor country may spend 80 percent of its income for food, whereas a rich one spends only perhaps 25 percent. It would hardly be sensible to say that the rich country "lags" in food spending.

And there is good reason to think that the same is true for education. A poor family certainly spends more for education as a proportion of its income than does a wealthy family, on average. Hence the data on proportion of income spent for education tell little about countries.

More generally – and very importantly – money expenditures on the education of a given student are a very dubious indicator of the quality of the education the student receives, if "quality" means some concept relevant to the preparedness of the student for adult life, economic and otherwise. One obvious defect is that teachers' salaries are a function of the income level of the society in which they teach; the "quality" of a teacher in the USA today, relative to that of a teacher in the USA one or two generations ago or of a teacher in Russia or South Korea or Israel, is not appropriately measured by the relative salaries. There are also other very great defects with this measure. The absence of evidence of a connection *within* the United States and *within* Great Britain between educational expenditures per student and the results of the education – including data on the performances of students schooled in parochial schools and at home (see Beisner, 1990 for a brief review and references), casts even more doubt on the meaningfulness of this measure. It is discussed here only to dispell wrong impressions left by unsound analysis of the data themselves.

If one is interested in how the United States compares against other countries, it would certainly be relevant to consider the total amount of education Americans have. The OECD book from which the above data come also includes a table of the educational levels of people 25–64 in various countries. Though this does not refer to current educational efforts, it does put the matter in perspective.

Given that most Americans finish high school eventually, the best measure of total education is the people who have a college education. These data show that the United States is far ahead of any other country. Figures 21.19a–c show that the US has a much more educated adult population than does any other country—suggesting that the USA does not "lag."

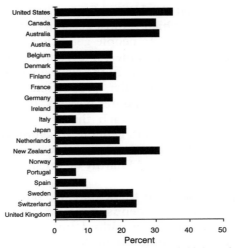

Figure 21.19a Percent of 25- to 64-year-old population with higher education, university and non-university
Source: Center for Educational Research and Innovation, Education at a Glance: OECD Indicators, 1992, 23.

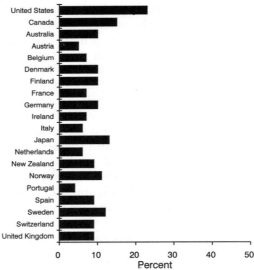

Figure 21.19b Percent of 25- to 64-year-old population with higher education, university
Source: Center for Educational Research and Innovation, Education at a Glance: OECD Indicators, 1992, 23.

SUMMARY AND CONCLUSIONS

The amount of education youths receive has been increasing rapidly in the United States, and in the rest of the world, over the periods for which we have data. This indicates an increase in people's capacity to produce economically, and therefore to have a higher standard of living for

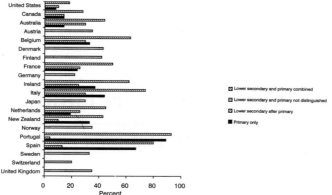

Figure 21.19c Percent of 25- to 64-year-old population with primary education only
Note: Austria, Denmark, Finland, Germany, Japan, Norway, Sweden, Switzerland, and the UK do not distinguish primary from lower secondary but lump them together; hence some persons counted as having "lower secondary" education in these countries will have had only the equivalent of primary education in the other countries. "Lower secondary after primary" includes only persons who have had both primary and lower secondary education. "Lower secondary and primary combined" includes persons, in countries that distinguish the two, who have had either primary alone or both primary and lower secondary education.
Source: Center for Educational Research and Innovation, Education at a Glance: OECD Indicators, 1992, 23.

themselves. The total amount of education has also been increasing, both because of the increase per person and also because of the larger populations. This implies an increase in the amount of knowledge created, which increases the standard of living for the community as a whole by increasing the productivity of individuals with given amounts of education. There is no obvious reason why these trends should not continue indefinitely.

NOTES

1   The assistance of Abbott Ferriss, Michael McEvoy, John Phelps, Ilya Somin, and George Washnis is much appreciated. Guenter Weinrauch did his usual excellent work as research assistant and prepared almost all of the graphs. Calvin Beisner gave us several valuable suggestions and worked on a few of the graphs. Many of the US graphs are built upon Abbott Ferriss's 1969 work.
2   Thomas Gray, *Elegy Written in a Country Churchyard*. In Joseph Auslander and Frank Ernest Hill, eds, *The Winged Horse Anthology*. New York: Doubleday, Doran, 1944, 223–4.

REFERENCES

Beisner, Calvin, "Here's the One Thing Public Schools Fear Most," *World*, March 17, 1990, p. 12.

Ferriss, Abbott, *Indicators of Trends in American Education*. (New York: Russell Sage Foundation, 1969).

22

Trends in Free Time

John P. Robinson

One of the hallmarks of a "modern" society is the amount of free time available to its members. Free time is usually inferred from simple estimates people report to government agencies about the weekly hours they spend at work. Decreases in the length of the workweek are looked on as one of the great advances in the quality of life in Western countries because it is thought to have increased free time.

As shown in figure 22.1, however, the last half-century has produced remarkably little change in the length of the American workweek. In contrast to the steady decline in the workweek of 2.7 hours per decade for nonagricultural employees between 1850 and 1940, less than a one hour per-week per-decade decrease has occurred since 1940, and virtually no decrease since 1950.

These aggregate data, however, conceal many important changes within the world of work:

- The proportion of women working has almost doubled since 1940.
- At the same time, there has been virtually no increase in the proportion of women aged 55–64 in the labor force.
- The proportion of men working has declined, particularly in the 55 to 64 age bracket (from 83 percent in 1970 to 67 percent in 1988). The average workweeks for those who work have declined as well.
- With advanced education, most individuals enter the labor force later in their lives.
- More people past age 65 have retired from work.
- More vacation days and holidays are available.

All of these factors have undoubtedly led to less overall time per lifetime spent working for those who live to age 65 or beyond, and they significantly decrease the ratio of people at work to the total adult population. These workweek figures, then, may exaggerate the prominence of time spend working in contemporary life.

The discrepancies between the figure 22.1 official workweek figures and "actual" time people spend working can be put into three categories:

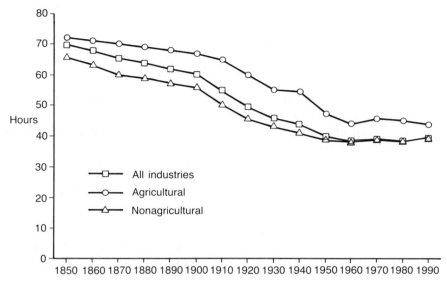

Figure 22.1 Length of average workweek, USA, 1850–1988
Sources: 1850–1930: Agricultural and nonagricultural industries from J. F. Dewhurst and Associates, *America's Needs and Resources: A New Survey*, New York: Twentieth Century Fund, 1955, appendix 20–4, 1073; all industries is average of other two columns weighted on the basis of percentage of gainfully occupied in agriculture and in nonagricultural industries as shown in *Historical Statistics of the United States, 1789–1945*, US Bureau of Census, 1949, Series D6–7, 63.

1940–52: Dewhurst and Associates, appendix 20–1, 1064–9.

1953–90: US Bureau of the Census, Current Population Reports: Labor Force, Series P-50, Nos. 59, 67, 72, 85, and 89; Bureau of Labor Statistics, Employment and Earnings, Annual Supplement, vol. 6, no. 11, May 1960, plus preliminary unpublished data for 1960 from DLS. The averages published by the Census were adjusted downward to reflect zero hours of work for those "with a job but not at work."

1. Decreasing numbers of *days* at work compared with numbers reported, due to failure of respondents to count holidays, sick days, and the like in their work-week estimates.
2. Decreasing numbers of *weeks* at work, due to the inability of analysts to include vacation periods in their estimates.
3. Decreasing numbers of *years* in the work force, due to delayed entry into the work force and earlier exit from the work force in the form of retirement and the like.

All of these factors add up to work comprising less of life than most workweek figures suggest, and they also lead to underestimates of available free time.

Such considerations led in 1985 to the first calculation of figures for the total years spent in one's work life by the Bureau of Labor Statistics, estimated at 38.8 years for men and 29.4 years for women. These figures work out to 55 percent of all time (per lifetime) as economically active for men and 38 percent for women as of 1980. No trend figures have been compiled since then, but there is good reason to expect that they are on the decline.

THE TIME DIARY

Do these work figures tell the whole story about free time? Data from three national time-diary studies suggest not. The time diary is a survey instrument that attempts to account for all the ways that people spend time. Each respondent is instructed to begin at midnight on a designated diary day and to report all their activities in the order in which they occurred across that entire 24-hour day.

In this way the diary exploits the many advantages of time as a measuring tool:

- All 24 hours are accounted for.
- If less time is spent on one activity, it must be picked up by another activity. Thus, trade-offs in time spent on various activities can be monitored.
- The sequence of activities are preserved in the order in which they occur during the day.
- Time-of-day of activities are recorded, so that the analyst can observe what aggregate proportions of the public are working (or watching TV) at 6 o'clock, midnight, or any other time of the day.

Because of its ability to capture all activity, and with minimal reporting burden, memory loss, and embarrassment to the respondent, time diary data are now regularly collected by several government agencies in European societies. In the United States, we have three national studies of time use (in 1965, 1975, and 1985, with 1994–5 in progress), each conducted by academic rather than government survey organizations.

In contrast to this diary, consider what is demanded of respondents when they try to give estimates of time expenditures (as in a typical workweek question) to interviewers in a government survey. Theoretically, they are expected to go through their last week (or more vaguely a "typical" week), and tally the work time that was spent on personal errands, health checks, long lunches, stops at the dry cleaners, car maintenance, and the like. Presumably these respondents will also remember to subtract sick days and holidays from their estimates. On the other hand, they should also remember to *include* time spent on work at home or their overtime work when it occurs. It seems rather optimistic of government analysts to expect that many respondents will take the time to perform these mental exercises in giving their estimates to interviewers. Moreover, as we find in our diary studies, a "typical" day of week is not typical, at least not when respondents in our surveys are asked to characterize the day on which they keep their diary.

The diary data themselves point out other problems with these questions. First, there is an important difference in the work time reported in diaries and in the estimates of their workweek that respondents give us. Across individuals, less time working is reported in the diaries than is estimated by these individuals, and that discrepancy is growing across time. In 1985, that discrepancy stood at about five hours per week (40-hour estimate vs. 35 hours in the diary); the gap was larger among working women (Robinson and Bostrom 1994). It was particularly high among those estimating longer work hours, has been found in eight other countries (Robinson and Gershuny 1994).

In time diary studies free time is defined as what is left over after subtracting the time people spend working and commuting to work, taking care of their families, doing housework, shopping, sleeping, eating, and doing other personal care activities. Put more actively, free time includes the time adults spend going to school, participating in clubs and other organizational activities, taking part in sports, recreational activities and hobbies, watching television, reading, and visiting with friends and relatives – as well as all related travel time.

There is a more dynamic element that comes into play when one looks at the actual relation between work and free time in the diary data. For each hour of reduced work time, only 60 percent of it was used for free time (Robinson, 1977). Many respondents with shorter workweeks spent more time doing housework and family care. Others sleep or eat more, or spend more time caring for children. Moreover, the same phenomenon is observed among retired people. They may gain 30 to 40 hours by retiring from work, but they only realize 15 to 20 hours more free times, because they increase their house care, sleep or eat more (not considered free time). Here again, an hour less work does not translate into an hour more free time.

We now look at trends in the diary data, and then how this free time gets divided – first across demographic groups in the population and second across specific activities.

GENERAL DIARY TRENDS

The Americans' Use of Time Project data in table 22.1 show that Americans of working age have more free time today than ever before. In 1985 men had 41 hours of free time a week, and women had 40 hours of free time. Both time and demographic factors lie behind this shift.

Time Factors

There are two activity-related reasons for the increase in free time. First, women are doing much less housework than they did several decades ago. Second, as noted above, the diary workweek is shorter today than it

Table 22.1 Differences in total weekly hours of free time, by year and demographic factors

	Men			Women		
	1965	1975	1985	1965	1975	1985
Total, 18–64	35	39	41	34	38	40
Age						
Age 18–35	40	42	43	36	40	39
Age 36–50	36	33	34	32	34	35
Age 51–64	34	40	44	35	39	44
Marital status						
Married	35	37	37	37	37	37
Not married	38	48	48	32	42	43
Children						
None	38	42	43	35	40	41
All aged 5+	36	33	39	33	35	38
At least one preschool	34	41	31	34	38	34
Employed						
No	63	55	56	40	44	47
Yes (10+ hours per week)	33	36	36	27	31	34

was in 1965 – not the "official" hours of work Americans report, but the work hours they record in their single-day diaries. While figure 22.1 indicates that official workweek figures have remained fairly constant over the past few decades, the number of work hours people record in their diaries has fallen significantly for both men and women between 1965 and 1985, as noted above.

Demographic Factors

The diary evidence is not the only data to suggest that Americans' free time should be expanding. There are two demographic trends that should result in more free time. First, fewer households have children. This means that today's adults spend less time involved in child care. Women with no children at home have three more hours of free time each week than women with older children, and seven more hours of free time each week than women with preschoolers – other factors being equal.

The second demographic trend that results in more free time is that Americans are spending more of their adult lives unmarried. Unmarried people have more free time than married couples. Unmarried women have six more hours a week of free time than married women. Unmarried men have 11 more hours of free time each week than married men.

The differences in the amount of free time available to married versus unmarried people, and to parents versus nonparents, has widened in the last two decades. But this change is relative. Most Americans have more

Table 22.2 How free time is distributed across activities, by year and gender (in hours per week)

	Total			Men			Women		
	1985	1975	1965	1985	1975	1965	1985	1975	1965
Total	40.1	38.3	34.5	41.1	38.6	35.6	39.6	38.3	34.4
TV	15.1	15.2	10.5	15.7	16.2	11.9	14.5	14.1	9.3
Visiting	4.9	5.5	6.6	5.0	5.1	5.9	4.8	5.7	7.5
Talking	4.3	2.2	2.6	3.5	1.9	1.6	5.1	2.7	3.6
Traveling	3.1	2.6	2.7	3.4	2.8	3.2	3.0	2.4	2.4
Reading	2.8	3.1	3.7	2.7	3.0	4.3	2.9	3.3	3.3
Sports/Outdoors	2.2	1.5	0.9	2.9	2.3	1.5	1.5	0.8	0.5
Hobbies	2.2	2.3	2.1	1.9	1.6	1.5	2.6	3.0	2.8
Adult Education	1.9	1.6	1.3	2.2	2.1	1.7	1.6	1.3	0.9
Thinking/Relaxing	1.0	1.1	0.5	1.2	1.0	0.2	0.9	1.2	0.6
Religion	0.8	1.0	0.9	0.6	0.8	0.8	1.0	1.3	1.0
Cultural Events	0.8	0.5	1.1	0.8	0.3	1.3	0.8	0.6	0.9
Clubs/Organizations	0.7	1.2	1.0	0.8	0.9	0.8	0.6	1.5	1.2
Radio/Recordings	0.3	0.5	0.6	0.4	0.6	0.7	0.3	0.4	0.4

free time today than they did 20 years ago, regardless of their living arrangements, as shown in table 22.1. Most of the gains in free time occurred between 1965 and 1975. Since 1975, the amount of free time people have has remained fairly stable.

However, the ages of such people in the sample needs to be taken into account. Table 22.1 shows that people aged 51 to 64 have gained the most free time since 1965, mainly because they are working less. As noted above, among people, in this age group, the proportion of men opting for early retirement increased considerably between 1965 and 1985, and the proportion of women in this age group who work has not increased at all since the mid-1960s.

Today, then, the 40-hour workweek is balanced by a 40-hour play week. But there is no denying the fact that many Americans are caught in a real time crunch. The diary data of working parents give evidence of severe time pressure. Like all averages, the diary data hide much individual variation. On balance, however, more people are gaining free time than losing it. How men and women are using their free time is shown in table 22.2.

SUMMARY AND CONCLUSIONS

Americans in general have more free time than 20 to 30 years ago. This comes both from increased life spans and its consequent increased years in retirement and from more free time for the younger (pre-retirement) work force. Signs are that this unofficial decrease in work time is also increasing. The lengthening lifespan will only accentuate that trend.

Unless more adults have children or stay married longer, or need to work longer hours to keep up their standard of living, it is doubtful the American population will experience any significant decrease in free time in the near future.

REFERENCES

Robinson, John P. (1977): *How Americans Use Time.* New York: Praeger.

—— (1994): *The Demographics of Time.* Ithaca, NY: American Demographics.

—— and Ann Bostrom (1994): "The Overestimated Workweek: What Time-Diary Studies Suggest." *Monthly Labor Review,* 117 (Aug.) 11–23.

—— and Jonathan Gershuny (1994): *Bulletin of Labor Statistics.* (Jan.)

EDITOR'S NOTE (JULIAN L. SIMON)

Figure 22.2 shows annual hours worked in selected countries from 1870 to 1989.

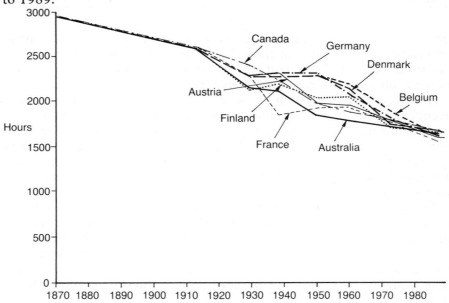

Figure 22.2   Annual hours worked per person, selected countries, 1870–1989

Source: 1850–1930: Agricultural and nonagricultural industries from J. F. Dewhurst and Associates, *America's Needs and Resources: A New Survey,* New York: Twentieth Century Fund, 1955, Appendix 20–4, 1073; all industries is average of other two columns weighted on the basis of percentage of gainfully occupied in agriculture and in nonagricultural industries as shown in *Historical Statistics of the United States, 1789–1945,* US Bureau of Census, 1949, Series D6–7, 63. 1940–52: Dewhurst and Associates, appendix 20–1, 1064–9. 1953–90: US Bureau of the Census, Current Population Reports: Labor Force, Series P–50, Nos. 59, 67, 72, 85, and 89; Bureau of Labor Statistics, Employment and Earnings, Annual Supplement, vol. 6, no. 11, May 1960, plus preliminary unpublished data for 1960 from DLS. The average published by the Census were adjusted downward to reflect zero hours of work for those "with a job but not at work."

23

Trends in Poverty in the United States

Rebecca M. Blank

We shall soon, with the help of God, be in sight of the day when poverty will be banished in the nation.

Herbert Hoover, 1928

The trend in poverty rates in the USA contains both good news and bad news. The good news, in a longer historical perspective, is that high levels of economic growth and income in this country have made poverty a relatively infrequent occurrence. Unlike any past age, poverty afflicts only a fraction of the nation's population, and even among those who are poor, few suffer the imminent threat of malnutrition and disease that still hovers over many impoverished persons in the world. Although Herbert Hoover's forecast was tragically wrong for the 1930s, in the years that followed that decade, poverty – while not banished – was certainly much reduced.

The bad news, however, cannot be ignored. Even in the midst of this country's recent and historically astounding affluence, a substantial number of Americans remain poor. The official poverty rate in 1990 – 13.5 percent of the US population – was almost identical to the rate two decades earlier. As we shall see, however, this aggregate rate hides some substantial changes among demographic groups. Over the past two decades, the elderly have seen major declines in poverty, while children's poverty rates have risen markedly; poverty rates among single-parent households and African-American households have declined only slightly.

HISTORICAL EVIDENCE ON POVERTY IN THE UNITED STATES

Attempts by social scientists in the USA to study poverty date back at least a century, starting perhaps with Jacob Riis's classic 1890 book, *How the Other Half Lives*. Robert Hunter attempted to estimate aggregate

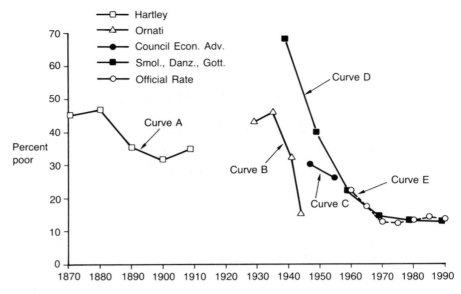

Figure 23.1   Historical trends in USA poverty rates
Sources: A, B, and C: Jeffrey G. Williamson and Peter H. Lindert, *American Inequality*, New
York: Academic Press, 1980, table 5.20. D: Eugene Smolensky, Sheldon Danziger and Peter
Gottschalk, "The Declining Significance of Age in the United States: Trends in the Well-Being
of Children and the Elderly Since 1939," table 3.1, in *The Vulnerable*, eds J. L. Palmer,
T. Smeeding, and B. Boyle Torrey, Washington, D.C.: Urban Institute Press, 1988.

poverty in the USA in 1904 based on a poverty line of $460 per year for
a family of five (slightly over $7,200 in 1989 dollars), concluding that
about 12 percent of Americans were poor. John Ryan, a turn-of-the-century reformer, argued that a working family with three children needed
between $600 and $800 per year (depending on where they lived) for a
"living wage." Others have estimated that close to 40 percent of wage
earners would have been poor using a $700 per year (slightly under
$11,000 in 1989 dollars) poverty line at that time. Adjusted for inflation,
both of these estimates are below the official US poverty line in 1989 of
$14,990 for a family of five. In contrast, a Brookings Institution report from
the early 1930s used a family need standard of $2,000 (close to $16,000 in
1989 dollars) as "sufficient to supply only basic necessities," estimating
that about 60 percent of Americans fell below this line in 1930.[1]

Figure 23.1 shows several recent attempts to estimate historical poverty rates. The estimates plotted in figure 23.1 are based on somewhat
different poverty line estimates and therefore are not directly comparable. As figure 23.1 indicates, Hartley (curve A) estimates that
poverty fell slightly during the late nineteenth century, using a poverty
line based on 1908 budget data. Ornati (curve B) shows a rise in poverty
during the depression, followed by a very fast decline from 1935 to 1944,
also using an estimate of a "minimally adequate" budget. Smolensky,

Danziger, and Gottschalk (curve D) also show a rapid decline in poverty between 1939 and 1959, but use a higher poverty line and therefore estimate much higher poverty rates. Their estimates, and those of the Council of Economic Advisors (curve C), are derived by backcasting official poverty lines from the 1960s using historical price data. Figure 23.1 also shows the official US poverty rate (curve E), available for more recent years.

POVERTY TRENDS OVER THE PAST THREE DECADES

In the mid-1960s the USA became one of the first countries to define an official poverty line. This definition has been used to calculate official US statistics on poverty from 1959 onward. A person is counted as poor if total cash income in his or her household is below the official US poverty line. This line is based primarily on the cost of a minimally adequate diet, multiplied up to reflect additional income needs. There is not a single poverty line; there are multiple lines that vary with household size, number of children, and whether the head of household is elderly or not. Thus, the poverty line in 1990 was $10,419 for a single parent with two children, $13,359 for a four-person family, and $6,268 for an elderly person.

The definition of poverty has not been changed since it was officially introduced in the mid-1960s, except for inflation adjustments. Because consumption patterns and expected standards of living have changed over recent decades, many argue that this definition needs to be re-thought and recalculated. Many alternative poverty definitions have been proposed, but there is little consensus on which alternative definition would be best.[2]

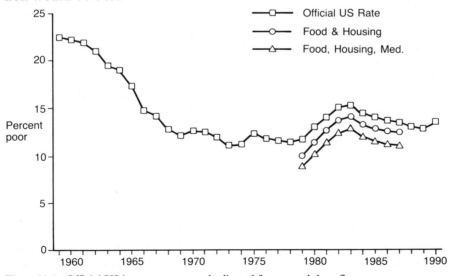

Figure 23.2 Official USA poverty rate, and adjusted for noncash benefits

Figure 23.2 plots official US poverty data since 1959. The poverty rate fell steadily through the 1960s, reaching a low of 11.1 percent in 1973, but rose again to a peak of 15.2 percent during the recession of the early 1980s. By 1990, the poverty rate stood at 13.5 percent, with 33.6 million Americans below the poverty line.

In 1990 the official poverty rate was at a slightly higher level than 20 years earlier. This lack of progress against poverty may seem particularly inexplicable because government transfer programs designed to help the poor expanded enormously over this time period. Federal spending on health, income security, and other social service programs more than tripled between 1969 and 1990, to almost $620 billion. In order to understand why poverty rates remained unchanged while government assistance increased, it is necessary to look more closely at the aggregate government poverty numbers.

First, a growing share of these government assistance dollars have been spent on what are called "in-kind" programs, that is, programs that provide specific goods or services rather than cash assistance. This includes housing subsidies, food stamps – which provide vouchers that can only be used to purchase food – and Medicare and Medicaid, which provide health coverage for Social Security recipients and certain low-income households, respectively. The effect of these programs does not show up in the poverty statistics, which are based only on a household's cash income.

Figure 23.2 includes recent estimates by the Bureau of the Census of what the poverty rate would have been in recent years if in-kind transfers were included. The top line in figure 23.2 indicates that in 1987, when the official poverty rate was 13.4 percent, poverty would have been 12.4 percent if food stamps and housing subsidies were included; the bottom line indicates poverty in 1987 would have been 11.0 percent if the health insurance provided by Medicare and Medicaid were also counted. Thus, progress against poverty has been somewhat greater than indicated by statistics based on cash income, because these statistics ignore noncash government assistance.

Second, the vast majority of increased government transfer dollars over the past two decades have gone to the elderly. Social Security and Medicare, which go almost entirely to the elderly, comprise almost 60 percent of government transfers. Since the mid-1970s, almost all the growth in government transfers has been in programs primarily benefitting the elderly. This means that the elderly have seen an enormous expansion in government assistance. Although the elderly have historically had higher poverty rates than the general population, figure 23.3 indicates that poverty among the elderly has fallen steadily throughout the past three decades, to historical lows. An elderly individual is now less likely to be poor than a nonelderly individual. Most of this decline in elderly poverty can be directly attributed to expanding government transfers.[3]

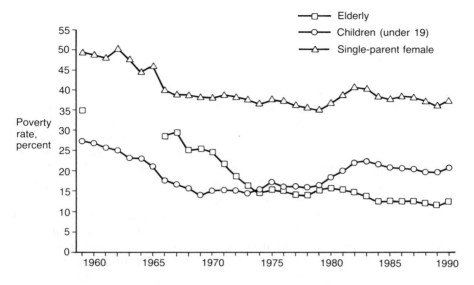

Figure 23.3 Poverty rates for elderly, children, and single female-headed households

Third, the expansion of government transfers over the past two decades has been offset by a slowdown in economic growth. The rapid declines in poverty during the 1950s and 1960s are largely attributable to strong economic growth over those decades. Beginning in the late 1960s, however, the economic situation worsened. Unemployment rates trended upward from the late 1960s through the early 1980s. Inflation was particularly high throughout the latter half of the 1970s. The early 1980s saw the longest and deepest recession since the 1930s. The expanding transfers to the poor in the 1970s basically offset the worsening economic opportunities. If transfers had not expanded, the worsening economy would have led to a substantial increase in poverty.[4]

Since the early 1980s there has been at least a short-term return to higher economic growth and lower unemployment rates, but poverty has not declined as much as in earlier periods of growth. This seems to be primarily due to declining real wages for unskilled workers. Although labor market involvement among low-income workers grew strongly over the expansion of the 1980s, this was offset by declining real wages.[5] The causes behind the widening wage distribution in the USA are still only partially understood, but they include changes in technology, changes in international markets that have produced changes in the skill mix of labor demand, and changes in the relative supply of more- and less-skilled workers relative to the rapidly growing demand for more-skilled workers by employers. The result for the poor has been that expanded employment opportunities brought about by the sustained economic expansion of the past decade have not translated into substantially higher family incomes. People worked more, but their work effort was repaid at a lower rate.

As poverty among the elderly has continued to decline, poverty among children has remained high. The middle line in figure 23.3 shows poverty trends among children. Currently over one child in five lives in a household below the poverty line, substantially above the poverty rate of about 15 percent among children two decades ago.

These high poverty rates among children are highly correlated with the increase in single-parent female-headed households. The proportion of the poor who live in single-parent female-headed households has doubled since 1959, from 18 percent to 37 percent. The top line in figure 23.3 shows the trend in poverty rates among these households. Although their poverty rates fell somewhat over the 1960s, throughout the 1970s and 1980s they remained between 35 and 40 percent. This contrasts with a poverty rate between five and eight percent among married-couple households over this same time period. Among black and Hispanic female-headed households, the poverty rate is over 50 percent. Children living in these families suffer disproportionately high poverty rates. Among children in female-headed households, over 50 percent are poor. Among children in black or Hispanic female-headed households, over two-thirds are poor.

Large differentials in poverty continue across racial and ethnic groups in the United States for all households. Figure 23.4 shows the trends in white, black, and Hispanic poverty rates. Poverty rates among African-Americans have not fallen as fast as among whites. Hispanic poverty rates have risen over the past decade. This means that a disproportionate share of poverty is borne by black and Hispanic families. Nonetheless,

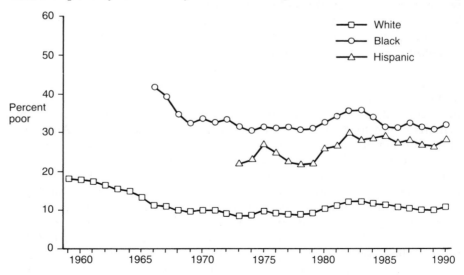

Figure 23.4 Official USA poverty rates, 1959–90, white, black, and Hispanic races

because so many more Americans are white, the majority of poor individuals – 66 percent in 1990 – are white.

In summary, the last three decades in the USA have brought generally lower poverty rates, but much of this progress occurred in the 1960s and not in more recent decades. Poverty rates among the elderly have fallen throughout this time period, primarily because the expanding government transfer system has been heavily targeted on the elderly. Although poverty rates among African-Americans and Hispanics, as well as among children and female-headed households, are lower in 1990 than they were in 1959, they have shown little downward trend since the mid-1970s.

OTHER MEASURES OF ECONOMIC NEED

The discussion has relied so far entirely on an income-based definition of poverty. As other chapters in this volume indicate, however, there are a wide variety of alternative measures of need that might also be considered. Adequate nutrition, housing, and medical care might be as important as income, per se. Thus, one may want to use measures of hunger, malnutrition, homelessness, housing standards, health, or household consumption to determine economic need.

While a detailed discussion of these issues is left to other papers in this volume, it is worth noting that the trends in income poverty in the USA over the past two decades give a more pessimistic picture of economic need than do most alternative measures. Housing conditions, food consumption, access to health services, and general health status have improved among poor individuals in this country over the past three decades, by virtually all measures.[6] Poverty indexes based on consumption rather than on income generally show a greater decline in need. This indicates the importance of looking at a broad range of information on economic and personal well-being, beyond income statistics.

There are, however, at least two frequently cited exceptions to the more optimistic information contained in nonincome statistics on household well-being that should be noted. First, the rise in homelessness since 1980, particularly in urban areas, indicates that for some groups among the poor, housing conditions may have significantly declined. Second, deterioration in certain urban neighborhoods, including increased crime and drug activity as well as declining labor market activity and declining school quality, may indicate that residents of some poor urban neighborhoods face higher barriers to escape poverty than they did in earlier time periods. This has evolved into a current debate on the potential implications of an urban "underclass" in concentrated areas of urban poverty. It is difficult to draw firm conclusions on either of these issues; data are scarce and there is substantial disagreement among researchers on the interpretation of currently available information.

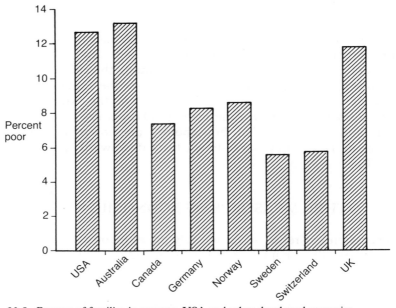

Figure 23.5   Percent of families in poverty, USA and other developed countries
Source: Timothy Smeeding, Barbara Boyle Torrey, and Martin Rein, "Patterns of Income and
Poverty: The Economic Status of Children and the Elderly in Eight Countries," table 5.2 in *The
Vulnerable*, eds J. L. Palmer, T. Smeeding, and B. Boyle Torrey, Washington, D.C.: The Urban
Institute Press, 1988.

HOW DOES THE USA COMPARE
WITH OTHER COUNTRIES?

There is little historical data available that provide comparable figures
on poverty rates in the USA and in other countries. Only recently have
researchers been able to define cross-country poverty rates, using similar
data and similar poverty definitions. Most comparisons that do exist
indicate that the USA does not do as well as many developed countries
in fighting poverty.

A recent study comparing the USA and Canada indicates that between 1969 and 1986, when poverty in the USA was largely stagnant,
poverty in Canada declined by 60 percent.[7] The result was that Canadian poverty, which had been substantially higher than in the USA in the
1970s, was substantially lower by the late 1980s. The research shows
that relative declines in poverty in Canada over the 1970s were the result
of a more robust macroeconomy, but the relative declines in the 1980s
resulted from an expansion in Canadian government transfers to poor
individuals, while the USA antipoverty transfer system (for the nonelderly) contracted. This seems to be evidence that the stagnation in US
poverty rates is not inevitable. Canada, a country with a very similar population and economic environment, has been able to make
continued progress against poverty.

Using the US definition of poverty, figure 23.5 indicates how US family poverty rates compare with similar rates from a selected group of other countries in the early 1980s. While cross-country calculations are never entirely comparable, these numbers are based on data explicitly constructed to provide equivalent cross-country income measures; poverty rates are calculated using identical poverty lines, adjusted for differences in purchasing power parity across countries. Australia shows a slightly higher poverty rate than USA, the United Kingdom is slightly lower, and Canada, Sweden, Germany, Norway, and Switzerland all have substantially lower poverty rates.

CONCLUSION

The news on poverty in the USA is clearly mixed. On the one hand, the nation has changed from a society in which 40 to 60 percent of the population struggled daily with economic survival, to a society in which only a fraction can be counted as poor. The elderly are the major success story of recent years. Historically, the elderly have always been poorer than other groups in society, but this is no longer true, largely because of an extensive government assistance network aimed at elderly households.

On the other hand, overall poverty rates have not declined in the past two decades. Changes in the demographic composition of the poor, particularly the growth in single-parent households, slower economic growth over the past two decades, and a contraction in the antipoverty transfer system for the nonelderly in the past decade have meant little overall progress against poverty. Poverty rates, particularly among certain groups, remain very high in this country, as comparisons with other developed countries indicate. The official poverty statistics, however, should not be used as the sole indicator of economic need or household well-being. Greater progress appears to have been made against poverty, when measured against consumption-based standards of food, housing, and medical care, than is shown by income measures alone.

NOTES

1   For a more extensive discussion of these early studies, see James T. Patterson, *America's Struggle Against Poverty, 1900–1980*. Cambridge: Harvard University Press, 1981. 1989 dollar values are calculated using the GNP deflator for personal consumption expenditures.

2   For a complete description of the US poverty line, and a discussion of its shortfalls, see Patricia Ruggles, *Drawing the Line*. Washington, DC: Urban Institute Press, 1990.

3   For evidence, see Daniel H. Weinberg, "Filling the 'Poverty Gap': Multiple Transfer Program Participation." *Journal of Human Resources*, 20, 1 (1985), 64–89.

4   See Rebecca Blank and Alan Blinder, "Macroeconomics, Income Distribution and Poverty," in *Fighting Poverty*, eds S. Danziger and D. Weinberg. Cambridge: Harvard University Press, 1986.

5   Rebecca M. Blank, "Why Were Poverty Rates So High in the 1980s?" in *Poverty and Prosperity in the USA in the Late Twentieth Century*, eds O. Papadimitriou and E. Wolff. London: Macmillan, 1993.

6   See Alan S. Blinder, "The Level and Distribution of Economic Well-Being," in *The American Economy in Transition*, ed. M. Feldstein, Chicago: University of Chicago Press, 1980; and Susan E. Meyer and Christopher Jencks, "Poverty and the Distribution of Material Hardship." *Journal of Human Resources*, 24, 1 (1989), 88–114.

7   Maria J. Hanratty and Rebecca M. Blank, "Down and Out in North America: Recent Trends in Poverty in the U.S. and Canada," *Quarterly Journal of Economics*, 107 (1), 233–54.

EDITOR'S NOTE (JULIAN L. SIMON)

In the years around 1690 in England, "nearly half the population [was] dependent upon alms or poor relief" (Colin Clark, *National Income and Outlay* [London: Macmillan, 1938], p. 217).

24

How "Poor" are America's Poor?

Robert Rector

In 1989, 31.5 million persons in the United States lived in poverty, according to the US Bureau of the Census. In 1989 the poverty income threshold for a family of four was $12,675.[1]

However, the official poverty report dramatically understates financial resources available to low-income Americans. Although some official data appear to show widespread poverty in the United States with little improvement over 25 years, unpublicized data reveal a different picture. "Poor" Americans today are better housed and fed and own more property than did average Americans throughout much of this century. In 1988, per capita expenditures, adjusted for inflation, by the lowest income fifth of the American population exceeded per capita income of the median American family in 1955.[2] Many Americans officially classed as poor would be considered as living in comfortable conditions by most Americans – and more than comfortable by most foreigners.

HOUSING CONDITIONS OF AMERICA'S "POOR"

The actual housing conditions of poor Americans contradict the general public image of poverty. According to the 1987 US Census *American Housing Survey*, 38 percent of US households officially defined as "poor" own their own homes. The median value of homes owned by poor persons in 1987 was $39,205, or roughly 60 percent of the median value of all owner-occupied housing.[3] Nearly 500,000 households defined as "poor" by the US Census owned homes valued over $100,000 in 1987; 1.25 million owned homes valued at over $60,000.[4]

As figures 24.1–24.4 show, the housing of these families, whether owned or rented, is spacious and includes excellent amenities by historic or international standards. By American standards, "crowded" housing means more than 1.5 persons occupy each room. Less than 2 percent of "poor" US households were "crowded" in 1987, according to this definition, and only 7.5 percent had more than one person per room. Officially poor US households, with 0.56 persons per room, are less

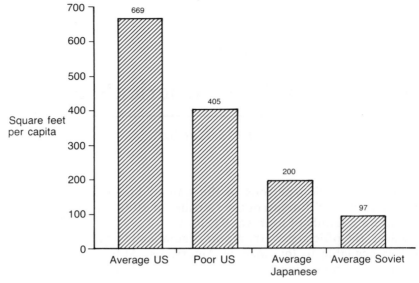

Figure 24.1   Housing space, square feet per capita (1987)
Sources: US Dept of Energy, Energy Information Administration, Housing Characteristics 1987 (Washington: Government Printing Office, 1989), 25. A. S. Zaychenko, "United States–USSR: Individual Consumption (Some Comparisons)," *World Affairs*, Summer 1989, 10. "The Affluent Japanese Household," *Business America*, March 23, 1981, 10.

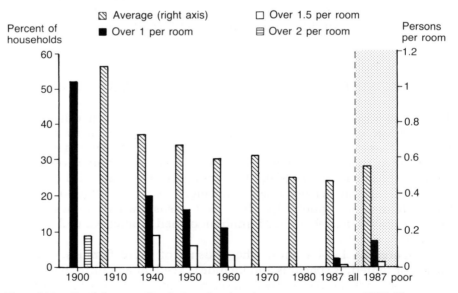

Figure 24.2a   Declining crowding in American housing, persons per room, 1900–87
Sources: Data for 1900–60: S. Lebergott, *The American Economy: Income, Wealth, and Want* (Princeton: Princeton University Press, 1976). Data for 1987: US Dept of Commerce and US Dept of Housing and Urban Development, *American Housing Survey for the United States in 1987*, Current Housing Reports, H-150-87.

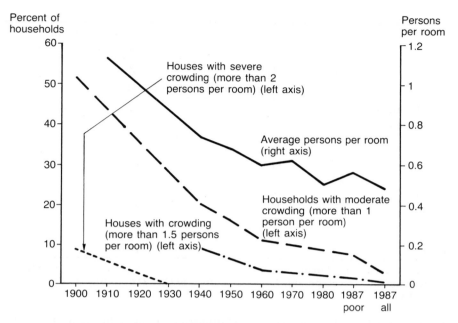

Figure 24.2b Declining crowding in American housing, persons per room, 1900–87
Sources: Data for 1900–60: S. Lebergott, *The American Economy: Income, Wealth, and Want* (Princeton: Princeton University Press, 1976). Data for 1987: US Dept of Commerce and US Dept of Housing and Urban Development, *American Housing Survey for the United States in 1987*, Current Housing Reports, H-150-87.

crowded today than average American households were in 1970 and average West European households in 1980.[6] By contrast, the average Japanese household has 0.8, the average Uruguayan 2.1, and the average Indian 2.8 persons per room.[7] (Note: all international comparisons in this chapter compare "poor" Americans to *average* persons in other nations, not to less affluent individuals in those nations.)

Figure 24.1 shows that poor Americans live in housing with an average twice as much living space per capita as average Japanese and four times as much as average Soviets.[8]

Nearly all officially poor US households, moreover, are equipped with basic modern plumbing, including hot and cold running water, indoor flush toilets, and indoor baths. While 30 percent of all Americans were without indoor toilets in 1950, less than 2 percent of poor Americans lacked them by 1987.[9] As table 24.1 shows, America's poor are less likely to lack indoor plumbing than the general population in Western Europe. The *average* Japanese is 22 times more likely to lack an indoor toilet than the *poor* American.[10]

The houses and apartments of America's "poor" are in far better condition than generally assumed. The median age of such housing units is only seven years greater than the median age of the overall US

Table 24.1 Households without modern amenities

	Percent lacking indoor flush toilet	Percent lacking fixed shower or bath
United States poor households	1.8	2.7
Other nations: all households		
United Kingdom	6	4
West Germany	7	11
Italy	11	11
Spain	12	39
France	17	17
Norway	17	18
Belgium	19	24
Ireland	22	26
Greece	29	–
Portugal	43	–
Japan	54	17

Source: European and Japanese data from 1980 census: Organization of Economic Cooperation and Development, *Living Conditions in OECD Countries* (Paris: OECD, 1986), 139.

housing stock.[11] The overwhelming majority of these housing units are in sound condition. According to the 1987 *American Housing Survey* of the US Census, only 2.4 percent of housing units owned or rented by

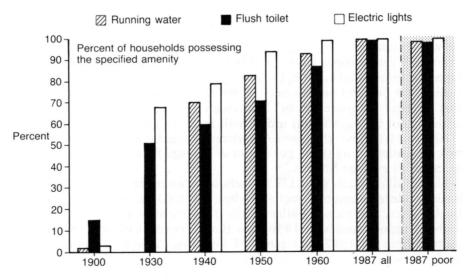

Figure 24.3a Amenities in American housing, 1900–87
Sources: Data for 1900–60: S. Lebergott, *The American Economy: Income, Wealth, and Want* (Princeton: Princeton University Press, 1976). Data for 1987: US Dept of Commerce and US Dept of Housing and Urban Development, *American Housing Survey for the United States in 1987*, Current Housing Reports, H-150-87.

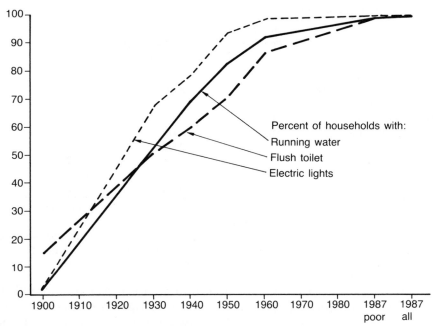

Figure 24.3b Amenities in American housing, 1900–87

Sources: Data for 1900–60: S. Lebergott, *The American Economy: Income, Wealth, and Want* (Princeton: Princeton University Press, 1976). Data for 1987: US Dept of Commerce and US Dept of Housing and Urban Development, *American Housing Survey for the United States in 1987*, Current Housing Reports, H-150-87.

households deemed poor had significant structural defects such as crumbling foundations or missing roof material.[12] Some 9 percent of the poor (roughly double the rate for the general population) reported being uncomfortably cold at least once during the previous winter due to inadequate insulation, inadequate heating capacity, or equipment failure.[13]

OWNERSHIP OF CONSUMER DURABLES AMONG POOR AMERICAN HOUSEHOLDS

Government data on the personal possessions of officially poor households show a steady improvement in the living standards of low-income Americans. These data starkly contradict the general public understanding of what it means to be poor (see table 24.2). For example, 62 percent of "poor" households own a car, truck, or van, and 14 percent own two or more cars,[14] nearly a third have microwave ovens.[15] According to government figures, 29 percent of all "poor" households own two or more color televisions.[16]

Table 24.2 Consumer durables owned by "poor" American households:[a] 1987

	All "poor" households	"Poor" owner-occupied households	"Poor" renter-occupied households
One or more automobiles[b]	62.2	77.9	52.4
Two or more automobiles	13.6	21.0	9.0
Air conditioning	49.0	55.8	44.7
Microwave oven	30.7	–	–
Washing machine	56.0	84.6	38.1
Dishwasher	17.0	23.2	13.2
Garbage disposal	18.9	15.1	21.3
Refrigerator	99.1	99.5	98.9
Telephone	81.3	91.4	75.0

[a] Figures represent percent of "poor" households that own the specified item.
[b] "Automobiles" includes personal trucks and vans.
Source: *American Housing Survey for the United States in 1987*, 40, 46, 90, 96, 108, 154. Energy Information Administration, US Department of Energy, *Housing Characteristics 1987* (Washington: US Government Printing Office, 1989), 87.

As figure 24.4 shows, officially poor households today are more likely to own common consumer durables such as refrigerators than the average family in the 1950s. In 1930, nearly two thirds of US households did

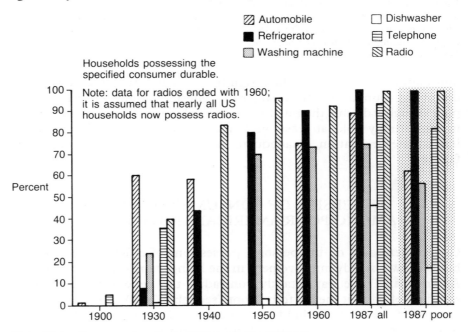

Figure 24.4a Consumer durables in US households, 1900–87
Sources: *The American Economy: Income, Wealth, and Want*, 281, 286–8, 355. *American Housing Survey for the United States in 1987*, 40, 46.

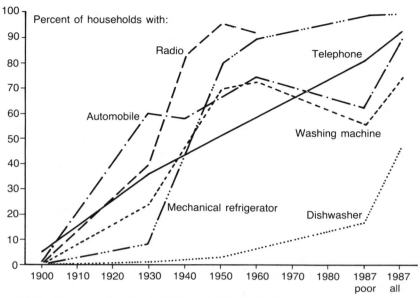

Figure 24.4b Consumer durables in US households, 1900–87
Sources: The American Economy: Income, Wealth, and Want, 281, 286–8, 290, 355. *American Housing Survey for the United States in 1987,* 40, 46.

not own a radio; over half had no form of refrigeration. Among the poor today fewer than 1 percent lack a refrigerator.[17]

Nearly 50 percent of officially poor households have air conditioning.[18] Seventeen percent of US households in poverty have automatic dishwashers, well above the rate for the general West European population in 1980.[19] Over 95 percent of all poor Americans own one or more televisions; a poor US household is nearly 50 percent more likely to own a television than is the average family in Ireland.[20] A poor American household is much more likely to own a *color* television set than is the average household in France, West Germany, or Italy.[21]

Among America's "poor" there are 344 cars per 1,000 persons.[22] This roughly equals the ratio for the total population of the United Kingdom. A poor American is 40 or 50 percent more likely to own a car than the average Japanese, twice as likely as the average Saudi Arabian, eight times as likely as the average Russian, 18 times as likely as the average Turk, and 215 times as likely as the average Indian.[23]

FOOD CONSUMPTION OF LOW-INCOME AMERICANS

Examination of actual food consumption by less affluent Americans also belies charges of widespread poverty (see table 24.3). On a per capita basis, the lowest income fifth of households in 1988 spent an average of

Table 24.3 Food consumption: low-income persons compared to upper-middle income persons (percent)

All meats	95
Steak	70
Poultry	109
Fish	114
Fresh vegetables	92
Fresh fruits	71

Note: Figures represent average per capita consumption of persons in the lowest income 20 percent of households compared to average per capita consumption in the most affluent 50 percent of households. Consumption is measured in pounds per week.
Source: Human Nutrition Information Service, US Department of Agriculture, *Food Consumption: Households in the United States, Spring 1977* (Washington: US Government Printing Office, 1982).

80 percent as much on food as the median American household,[24] and 32 percent of that was spent in restaurants.[25] This is scarcely a pattern to be expected from individuals suffering chronic hunger.

Surveys conducted by the Department of Agriculture show relatively little difference in overall food consumption between high- and low-income households. Though the food purchased by low-income households normally is of lower quality and less expensive than that consumed by the upper middle class, there is little evidence of material shortages. For instance, the average low-income person eats 95 percent as much meat as the average person in the upper middle class. Measured in pounds of food consumed per week, low-income persons actually consume 14 percent more poultry, 9 percent more fish, and only 8 percent less fresh vegetables than upper middle class persons.

Table 24.4, derived from studies conducted by the Human Nutrition Information Service of the US Department of Agriculture, shows the average nutritional status of persons from three income groups:

1   "Low-income" persons, from the least affluent 20 percent of the population.
2   Persons receiving food stamps, and those whose income is low enough to be eligible for food stamps but who did not actually receive them.
3   Persons from the most affluent half of the population.

Table 24.4 compares average food consumption in each of the three income groups to USDA recommended nutritional standards.[26] For all three groups food consumption exceeds the standards in almost every nutritional category. Differences between the middle class and poor in almost all cases are quite modest.

Historic Diet Trends

The diets of low-income Americans have shown considerable improvement over time. Consumption of vegetables among low-income persons

Table 24.4 Average nutrients consumed as percentage of 1980 recommended dietary standards

	Persons eligible for food stamps	*Low-income persons*	*Upper middle income persons*
Protein	169	156	168
Calcium	89	79	89
Iron	99	100	101
Magnesium	87	78	86
Phosphorus	131	126	139
Vitamin A	124	144	129
Thiamin	119	113	111
Riboflavin	135	130	133
Niacin	118	120	125
Vitamin B$_6$	75	71	77
Vitamin B$_{12}$	142	178	171
Vitamin C	137	134	156

Source: Human Nutrition Information Service, US Department of Agriculture, *Nutrient Intakes: Individuals in 48 States, Year 1977–78*, Nationwide Food Consumption Survey 1977–78, Report No. I–2 (Washington: US Government Printing Office, May 1984), 254, 251. Human Nutrition Information Service, US Department of Agriculture, *Food and Nutrient Intakes of Individuals in 1 Day, Low-Income Households, November 1979–March 1980*, Nationwide Food Consumption Survey 1977–78, Preliminary Report No. 13 (Washington: US Department of Agriculture, 1982), 126.

today exceeds average consumption among the general population in the mid-1950s.[27] Low-income Americans today consume about 20 percent more meat, poultry, and fish than did the average citizen in 1955[28] and nearly 70 percent more than the average urban dweller in 1948.[29] Beef consumption, often considered a symbol of diets of the prosperous middle class, has shown a similar increase. Low-income households consume 35 percent more beef than average households consumed in 1955,[30] and the lowest 20 percent of households in income consume nearly a third more beef than the top 5 percent of urban households in 1948.[31]

Consumption of frivolous items like soft drinks is 100 percent greater among low-income persons today than among the general population in the mid-1950s.[32] By contrast, cheap, starchy goods like potatoes, bread, and flour are consumed by low-income individuals today at rates 25 to 50 percent below that of the general population in the 1950s.[33]

International Comparisons

Rich and poor Americans alike typically eat rich diets compared with the rest of the world. The item most associated with an expensive diet is meat; as income increases, meat consumption increases sharply.

Table 24.5 Meat consumption by average citizens in various nations as percentage of consumption by low-income Americans

Low-income persons	
United States	100
Average persons	
West Germany	75
France	70
Italy	62
Great Britain	57
Soviet Union	56
Romania	40
Mexico	39
Japan	39
Venezuela	31
Brazil	27

Source: Food Consumption: Households in the United States, Spring 1977, and unpublished data provided by the US Department of Agriculture.

Table 24.5 shows that the lowest income fifth of American households consume more meat per person than *average* citizens in various other countries.[34]

Food Consumption of Poor Children

Many critics of US society have expressed concern about malnutrition caused by poverty among American children. In 1985 the Department of Agriculture conducted a thorough study of the food consumption and nutritional status of preschool children and young mothers. This study showed very little difference in the nutritional content of food consumed by low-income as compared to affluent Americans. Children from families with income below 75 percent of the poverty level consumed 54.4 grams of protein per day compared to 53.6 grams for children in families with incomes above 300 percent of poverty (roughly $33,000 for a family of four in 1985).[35] Black preschool children consumed 56.9 grams of protein compared to 52.4 grams for white children.[36] Surprisingly, protein and calorie consumption were slightly higher among children in the central cities than in the suburbs.[37]

Average consumption of nutrients was very high for preschool children of all income classes. Average protein consumption among children living in families with incomes below 75 percent of the poverty level equalled 211 percent of recommended USDA standards.[38] Consumption of essential vitamins and minerals among both high- and low-income children generally exceeded USDA standards, often by as much as 50 to 100 percent. Shortfalls in the average consumption of iron and zinc were unrelated to income class or race.[39]

Poverty and Malnutrition.

Malnutrition and hunger caused by poverty are virtually nonexistent in the United States. The most expensive factors in a diet are the overall levels of protein and calorie intake. But, in its extensive surveys, the US government has found no evidence of significant caloric or protein deficiencies among the poor.[40] Indeed, being overweight is the number one dietary problem of both rich and poor Americans.[41]

Nor is it true that poor Americans are forced by economic necessity to eat low-quality food that is damaging to their health. In fact, poor persons have lower levels of serum cholesterol than nonpoor persons of the same age, sex, and race.[42] Moderate deficiencies of certain vitamins and minerals such as vitamin B_6 and zinc occurring in part of the US population are unrelated to income class.[43] Moderate calcium and iron deficiencies do occur more frequently among poor women than nonpoor women, but they normally result from the type of food consumed rather than the amount of money spent on food; simply raising the income of poor women would have little bearing on the problem. A more effective response would be to distribute inexpensive vitamin and mineral supplements to adult women receiving assistance through the Women, Infants, and Children (WIC) or Food Stamp programs.

WHAT IS WRONG WITH THE OFFICIAL POVERTY MEASUREMENT?

It is difficult to reconcile these data on the actual living standards of less affluent Americans with official government reports showing widespread and unremitting poverty. In part, this difficulty is a product of the very high levels of US poverty income thresholds. When initially established in the early 1960s, the US poverty income threshold of $3,165 for a family of four ($12,675 in 1989 dollars) was nearly 30 times greater than the world median per capita income.[44] In historical terms, the comparison is equally striking: an American family of four may be currently defined as poor while having an income, measured in constant dollars, more than twice the annual earnings of the average American worker at the beginning of this century.[45]

Official government figures also significantly undercount financial resources available to lower income Americans. Three factors contribute to this undercount.

1. Census officials acknowledge that lower income Americans do not distinguish between the Census Bureau and other government agencies. Therefore they are unlikely to reveal to the Census sources of income that they have concealed from the IRS or welfare agencies. The Census Bureau's survey of consumer expenditures, in fact, shows that low-income households spend $1.94 for every $1.00 in income reported to the Census. If poverty estimates were based on the

actual expenditures of poor households rather than reported income, the number of officially poor Americans would be far lower.[46]

2. The Census Bureau ignores almost the entire US welfare system when calculating the incomes of poor Americans. Official poverty reports include only a portion of current cash welfare assistance. Noncash welfare is excluded, despite the fact that the overwhelming majority of welfare spending in the USA is in the form of noncash benefits such as Food Stamps and Medicaid. Entire multibillion dollar welfare programs such as public housing and school lunch subsidies are excluded in the normal Census Bureau estimates of the living standards of low-income Americans.[47] Total means-tested welfare spending in the USA came to $155.6 billion in 1988. Of this the Census poverty report counted only 17.3 percent as income received by US households.[48] The welfare spending ignored in the official poverty report comes to $128.7 billion, or $9,058 for each "poor" household.

3. The official poverty measure includes only reported income and ignores all household assets. Thus, elderly families living off savings or middle class families that have undergone a temporary dip in income will be counted as "poor" while their actual living conditions may be quite affluent.

For these reasons, the official US poverty data do not provide useful information about living standards or economic stratification. But the political utility of the official poverty reports remains enormous. Consequently, we can expect a continuing flow of official reports revealing large-scale and unchanging poverty for the foreseeable future, regardless of actual changes in living standards.[49]

CONCLUSION

The official US measurement of poverty is a Potemkin village in reverse. Based on inaccurate estimates of the cash incomes of US households, it gives a false impression of living conditions in the USA. A more accurate picture can be obtained by examining the food consumption, property ownership, and housing conditions of individuals defined as "poor" by the government.

These measures show that the real living standards of "poor" Americans differ widely from the normal image of poverty. The living standards of poor Americans show continuing improvement and are quite high by any historic standard; many if not most of today's poor Americans have a material standard of living that would have been considered middle class or better by Americans earlier in the century. In many respects they compare favorably to the current general population of Western Europe and would be considered affluent in most parts of the globe.

NOTES

1   The original poverty income thresholds were developed in 1963 by the US Department of Health Education and Welfare. Annual poverty counts had

quasi-official status until 1969 when the Budget Bureau designated them as an official US government measure to be published regularly by the Census Bureau.

2 US Department of Labor, Bureau of Labor Statistics, "Consumer Expenditures in 1988," *USDL Press Release Number 90–96*, February 26, 1990, table 1. US Department of Commerce, Bureau of the Census, *Historical Statistics of the United States, Part I* (Washington, DC: US Bureau of the Census, 1975), 297 and 301.

3 Data are from the US Department of Commerce and the US Department of Housing and Urban Development, *American Housing Survey for the United States in 1987*, Current Housing Reports H-150-87 (Washington, DC: US Government Printing Office, 1989), 34, 84, 114, and 304. Fifty-eight percent of "poor" owner-occupied households are nonelderly. The average value of homes owned by both poor and nonpoor households was established by self-reporting of the owners and therefore, on average, is probably an underestimate, particularly in the case of older dwellings.

4 Ibid., 114.

5 Ibid., 38.

6 Organization for Economic Cooperation and Development, *Living Conditions in OECD Countries* (Paris: OECD, 1986), 133. US average was computed from *American Housing Survey* data, 1987.

7 Department of International Economic and Social Affairs Statistical Office, United Nations, *Compendium of Human Settlement Statistics: 1983* (New York: United Nations, 1985), 251–61.

8 US Department of Energy, Energy Information Administration, *Housing Characteristics 1987* (Washington, DC: Government Printing Office, 1989), 25. A. S. Zaychenko, "United States–USSR: Individual Consumption (Some Comparisons)," *World Affairs* (Summer 1989), 10. "The Affluent Japanese Household," *Business America*, March 23, 1981, p. 10.

9 Stanley Lebergott, *The American Economy: Income, Wealth and Want* (Princeton, NJ: Princeton University Press, 1976), 272. *American Housing Survey for the United States in 1987*, op. cit., 38, 44.

10 *Living Conditions in OECD Countries*, op. cit., 139. *American Housing Survey for the United States in 1987*, op. cit., 38, 44.

11 *American Housing Survey for the United States in 1987*, 34.

12 Ibid., 36.

13 Ibid., 44.

14 *American Housing Survey for the United States in 1987*, 46.

15 US Department of Energy, Energy Information Administration, *Housing Characteristics 1987* (Washington, DC: Government Printing Office, 1989), 87.

16 US Department of Energy, Energy Information Administration, *Housing Characteristics 1990* (Washington, DC: US Government Printing Office, 1991), 115.

17 Lebergott, op. cit., p. 282. *American Housing Survey for the United States in 1987*, 40.

18 *American Housing Survey for the United States in 1987*, 40.

19 Organization for Economic Cooperation and Development, *Living Conditions in OECD Countries* (Paris: OECD, 1986) 126–7.

20 Over 95 percent of poor US households owned one or more televisions in 1987; in comparison only 65 percent of Irish households owned televisions when surveyed in 1980. Data are from unpublished US Department of Energy figures and *Living Conditions in OECD Countries*, 125.

21 *Housing Characteristics 1987*, 87. US Bureau of the Census, *Statistical Abstract of the United States: 1989* (Washington, DC: US Government Printing Office, 1989), table 1417.

22 *American Housing Survey 1987*, 46, 50.

23 Ibid., 46, 50. US Department of Commerce, Bureau of the Census, *Statistical Abstract of the United States: 1989*, table 1418. Comparison based on cars per 1,000 persons.

24 Bureau of Labor Statistics, "Consumer Expenditures in 1988," *USDL Press Release*, Feb. 26, 1990, op. cit., 90–6.

25 Ibid. Throughout this chapter the term "low-income" shall be used in reference to the one-fifth of households with the lowest income in a given year, usually termed the bottom quintile. The term "upper middle class" shall be used to refer to households in the top 50 percent of the income distribution.

26 Human Nutrition Information Service, US Department of Agriculture, *Nutrient Intakes: Individuals in 48 States, Year 1977–78, Nationwide Food Consumption Survey 1977–78*, Report No. I-2 (Washington, DC.: US Department of Agriculture, May 1984), 251, 254. This 1978 study represents the latest survey available on food consumption in the general US population; subsequent studies were undertaken in the late 1980s, but the data are not yet publicly available. Human Nutrition Information Service, US Department of Agriculture, *Food and Nutrient Intakes of Individuals in 1 Day, Low-Income Households, November 1979–March 1980*, Nationwide Food Consumption Survey 1977–78, Preliminary Report No. 13, (Washington, DC: US Department of Agriculture, 1982), 126.

27 The comparison is based on total vegetable consumption excluding potatoes. Agricultural Research Service, US Department of Agriculture, *Food Consumption of Households in the United States*, Household Food Consumption Survey 1955, Report No. 1 (Washington, DC: US Government Printing Office, December 1956), passim. *Food Consumption: Households in the United States, Spring, 1977*, passim.

28 Agricultural Research Service, US Department of Agriculture, *Food Consumption of Households in the United States: Spring 1965*, Household Food Consumption Survey 1965–66, Report No. 1 (Washington, DC: US Government Printing Office, 1968), 3. *Food Consumption: Households in the United States, Spring 1977*, passim.

29 Agricultural Research Administration, US Department of Agriculture, *Nutritive Value of Diets of Urban Families: United States, Spring 1948 and Comparison with Diets in 1942*, 1948 Food Consumption Surveys, Preliminary Report No. 12, November 30, 1949 (Washington, DC: US Department of Agriculture, 1949), 13. The 1948 survey was limited to households of two or more persons in urban areas.

30 *Food Consumption: Households in the United States, Spring 1977*, 27. *Food Consumption of Households in the United States*, Household Food Consumption Survey 1955, 66.

31 *Nutritive Value of Diets of Urban Families: United States, Spring 1948 and Comparison with Diets in 1942*, 13.

32 *Food Consumption: Households in the United States, 1977*, 58. *Food Consumption of Households in the United States*, Household Food Consumption Survey 1955, Report No. 1, 158.

33 Ibid. passim.

34   There are three methods to measure consumption of a given food item. (1) *Food supply data*. These data are based on aggregate economic production and consumption figures (for meat this measure is usually made on a "carcass weight equivalent" basis). (2) *Household food consumption surveys*. These surveys determine the amount of food consumed in a household over a given time. (3) *Individual intake surveys*. These measure the food consumption of a specific household member over time. When comparing per capita meat consumption between nations, it is important that a consistent measurement method be used. In compiling the data presented in table 24.5 a two-stage process was used to provide cross-national comparability. First, average per capita US consumption was compared to average per capita consumption in other nations using internationally compatible data on a carcass weight equivalent basis provided by US Department of Agriculture. Second, the ratio of per capita low income consumption to average per capita US consumption was determined using a separate set of internally consistent household survey data provided in *Food Consumption: Households in the United States, Spring 1977*, 9 and 12. Because low-income per capita consumption is not available on a carcass weight basis, the household survey ratio of low-income US consumption to average US consumption was used to estimate low-income US consumption on a carcass weight basis. Low-income US consumption on a carcass weight basis was then compared to average carcass weight consumption in other nations. Since there is very little difference between average and low-income meat consumption in the USA according to the household survey data, the second stage in this process had little effect on the overall totals presented in table 24.5.

35   Human Nutrition Information Service, US Department of Agriculture, *Low Income Women 19–50 Years and Their Children 1–5 Years, 4 Days: 1985*, CSF II Report 85-5 (Washington, DC: US Department of Agriculture, 1988), 50. Human Nutrition Information Service, US Department of Agriculture, *Women 19–50 Years and Their Children 1–5 Years, 4 Days: 1985*, CSF II Report No. 85-4 (Washington, DC: US Department of Agriculture, 1987), 42.

36   *Women 19–50 Years and Their Children 1–5 Years, 4 Days: 1985*, 42.

37   Ibid.

38   *Low Income Women 19–50 Years and Their Children 1–5 Years, 4 Days: 1985*, 72.

39   *Low Income Women 19–50 Years*, 73. *Women 19–50 Years*, 65.

40   Life Sciences Research Office, Federation of American Societies for Experimental Biology, *Nutrition Monitoring in The United States: An Update on Nutrition Monitoring*, Report prepared for the US Department of Agriculture and the US Department of Health and Human Services (Washington, DC: US Government Printing Office, 1989), 51.

41   Ibid., 73.

42   Ibid., II 72–4.

43   Ibid., II 136–42.

44   This international comparison was made by contrasting US poverty income thresholds for 1959 against global income data from 1958 provided in Simon Kuznets, *Modern Economic Growth: Rate, Structure, and Spread* (New Haven and London: Yale University Press, 1967), 359–90. The comparison uses official currency exchange rates which are sometimes regarded as underestimating relative purchasing power in less developed nations; if the comparison were made instead on the basis of purchasing power parity the differential between

the US poverty income thresholds and world median income might be reduced by 50 percent.

45　See *Historical Statistics of the United States: Colonial Times to 1970*, Part 1 (Bureau of the Census, 1972), series E135 and D722.

46　See Robert Rector, Kate Walsh O'Beirne, and Mike McLaughlin, "How Poor are America's Poor?" *Heritage Foundation Backgrounder #791*, September 21, 1990. This gap between expenditures and income is even more remarkable given that the Census Bureau excludes most noncash welfare assistance from the consumer expenditure totals. While food purchased with food stamps is included as a household expenditure, other food programs, medical assistance, and housing subsidies are not. Thus the undercount of consumer expenditures is largely separate from and in addition to the undercount of welfare benefits discussed in the succeeding paragraph of the text.

47　Ibid. The Census Bureau also provides some experimental poverty measures that include a partial valuation of some noncash welfare benefits, but even with these measures the overwhelming bulk of welfare expenditures are excluded. See Robert Rector and Kate Walsh O'Beirne, "Dispelling the Myth of Income Inequality," *Heritage Foundation Backgrounder #710*, June 6, 1989.

48　This figure excludes means-tested welfare expenditures on persons residing in institutions such as nursing homes because these individuals are excluded both from the Census Bureau poverty count and its annual count of the general population. Total means-tested expenditures by federal, state, and local governments, including expenditures on institutionalized persons, came to $184 billion in 1988. See Rector, O'Beirne, and McLaughlin, "How Poor are America's Poor?"

49　One irony is that while most welfare programs are not counted as income by the Census, these programs do induce low-income Americans to reduce wages – which are counted as income by Census. Thus, most welfare spending actually causes a decrease in measured income and an increase in poverty according to the Census Bureau.

25

Homelessness in America

Randall K. Filer

Few issues have generated as much discussion during the 1980s as the specter of homelessness in America. The sight of men and women huddled over heating vents on a winter's night raises obvious social policy issues. Unfortunately, few problems have proven as difficult to quantify and understand. We begin with the most simple question – how many homeless are there in the United States?

Despite claims of up to four million homeless Americans (see, for example, Maitland, 1989), most estimates place the number substantially lower. Table 25.1 shows estimates for several years during the 1980s. These estimates, derived from a variety of sources and using different techniques, show remarkable conformity. It seems reasonable to conclude that there are somewhere between 250,000 and 500,000

Table 25.1 Estimated total number of homeless in the USA

Year	Number of homeless	Source	
1983	279,000	Freeman and Hall[b]	
1984	254,000	HUD	Extrapolation from local interviews
	353,000	HUD	Extrapolation from shelter operators
	192,000–267,000	HUD	Shelter population plus extrapolation from street counts
1987	567,000–600,000[a]	Urban Institute	
	355,000–445,500[a]	Burt	Urban Institute data revised to account for new information and remove biases
1990	229,000	Census Count	

[a] Estimated on the basis of service users (including soup kitchens). Estimate is for a seven-day period and thus will overstate the number in any given night.
[b] Richard B. Freeman and Brian Hall, 1987.

homeless persons in the United States and that this number has been roughly stable for the past decade.

Aside from the obvious difficulty of counting people who are defined as *not* being somewhere (in a home), what can account for the vast discrepancy between these numbers (between 0.1 percent and 0.2 percent of the population) and the figures an order of magnitude greater often cited in the popular debate? Although any attempt to count the homeless will find it difficult to identify and include everyone, in view of the care, attention, and diversity of techniques with which the estimates reported in table 25.1 were developed, it does not seem likely that they missed nine out of every ten homeless persons.[1] Rather, two other factors are likely to account for the wide discrepancy. With the exception of the 1987 figures, the estimates in table 25.1 are for a single night. People may move in and out of periods of sleeping in shelters over the course of a year as their economic and personal situations change. Thus, the cumulative incidence of experiencing a spell of homelessness will be greater than the single night probability. Second, all of the estimates of homelessness in table 25.1 define someone as "homeless" if he or she slept in either a public shelter or in a place not typically used for sleeping (such as on the street or in doorways, bus terminals, subways, abandoned buildings, or other such places). Higher estimates typically include among the homeless anyone who does not have a permanent place of abode *in their own name* (or in their partner's name). Such a definition would include large numbers of individuals and families sharing housing with friends or relatives. While the phenomenon of shared housing raises a number of important public policy issues, the determination of whether such adaptation is desirable or not, as well as where to draw the line regarding who is to be included among the homeless, argues for the observable definition conventionally adopted (see Filer, 1991). As an illustration, consider the case of a teenaged mother who would like to escape the rules and regulations of her parents' house but who cannot find or afford her own apartment. Should such a mother be considered "homeless"? Would it even be in society's interest for her to have her own home?

No matter what the overall level of homelessness in the USA, there is substantial variation in this level across cities. Table 25.2 shows the estimated number of homeless in 30 large cities for 1984 and 1990. With certain notable exceptions, there is a strong degree of congruency in the relative levels of homelessness in these estimates, even though they were made six years apart and used vastly different techniques.[2] Pairs of cities with approximately the same population have vastly different rates of homelessness. For example, homelessness is between four and 12 times as common in San Francisco as in Baltimore, and approximately five to six times as common in Washington, DC as in Milwaukee, in Seattle as in Cleveland, in Atlanta as in Kansas City, or in Newark as in Charlotte. This suggests that whatever the national situation, there is

Table 25.2 Estimated number of homeless in selected cities[a]

City	1984 HUD Estimate[b]	1990 Census estimate
New York	29,000	33,830
Los Angeles	32,550	7,706
Chicago	19,850[c]	6,764
Houston	6,350	1,931
Philadelphia	3,600	4,485
San Diego[d]	3,000	4,947
Detroit	7,500	1,311
Dallas–Ft Worth[d]	7,000	2,724
Phoenix	1,075	1,986
Baltimore	690	1,531
San Francisco	8,250	5,569
Washington DC	4,700	4,813
Milwaukee[d]	1,000	566
Boston	3,200	2,463
New Orleans[d]	1,600	892
Cleveland	410	694
Seattle	3,175	2,539
Denver[d]	2,500	1,269
Kansas City	370	593
Atlanta[d]	2,000	2,491
Portland OR	1,550	1,702
St Louis[d]	5,000	827
Tucson[d]	1,000	539
Albuquerque[d]	1,000	370
Pittsburgh	888	760
Miami	5,950	1,164
Cincinnati	875	1,012
Charlotte	275	638
Minneapolis–St Paul	1,010	1,589
Newark[d]	3,000	2,816

[a] Sample consists of the 30 largest cities for which estimates are available in both years.

[b] HUD estimate is the midpoint of the "most reliable range" determined by HUD from its interviews with local officials.

[c] The accuracy of the HUD estimate for Chicago is dubious. Rossi (1989) reports an estimate for 1985 to 1986 of 2,020 to 2,722.

[d] 1984 figures are from Tucker (1987) since the city was not in the HUD study. Tucker (personal conversation) asserts that his estimates were derived in a manner analogous to that used by HUD.

considerable room for local conditions and policies to influence home-lessness levels. Indeed, Honig and Filer (1993) are able to explain over 80 percent of the cross-city variation in 1984 homelessness rates (which ranged from 6.8 per 100,000 to 412 per 100,000) as a function of local labor market conditions, rents at the low end of the housing market, and state government policies with respect to hospitalization of the mentally ill and the tightness of AFDC and other public assistance program rules.

Although the attention paid to homelessness may have increased in the 1980s, it is clear that the problem itself is not a new one. There are no data that enable even rudimentary comparisons of the number of homeless in the United States at various times. Indeed, it is not possible to obtain anything more than suggestive numbers from individual cities. Perhaps the most extensive documentation is available for New York City, which has had both a long tradition of public assistance and a larger than typical homeless population. As early as 1816, a census of public institutions reports that 2,600 people out of a population of around 110,000 were being housed by the city[3] (*New York Evening Post*, 1816). By 1826, the number housed by the city had increased to over 3,400 (including 1,742 in the almshouse and 340 in the debtor's prison and the workhouse) out of a population of around 165,000 (*New York Commercial Advertiser*, 1826).

By the early twentieth century, New York provided temporary shelter to homeless persons (overwhelmingly men) in a Municipal Lodging House as well as an annex to this house located on an East River Pier. In January 1915, these two facilities housed an average of 2,096 persons each night. In addition, several hundred persons were housed by the federal government in excess beds on Ellis Island, and an additional several hundred were housed by various private organizations such as the Salvation Army. Even so, these facilities could not begin to provide shelter for the estimated 26,000 homeless, unemployed men in the city out of a population of slightly less than four million (Schneider and Deutsch, 1941, p. 214). These figures point out the difficulty in comparing estimates of the number of homeless persons at various times. What little data that are available almost exclusively report the number of people in public shelters, yet the fraction of the homeless receiving such shelter has increased dramatically over the century. Thus, the current situation will appear relatively worse than it actually is when compared with earlier periods when far more of the homeless lived outside the formal shelter system and were not, therefore, included in the available counts.

The reasons for this shift toward public shelters over time are an important public policy issue. While some homeless persons were traditionally sheltered by public agencies and charities outside of the formal shelter system,[4] by far the majority of transient or homeless persons were accommodated by private market alternatives. These ranged from rooming houses or SRO hotels at the top end of the market (costing between $0.50 and $1.00 per night in the early twentieth century) down through cage or cubicle hotels (where semiprivate rooms could be rented for $0.15 to $0.25 a night) to dormitories where beds were available for as little as $0.10 a night. Hoch and Slayton (1989) provide convincing evidence that homeless persons preferred these private market alternatives over municipal shelters and missions for reasons of privacy, autonomy, and self-respect. Unfortunately, a series of public policy initiatives directed at slum clearance and urban renewal as well as imposing higher health, safety, and quality standards on the private housing market have made it uneconomical to continue to provide such housing in most American cities. Since these policies have not improved the ability of the very poor to afford high-quality housing, they have had the perverse effect of worsening the quality of the actual shelter used by homeless persons by eliminating their preferred options and forcing them to rely on municipal shelters.

By the great depression, shelter capacity in New York had increased substantially. In January 1932, 8,668 men (and 246 women) were reported as receiving temporary shelter from the city and various private agencies. The numbered sheltered rose to a peak of 18,719 men (and 420 women) in March 1935, then gradually declined for the next six years (with a temporary upturn during the 1938 recession). In January 1940, 10,178 men and 216 women were being sheltered, while in January 1941, 7,388 men and 171 women were in public homeless shelters. With the start of World War II, the number of homeless fell rapidly, so that by August 1944, only 537 men and 72 women remained. After the war, the number of homeless sheltered in New York increased somewhat to slightly less than 2,000 a night during the winter months of 1946 and between 2,500 and 3,000 a night during the winters of 1947 and 1948 (Welfare Council of New York City, 1949).

During the 1980s, the number of "traditional" homeless (i.e., single unattached individuals) sheltered by the city and various private agencies rose from a daily average of 2,804 in 1981 to a peak of 12,506 in the 1989 fiscal year (New York City Human Resources Administration, 1989). What is unique about the 1980s in New York is that single men (who have traditionally comprised the vast majority of the homeless) are now a minority of the homeless. In March 1971, there were 1,350 families receiving temporary shelter from the city of New York. Placements into apartments reduced this number to about 300 by the end of 1971. It remained below 1,000 until well into 1982, when it began to rise. Figure 25.1 shows the number of homeless families in city and private shelters (or provided with hotel rooms) for each month between January 1983 and August 1992. Given that the average homeless family has approximately three members (one adult and two children), it is clear that during the mid- and late 1980s, New York City had more homeless family members than single adults.

The emergence of homelessness among families in the 1980s once again points out the difficulty of obtaining an accurate picture of trends in homelessness over time. There is little evidence of significant numbers of homeless families in major urban areas prior to the current decade. It might be expected that increases in public benefits such as AFDC should create lower rates of homelessness among families now

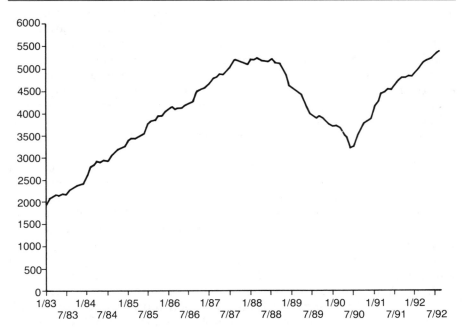

Figure 25.1 Number of homeless families in New York City

than in the past. Yet this incidence appears to have increased substantially. In addition, there is substantial variation in the extent of family homelessness across various cities. A 1987 survey of officials in 26 cities indicated that the fraction of homeless persons in families ranged from 13 percent in Cleveland to 65 percent in Providence (US Conference of Mayors, 1987). The mean fraction of homeless persons living in families across the cities in this survey was reported to be 30 percent (with a standard deviation of 15 percent). A similar survey of most of the same cities a year later (US Conference of Mayors, 1989) claimed that the fraction in families was 33.5 percent (with a standard deviation of 16.8 percent).[5] Filer (1990) suggests that family homelessness should be viewed in the context of public housing policies in general and may be largely a response to city policies that favor those currently in shelters over other poor residents in allocating assisted housing units.

It is clear that there are, in reality, two very different homeless populations – those in families and single individuals. Survey results in various cities present a remarkably consistent picture of the characteristics of each of these two groups. Results with respect to some of the major characteristics are presented in table 25.3.[6]

Homeless families are almost exclusively unmarried women with children rather than conventional two-parent families. Almost all of the women heading these families are not now working and very few have significant prior labor market experience. Very few in any city are

Table 25.3 Characteristics of homeless persons estimated in various local studies

City or State	Alabama	Baltimore	Boston	Boston	Chicago
Author	La Gory et al. (1989)	Breakey et al. (1989)	Mulkern et al. (1985)	Schutt (1988)	Rossi (1989)
Families included	Yes	No	No	Yes	Yes
Shelter only	No	No	No	Yes	No
Male	75%	n.a.	81%	71%	76%
Mean age	<40	40 (m) 33 (f)	38	31[b]	40
Nonwhite	30%	63% (m) 70% (f)	30%	34%[c]	69%
Nonwhite in area	26%	56%	30%	30%	50%
High School graduates	n.a.	34% (m) 46% (f)	52%	46%	55%
Employed	n.a.	18% (m) 16% (f)	11%	10%	25%
Public assistance	n.a.	40% (m) 53% (f)	37%	32%	n.a.
Never married	n.a.	98%[d] (m) 93%[d] (f)	61%	61%	57%
Criminal record[e]	n.a.	72% (m) 32% (f)	n.a.	28%	33%
Alcohol abuse[j]	n.a.	68% (m) 32% (f)	38%	38–49%	n.a.
Drug abuse[j]	n.a.	22% (m) 17% (f)	n.a.	35%	n.a.
Alcohol or drug abuse[l]	n.a.	75% (m) 38% (f)	n.a.	75%	38%
Mental illness: Current[n]	n.a.	42% (m) 49% (f)	29%	n.a.	n.a.
Prior Hospitalization	n.a.	n.a.	45%	38%	23%

City or State	Colorado	Detroit	Los Angeles	Milwaukee	Minneapolis
Author	James (1989)	Mowbray et al. (1986)	Farr et al. (1986)	Rosnow et al. (1985)	Piliavin et al. (1987)
Families included	Yes	No	No	No	Yes
Shelter only	No	Yes	No	No	No
Male	80%	71%	96%	87%	85%
Mean age	<40	35	35[b]	35[b]	32
Nonwhite	42%	74%	73%	40%	57%
Nonwhite in area	11%	65%	38%	26%	12%
High School graduates	n.a.	43%	52%	n.a.	32%
Employed	31%	12%	15%	1%	36%
Public assistance	n.a.[d]	29%	20%	31%	>55%
Never married	81%[d]	51%	59%	36%	53%

Criminal record[e]	7%[f]	28%	36%[g]	56%	21–54%
Alcohol abuse[i]	n.a.	31%	27%[g]	n.a.	n.a.
Drug abuse[j]	n.a.[f]	11%	10%	n.a.	n.a.
Alcohol or drug abuse[l]	13%[f]	n.a.	31%	24%	n.a.
Mental illness: Current[n]	n.a.	n.a.	28%	48%	n.a.
Prior Hospitalization	6%[f]	26%	27%	42%	19%

City or State	New York	New York	New York	New York	Ohio
Author	Burt and Cohen (1988)	Crystal et al. (1986)	Cuomo Commission	Freeman and Hall (1987)	Roth et al. (1985)
Families included	No	No	Yes	No	No
Shelter only	No	Yes	Yes	No	No
Male	84%	78%	54%	n.a.	81%
Mean age	n.a.	33	30–39[b]	n.a.	33
Nonwhite	85%	83%	93%	64%	41%
Nonwhite in Area	39%	39%	39%	39%	32%
High School graduates	45%	47%	57%	47%	45%
Employed	12%	4%	11%	n.a.	25%
Public assistance	>21%	18%	70%	12%	44%
Never married	79%	69%	57%[d]	n.a.	45%
Criminal record[e]	39%	39%	17%[g]	39%	59%
Alcohol abuse[i]	n.a.	23%	29%	29%	27%
Drug abuse[j]	n.a.	17%	47%	14%[k]	8%
Alcohol or drug abuse[l]	41%	32%	29–65%[m]	n.a.	n.a.
Mental illness: Current[n]	n.a.	20%	n.a.	n.a.	31%
Prior Hospitalization	25%	22%	20%[o]	n.a.	30%

City	Phoenix	St. Louis
Author	Brown et al. (1983)	Morse et al. (1985)
Families included	Yes	No
Shelter only	?	Yes[a]
Percent male	86%	n.a.[a]
Mean age	37	29[b]
Nonwhite	39%	65%
Nonwhite in Area	15%	46%

Table 25.3 (Continued)

City or State	Phoenix	St. Louis
Author	*Brown et al. (1983)*	*Morse et al. (1985)*
High School graduates	60%	57%
Employed	18%	9%
Public assistance	>6%[d]	31%
Never married	87%[d]	52%
Criminal record[e]	37%[h]	22%
Alcohol abuse[i]	n.a.	36%
Drug abuse[j]	25%	19%
Alcohol or drug abuse[l]	n.a.	n.a.
Mental illness: Current[n]	n.a.	46%
Prior Hospitalization	17%	20%

[a] St Louis study drew a nonrandom sample with excess females. Since no population weights are available, all resulting averages will be biased with too much weight given to female values.

[b] Median age

[c] Excluding Asians

[d] Not currently married. New York (Cuomo Commission also excludes those currently "living with someone".

[e] Felony conviction or time in prison [Chicago, Detroit, Minneapolis (21%), New York (Burt and Cohen, 1988)]. Otherwise time in jail.

[f] Within the past month only.

[g] Within the past 12 months only.

[h] Within the past 24 months only.

[i] Baltimore – diagnosed dependency, Boston – 38% treated, 49% use several times a week, NY (Crystal) – "use regularly," NY (Cuomo) – "willing to accept rehabilitation treatment," others treated in detox program.

[j] Baltimore – diagnosed dependency, Boston – use several times a week, Detroit – recent use, NY (Crystal) – "use regularly," NY (Cuomo) – positive urine test (note that only 33% reported use in the past year), Phoenix – "use Street drugs," others treated in detox program.

[k] Excludes marijuana.

[l] As in notes e and f, if not mentioned, treated in a detox program for one or both of alcohol or drugs.

[m] Twenty-nine percent for heads of families, 65% for residents of singles shelters.

[n] Baltimore – diagnosis of "major mental illness" according to APA guidelines by clinicians, Boston – "high score on mental illness scale," Los Angeles – Diagnostic Interview Schedule (DIS) diagnosis of "major mental illness" within past six months, Milwaukee – two or more self-reported symptoms such as "hearing voices" or "obsessive thoughts." New York (Crystal) – case workers' judgment, Ohio – Psychiatric Severity Index score greater than one, St Louis – elevated score (above 0.72) on the Global Severity Index on the Brief Symptom Inventory (BSI).

[o] Also includes prior reported treatment in "clinic" or with long-term drug therapy.

reported as unsheltered (i.e., living on the street or in public places). Many move in and out of spells of residence in shelters and very few are homeless for extended periods. In New York City, for example, data from late 1990 show an average spell length to date of less than six months. Most families in New York are placed in permanent apartments within a year of entering the shelter system and spend at most a month in group environments before being placed in facilities with private rooms, baths, and (often) kitchens.

The data with respect to homeless singles are similar in every city for which studies exist. On average, 90 percent are men. Only a small minority are employed and only slightly more are receiving public assistance. A disproportionate number are black and a majority have never married. Many have substance abuse or mental problems. The widely cited figure of one third suffering from a mental illness and one third who are drug or alcohol abusers represents a typical figure for the percentage who claim to have been hospitalized for mental illness or treated in an in-patient substance abuse rehabilitation program in the recent past. Given the shortage of treatment slots in rehabilitation programs and the trend toward nonhospitalization, this estimate must be a lower bound on the actual incidence of these pathologies. Perhaps the best data come from the study of Baltimore homeless included in table 25.3. When clinicians from the Johns Hopkins University Medical School actually performed standard diagnostic tests on a sample of homeless persons, they found 91 percent of the men and 80 percent of the women suffered from a psychiatric disorder according to the diagnostic scheme of the American Psychiatric Association's *Diagnostic and Statistical Manual of Mental Disorders*. (In this framework, substance addiction is considered a form of psychiatric disorder.) A recent study of shelter residents in New York City found that 29 percent of the heads of families and 65 percent of single residents who agreed to a voluntary urine test had recently used one or more drugs (primarily cocaine; New York City Commission on the Homeless, 1992).

In summary, although we do not know the causes of homelessness with as much certainty as we might, the list of factors that appear to be related includes: mental illness, substance abuse, lack of employment opportunity and/or interest, lack of family ties, tight housing markets, and incentives in public policies. Lest we think that we have come far in our understanding of homelessness, consider the "First Annual Report of the Managers of the Society for the Prevention of Pauperism in the City of New York" which, in 1818, in studying the use of city shelters, listed the causes of such reliance as (in order, with selected comments):

1ST. IGNORANCE, arising either from inherent dullness, or from want of opportunities for improvement.

2ND. IDLENESS. A tendency to this evil may be more or less inherent.

3RD. INTEMPERANCE IN DRINKING. This most prolific source of mischief and misery drags in its train almost every species of

suffering which afflicts the poor. This evil, in relation to poverty
and vice, may be emphatically styled the *Cause of Causes.*

4TH. WANT OF ECONOMY

5TH. IMPRUDENT AND HASTY MARRIAGES[7]

6TH. LOTTERIES

7TH. PAWNBROKERS

8TH. HOUSES OF ILL FAME

9TH. THE NUMEROUS CHARITABLE INSTITUTIONS OF THE
CITY. . . . Is not the partial and temporary good which they accomplish . . . more than counterbalanced by the evils that flow
from the expectations they necessarily excite; by the relaxation of
industry, which such a display of benevolence tends to produce; by
that reliance upon charitable aid, in case of unfavorable times,
which must unavoidably tend to diminish . . . that wholesome
anxiety to provide for the wants of a distant day, which alone can
save [the laboring classes] from a state of absolute dependence, and
from becoming a burden to the community?

What is clear is that the past decade has seen a remarkable amount of
attention paid to a serious social problem without a great deal of understanding. We now have a reasonable picture of how many homeless
there are and what the characteristics of the homeless population are.
We know very little about whether the problem has been getting worse
over time or simply whether we have become more aware of it. Finally,
with very few exceptions, work on understanding the causes of homelessness and how it might respond to various public policy interventions
has only just begun.

NOTES

1   Indeed, they may contain errors tending to overstate the number of homeless
persons. Rossi (1989) reports that only 9 percent of those approached on the
street between 1 a.m. and 6 a.m. were actually homeless. Census methodology
attempted to exclude those who were engaged in "obvious economic activity"
but made no effort to determine whether those in shelters or on the streets on
the enumeration night had a residence where they could have slept. Similarly,
every study to date has included all persons in shelters as "homeless" without
attempting to identify those who had alternative housing options where they
could have slept. While it is likely that the undercount from missed individuals
exceeds the overcount from improperly classified ones, such a conclusion must
be based on intuition rather than firm evidence.

2   The HUD estimates for 1984 were based primarily on "key-person" interviews,
while the 1990 census estimates were based on actual counts of those in shelters
combined with an attempt to enumerate visually those in public places on a given
night. The obvious discrepancies involved the cities of Los Angeles, Chicago,
Houston, Detroit, Dallas, St Louis, and Miami. In each of these cases, local
estimates of the problem in 1984 were far greater than actual counts in 1990.

Perhaps the situation has gotten substantially better in these cities since 1984, or perhaps the census count was particularly understated in them. A more likely possibility is that local officials, for political or other reasons, may have misperceived the extent of the problem in the earlier year as they confronted increased awareness of the issue and demands for action. William Tucker, who provided the estimate for St Louis, reports that it was heavily influenced by his interview with one local official who claimed a problem of enormous size (personal conversation). Similarly, Rossi (1989) reports estimates for Chicago an order of magnitude lower than those provided to HUD by local officials in 1984.

3   About 950 of these were in the state prison or the city hospital. However, 1,242 were in the almshouse, which was at the time of its opening in 1816 the largest building in the city. About half of these were adult men, the rest were women or children. An additional 300 men were incarcerated in the debtor's prison and the "bridewell" or workhouse where individuals convicted of vagrancy or similar crimes were sent.

4   Perhaps the most important informal source of shelter in the late nineteenth and early twentieth centuries was the local police station. Figures for 1890 indicate that police stations in New York were providing approximately 150,000 lodgings a year.

5   This data should be interpreted with great caution. It represents simply the "guess" of a local official answering a survey form and is remarkably variable for certain cities. For example, in 1987 Norfolk reported that 30 percent of its homeless were family members, while a year later it claimed that 80 percent were. Conversely, in 1987 Providence reported 65 percent as family members, while in 1988 this figure had fallen to 30 percent.

6   This table is adapted from Filer and Honig (1990) and the reader is referred there for a more extensive discussion of the characteristics of the homeless in these and other studies.

7   Today, perhaps, the lack of marriage at all.

REFERENCES

Breakey, William R., Pamela J. Fischer, Morton Kramer, Gerald Nestadt, Allan J. Romanoski, Alan Ross, Richard Royall, and Oscar C. Stine (1989): "Health and Mental Health Problems of Homeless Men and Women in Baltimore." *Journal of the American Medical Association* (262) September 8.

Brown, Carl, Steve MacFarlane, Ron Paredes, and Louisa Stark (1983): "The Homeless of Phoenix: Who Are They? And What Should Be Done?" Prepared for the Consortium for the Homeless (Phoenix: South Community Mental Health Center).

Burt, Martha R. (1990): "Developing the Estimate of 500,000–600,000 Homeless People in the United States in 1987," presented at the Conference on "Enumerating Homeless Persons: Methods and Data Needs." US Census Bureau, Department of Housing and Urban Development and Interagency Council on the Homeless. November.

——— and Barbara E. Cohen (1988): *Feeding the Homeless: Does the Prepared Meals Provision Help?* Vols. I, II Washington, DC: The Urban Institute.

Crystal, Stephen, Susan Ladner, and Richard Towber (1986): "Multiple Impairment Patterns in the Homeless Mentally Ill." *International Journal of Mental Health* (14), 4.

Farr, Rodger, Paul Koegel, and Audrey Burnam (1986): *A Study of Homelessness and Mental Illness in the Skid Row Area of Los Angeles.* Los Angeles County, CA: Department of Mental Health.

Filer, Randall K. (1990): "What Really Causes Family Homelessness?" *NY: The City Journal,* (Autumn) 31–40.

—— (1991): "Tracking Down the Hidden Homeless." *NY: The City Journal* (Summer) 10–12.

—— and Marjorie Honig (1990): *Policy Issues in Homelessness: Current Understanding and Directions for Research.* The Manhattan Institute for Policy Research, March.

Freeman, Richard B., and Brian Hall (1987): "Permanent Homelessness in America?" *Population Research and Policy Review* (6).

Hoch, Charles, and Robert A. Slayton (1989): *New Homeless and Old.* Philadelphia: Temple University Press.

Honig, Marjorie, and Randall K. Filer (1993): "Variation in the Extent of Homelessness Across American Cities." *American Economic Review,* March, 248–55.

James, Franklin J. (1989): "Factors Which Shape the Risks of Homelessness: Preliminary Observations From Colorado." Denver: University of Colorado Graduate School of Public Affairs, September.

La Gory, Mark, Ferris J. Ritchey, Timothy O'Donoghue, and Jeffrey Mullis (1989): "Homeless People in Alabama: A Variety of People and Experiences." In J. Momeni, ed. *Homelessness in America.* Westport, CT: Greenwood Press.

Maitland, Leslie (1989): "Plan for Homeless is Called Modest." *New York Times,* November 26.

Morse, Gary, Nancy M. Shields, C. R. Hanneke, R. J. Calsyn, G. K. Burger, and B. Nelson (1985): *Homeless People in St. Louis: A Mental Health Program Evaluation, Field Study, and Follow-up Investigation, Vol. I.* Jefferson City, MO: State of Missouri Department of Mental Health.

Mowbray, C. V. Johnson, A. Solarz, and C. Combs (1986): "Mental Health and Homelessness in Detroit." Lansing, MI: Michigan Department of Mental Health.

Mulkern, V., V. Bradley, R. Spence, S. Allein, and J. Oldham (1985): "Homelessness Needs Assessment Study: Findings and Recommendations for the Massachusetts Department of Mental Health." Boston: Human Services Research Institute.

New York City Commission on the Homeless, Andrew M. Cuomo, Chairman (1992): "The Way Home: A New Direction in Social Policy." New York: City of New York, February.

New York City Human Resources Administration (1989): "Progress Report on the Five-Year Plan for Housing and Assisting Homeless Families." New York: City of New York, February.

New York Commercial Advertiser, January 19, 1826.

New York Evening Post, June 12, 1816.

Piliavin, Irving, Michael Sosin, and Herb Westerfeldt (1987): *Conditions Contributing to Long-Term Homelessness: An Exploratory Study.* Madison, WI: Institute for Research on Poverty.

Rosnow, M. T. Shaw, and C. Concord (1985): "Listening to the Homeless: A Study of Homeless Mentally Ill Persons in Milwaukee." Milwaukee, WI: Human Services Triangle.

Rossi, Peter H. (1989): *Down and Out in America: The Origins of Homelessness*. Chicago: University of Chicago Press.

Roth, Dee, Jerry Bean, Nancy Lust, and Traian Saveanu (1985): *Homelessness in Ohio: A Study of People in Need*. Columbus, OH: Department of Mental Health.

Schneider, David M., and Albert Deutsch (1941): *The History of Public Welfare in New York State: 1867–1940*. Chicago: University of Chicago Press.

Schutt, Russell K. (1988): "Boston's Homeless, 1986–87: Change and Continuity." Boston: University of Massachusetts Department of Sociology, mimeo.

Society for the Prevention of Pauperism in the City of New York (1818): "First Annual Report of the Managers."

Tucker, William (1987): "Where Do the Homeless Come From?" *National Review*, September 25, 1987.

US Conference of Mayors (1987): *The Continuing Growth of Hunger, Homelessness and Poverty in America's Cities: 1987*. Washington, DC: US Conference of Mayors.

——— (1989): *A Status Report On Hunger and Homelessness in America's Cities: 1988*. Washington, DC: US Conference of Mayors.

US Department of Housing and Urban Development (1984): *A Report to the Secretary on the Homeless and Emergency Shelters*. Washington, DC: Office of Policy Development and Research.

Welfare Council of New York City, Project Committee on Homeless Men (1949): "Homeless Men in New York City" mimeo.

26

The Recent US Economy

Alan Reynolds

Many people believe the US economy ended the decade of the 1980s in worse shape than it began, requiring many future years of austerity and suffering to atone for the "excessive greed" and "debt binge" of the 1980s. The data indicate otherwise.

The 1980s began with four terrible years of "stagflation" – 1979–82 – and ended with a brief recession from July 1990 to March 1991. Productivity, for example, actually fell from 1978 to 1982, but then rose relatively briskly for six years (figure 26.1).

Actually, the usual figures greatly understate industrial productivity gains because of familiar problems in the financial sector (including savings and loan failures). While output per hour rose by 9.4 percent between 1980 and 1991 for all nonfarm businesses, it rose by 16.4 percent among nonfinancial business – doubling the 8 percent increase from 1970 to 1980. In manufacturing, the total productivity increase was 41.4 percent from 1980 to 1990, more than 4 percent per year.

The cumulative gains in the level of productivity were particularly impressive since, unlike much of the 1960s, they were also combined with rapid growth of employment. From 1983 to 1989, US employment grew by 2.4 percent per year – twice as fast as in Japan and four times as fast as in Germany. Increases in the percentage of the population employed (to a record 63.3 percent by 1989) were well above the postwar trend (figure 26.1). This unprecedented increase in the number of people willing and able to work reflected improved job opportunities and labor incentives arising from (1) reduced marginal tax rates, (2) no increases in the minimum wage from 1981 to 1990, and (3) a large increase in the earned income tax credit for low-wage workers.

The addition of 18.6 million jobs from 1980 to 1990 is sometimes thought to be the result of wives being forced to work outside the home to maintain the same living standards. But most income gains were not, in fact, created by turning one-earner families into two-earner families. Participation in the labor force among married women increased less rapidly on a proportional basis in the 1980s (from 49.8 percent to 58.4

Figure 26.1 Employment, productivity, and output, USA, 1954–92
Source: Bureau of Labor Statistics; Federal Reserve Board.

percent) than in the 1970s (from 40.5 percent to 49.8 percent). And among those who were two-earner families in both decades, real median income increased by 11.9 percent from 1980 to 1989, compared with a 3.1 percent increase over the previous decade.

Production increases were not confined to service industries. After a period of weakness from 1979 through 1982, manufacturing output surged nearly 40 percent by 1990 and declined very little in the 1990–1 recession, despite large defense cuts (figure 26.1).

Even if the 1981–2 recession is included, the growth of US manufacturing output in the 1980s was much stronger than in Germany, and also much stronger than that during the 1970s (figure 26.2). Japan continued to experience rapid manufacturing growth until 1991 but then suffered a relatively severe downturn. By October 1992, for example, industrial production in Japan was 6.4 percent lower than a year before, though rising in the USA.

Increases in real output brought increases in real income. Measured in 1990 dollars, median family income fell from $33,370 in 1973 to $33,037 in 1982 but rebounded by 12.6 percent to $36,062 by 1989. The alternative "mean" average of family income rose even more dramatically in the 1980s (figure 26.3). Between the cyclical peaks of 1979 and 1990, real after-tax income *per capita* rose by 17.6 percent compared with 9.3 a percent rise between the comparable peaks of 1973 and 1979.

The increase in incomes has often been obscured by looking at average wages, which were diluted by larger numbers of young and part-time

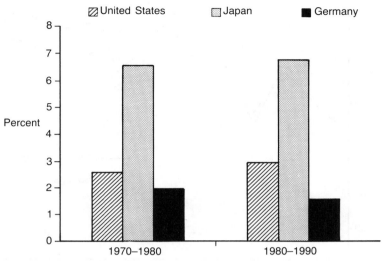

Figure 26.2   Annual increase in manufacturing output, average over 10-year period
Source: US Bureau of Labor Statistics.

workers. Wage statistics also exclude health care and other fringe benefits (which were a rising portion of compensation) and include only
nonsupervisory workers, though many previous wage-earning jobs were
upgraded to salaried lower management positions. Many people also
stopped receiving either wages or salaries in the 1980s and instead
started their own businesses. Total real income of nonfarm proprietors

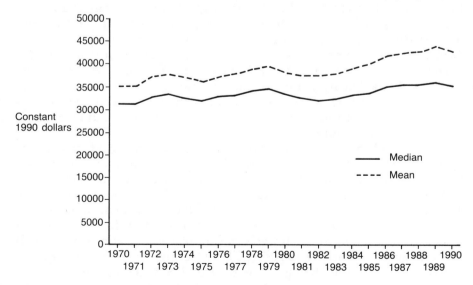

Figure 26.3   Average real family income, USA, 1970–90
Source: Bureau of the Census, P-60, no. 174.

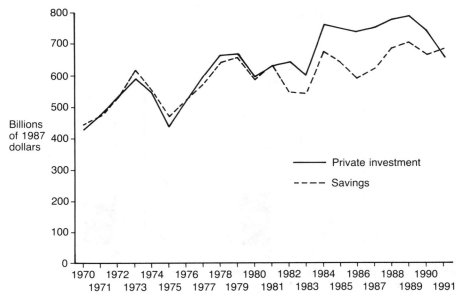

Figure 26.4   The gap between investment and savings

Notes: Adjusted by deflator for fixed investment. Figures imply zero government investment.
Source: Dept of Commerce.

rose by 45 percent from 1980 to 1990 and was 59 percent as large as all wages and salaries in manufacturing by 1990.

Lower inflation in the 1980s makes it particularly essential to adjust for inflation. In such "real" terms, the gap between savings and investment – which was filled by foreign investment – looks quite different. This is true despite conceptual problems with conventional measures (which do not count education or FBI computers as investment, for example, and do not count home equity or the near-quadrupling of the stock market as savings). Looking at this issue in terms of real absolute amounts of investment and savings, rather than as percentages of (rising) income, shows that annual savings actually did rise – but investment rose much more (figure 26.4).

As the USA became a new "tax haven," the improved investment opportunities attracted foreign capital, both real and financial. In fact, the USA virtually imported entire factories that are now starting to export things like the Honda Accord and the Mazda-built Ford Probe and Explorer. This capital surplus accounted for current account (trade) deficits, even though real exports doubled from 1986 to 1992. Instead of being a net lender to such unpromising places as Latin America and Eastern Europe, as in the 1970s, the USA became, by some measures, a net debtor. Far from being a measure of US weakness, becoming a magnet for world capital was both an effect and a cause of renewed economic vitality.

The common belief that budget deficits were due to tax cuts rather than spending increases rests on a confusion between marginal tax rates on extra income (such as overtime or promotions) and average tax rates (taxes actually collected, as a percent of total income). Although the highest marginal tax rate was reduced from 70 percent to 31 percent, average federal taxes actually rose – to 19.2 percent of GNP in 1989, from an average of 18.2 percent in 1971–9. Many families earned more money at the lower tax rates and had less incentive to hide that fact from the IRS. Measured in constant 1987 dollars, *real* federal revenues rose by 27 percent from 1979 to 1989, compared with less than 19 percent in the previous ten years. One reason for the sharp increase in federal revenues was the large capital gains resulting from tripling stock prices and cutting interest rates in half (thus raising the value of bonds). The lower capital gains tax, prior to 1987, ensured that more of such gains would be realized through frequent asset sales, thus contributing to revenue.

Although real revenues rose rapidly in the 1980s, so did real federal spending (figure 26.5). Much of this spending was for long-lasting capital goods, though, particularly for defense. Measured in 1982 dollars, annual investments in military hardware rose from $32.2 billion in 1976 to $87.4 billion in 1989. Despite this near tripling of real outlays on defense capital, the budget deficit was down to 2.9 percent of GNP in fiscal 1989 (before temporary outlays to acquire the assets of failed thrifts were added to the budget), compared with 4.3 percent in 1976. Unlike previous decades, the deficits in federal budgets in the 1980s

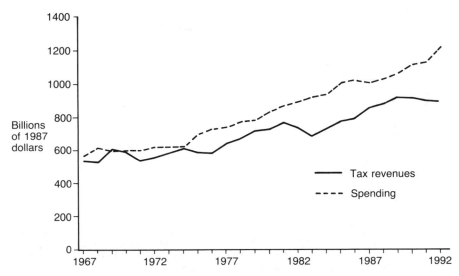

Figure 26.5 Real federal receipts and spending, 1967–92
Source: Budget of the United States, 1993, table 1.3.

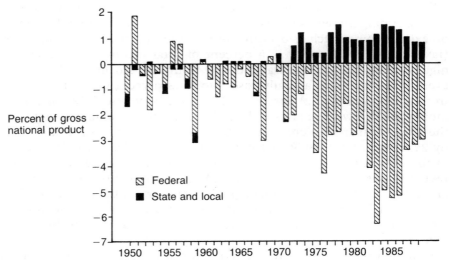

Figure 26.6 Budget deficits (–) and surpluses
Source: Office of Management and Budget.

were also partly offset by surpluses in state and local budgets (figure 26.6).

Many worry about rising household debt. This neglects the even greater increase in the value of household assets, such as stocks and

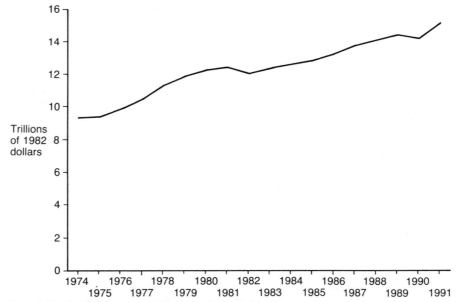

Figure 26.7 Real private net worth, households and businesses
Note: 1991 estimated.
Source: Federal Reserve Board.

bonds (see the article by Robert Rector in this volume). Housing construction was strong from 1983 through 1988 (as indicated in the article by Richard Muth in this volume), and more and better houses mean more mortgages. A long period of strong auto sales likewise meant more auto loans. More students going on to college meant more student loans. However, the added value of both real and financial assets substantially exceeded the added debts, so that household net worth grew much larger, not smaller. The measured increases in net worth from 1982 through 1989 largely reflected higher values for US stocks and bonds, unlike increases from 1975–80, which were largely dependent on inflation in house prices that exceeded the general inflation rate (figure 26.7).

Every generation "mortgages the future" – that is, issues bonds and stocks and mortgages to help finance investments that are expected to increase the quantity and quality of goods and services in the future. Americans in the next decade or two will indeed inherit such financial obligations from the 1980s. But they will also inherit the real assets that those debts (and equity) financed – from computers and robots to missiles and college degrees.

The 1980s were a period of rapid gains in employment and output combined with moderate inflation and renewed confidence in US stocks and bonds. In a longer-term perspective, the 1980s actually scored a number of impressive achievements, not the least of which was a substantial increase in the quantity and modernity of the nation's physical and human capital that will increase real output and income in the decades ahead.

PART III

Natural Resources

27

Long-Term Trends in Energy Prices

William J. Hausman

Over the long course of history, successfully developing nations have exhibited great adaptability in their choices regarding energy utilization. In the preindustrial period, wood, wind, water, and animate power were the major sources of energy for both household and industrial needs. With the advent of industrialization, there was a tremendous increase in the demand for power, and these traditional sources of energy were relatively quickly displaced by fossil fuels. The three accompanying graphs draw on British and American experience to illustrate a few basic points about long-term trends in energy prices and energy choices. First, there is little evidence that the price of "energy" has risen dramatically over the last 550 years, although the energy shock of the 1970s, evident in the data, is still playing itself out. Second, although the price of some fuels in some areas (British coal in the twentieth century, for example) may have risen, the same fuel in other areas, or other fuels, have replaced those that have become particularly scarce. Finally, in the United States in the twentieth century there has been a definite movement away from reliance on a single source of fuel and toward use of a diversity of fuel sources (a movement found in virtually all developed nations of the world).

LONG-TERM TRENDS IN WOOD AND COAL PRICES, 1450–1989

Long-term trends in the real price of wood in Britain and the real price of coal in Britain and the United States are presented in figure 27.1. The prices underlying the indices represent the primary price of the fuel at or close to the source of supply and thus exclude all transport and marketing costs as well as any taxes. The price indices are superimposed for at least part of this time period, and this procedure should be justified. In the case of coal, the justification is straightforward. After adjusting for the exchange rate, British and US nominal (pithead) coal prices were approximately equal in 1880 (assuming that the respective energy contents were approximately equal); thus, both real coal indices were set equal to one at 1880. Although it would have been desirable to standardize the British wood and coal indices for energy content, there is insufficient information to determine the price per energy unit for wood versus coal for the late medieval and early modern time period. However, it has been argued that "at any time in the 16th and 17th centuries coal was less expensive than charcoal per unit of heat output."[1] The wood and coal indices were arbitrarily joined at 1600, which, if there is a bias, understates the real price of wood relative to coal over the length of the index.

Several features of figure 27.1 stand out. First, divergent trends in wood versus coal prices in seventeenth-century Britain are evident. This period of British history has been identified as containing the first "energy crisis" of the modern era. Second, the real price of coal in Britain, after remaining reasonably stable for about two and a half centuries (1600–1850), began to rise some time after the middle of the nineteenth century. By the beginning of the twentieth century, its price had risen much faster than the price of coal in the USA, which had remained relatively stable. It can be said that the British coal industry faced a twentieth-century "crisis," with which it is still dealing.

Economic historians have argued that Britain's response to the seventeenth-century "energy crisis," which was to vigorously develop its coal resources, propelled the nation into an industrial revolution that ultimately provided the British people with the highest standard of living in

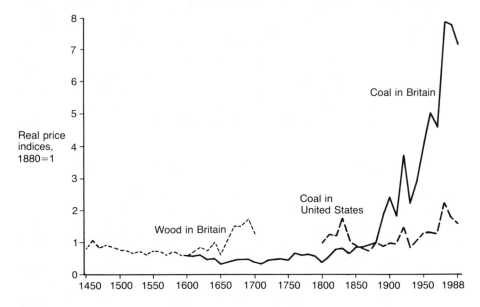

Figure 27.1 Long-term energy prices: wood and coal in Britain and the USA

the world and led the way for developing nations in the rest of the world. As E. A. Wrigley has argued recently:

> Something more than modernization or capitalism in the usual meaning of the term was needed to enable England to escape the same fate as Holland and to become the first country in which the age-old constraints on growth lost their force, the country where in spite of the forebodings of some contemporaries, it became clear that poverty was not the inevitable fate of the bulk of the population, nor a sweating brow the precondition of the daily loaf. The transition that produced the change in prospects was the move from an advanced organic to a mineral-based energy economy.[2]

Throughout the seventeenth century, the British economy still relied predominantly on traditional sources of energy, but the transition had begun. Coal production reached three million tons per annum by 1700. The next two centuries witnessed the almost continuous expansion of Britain's coal industry and coal-based economy. Average annual coal production rose to around 30 million tons by 1830 and reached a peak of nearly 290 million tons in 1913. After the middle of the nineteenth century, the price of coal began to rise. There is disagreement among historians over the causes of this rise, although most explanations are based on deteriorating geological conditions, declining labor productivity, or entrepreneurial sluggishness and mismanagement.[3] Such conditions did not occur to the same extent, or were overcome, in the USA, as is evident from figure 27.1. By the end of the nineteenth century, the USA was producing more coal than Britain. The British coal industry was becoming increasingly uncompetitive in the world market, and it faced particularly severe difficulties between the world wars. The industry was nationalized in 1946 with little opposition from mineowners or consumers, but this has not solved its problems. Britain, once the dominant world leader, is no longer among the larger coal producers of the world. The US coal industry, by contrast, remains vigorous, extracting nearly a billion tons per annum (a portion of which is exported) at a real pithead price not much different from that obtaining nearly 200 years ago.

ENERGY IN THE UNITED STATES, 1800–1989

Figures 27.2 and 27.3 summarize fuel prices and consumption patterns for energy in the United States over the past two centuries. For the first three quarters of the nineteenth century, wood was the predominant fuel for all basic uses (although water wheels were important for some specialized uses). Although the steam engine had been in use in the USA from the beginning of the nineteenth century, it did not have a large impact on total energy consumption. Coal was used primarily for specialized purposes in particular areas, such as for the smelting of iron in the Pennsylvania anthracite region. It was not until after the Civil

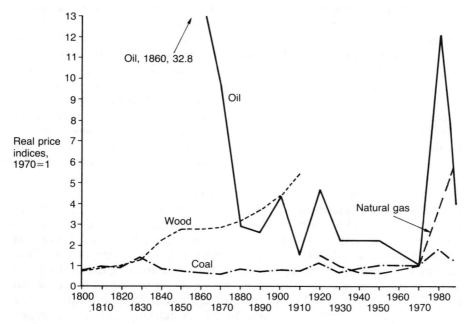

Figure 27.2 US energy prices, 1800–1988

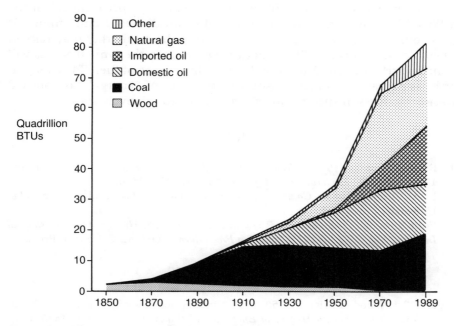

Figure 27.3 US energy consumption by type of fuel, 1850–1989

War, with the development of a national rail network and the coming of the "second industrial revolution," that coal came to be used more widely. Unfortunately, there is no extant price series for wood fuel in the United States (nor could it be said that there was a national market for wood fuel). The price index for wood in figure 27.2 is composed of the price of timber, which likely is a good proxy for wood fuel. The index exhibits a fairly strong upward trend for most of the nineteenth century.

Between 1870 and around 1910, there was a tremendous increase in energy consumption in the USA and a dramatic shift from wood to coal as the dominant fuel. What is remarkable is that the expansion in the coal industry occurred with virtually no upward price pressure, and the same is true of its period of relative decline.

During the course of the twentieth century, oil and natural gas became major sources of energy in the United States. By 1970, the prices per unit of energy (BTU – British thermal unit) of oil, coal, and natural gas had nearly converged (which is why the indices in figure 27.2 were set equal to one in 1970). Oil has had a particularly interesting history. One feature of figure 27.2 that stands out is the dramatic decline in the price of oil in the first two decades of its existence. At the time, however, oil provided an infinitesimal proportion of US energy consumption, so the impact of this decline was negligible. It is actually difficult to make generalizations about the long-term trend in oil prices (and it certainly cannot be said that there has been a secular upward trend). Perhaps the most remarkable feature of the oil industry has been its relative price instability, which is especially evident in the energy price shocks of the 1970s. We remain heavily reliant on oil (both domestic and imported) as a source of energy. It is conceivable that the oil market may remain unstable, and thus it is likely that energy users in the United States will continue to move toward more diversity in sources of fuel. This adaptability is perhaps the major reason why the price of "energy" (as opposed to specific fuels in specific areas) has remained relatively stable.

NOTES

1 Brinley Thomas (1986): "Was There an Energy Crisis in Great Britain in the 17th Century?" *Explorations in Economic History*, 23, 125–6.
2 E. A. Wrigley (1988): *Continuity, Chance and Change*. Cambridge: Cambridge University Press, 104.
3 For a thorough discussion see Roy Church (1986): *The History of the British Coal Industry, Vol. 3, 1830–1913: Victorian Pre-eminence*. Oxford: Clarendon Press.

FIGURE SOURCES

Figure 27.1: Wood in Britain: 1450–1649, timber price index from Joan Thirsk, ed. (1967): *The Agrarian History of England and Wales, Vol. IV, 1500–1640*. London: Cambridge University Press, 846–50; 1650–1701, charcoal price at Eton

College from William Beveridge (1939): *Prices and Wages in England from the Twelfth to the Nineteenth Century.* London: Longmans, 143–6.

Coal in Britain: 1601–90, estimates of f.o.b. Newcastle price based primarily on John U. Nef (1932): *The Rise of the British Coal Industry, Vol. II.* London: Routledge, 396–7, with interpolations based on prices at Eton and Westminster schools from William Beveridge (1939): *Prices and Wages in England from the Twelfth to the Nineteenth Century.* London: Longmans, 145–6, 193–4; 1691–1829, f.o.b. Newcastle price described in William J. Hausman (1987): "The English Coastal Coal Trade, 1691–1910: How Rapid was Productivity Growth?" *Economic History Review*, 40, 591; 1830–1913, pithead coal price index from Roy Church (1986): *The History of the British Coal Industry, Vol. 3, 1830–1913: Victorian Pre-eminence.* Oxford: Clarendon Press, 54; 1914–46, pithead selling value from Barry Supple (1987): *The History of the British Coal Industry, Vol. 4, 1913–1946: The Political Economy of Decline.* Oxford: Clarendon Press, 8–9; 1947–82, colliery costs per tonne from William Ashworth (1986): *The History of the British Coal Industry, Vol. 5, 1946–1982: The Nationalized Industry.* Oxford: Clarendon Press, 682–5; 1982–8, coal manufacturing price index from Great Britain Central Statistical Office (1986, 1990): *Annual Abstract of Statistics.* London: HMSO.

Coal in the United States: 1800–79, delivered price of anthracite coal from US Department of Commerce (1975): *Historical Statistics of the United States*, Part 1. Washington: US GPO, 208–9; 1880–1970, avearge value of bituminous coal, f.o.b. mine, from US Department of Commerce (1975): *Historical Statistics of the United States*, Part 1. Washington US GPO, 589–90; 1971–87, average value of bituninous coal, f.o.b. mini, from Energy Information Administration (1989): *Coal Data: A Reference.* (Washington: US GPO, 1989), p. 65.

British Price Index: 1450–1780, E. H. Phelps Brown and Sheila Hopkins (1956): "Seven Centuries of the Price of Consumables, Compared with Builders' Wage Rates." *Economica*, 23, 296–314; 1781–1850, Peter Lindert and Jeffrey Williamson (1983): "English Workers' Living Standards during the Industrial Revolution: A New Look." *Economic History Review*, 35, 468–73, with interpolations from Peter Lindert and Jeffrey Williamson (1985): "English Workers' Real Wages: A Reply to Crafts." *Journal of Economic History*, 45, 145–53; 1851–1911, E. H. Phelps Brown and Sheila Hopkins (1956): as above 1912–84, retail price index based on official statistics taken from David Butler and Gareth Butler (1986): *British Political Facts, 1900–1985.* New York: St Martin's 380–2; 1985–8, retail price index from Great Britain Central Statistical Office (1990): *Annual Abstract of Statistics.* London: HMSO.

US Price Index: 1800–1970, consumer price index from US Department of Commerce (1975): *Historical Statistics of the United States*, Part 1. Washington: US GPO, 210–11; 1971–88, *Economic Report of the President*, 1990. Washington: US GPO, 359.

Figure 27.2: Wood: 1800–1910, wholesale lumber index from US Department of Commerce (1975): *Historical Statistics of the United States*, Part 1. Washington: US GPO, 548.

Coal: See sources for figure 27.1.

Oil: 1860–1949, average value at well from US Department of Commerce (1975): *Historical Statistics of the United States*, Part 1 Washington: US GPO, 593–4; 1950–84, estimated actual world transactions price in US dollars from James M. Griffin and Henry B. Steele (1986): *Energy Economics and Policy.* Orlando:

Academic Press, 16; 1985–9, *Monthly Energy Review, May 1990.* Washington: US Energy Information Agency, 92.

Natural Gas: 1920–89, *Monthly Energy Review, August 1990.* Washington: US Energy Information Agency, 109.

Figure 27.3 1850–1970, US Department of Commerce (1975): *Historical Statistics of the United States*, Part 1. Washington: US GPO, 587–8; 1973–89, *Monthly Energy Review, August 1990.* Washington: US Energy Information Agency, 9,11.

28

Trends in the Price and Supply of Oil

Morris A. Adelman

Figure 28.1 is an index (1982 = 100) of the price of gasoline at the pump since 1920. Nominal prices have been deflated by the Consumer Price Index.[1] Most of the price consists of refining, marketing, transportation, and taxes. Only a minor portion consists of the mineral crude oil, extracted from a supposedly "limited and increasingly scarce resource."

Figure 28.2 shows the inflation-adjusted price of crude oil over a longer period. Before World War II, only the United States had publicly known "arm's length" crude oil prices. Moreover, the United States was a substantial exporter. Hence, its price was the world price. After the war, the Persian Gulf price became the world price. The US price was considerably higher than the world price before 1972 because of informal, then mandatory, import quotas. After 1972, it was lower than

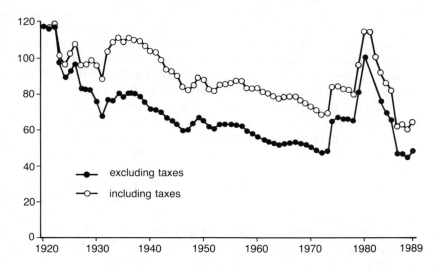

Figure 28.1   Real gasoline prices, 1920–89; index: 1982 =100

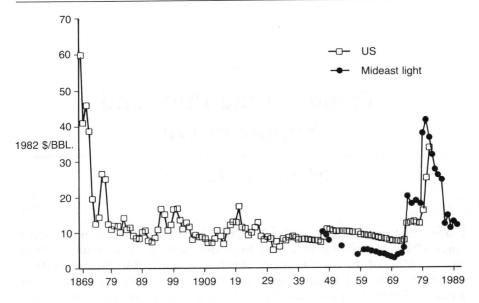

Figure 28.2   Crude oil prices, 1869–1990

the world price because of price ceilings. But by 1981, the two series merged, and for later years the figure shows only the Persian Gulf price.

There is no long-term, gradual increase in price, contrary to expectations for a gradually exhausted resource. Indeed, the trend of the Persian Gulf price was strongly downward through 1970. This does not necessarily imply increasingly plentiful supply, although that may have been the case. The change in scarcity, if any, was overlain by the loss of market control. The multinational oil companies could not keep the price from eroding toward (but not all the way to) the competitive level.

The sharp price spikes in 1973–4, 1979–81, and 1990 were all occasioned by deliberate supply curtailment by some or most of the governments comprising the Organization of Petroleum Exporting Countries (OPEC). The price declined mildly after the 1974 peak; it sank and collapsed from the even higher 1981 peak; another decline was to be expected after November 1990, as this paper was completed. [In May 1992, the Persian Gulf indicator price we use (Dubai Fateh 32°) was $17.60 per barrel. In 1982 dollars (US gross domestic product implicit price deflator), this was $12.14.]

It is widely believed that the price of a mineral must increase over time because the limited stock is being exhausted. Figure 28.3 shows that through 1973, additions to reserves in the United States outstripped production; hence reserves continued to grow. Thereafter, as cost increases outstripped price increases, less was added by exploration and development than was produced. Figure 28.4 explains *why* the turnabout occurred in the United States: in the rest of the world, additions

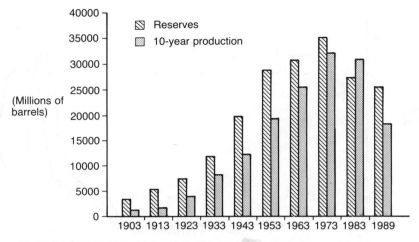

Figure 28.3 Crude reserves and 10-year production, USA, 1903–89

to reserves continued to far exceed production. The constant growth in cheaper non-US reserves has pre-empted the demand for additional US reserves, causing the latter to decline.

If the price of any product is expected to rise in the future, the value of an inventory of that product must rise accordingly. Figure 28.5 shows that the value of reserves held in the United States showed no tendency to rise after World War II, until the price explosion of 1973. Figure 28.5 also shows that the value of a barrel in the ground paralleled the cost of developing the barrel into a reserve. This is necessary because developing a barrel is a substitute for buying one. Figure 28.6 shows that the

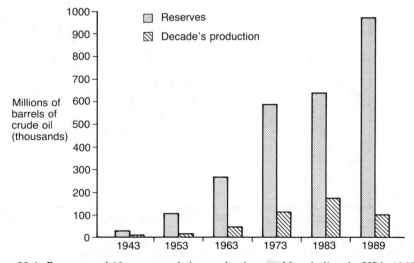

Figure 28.4 Reserves and 10-year cumulative production, world excluding the USA, 1943–89

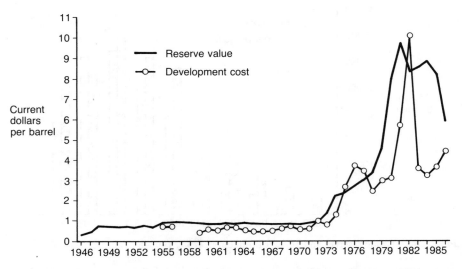

Figure 28.5 Crude oil reserve value, 1946–86, and post-tax development cost, 1955–86

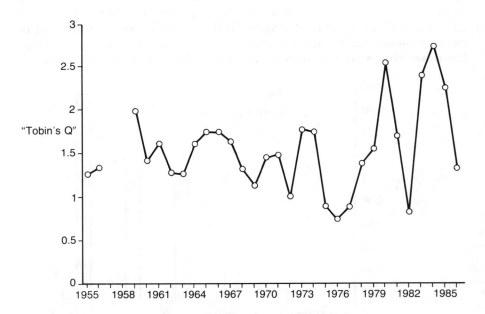

Figure 28.6 Ratio, reserve value/development cost, crude oil reserves, USA, 1955–86 (development cost is post-tax)

ratio of reserve value to capital cost, or "Tobin's Q" stayed around 1.5 and was little affected even by the turbulence of the 1970s and 1980s, although there were great year-to-year fluctuations.

Figure 28.7 is a supply curve for the non-communist world outside the United States, Canada, and Europe. The length of each horizontal line segment shows the capacity of those countries in the given year. For non-cartel countries, this is simply current production. For cartel countries, it is the highest recorded past output. The height of the line on the vertical axis shows the thousands of dollars spent to establish, not a barrel in the ground (as in figure 28.4), but a barrel of additional daily capacity, according to data on expenditures, well completions, and well productivity for the given year. The lines are arrayed in rising order of expenditures per incremental capacity unit. There is a bias, because 1985 costs per foot drilled are applied to earlier years, when technology was less efficient; the true starting curve would therefore be farther to the left.

As calculated in figure 28.7, the curve moved sharply to the right from 1955 to 1975, a period when capacity expanded as much as or more than consumption. But it was stable through the next 10 years as consumption stagnated. The dashed line shows the capacity that would exist if each country produced 5 percent of its 1985 proven reserves per year. (The industry rule of thumb is 7 percent; in the United States, it is about

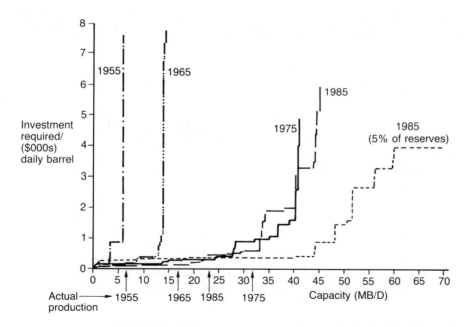

Figure 28.7 Non-communist world supply curve (excluding N. America and W. Europe)
Source: Adelman and Shahi 1989.

10 percent.) If we used not 1985 but 1989 reserves, the dashed line would be much farther to the right.

It seems impossible to reconcile these data with any theory or vision that oil is a "limited exhaustible resource," becoming ever more scarce and expensive. What we observe is the net result of two contrary forces: diminishing returns, as the industry moves from larger to smaller deposits and from better to poorer quality, versus increasing knowledge – of science and technology generally, and of local geological structures. So far, increasing knowledge has won, but nobody can say how long this will continue. However, figures 28.4 and 28.5 show how the process can at least be monitored: by measuring the investment needed at the margin to install an additional unit of reserves or capacity, and by measuring the current values of existing reserves. A persistent substantial increase would be the glowing red light of increasing scarcity.

NOTE

1   The data in this paper (whose preparation was much helped by Rachel E. Obstler) are from a research project embodied in two books published by MIT Press: *The Economics of Petroleum Supply* (1993) and *The Genie Out Of The Bottle: World Oil Since 1970* (1995).

EDITOR'S NOTE (JULIAN L. SIMON)

Figure 28.8 traces the known world reserves of crude oil compared with annual world production, thus showing the number of years' consumption,

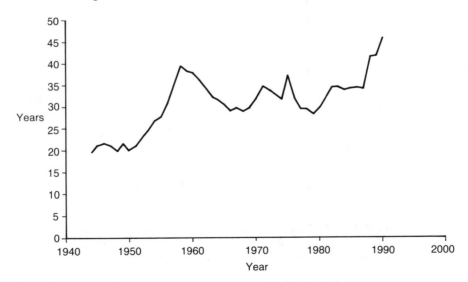

Figure 28.8   Crude oil, world known reserves/annual world production
Source: J. L. Simon, G. Weinrauch, and S. Moore, "The Reserves of Extracted Resources: Historical Data", *Non-Renewable Resources*, summer, 1994.

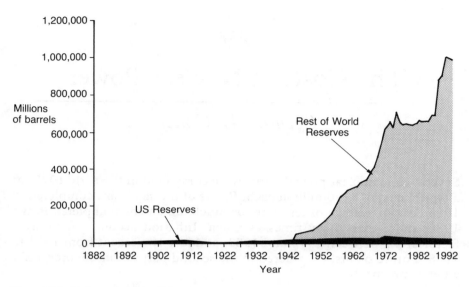

Figure 28.9 Estimated crude oil reserves
Source: See figure 28.8.

at current rates, represented in each year by that year's known reserves. Figure 28.9 traces estimated US and world crude oil reserves in millions of barrels.

29

The Costs of Nuclear Power

Bernard L. Cohen

Several large nuclear power plants were completed in the early 1970s at a typical cost of $170 million each. Plants of the same size completed in 1983 cost an average of ten times as much, and some completed in the 1980s more than thirty times as much. Inflation accounts for only a factor of 2.2 between 1973 and 1983, and just 18 percent from 1983 to 1988. Future plants, considered in the second part of this chapter, offer a better prospect.

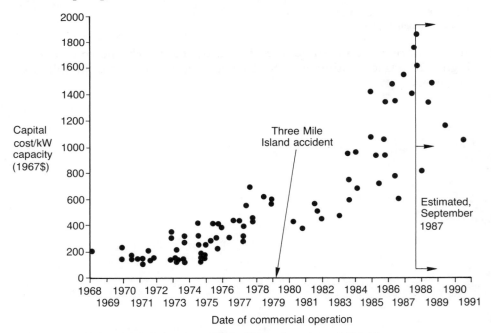

Figure 29.1 Estimated capital cost per kW for US nuclear plants, 1968–90
Notes: Adjustments for inflation are based on the Consumer Price Index mid-way between the dates of construction start and commercial operation. In a few cases, costs are averaged over two or three units at the same site. Dates after mid-1987 are estimates projected at that time.
Source: Data are from Tennessee Valley Authority, "US Nuclear Power Plants' Cost Per Kilowatt Report," Sept. 1987.

REASONS FOR CONSTRUCTION COST INCREASES

The construction costs of various plants, adjusted for inflation, are plotted versus their date of initial commercial operation in figure 29.1. There is about an 800 percent increase over and above inflation during this fifteen-year period.

For example, Commonwealth Edison, the utility serving the Chicago area, completed its Dresden nuclear plants in 1970–1 for $146 per kilowatt (or kW, the unit of electrical power used to measure generating capacity), its Quad Cities plants in 1973 for $164/kW, and its Zion plants in 1973–4 for $280/kW. But its LaSalle nuclear plants, completed in 1982–4, cost $1,160/kW, and its Byron and Braidwood plants, completed in 1985–7, cost $1,880/kW – a thirteen-fold increase over the seventeen-year period. Northeast Utilities completed its Millstone 1, 2, and 3 nuclear plants, respectively, for $153/kW in 1971, $487/kW in 1975, and $3,326/kW in 1986, a 22-fold increase in fifteen years. Duke Power, widely considered one of the most efficient utilities in the nation in handling nuclear technology, finished construction on its Oconee plants in 1973–4 for $181/kW, on its McGuire plants in 1981–4 for $848/kW, and on its Catauba plants in 1985–7 for $1,703/kW, a nearly ten-fold increase in fourteen years. Philadelphia Electric Company completed its two Peach Bottom plants in 1974 at an average cost of $382 million, but the second of its two Limerick plants, completed in 1988, cost $2.9 billion – 7.6 times as much. A long list of such price escalations could be quoted, and there are no exceptions.

Understanding Construction Costs[1]

The Philadelphia office of the United Engineers and Constructors (hereafter called "United Engineers"), under contract with the US Department of Energy, makes frequent estimates of the cost of building a nuclear power plant at the current price of labor and materials. These are called the EEDB (energy economic data base). They include the median experience (ME) for all plants under construction at any time, and the best experience (BE), based on a small group of plants with the lowest costs.

There is little difference between BE and ME plants with regard to materials. They purchased the same items from the same suppliers for the same prices. Incidentally, the equipment for generating electricity is purchased from vendors and represents only a small part of the materials cost – 24 percent for the nuclear steam supply system, which includes the reactor, steam generators, and pumps, and 16 percent for the turbine and generator. They represent only 7.4 percent and 5 percent, respectively, of the total EEDB cost. The rest of the cost is for concrete, brackets, braces, piping, electrical cables, structures, and installation.

While there is little difference in materials cost, the difference in labor costs between ME and BE plants is spectacular. The comparison is broken down in table 29.1. About half of the labor costs are for professionals. It is in professional labor, such as design, construction, and quality control engineers, that the difference between BE and ME projects is greatest. It is also for professional labor that the escalation has been largest – in 1978 it represented only 38 percent of total labor costs; in 1987, 52 percent. However, essentially all labor costs are about twice as high for ME as for BE projects.

The total cost of a power plant is defined as the total amount of money spent up to the time it goes into commercial operation. In addition to the cost of labor and materials, there are two other very important factors:

First is the cost escalation factor, which takes into account the inflation of costs with time after project initiation. Inflation for construction projects has been about 2 percent per year higher than general inflation as represented by the consumer price index (CPI).[2] For example, between 1973 and 1981, the average annual price increase was 11.5 percent for concrete, 10.2 percent for turbines, and 13.7 percent for pipe, but only 9.5 percent for the CPI.[3] If the assumed inflation for construction projects is 12 percent per year, which was typical of the late 1970s and early 1980s, the cost of equipment not purchased until near the end of the ten-year construction period will have tripled by that time.

Second is the interest charges on funds used during construction. All money used for construction must be borrowed or obtained by some equivalent procedure. Hence the interest paid up to the time the plant goes into operation is included in the total cost of the plant. For example, the basic engineering may involve salaries paid twelve years

Table 29.1 Breakdown of labor costs for nuclear power plants and coal-burning plants from the 1987 EEDB

Type of labor	Median experience plant	Best experience plant	Median/ Best	Coal-burning plant
Structural craft	150	91	1.60	76
Mechanical craft	210	100	2.10	180
Electrical craft	80	48	1.70	52
Services (indirect costs)	170	86	2.00	38
Engineering	410	170	2.30	56
Field supervision	320	65	4.90	50
Other professional	580	27	2.10	6
Insurance taxes	115	65	1.80	65
Total	1520	660	2.30	520

Source: United Engineers and Constructors, Inc., "Phase IX Update (1987) Report for the Energy Economic Data Base Program," July 1988; also, "Phase X Update Report," September 1989.

before the plant becomes operational. If the annual interest rate is 15 percent, its cost is therefore multiplied by $1.15^{12} = 5$.

These two factors depend on the length of time required for construction and the rate of inflation and interest rates. Without inflation, or if plants could be built very rapidly, these factors would be close to 1.0, having little impact on the cost.

The product of these two factors – cost escalation and interest charges – has been estimated by United Engineers for various project initiation dates. It was only 1.17 in 1967, when construction times were 5.5 years and the inflation rate was 4 percent per year; it increased to 1.45 in 1973, when construction times reached eight years but inflation was still only 4 percent; it went up to 2.1 in 1975–8, when construction times lengthened to ten years and the inflation rate averaged about 7 percent; and it jumped to 3.2 in 1980 when construction times reached twelve years and inflation soared to 12 percent. That is, the cost of a plant started in 1980 would have been more than triple the EEDB labor and materials cost; 69 percent would have been for inflation and interest.

This analysis reveals two important reasons, besides increased labor prices, why costs of nuclear plants completed during the 1980s were so high: their construction times were much longer than in earlier years, and they were built during a time of high inflation. Why did they take so long to build?

Regulatory Ratcheting

The Nuclear Regulatory Commission (NRC) and its predecessor, the Atomic Energy Commission Office of Regulation, as parts of the United States Government, must respond to public concern. Starting in the early 1970s, the public grew concerned about the safety of nuclear power plants (see the chapter on that subject in this volume): the NRC therefore tightened regulations and requirements for safety equipment.

This process came to be known as "ratcheting." Like a ratchet wrench that is moved back and forth but always tightens and never loosens a bolt, the regulatory requirements were constantly tightened, requiring additional equipment and construction labor and materials. This also increased costs by extending the time required for construction. Total construction time increased from seven years in 1971 to twelve in 1980, roughly doubling the final cost of plants. In addition, labor and materials costs, corrected for inflation, approximately doubled during that period. Thus, regulatory ratcheting, quite aside from the effects of inflation, quadrupled the cost of a nuclear power plant.

Regulatory Turbulence

The escalating labor costs for construction of nuclear plants were not all directly the result of regulatory ratcheting. This ratcheting, applied to plants under construction, had more serious effects. As new regulations were issued, designs had to be modified to incorporate them. In some instances, the walls of a building were already in place when new regulations appeared requiring substantial amounts of new equipment to be included within them. In some cases this proved nearly impossible, and in most cases it required great added expense for engineering and repositioning of equipment, piping, and cables already installed. Constructors, attempting to avoid such situations, often included features not required in an effort to anticipate rule changes that never materialized. This also added to cost.

Changing plans in the course of construction is a confusing process that can easily lead to costly mistakes. The Diablo Canyon plant in California was ready for operation when such a mistake was discovered, necessitating many months of delay. Delaying completion of a plant typically costs more than a million dollars per day.

Another source of cost escalation in some plants was delays caused by opposition from well-organized local or national groups that took advantage of hearings and legal strategies to delay construction. The Seabrook plant in New Hampshire suffered two years' delay[4] due to vocal opponents' concern with the plant's discharges of warm water (typically 80°F) into the Atlantic Ocean. The utility eventually provided a large and expensive system for piping this warm water two-and-a-half miles out from shore before releasing it. Three years' additional delay, costing about another billion dollars, occurred after completion when Massachusetts Governor Michael Dukakis, in deference to the plant's opponents, refused to cooperate in emergency planning exercises for a part of his state included in the emergency planning zone. The NRC ruled in 1990 that the plant could operate without that cooperation.

In summary, there are many reasons why the costs of nuclear plants begun in former years were far higher than originally estimated. Nearly all – other than unexpectedly high inflation rates – are closely linked to regulatory ratcheting and turbulence.

THE NEXT GENERATION OF NUCLEAR POWER PLANTS[5]

The nuclear power plants in service today were conceptually designed and developed during the 1960s. At that time, it was deemed necessary to achieve maximum efficiency and minimum cost in order to compete successfully with coal- and oil-burning plants. The latter were priced at 15 percent of their present cost and used fuel that was cheap by current standards. In order to maximize efficiencies in nuclear plants, temperatures, pressures, and power densities were pushed up to their highest practical limits.

As the NRC tightened its safety requirements, new systems were added to meet them. Eventually, the amount of labor and materials needed for these add-ons exceeded that for the plant as originally conceived.

New Plant Designs

By the early 1980s it became apparent that a new conceptual design of nuclear reactors was called for. The cost of electricity from coal and oil had risen to make competing with them possible at less than maximum efficiency. Furthermore, the added efficiency achieved by pushing temperatures, pressures, and power densities to their limits was overshadowed by the efficiency lost due to shutdowns when these limits were exceeded. But above all it would be much easier to satisfy the public's demand for super safety by starting over with a new conceptual design than by using myriads of add-ons to a design originally targeted at different goals. In the mid-1980s, several reactor vendors undertook these new designs.

In attempting to obtain maximum performance per unit of cost, designers nearly always find it advantageous to build plants with higher power output. It costs substantially less to build and operate a 1,200,000-kW plant than two 600,000-kW plants of the same basic design. However, if safety is the primary goal, as it is in the United States, it is much easier to assure adequate cooling if only half as much heat must be dissipated. In a 600,000-kW reactor of appropriate design, simple gravity flow of water with natural convection is adequate. Unlike pumps, which can fail or stop from power outage, gravity never stops working. That makes such a 600,000-kW reactor inherently safer. This is called "passive stability."

Since construction time is so important, another consideration is that many more parts of a smaller reactor than of a larger one can be produced and assembled in a factory. Operations are much more efficient in a factory than at a construction site because permanently installed equipment and long-term employees can produce many copies of the same item for many plants.

The 600,000-kW size range is now considered optimum for US utilities, and domestic and foreign reactor vendors are competing to provide it. Coal- and oil-burning plants of this size have always been available and are the most widely used in the United States.

Another aspect of the new design philosophy is to favor a larger tolerance to variation in operating conditions over optimizing efficiency. To take just one example, a reactor must be shut down for safety reasons if the power density – the energy per second produced by each foot of fuel assembly – exceeds a predetermined limit. Present reactors operate at 5.5 kW per foot, but in the new reactors this is reduced to 4.0 kW per foot. This change reduces efficiency but means fewer shutdowns.

Another change in the new design is to favor simplicity over complexity. Adding complex equipment and operations may improve the efficiency of a reactor, but it also introduces more possibility for equipment failures and operator mistakes. In earlier reactors, accepting these problems was deemed a worthwhile sacrifice, but not under the new design philosophy.

In addition to changes in design philosophy, there have been many lessons learned from experience that can be incorporated into the new reactors. These include systems for adding small quantities of various chemicals to the water to reduce corrosion, employing different materials for certain applications, and using new methods for construction.

Licensing Reform

Licensing of reactors has been a major problem for US utilities. They must obtain a construction permit to start building, and when the plant is nearly complete they must apply separately for an operating license. These are long, expensive processes that can cause disastrous delays. To avoid this sort of fiasco in future plants, new rules have been proposed under which both a construction permit and an operating license will be obtained in a single procedure before construction begins. When the plant is complete, the utility need only show that it was constructed in accord with the plans originally approved.

Even more important than the licensing problem is standardization. Almost every US reactor has been custom designed and custom built. This is no longer necessary. French experience, among other examples, shows that a standardized design approach is very efficient and successful, resulting in high-quality plants built quickly and cheaply. For the United States, with its current tumultuous licensing procedures, standardization will be enormously helpful.

Cost per Kilowatt-Hour

Table 29.2 breaks down the cost per kilowatt-hour for different types of power plants. The costs of operating and maintaining a facility, and of providing it with fuel, are easy to understand. The contribution of the capital used in constructing it is calculated as the amount of money per kilowatt-hour needed to purchase an annuity that will pay off all capital

Table 29.2 Cost per kilowatt-hour for various types of power plants (in mills [0.1 cents] of 1987 dollars, for plants going into operation in the year 2000)

Type	Capital	Operation and maintenance	Fuel	Decommissioning	Total
Median experience	56	13.0	7.2	0.5	77
Best experience	28	9.1	6.4	0.5	44
APWR	22	9.1	6.4	0.5	38
New-600 (2 units)	22	10.4	6.4	0.7	40
Coal (2 units)	21	5.9	21.0	0.1	48

costs and interest before the end of the facility's life. The decommissioning charge is the amount of money per kilowatt-hour that must be set aside so that, including the interest it has earned, there will be enough money to pay for decommissioning the plant when it is retired. There are many uncertainties in these calculations, but both the US Government and the utilities employ groups of experts who have developed standard procedures that are normally applied and widely accepted.

In table 29.2, the APWR (advanced pressurized water reactor) refers to present-type reactors with various upgrades for safety, a standardized design, and the streamlined licensing procedures. The New-600 represents the new generation of reactors that fully use passive safety features. Both the New-600 and the coal-burning plant are of half the capacity of the present generation reactors and the APWR, and therefore table 29.2 considers two of these plants at the same site in drawing comparisons. Note that having two plants rather than one increases the operating, maintenance, and decommissioning costs and makes electricity from the New-600 slightly more expensive than from the APWR. That is part of the price we pay for super safety, but it is hardly significant.

The most important point in table 29.2 is that electricity from the new-generation reactors is about 20 percent cheaper than that from a coal-burning plant. The coal-burning plant is cheaper to build, maintain, and decommission, but its fuel cost is more than three times higher.

Of course, this conclusion depends heavily on the assumptions of standardized design, streamlined licensing, and no regulatory turbulence. These would reduce construction times to five or six years, according to Department of Energy analysts, and that fact alone would account for a large fraction of the cost saving. Industry planners are hoping for even shorter construction times, especially after a few of these standardized plants have been built.

In Western Europe, where there has been no regulatory turbulence comparable to that in the United States, most nuclear power plants have been relatively cheap. A 1982 study by the European Economic Community[6] estimated that for projects started at that time, electricity from coal-fired plants would be more costly than that from nuclear plants by 70 percent in France and Italy, 40 percent in Germany, and 30 percent in Belgium. These results were based on the slowest rise in the price of coal that was considered reasonable to expect and did not include scrubbers to remove sulfur, which are usually required on new coal-burning plants in the United States and add substantially to their cost.

NOTES

1   United Engineers and Constructors, Inc., "Phase IX Update (1987) Report for the Energy Economic Data Base Program," July 1988. Also, "Phase X Update Report," September 1989.

2   A. Reynolds, "Cost of Coal vs. Nuclear in Electric Power Generation," US
    Energy Information Administrative Document (1982).
3   W. W. Brandfon, "The Economics of Nuclear Power," American Ceramic
    Society, Cincinnati, 1982.
4   M. R. Copulos, *Confrontation at Seabrook* (Washington: The Heritage Founda-
    tion, 1978).
5   J. J. Taylor, "Improved and Safer Nuclear Power," *Science* 244, 318 (April 21,
    1989); J. J. Taylor, K. E. Stahlkopf, D. M. Noble, and G. J. Dau, "LWR
    Development in the U.S.A.," *Nuclear Engineering and Design* 109, 19 (1988);
    J. Catron, "New Interest in Passive Reactor Designs," *EPRI Journal* 14, 3–4
    (April 1989); R. Livingston, "The Next Generation," *Nuclear Industry*, July
    1988; K. E. Stahlkopf, J. C. DeVine, and W. R. Sugnet, "U.S. ALWR Pro-
    gramme Sets Out Utility Requirements for the Future," *Nuclear Engineering
    International*, November 1988, 16; R. Vijuk and H. Bruschi, "AP600 Offers a
    Simpler Way to Greater Safety, Operability, and Maintainability," *Nuclear En-
    gineering International*, November 1988, 22; K. Hannerz, "Applying the PIUS to
    Power Generation," *Nuclear Engineering International*, December 1983; K. Han-
    nerz, "Making Progress on PIUS Design and Verification," *Nuclear Engineering
    International*, November 1988, 29; "Nuclear Power: The New Generation,"
    IAEA (International Atomic Energy Agency) *Bulletin* 331:3 (1989), *passim*.
6   *Nuclear News*, July 1982, 48.

30

Trends in Availability of Non-Fuel Minerals

John G. Myers, Stephen Moore, and Julian L. Simon

Prices of minerals can be thought of in (at least) two ways. Minerals are used as an input to production by combining them with other inputs, labor in particular, to produce products; if the relative price of a mineral rises, it serves as a constraint on production and limits output. In order to determine whether a mineral has served as a constraint on production over the long term, we examine the movements of mineral prices relative to wage rates.

Alternatively, we ask whether the quantity of other commodities required to exchange for a unit of a mineral has risen over the long term. The answer to this question can be found in the relative price of the mineral, which is the nominal price of the mineral divided by a general price index (here an index of consumer prices).

The sample examined in this paper includes 13 minerals. The mineral with the longest series of available prices is copper, for which we have data from 1801. This metal has been the object of much study because of its special qualities, particularly in electrical uses, and the large value of its annual consumption. As may be seen in figure 30.1, the price of copper relative to the wage rate fell sharply and fairly steadily from 1801 to about 1930; it has drifted downward since then. Relative to consumer prices, the decline in the copper price has been less steep and has shown little trend since 1930. The difference between the two long-term patterns results from the fact that real wages (the wage rate divided by the index of consumer prices) rose during most of the 190 years; this is shown in figure 30.2. The same pattern, the mineral price falling more (or rising less) relative to the wage rate than to the consumer price index, will appear for each mineral we examine.

The second longest price series is for lead, which begins in 1819. As shown in figure 30.3, there is a strong downward trend in the price–wage ratio, which extends over the entire 17 decades. The movement of the price of lead relative to consumer prices is also downward, but more

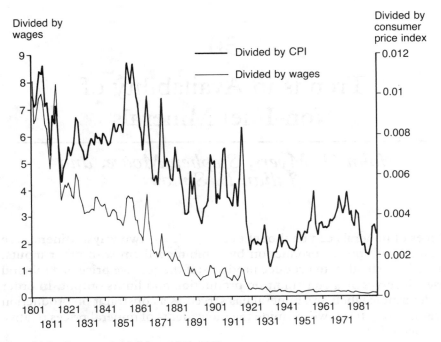

Figure 30.1 Copper price indexes, 1801–1990
Sources: US Bureau of Mines (1987), *Minerals Yearbook, 1985, volume I, Metals and Minerals,* Washington: US Govt Printing Office; id. (1990), *Nonferrous Metal Prices in the United States through 1988, Washington: US Bureau of Mines.*

gradual, with several interruptions. The value of lead consumed in the United States is not great in comparison with that of copper, for example, but it is used extensively in automotive manufacturing and in batteries.

Nickel prices are available from 1840, and reveal a strong downward trend relative to wages up to about 1920, but do not change substantially thereafter (figure 30.4). A similar pattern is revealed for the ratio of the price of nickel to the index of consumer prices. The value of the consumption of nickel in the United States is quite high, but it is all imported (other than recovered scrap). The principal use in this country is for stainless and heat-resisting steel; other important uses are in the manufacture of nonferrous alloys and in electroplating.

Mercury prices have fluctuated widely during the last 140 years; the general movement relative to wages was downward, but with several reverses of direction. As may be seen in figure 30.5, the pattern relative to consumer prices is less clear; fluctuations in relative prices have become larger and larger over time; the last ten years of the period were well below the first ten, however. Mercury is most heavily used today in batteries, chemicals, electrical equipment, and paint.

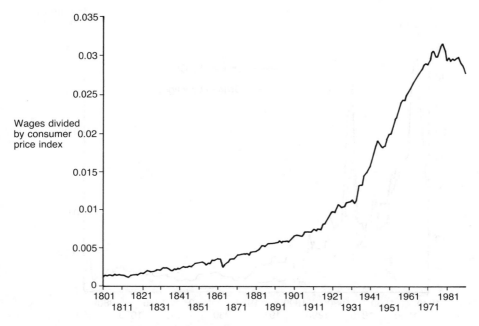

Figure 30.2   Change in real wages over time

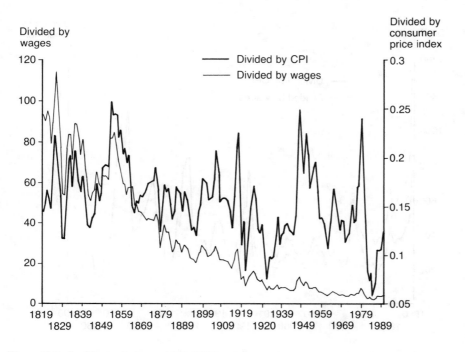

Figure 30.3   Lead price indexes, 1801–1990

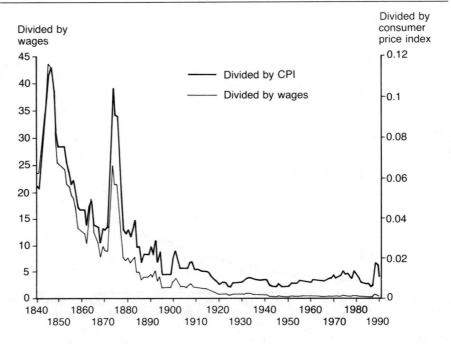

Figure 30.4 Nickel price indexes, 1840–1990

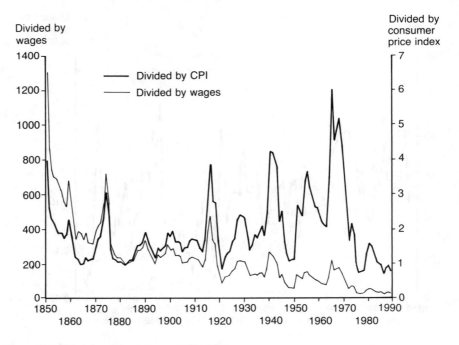

Figure 30.5 Mercury price indexes, 1850–1990

Silver (figure 30.6) was subject to extensive market manipulation in 1979–80, which raised the price to record levels. The end of that episode saw the price return to the range that had prevailed from the turn of the century to 1974. The price of silver relative to the wage rate fell steeply from the 1850s to the 1950s; there has been little net change since then. The trend of silver prices relative to consumer prices was down through the 1930s; and during the late 1980s, the relative price returned to the level of the 1960s. Monetary use of silver is insignificant in the United States today; photography is the main market, followed by electrical uses.

Zinc is an important metal in the United States, although only one-fifth of consumption is domestically produced. The principal uses are in galvanizing, die-casting, and brass. Zinc prices indexed by wages fell sharply from 1853 through the 1930s (figure 30.7) and fell more slowly thereafter. The trend of zinc prices relative to consumer prices is less steep, but clearly downward.

From 1880 to about 1918, the price of platinum rose sharply relative to both wages and consumer prices, as shown in figure 30.8. Thereafter it fell to about 1940, but the price during the 1980s was higher than during the 1880s, relative to wages or prices. Platinum has gained wide use in recent years in the automotive industry, primarily in emission control; this is in addition to its traditional uses in jewelry and electrical equipment, and as a store of value.

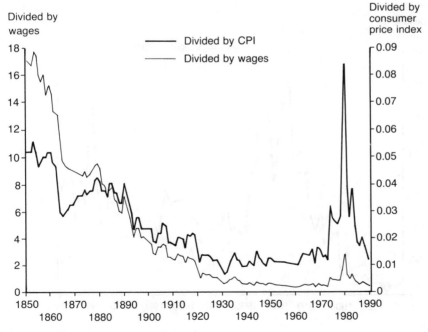

Figure 30.6   Silver price indexes, 1850–1990

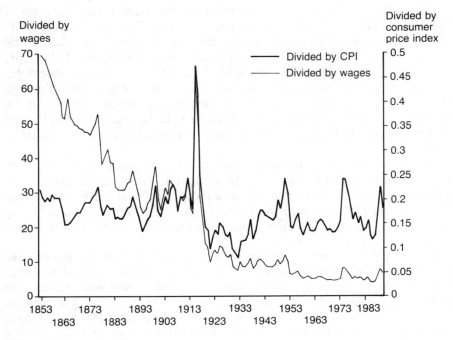

Figure 30.7   Zinc price indexes, 1853–1990

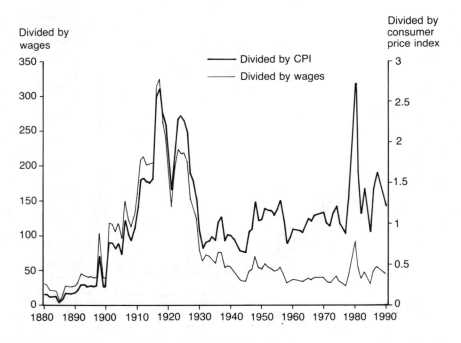

Figure 30.8   Platinum price indexes, 1880–1990

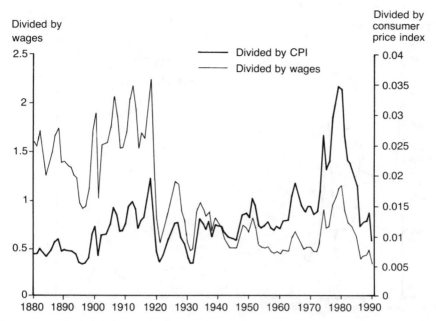

Figure 30.9 Tin price indexes, 1880–1990

Tin is one of the oldest metals used by humans. The long-term pattern of tin prices during this century has been affected by two special features: a large US government stockpile, which is often used in an attempt to stabilize the domestic price; and a long history of international agreements, also designed for price stability. The uses of tin in the United States have also changed substantially over the years as aluminum replaced steel and tin in many containers, while tin chemicals experienced solid growth; solder has remained the principal market for tin. Relative to the wage rate, the price of tin fell substantially from 1880 to the present, as may be seen in figure 30.9. Tin has risen relative to the consumer price index over the same period, however. This result underscores the importance of the choice of deflator in studying long-term trends in commodity prices.

Aluminum is a more recent addition to the list of metals in use. Quoted prices are available from 1895 and reveal a sharp decline to about 1945, as improved production techniques were applied, made possible by a very rapid growth in demand (figure 30.10). The price continued to decline during the next 45 years, relative to both wages and consumer prices. The principal uses of aluminum in the United States are in containers, construction, and the automotive industry.

Another mineral of recent use is antimony, which has a very wide range of markets, including plastics, ceramics, glass, and pigments. The total value of antimony is small compared with such major nonferrous

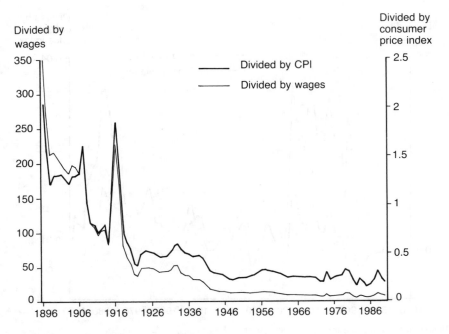

Figure 30.10 Aluminum price indexes, 1896–1990

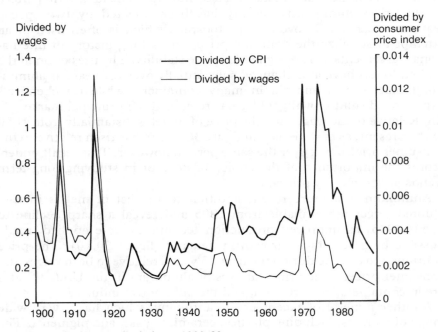

Figure 30.11 Antimony price indexes, 1900–89

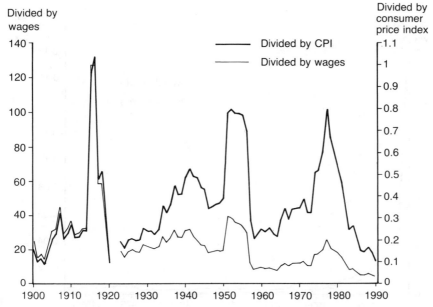

Figure 30.12   Tungsten price indexes, 1900–90

metals as aluminum or copper, however. This mineral has experienced price trends similar to those of tin during the last 90 years, falling relative to the wage rate but rising relative to the index of consumer prices (figure 30.11).

Tungsten prices are available since the turn of the century and reveal the same pattern as tin and antimony, falling relative to wages and rising relative to consumer prices (figure 30.12). The metal is used primarily in cutting and wear-resistant alloy materials such as machine tools and oil well drilling bits.

Prices for magnesium are available only since 1915. They reveal a sharp drop from 1922 to 1949, relative to both wages and consumer prices, as may be observed in figure 30.13. This early pattern is similar to that of aluminum, reflecting the implementation of improved techniques in production made possible by the rapid growth in demand for this new metal, which was attractive because of its high strength-to-weight ratio. Subsequent years have been dominated by conflicting forces, such as strategic stockpile-building and then -releasing, a great diversification of civilian uses, and energy price increases, which raised production costs. Relative prices have fluctuated in a narrow range during the last 45 years.

Manganese is the mineral with the shortest price series in our study, only 37 years (1954–90). This metal has two main uses, in iron- and steel-making, and in dry cell batteries. It is classified as a strategic material and is stockpiled by the US government. Relative to both the

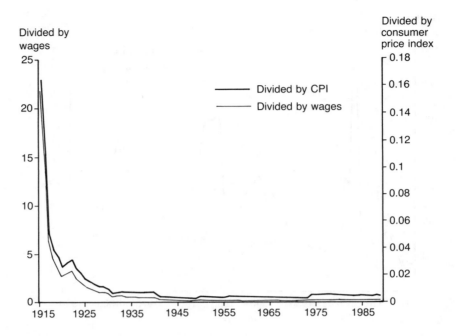

Figure 30.13 Magnesium price indexes, 1915–89

wage rate and the index of consumer prices, the price trend of manganese was generally downward for the available period, with some upturn during the last three years for which we have data.

The results of this survey of mineral prices can be summarized in a few words: 12 of the 13 fell relative to the wage rate, and 9 of the 13 fell relative to the index of consumer prices. Only the price of platinum rose relative to the wage rate, while the prices of platinum, tin, antimony, and tungsten rose relative to consumer prices. As the preceding paragraphs indicate, each of these minerals has its own history and peculiarities. It seems warranted to conclude, nevertheless, that there is practically no indication that scarcity of minerals served as a constraint on production over the long term, and scant evidence of increased costs of minerals, in the sense that more of other commodities must be exchanged for them.

31

Trends in Nonrenewable Resources

H. E. Goeller

Over the last twenty-five years a number of studies have questioned the long-term adequacy of the world's nonrenewable resources, chiefly fossil fuels and the chemical elements. An early and notable study among these, Meadows et al. (1972), suggested that world aluminum resources might be exhausted by 2003, about ten years from now. In reality, aluminum will always be recoverable at only marginally increased costs from unlimited supplies of clays and anorthosite using already developed processes.

This chapter is concerned with the world's supply of chemical elements, leaving questions of fossil fuels and energy in general to other chapters in this volume. I draw here on two earlier papers (Goeller and Weinberg, 1976; Goeller and Zucker, 1984) which indicate that the world is unlikely to run short of any elements before about 2050 and that new technology will almost certainly provide alternative sources and substitutes to meet the continuing requirements of civilization in the centuries to come. In an "age of substitutability," perhaps two centuries from now, humanity should be able to subsist permanently on renewable resources of energy, plus substitutes for some chemical elements, and nonrenewable resources of about 36 elements that will be in effectively infinite supply.

We can expect to reach this eventual, permanently sustainable state of humanity in three stages. First, for the next 30 to 50 years, there should be much concerted effort on the development of processes for lower grade and alternative sources of plentiful elements and research on providing substitutes for limited ones. The second stage will entail further research and development and implement the processes and substitutes developed in the first stage. Stage three will complete the transition to this sustainable state of the world. A great amount of time is available in which to do the requisite research and development, much of which will depend on future science and engineering not yet envisioned.

Long-term forecasting over such an extended time can, at best, be only educated speculation. Nevertheless, I believe that this has much greater plausibility and is more realistic than the simple manipulation of "current" resource data, which is essentially what is done by those who see the imminent exhaustion of resources. Who in 1850, for instance, could possibly have given a very accurate idea of lifestyles, science, and technology in the latter part of the twentieth century? Perhaps science fiction writers like Jules Verne and H. G. Wells have come the closest.

In support of this contention I use current reserve and resource data sparingly and only with strict qualifications. Thus much of what is said here is more qualitative than quantitative in nature. The nonrenewable materials of concern here are the 81 stable (nonradioactive) chemical elements which can neither be created nor destroyed except in minute amounts by nuclear processes at totally unrealistic and impractical costs.

Many readers may assume that these elements are obtained only by mining the earth's crust, which is a very heterogeneous medium. However, it turns out that of the 81 chemical elements a dozen or so that are consumed in the greatest amounts are already produced from the atmosphere and the oceans, both of which are homogeneous sources of infinite extent for all intents and purposes. In addition, some of the most used elements are also found in near infinite supply in terrestrial sources as well.

In the material which follows, I take a world rather than a national perspective, because national economies are rapidly becoming part of a single world economy. Most of the material in the text and all of the data in figure 31.1 and the tables are derived from US government sources. Information on reserves and resources are from US Geological Survey (USGS) references and data on demand and consumption are from US Bureau of Mines (USBM) sources. Resources in these references nearly always refer to presently conventional or near conventional sources recoverable with existing processes at or near current costs. The USGS has developed a very specific set of definitions for reserves and resources which is given in detail in USGS Survey Circular 831, 1980 and is well summarized in the annual USBM Mineral Commodity Summaries series. In many cases new processes are already available for both conventional resources at higher costs and nonconventional resources as already pointed out for aluminum.

STAGE 1   PROCESS DEVELOPMENT

Figure 31.1 provides a very schematic diagram of the world flow of material from resources to use to ultimate disposal and loss, using zinc as an example from 1880 to the present. As shown, the in-use-pool is only a small part of either resources, production, or losses over the long term. Note that extended resources recoverable at only moderately increased costs are 15 times larger than current estimates of reserves and resources.

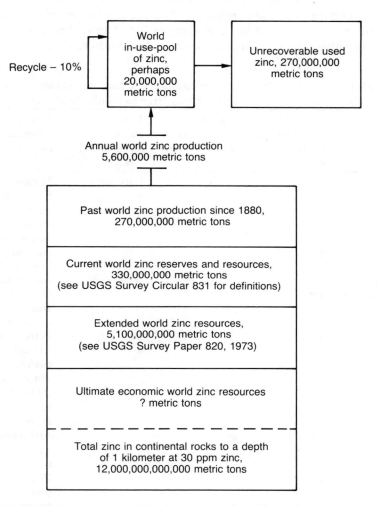

Figure 31.1 World zinc resources, production use, recycle and losses since 1880

It is impossible to quantify ultimate resources. As the geologist Frasché said in 1962, "total exhaustion of any mineral resource will never occur." Some resources will always remain and as we are forced to use ever leaner ores the costs of recovery will increase, generally gradually, until cheaper substitutes, where feasible, take over.

Finally, figure 31.1 indicates that about 10 percent of annual zinc production is recycled material. Although recycling is important, it should not be overemphasized, because unless it is very high it makes little difference in the long run. If it is 10 percent on the first recycle it is only 1 percent on the second and 0.1 percent on the third. However, in the case of aluminum and magnesium it provides for great energy savings.

Table 31.1 is derived primarily from Goeller and Zucker (1984). It provides consumption and resource data for a number of the chemical elements and is arranged in decreasing order of annual world consumption. The table also indicates resource to demand (R/D) ratios for both

Table 31.1 Comparison of current consumption and current and ultimate resources of elements expected to eventually have very large to infinite resources

Element	Annual World Consumption (million metric tons)	Current Resource to Demand Ratio (Years)	Ultimate Resource to Demand Ratio (Years)	Present/ Ultimate Sources
Iron	480	415	Very large ∞	Magnetite, hematite, magnetic taconites, nonmagnetic taconites, laterites, basalt
Nitrogen	140	∞	∞	Atmosphere
Calcium[a]	100	∞	∞	Gypsum, anhydrite, limestone
Carbon	100(?)	4	Very large ∞	Petroleum, natural gas/coal, tar sands, oil shales, C from limestone and H from water
Oxygen	93	∞	∞	Atmosphere/ water electro-lysis
Sodium	77	∞	∞	Rock salt, seawater
Sulfur	53	53	∞	Native sulfur, pyrites/gypsum, anhydrite
Chlorine	33	∞	∞	Rock salt, seawater
Hydrogen	24	Very large	∞	Fossil fuels/water electrolysis
Potassium	22	5450	∞	Sylvite/seawater
Aluminum	17	50	∞	Bauxite/clays, anorthosite
Phosphorus	17	1060	Very large	Phosphate rock/lower grade P.R.

Element	Annual World Consumption (million metric tons)	Current Resource to Demand Ratio (Years)	Ultimate Resource to Demand Ratio (Years)	Present/ Ultimate Sources
Manganese	8.7	320	Very large	Current ores/seafloor nodules
Magnesium[a]	51	∞	∞	Magnesite, brines, seawater
Chromium	3.2	3125	Very large	Current ores/ultramatic rocks[e]
Silicon	2.6	∞	∞	Quartz, sandstone
Argon[b]	1.9	∞	∞	Atmosphere
Titanium[a]	1.7	418	Very large	Current ores/titaniferous magnetites
Nickel	0.71	296	Very large ∞	Current ores/ultramatic rocks[e]
Boron	0.42	643	∞	Current ores/seawater
Bromine	0.30	∞	∞	Seawater
Strontium	0.047	240	∞	Current ores/seawater
Cobalt	0.023	270	Very large ∞	Current ores/sea floor, ultramatic rocks[e]
Iodine	0.012	375	∞	Brines, Chile saltpeter, seawater
Lithium	0.0064	1270	∞	Current ores/seawater
Platinum[c]	0.00019	500	Very large ∞	Current ores/ultramatic rocks[e]
Gallium	<0.00001	?	∞	Bauxite/clays
Rubidium	<0.00001	?	∞	Current ores/seawater

Notes:

[a] Some elements, such as calcium, magnesium, and titanium, are used in both elemental (usually metallic) form and as compounds; consumption values include both forms.

[b] Also neon, krypton, xenon, but not helium.

[c] Also palladium, rhodium, ruthenium, osmium, and iridium.

[d] Depends on split between fossil fuels for fuels versus petrochemicals.

[e] About 4 percent of all surface igneous rocks are ultramatic rocks.

current resources and our estimates of ultimate resources which, in all cases, are judged eventually to be very large to infinite in extent. The table also indicates current and ultimate sources of each element. The annual world consumption of the 28 listed elements is about 1,200 million metric tons; consumption of the remaining 36 elements, shown in table 31.2a, b, is only 24 million tons or only 2 percent of the total.

Table 31.2a Elements expected to always have limited resources listed in order of decreasing resource to demand (R/D) ratios based on *current* estimates of reserves and resources

Element	Extended to current resource ratios	Element	Extended to current resource ratios
Scandium		Tin	
Hafnium		Tantalum	
Beryllium		Selenium	
Rhenium		Barium	2.14
Vanadium		Fluorine	21.5
Cesium		Copper	
Niobium		Tungsten	
Helium		Molydenum	49.8
Lanthanum[a]		Thallium	
Tellurium		Zirconium	
Arsenic	1.95	Mercury	
Yttrium		Germanium	
		Lead	5.27
		Cadmium	
		Antimony	
		Silver	
		Zinc	15.5
		Gold	
		Bismuth	
		Indium	

Table 31.2b Extended to current resource ratios for elements with currently limited resources

Cobalt	4.20	Nickel	4.34
Iron	1.56	Phosphorus	1.89
Manganese	6.67	Titanium	2.68

Note:
[a] Plus the 14 rare earth elements.

Proceeding now to the 38 elements which I expect will always be in limited supply, table 31.2, lists the elements in decreasing order of depletion under growing demands based on *current* resource estimates; thus elements at the end of the list would be depleted first. If current resources were to remain constant, elements on the right in table 31.2 might be depleted before 2100; those on the left sometime after 2100.

Thus, the table sets a *lower limit* for depletions with the assurance that ultimate depletions would be delayed by expanding resources through new technology, through further exploration as times goes by and by gradual use of substitutes.

Table 31.2 also gives the ratio of extended to current resources for a dozen selected elements. The ratios for which data are available were derived from a careful reading of Brobst and Pratt (1973) and vary from 1.56 for iron to nearly 50 for molybdenum. Meadows et al. (1972) assumes that ultimate resources of all elements will be five times known resources. The wide range of values in table 31.2 demonstrates that this assumption is a gross oversimplification.

For just such reasons I purposely refrain from providing numerical values for rates (and dates) of depletion in table 31.2, since these data are beset with enormous uncertainty over longer time frames. Instead, I prefer to think of estimated depletion rates as planning guides for ordering new process research and development and research on substitutes for limited materials.

Use of fertilizers and soil additives in agriculture is perhaps the most nonsubstitutable use of materials. Of the major additives (nitrogen, phosphorus and potassium), only phosphorus is in limited supply. More recently, more optimized use of phosphates, new methods such as trickle irrigation, and a limited but growing return to organic farming have reduced demands, and new genetically engineered plants are on the horizon. Further, Emigh (1972) has shown that phosphate resources are much more extensive than formerly realized. A number of trace elements are also needed in agriculture and animal husbandry. However, even in the long-term future the amounts involved will probably be a small but growing part of overall demands for these elements. The 13 elements involved include fluorine, silicon, vanadium, chromium, manganese, iron, cobalt, copper, zinc, selenium, molybdenum, tin, and iodine.

Since iron and steel constitute over 90 percent of the demand for all metals it is important that iron resources continue to be available at reasonable costs. It appears that current resources of magnetite, hematite, and magnetic taconites will be augmented later by new technology for nonmagnetic taconites and laterites, and ultimately basalt, which averages 8 percent iron. In addition, as supplies of coking coal are depleted it will be feasible to turn to infinite supplies of hydrogen as an iron ore reductant. Although little used now, the H Iron process is already fully developed. It involves passing hydrogen through heated beds of iron ore to reduce the iron oxide to metallic iron.

Although fossil fuels are used predominately as fuels, significant amounts provide the basis for the large petrochemical industry that produces organic plastics, solvents, fibers, and a myriad of other products. As natural gas, oil, and coal resources become depleted, it will be possible to use tar sands, oil shales, possibly organic soils and, ultimately, carbon from limestone and hydrogen from water electrolysis as resources. Processes already exist but are currently uneconomic and increasingly energy intensive.

Resource estimates are often given as amounts recoverable at varying costs. As an extreme example, the US Bureau of Mines (*Mineral Yearbook, 1964*) estimated domestic resources of mercury as 1,600 metric tons recoverable at $2,900 per ton or 50,000 tons at $43,500 per ton; over the past 38 years prices have actually ranged from a low of $4,000 per ton to a high of $34,000 per ton in terms of constant dollars as the ratio of demand to production has varied.

Problems of disruptions of supply, having nothing to do with actual resources, can drastically affect near-term availability of materials. A good example in recent years was the "cobalt crisis," more perceived than real, when internal conflicts threatened cobalt supplies from Zaire in 1978 (*Mineral Yearbook 1978–79*). About one-third of cobalt resources and two-thirds of world production are in Zaire and Zambia (but mostly in Zaire). Panic buying forced a ten-fold increase in prices in some cases. South Africa, which has the preponderance of world chromium, gold, and platinum resources and production, provides an even more serious potential threat. As more plentiful elements substitute for more limited ones in the future, many such threats will surely abate.

It is also important to note that many elements are recovered by processes that include two or more steps and that, though costs of the first step may increase as lower grade or alternative resources are utilized, costs of the secondary steps may not. For example, in aluminum production, alumina is produced first, which is then reduced to aluminum. In this case only the cost of alumina production is increased.

STAGE 2 RESEARCH AND DEVELOPMENT TOWARD SUBSTITUTION

As noted earlier, new processes and/or the use of new minerals can also greatly expand resources. The successful development of flotation technology about 1916 permitted the use of formerly unusable porphyry copper ore deposits which expanded copper resources at least ten-fold. More recently, the discovery of extensive exploitable rare-earth-containing bastnaesite deposits in California required a significant increase in the average natural abundance value for the 14 rare earth elements. I believe that in the longer term, other such breakthroughs will be achieved.

Seldom is any material fully recovered from its ore. The extent of recovery is generally based on minimizing present recovery costs. As ores become leaner, process wastes (tailings) may eventually become the equivalent of new ore and be reworked by further processing. Some phosphate rock operations recover only half the contained phosphate the first time. It is important, therefore, to prevent tailings from being abandoned or dissipated since they will then be lost forever to future use. This is especially important for multi-element ores such as copper–zinc and lead–zinc–silver ores when all the elements contained are not recovered.

In particular, this is very important for those elements obtained mainly or only as byproducts of more plentiful elements. These byproduct elements usually end up in waste streams such as anode slimes, electro-precipitator dusts, or baghouse dusts, which should be rigorously se-questered for future use. A list of such elements is given in table 31.3.

Before discussing substitution, it seems desirable to at least mention other methods that can extend resources. We have already mentioned that recycling, though helpful, is relatively ineffective unless very efficient. Resources can also be extended by miniaturization of products, as already done for automobiles and electronic products; by increasing product longevity, and by reducing aesthetic applications as in jewelry, which was done in both world wars. Finally, in many instances, the materials costs are only a small fraction of final product costs.

STAGE 3   THE AGE OF SUBSTITUTABILITY

Most materials are used solely for their physical or chemical properties. Any alternative material with similar properties will suffice. Further, although a certain material may be optimum for a given use, other materials may be quite adequate. Many examples of substitutes are given in Goeller and Weinberg (1976) and Goeller and Zucker (1984).

Until recently, substitutes have generally been found by trial and error and by empirical means. Now, however, we are rapidly entering a time when new and substitute materials are being "made to order" (Phillips et al., 1982; McCarroll, 1990; Brauman, 1992). Many new plastics,

Table 31.3 Byproduct elements and their sources

Element	Byproduct of
Cadmium	Zinc
Thallium	Zinc
Germanium	Zinc
Indium	Zinc
Arsenic	Copper
Selenium	Copper
Tellurium	Copper
Antimony	Lead
Bismuth	Lead
Hafnium	Zirconium
Gallium	Aluminum
Rhenium	Molybdenum

composite materials, ceramics, gels, metal alloys, and other substances not even dreamed of only a few years ago are rapidly becoming commercially available, and no end is in sight for these exciting developments.

Unfortunately, research and development in the domestic mining and metals sector of our economy appears inadequate at this time, as evidenced in a recently completed two and a half year study by the National Academy of Science (NAS, 1990a, 1990b). Thus, we do not argue that our predicted age of substitutability will be easily achieved; only that the landfall, if we arrive safely at stage 3, should be surprisingly better than the catastrophisists have predicted.

REFERENCES

Brauman, John I. (1992): "Frontiers in Materials Science." *Science* 255, 1049, 1077–112 (February 28).

Brobst, D. A., and Pratt, W. P., eds (1973): "United States Mineral Resources." Geological Survey Professional Paper 820.

Emigh, G. D. (1972): *Engineering and Mining Journal*, 173, 90.

Frasché, D. F. (1963): NAS-NRC *Publication 1000-C*, 18.

Goeller, H. E., and Weinberg, A. M. (1976): "The Age of Substitutability." *Science*, 191, 683.

—— and Zucker, A. (1984): "Infinite Resources: The Ultimate Strategy." *Science*, 223, 456.

McCarroll, Thomas (1990): "Solid as Steel, Light as a Cushion." *Time*, 94 (November 26).

Meadows, D. H., Meadows, D. L., Randers, J., and Behrens, W. W. III (1972): *The Limits to Growth*. New York: Universe Books.

Mineral Yearbook 1964 (1964): Washington: US Government Printing Office, 766.

—— *1978–79* (1980): Washington: US Government Printing Office, 256.

National Academy of Science (1990a): *National Research Council News Report*. 40(10), 2.

NAS (1990b): *Competitiveness of the Minerals and Metals Industry*. USA: National Academy Press.

Phillips, J. C., Marvin Cohen, and Volker Heine (1982): "The Quantum Mechanics of Materials." *Scientific American* 246, 82–102 (June).

32

Trends in Availability and Usage of Outdoor Recreation

Robert H. Nelson

The concept of "outdoor recreation" as an item of consumption is a recent development in Western history, a product of the revolutionary social changes of the past two centuries. For a primitive tribe the outdoors was virtually the entire world; both work and play largely occurred in forests and fields. Even when technology and skills were developed to erect buildings that made indoor living possible, the outdoors was usually near at hand. When the Pilgrims arrived at Plymouth, the outdoors did not appear as a recreational opportunity but as a place and a resource to be tamed and used for the very survival of the community.

Following the Industrial Revolution, however, the outdoors took on a new meaning. Residents of congested cities, with their numerous factory and office workers, might go long periods without entering a secluded wood or seeing a running stream. Rising incomes and improved transportation brought travel to distant locations within reach of large parts of the population. Spurred by large capital investments and technological advances, the productivity of the labor force grew rapidly and made possible growing amounts of leisure time for the pursuit of recreation and other enjoyments.[1] To trace the history of outdoor recreation is, in short, to develop a record of the spread of urban and industrial living arrangements, growing wealth, and rising leisure in the modern age.

Governments have responded to growing public demands for outdoor recreation by creating new park and other recreational land systems. As shown in figure 32.1, there has been an explosive growth in the amount of land set aside in the twentieth century for this purpose. The size of the National Park System has increased from 3.3 million acres at the turn of the century to 76.2 million acres in the late 1980s. Since 1940, the total land area of state park systems has more than doubled. The acreage within the National Wildlife Refuge System grew by a factor of five from 1950 to 1988 alone (although much of the increase represented lands in Alaska that are not now easily accessible for recreation).

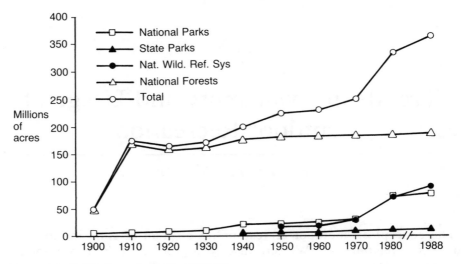

Figure 32.1   Public recreation lands, 1900–88

Legislation to create the national forest system was enacted in 1891. By 1900, the national forests already covered 46.5 million acres. Today, these forests comprise a huge domain of 186.3 million acres, almost twice the land area of California, and most of it available for outdoor recreation.

As shown in figure 32.2, the growth in usage has more than matched the increasing areas of land available for recreation. By 1988, there were

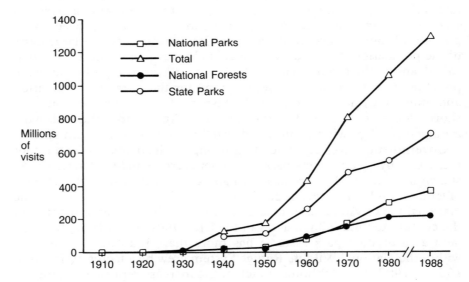

Figure 32.2   Recreational use of public land systems, 1910–88

368 million visits to the 354 units of the National Park System (including national parks, battlefields, historic sites, seashores, and other park system designations). State parks received 710 million visitors, averaging almost three visits per person for the US population as a whole. From 1950 to 1988, the number of visitor days spent in the national forests increased by a factor of six, reaching a total of 242.3 million.

There are also more than 300 million acres of private land available for public recreational use.[2] In 1986 the private sector provided 55 percent of the total of 17,000 campgrounds in the United States and 70 percent of the total 1.3 million campsites. Lease and fee arrangements are found on more than 50 million acres of private lands, as more and more owners seek to obtain recreational revenues from their lands. The overall availability of private land for general public use has nevertheless declined somewhat in recent years – a trend partly attributable to private landowner fears of liability suits.

Overall, public participation in outdoor recreation activities in the United States is estimated to have increased at the rate of around 10 percent per year from the end of World War II into the 1970s. While continuing to rise, the rate of growth has since slowed to around 3 to 4 percent per year.[3] Factors contributing to slower growth have included a longer life span and growing numbers of elderly; a lower birth rate and fewer families with children; rising travel costs (especially in the late 1970s and early 1980s); growing labor force participation by women (reducing available leisure time and scheduling discretion); and greater travel by US residents to other countries to pursue outdoor recreation activities.

From 1960 to 1982, the percent of the population participating in several forms of outdoor recreation declined in absolute terms. Less favored activities included picnicking, driving for pleasure, and motor boating. However, there were significant increases in participation rates (the percent saying that they participated at least once in a given year in a given activity) for bicycling, walking for pleasure, attending sporting events, and swimming. New technologies have contributed to the emergence of several new forms of outdoor recreation, including running rivers in inflatable rafts, snowmobiling, trail riding, hang gliding, and sail boarding. In addition to slower growth, other basic trends in recent years have included a shift toward more physically demanding forms of outdoor recreation and a greater diversity in the types of recreational activities pursued.

Hunting and fishing remain popular forms of outdoor recreation. The US participation rate for big game hunting rose from 4.6 percent in 1965 to 6.4 percent in 1985.[4] Populations of deer, elk, antelope, and several other big game species increased significantly over this period, resulting in sharp increases in harvest levels. For example, deer harvests in 20 northern states rose by almost 100 percent; elk and pronghorn antelope harvests in the Rocky Mountain states grew by 58 percent and 104 percent, respectively. Participation rates for small game (grouse, rabbit, pheasant, etc.) hunting, however, declined by about 2 percent to reach 5.7 percent in 1985. Populations of ducks recorded on the major US flyways typically held steady or rose from 1965 into the 1970s and then declined significantly, falling for the nation as a whole well below 1965 levels.

The first wilderness area was created in 1924 in the Gila National Forest in New Mexico, partly at the suggestion of Aldo Leopold, then a Forest Service employee. Acting under its own administrative authority, the Forest Service later added a number of other areas to the wilderness system. In 1964, Congress formally established the national wilderness system in legislation, designating 9.1 million acres for immediate inclusion. The system has now expanded to contain more than 90 million acres, including 36.8 million acres administered by the National Park Service and 19.3 million acres by the Fish and Wildlife Service. Wilderness recreation has a glamour in the public eye that exceeds its share of direct recreational use. In 1986, there were about 15 million visitor days spent in wilderness areas, less than 3 percent of total recreation on federal lands, although wilderness areas comprise some 25 percent of such lands.

In recent years there has been a growing tendency to see nature as a setting where outdoor recreation represents a foreign presence. Indeed, human impacts in general are currently seen by influential segments of the population as detractions from a higher goal, which is to preserve nature in an original condition that existed prior to the arrival of Western civilization.[5] In the past, the government sought to build roads, construct lodges, and otherwise ease the way for the direct experience of nature by millions of Americans. Today, many such projects built in the past would not be permitted, because they would be seen as an intrusion on nature. Instead of increasing public access, government policy is often more concerned with controlling and limiting access to nature, perhaps even reducing it substantially from current levels in some areas. These new public attitudes have contributed to the slowing of the rates of growth of participation in outdoor recreation activities that have been seen in the past 15 years.

In summary, trends in outdoor recreation are a kind of litmus test for the changes taking place at any given moment in American life. Following World War II, American society was characterized by rapid growth in family income, a high birth rate, growing leisure time, improving transportation, and other factors that yielded an explosive growth in participation in outdoor recreation. The National Park Service, transportation departments, manufacturers of recreational equipment, and many other suppliers responded to facilitate this growth. In the past two decades the population has aged, pressures for use of scarce leisure time have intensified, growth of family income has slowed, more women have entered the labor force, and other developments have caused a decline from previous very high rates of growth. Current trends thus suggest that in the future the areas of greatest expansion in outdoor recreational activity will be closer to home, less time intensive, less land intensive, and further separated from primitive nature. Participation in outdoor recreation will continue to grow, but at somewhat slower rates than in the past and with some shifts in emphasis.

NOTES

1 See Douglas M. Knudson, *Outdoor Recreation*, New York: Macmillan, 1984.
2 The data on outdoor recreation presented in this paper draw heavily on: Forest Service, US Department of Agriculture, *An Analysis of the Outdoor Recreation and Wilderness Situation in the United States, 1989–2040: A Technical Document Supporting the 1989 USDA Forest Service RPA Assessment*, General Technical Report RM-189, Washington, DC: Apr. 1990.
3 See the President's Commission on Americans Outdoors, *Working Papers*, Washington, DC: Dec. 1986, 3.
4 See Forest Service, US Department of Agriculture, *An Analysis of the Wildlife and Fish Situation in the United States, 1989–2040: A Technical Document Supporting the 1989 USDA Forest Service RPA Assessment*, General Technical Report RM-178, Washington, DC: Sept. 1989.
5 See Robert H. Nelson, "Unoriginal Sin: The Judeo-Christian Roots of Ecotheology," *Policy Review* (Summer 1990).

33

Global Forests Revisited

Roger A. Sedjo and Marion Clawson

Over a decade ago we addressed the notion that the world's forests, particularly the tropical forests, "are disappearing at alarming rates as growing numbers of people seek land to cultivate, wood to burn, and raw materials for industry" (Sedjo and Clawson, 1983). Using data newly released by FAO (1983), we showed that the actual rates of tropical deforestation, while significant at about 0.6 percent per annum, were considerably below rates often cited in reports of the time and used in the 1980 *Global 2000* report. In that paper we found that although significant deforestation was occurring in the tropics, the temperate forests were experiencing stability and even expansion. We concluded that industrial forest plantations were likely to become more important and that timber supply problems were unlikely to be a major global problem into the indefinite future. However, problems of deforestation, environmental spillovers, and timber availability might exist at some times on the local level.

We also noted that deforestation is not identical to logging or even land clearing *per se*. Deforestation involves land conversion in which the forest cover is permanently removed and the land used for another purpose, e.g., agriculture. Despite our cautiously optimistic assessment a decade ago, we also stressed that weak forest tenure in the tropics was giving rise to the open access common property type problems that could lead to excessive rates of deforestation. We expressed concern over the possibility of excessive destruction of genetic resources and stated that "the potential also exists for serious depletion of the world's genetic resources, especially in the moist tropical forests" (p. 167).

In many respects we find conditions today quite similar to those of a decade ago. The important role that plantation forestry is playing in providing adequate timber is now well recognized and concerns over long-term timber availability have, in general, continued to decline. However, tropical deforestation has continued and current data indicate that over the past decade the rate of deforestation has increased significantly to about 15.4 million hectares annually or about 0.8 percent (table 33.1).

Evidence accumulated over the decade suggests that the forces that are promoting deforestation have remained largely unaltered. Land tenure problems for tropical forests are still common and give rise to socially excessive rates of deforestation. This tendency is exacerbated by misdirected government policies which often subsidize deforestation (Repetto and Gillis, 1989; Biswanger, 1990).

FORESTS IN TRANSITION

The production of wood today is experiencing a transition similar to that experienced two or three thousand years ago in agriculture. Just as agriculture long ago evolved from hunting and gathering to cropping

Table 33.1 Estimates of tropical forest area and rate of deforestation by geographical subregions

Continent	Number of countries	Total[a] land area	Forest area 1980	Forest area 1990	Annual deforest. 1981–90	Rate of change 1981–90
			thousands of ha			(percent per annum)
Africa	40	2,236.1	568.6	527.6	4.1	−0.7
W. Sahelian Africa	9	528.0	43.7	40.8	0.3	−0.7
E. Sahelian Africa	6	489.7	71.4	65.3	0.6	−0.8
West Africa	8	203.8	61.5	55.6	0.6	−1.0
Central Africa	6	398.3	215.5	204.1	1.1	−0.5
Tropical Southern Africa	10	558.1	159.3	145.9	1.3	−0.8
Insular Africa	1	58.2	17.1	15.8	0.1	−0.8
Asia	17	892.1	349.6	310.6	3.9	−1.1
South Asia	6	412.2	69.4	63.9	0.6	−0.8
Continental S.E. Asia	5	190.2	88.4	75.2	1.3	−1.5
Insular South East Asia	5	244.4	154.7	135.4	1.9	−1.2
Pacific Islands	1	45.3	37.1	36.0	0.1	−0.3
Latin America	32	1,650.1	992.2	918.1	7.4	−0.7
Central America Mexico	7	239.6	79.2	68.1	1.1	−1.4
Caribbean	19	69.0	48.3	47.1	0.1	−0.3
Tropical S. America	7	1,341.6	864.6	802.9	6.2	−0.7
Total	90	47,783	1,910.4	1,756.3	15.4	−0.8

[a] Totals may not tally due to rounding.
Source: FAO (1993).

and livestock raising, in many places wood production has evolved – or is evolving – from gathering natural inventories to cropping forest plantations. Although forest output use in much of the tropical forests remains a matter of searching for valuable trees and harvesting them with little or no concern for regeneration, managed forestry is becoming increasingly important in many areas where it was unknown a generation ago. Industrial forest management and plantation forestry are important in the temperate world and beginning to emerge in Latin America, the Pacific basin, and elsewhere outside the northern temperate zone. Similarly, increased attention is being given to the sustainable management of native tropical forests for timber and other outputs.

Of all the natural resources examined by Potter and Christy (1962) and Manthy (1978), only wood experienced a continuous long-term upward trend in its real price in the United States (figure 33.1). This rising real price trend is reflected in the price of lumber (Ulrick, 1988).

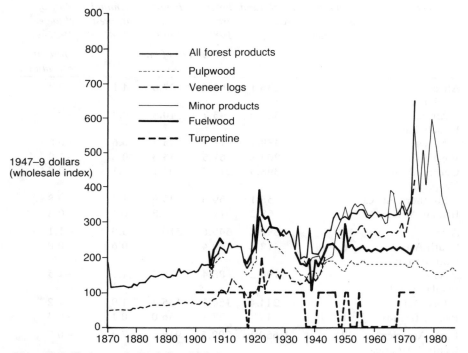

Figure 33.1 Forest products, deflated prices

Notes: Sawlogs price is $/m board feet, × 10, Douglas fir, average. Pulpwood price is $/cord, × 10, Southern pine, Southeast. Veneer logs price is $/m cubic feet, Douglas fir, average. Minor products price is $/m cubic feet. Fuelwood price is $/m cubic feet. Turpentine price is $/100 gallons. Rosin price is $/1,000 pounds. All forest products shows total expenditures, 10 millions of dollars.

Sources: 1870–1973: Manthy (1978); 1974–87: "US Timber Production, Trade, Consumption, and Price Statistics 1950–87"; A. H. Ulrich (1988, 1989). Washington: USDA Forest Service, Miscellaneous Publications nos. 1471, 1460.

Evidence indicates that a similar trend was experienced in Europe (Sivonen, 1979). This trend suggests that wood (or at least certain types of wood) has become increasingly scarce in an economic sense, and this has created financial incentives that have been manifest in the rapidly increasing rate of investments in tree planting and forest management.

However, the rising real price trend moderated considerably in the post-1950 period, although this period has been characterized by a great deal of price variability. Figure 33.1 shows that real prices for most forest resources in the late 1980s were no higher than they had been in the earlier 1950s. This general observation also applies to lumber price trends generally. The lack of a rising long-term real price trend in recent years is consistent with the notion that supply and demand have achieved long-term balance in the post-1950 period, with both growing at about the same rate. In 1992 and 1993, however, lumber prices in particular rose substantially in North America. We would attribute this rise to large reductions in timber harvests on public lands in the west, the result of widespread environmental concerns. This reduction in supply has been reinforced by simultaneous declines in timber harvests in western Canada, again driven by environmental concerns.

THE EXPERIENCE OF US FORESTS

The experience of US forests is in many respects an excellent example of the generalized history of the world's temperate forests. As recently as 400 years ago the United States had extensive areas of mature forests, inherited from the operation of natural forces over centuries (Clawson, 1979). Clearing of forests proceeded slowly for many decades and as late as 1800 had taken only a few million hectares out of the original forest. The rate of clearing accelerated during the nineteenth century, especially the latter part. The peak probably occurred around 1905. The rate of clearing, or the shrinkage in the original forest area, was almost exactly matched by an expansion of crop agriculture until prairies and plains were reached, after which crop acreage could be increased at the expense of grazing land rather than at the expense of forest land.

By about 1920 the process of forest clearing had largely run its course. Since then the area of forest land has remained relatively stable in the United States, even as the forest stock has increased dramatically, reflecting reforestation, fire control, and other factors (Sedjo, 1991). While abandonment of farm land and reversion to forest began in New England and the Middle Atlantic States before the Civil War, the really large reversions came after World War I, in the South particularly but also in the Lake States. The 1920 Census of Agriculture records both the highwater mark of cropping in much of the South and the low point for the nation's forests.

The dramatic changes in US forestry in the past 70 years have been not in land area but in annual growth of wood. Average annual growth

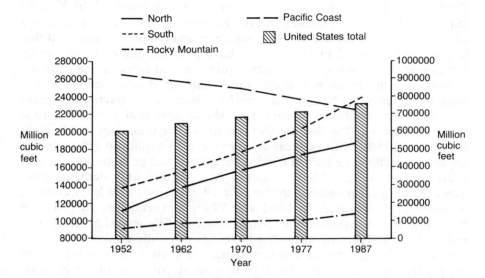

Figure 33.2 Growing commercial timber in the USA, net volume, all species

Notes: Net volume of growing stock on commercial timberland in the US, by section, 1952, 1962, 1970, 1977, and 1987; million cubic feet. Data may not add to totals because of rounding.

Source: *An Analysis of the Timber Situation in the United States 1952–2030, 1989–2040*; Part I: The Current Resource and Use Situation. Washington: US Dept of Agriculture, Forest Service, Forest Resource Report no. 23, Dec. 1982, p. 133; draft 1989.

today is three-and-a-half times what it was in 1920. Each successive timber inventory and appraisal by the Forest Service has shown a substantially higher total annual wood growth than its predecessor, and between 1952 and 1987 the total forest inventory (growing stock) increased 25 percent (figure 33.2). In a series of major studies since 1900 and more particularly since 1930, the Forest Service has estimated future wood growth. Every such estimate has been too low by a substantial amount (Clawson, 1979).

Although ecological and environmental conditions are different in countries less developed than the USA, and hence one would not want to overdraw the parallel between the experience of the USA and that of the countries of the developing world, it is clear that dynamic growing societies will generate pressures on and changes in the forest resource base. The experience of the temperate world indicates, however, that forests and forest lands need not inevitably continue the process of destruction and reduction.

OVERALL WORLD FOREST SUMMARY

Forests cover about 30 percent of the world's land surface; most (66 percent) of this forested area is closed forest, meaning that there exists a substantially complete cover of trees over the whole surface of the land. Closed forests vary greatly, from totally closed forests, where very little sunlight reaches the ground, to less dense forests in many temperate zones. The open forests are also highly variable from region to region, varying from tree stands which at a distance seem both continuous and pervasive on the landscape, to areas where trees or large shrubs occur at wide intervals with truly open areas between the individual trees.

For the world as a whole, 60 percent of the forest land area, both closed and open, is classified as hardwood and the remainder as softwood. This varies greatly by major parts of the world. In the northern hemisphere by far the greater part of the total forest land area is composed of softwood species; in the southern hemisphere and the tropics, the reverse is true. There are, of course, exceptions to this generalization, both within hemispheres and within countries. About 83 percent of the softwood (coniferous) closed forests are located in the former USSR and North America, while about 57 percent of the hardwood (nonconiferous, or broadleaved) closed forest land is found in South America and Asia (table 33.1). In general softwoods are more readily usable for commercial purposes.

For standing volume of timber, the current overall picture is generally similar. While timber volumes vary from region to region and site to site, the aggregate regional timber volumes correspond roughly with the forest land area. About 44 percent of the total production is produced in the temperate regions of North America, Europe, the former USSR, and Oceania alone. If fuelwood is excluded, this jumps to 74 percent.

FORESTS IN THE TEMPERATE REGIONS

Over half of the world's forest lands are located in the temperate climate regions. These include the forests of North America, Europe, the former USSR, China, and Oceania. While these regions have large areas of both hardwood and softwood, the softwoods are predominant. Figures 33.3a and 33.3b present FAO data on the forested area for the temperate zones between 1950 and 1988. Recently, the ECE/FAO has done a major assessment of the timber situation in Europe (1986), the former USSR (1989), and North America (1990). The results indicate that the timber volumes have risen dramatically in recent decades. One particularly impressive feature of these results is the pervasiveness of the expansion of standing timber volume in the temperate climate forest, including the former Soviet Union and Canada. Although these findings are contrary to common perceptions that forests are declining in these countries, other data tend to confirm the ECE/FAO results (e.g., Holowacz, 1985; Honer et al., 1990; Kauppa et al., 1992; WRI, 1986). These results indicated that timber volumes in the world's temperate climate

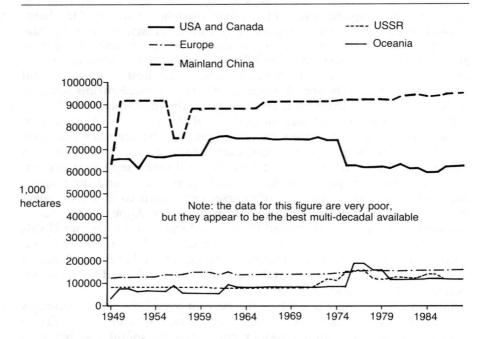

Figure 33.3a Forested area in temperate regions, selected areas
Notes: North America includes USA (including Alaska and Hawaii) and Canada. Forest data are collected and updated only intermittently. Extended periods in series without change reflect this procedure.
Source: United Nations Food and Agriculture Organization Production Yearbooks. The data given for a particular year are the data given for that particular yearbook; they are not the result of forest surveys taken in that particular year.

forests are expanding at rather rapid rates, with net increases estimated to be approaching 700 million m³ annually.

PLANTATION FORESTRY

While less than 1 percent of the forests of Latin America are industrial plantations, about one third of the region's industrial wood output comes from industrial forest plantations. Furthermore, the total area in industrial forest plantations is projected to increase 300 percent between 1979 and the year 2000. In addition, other forest plantations, such as protection and fuelwood, are also expected to increase in number and land area. By the year 2000 it is expected that more than half of the greatly expanded industrial wood production of Latin America will be produced from plantation forests (IDB, 1982, p. 17).

By the mid-1980s, the area of artificially regenerated forest worldwide totalled about 90 million hectares or roughly 3 percent of the world's closed forest area. The majority of these forests were located in Europe, North America, the USSR, and China, with nearly one-fifth of the

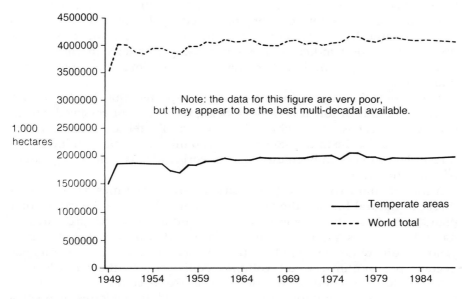

Figure 33.3b World forested area
Notes: See figure 33.3a.
Source: See figure 33.3a.

world's plantation area found in China. (Opinions about the success of China's reforestation efforts vary considerably. A World Bank assessment [1982] of China's environmental problems presents a pessimistic appraisal. This estimate is obtained by summing the separate estimates of artificially regenerated forests by regions as presented in World Wood [1982].) It is only in very recent times that plantation forestry has been a factor affecting the world's forests. Although some conscientious tree planting took place in parts of Europe and the Far East before the twentieth century, the vast majority of reforestation occurred through natural regeneration. However, the incidence of artificial regeneration has increased dramatically since World War II and particularly after 1960.

While the data indicate that the majority of forest plantations are situated in the northern hemisphere temperate climate regions, in recent years increased attention has been given to plantation activities in the tropics and subtropics and in the southern hemisphere temperate climate regions. Today major forest plantation activities are under way in Brazil, Chile, Venezuela, South Africa, India, Indonesia, the Philippines, Australia, New Zealand, and a host of other tropical and southern hemisphere countries. A study of forest plantations in the tropics (Lanly and Clement, 1979) estimated that the tropic and subtropic regions of Central and South America, Africa, and Asia had about 11.8 million hectares of plantation forest in the mid-1970s. Of this, about 6.7 million hectares were industrial forest plantations. By the 1980s, only five years later, the industrial forest plantations were projected by Lanly and Clement to have increased by 36 percent to 9.1 million hectares, and projections for the year 2000 indicated the industrial plantations in that region will cover over 21 million hectares, or three times the area of the mid-1970s.

The World Resources Institute (1986) reported that by 1980 14.5 million hectares of forest land were being reforested or renewed world-wide every year (p. 66), while Worldwatch (Postel and Heise, 1988) reported that by the mid-1980s almost 100 million hectares of industrial forest plantations had been established worldwide, as well as large areas of protection forests.

Although industrial forest plantations are sometimes viewed as a threat to natural forests, the opposite is more likely. To the extent that plantation forests meet society's needs for industrial wood, pressures on natural forests as a source of industrial wood will be reduced. Given yields readily obtainable in plantation forests (10 to 20 cubic meters per hectare per annum), wood volumes equivalent to the world's pro-jected industrial wood needs for the year 2000 could be met by the sustained yield production of just 100–200 million hectares of planta-tion forests – or only 3.5 to 7.0 percent of the world's closed forest land area (Sedjo, 1983). In addition, to the extent that nonindustrial protec-tion forests are being artificially established, environmental damage that was being generated in the absence of protective forest is being reduced.

TROPICAL FORESTS AND DEFORESTATION

Lying on each side of the equator around the world is an immense area of tropical moist forests that have special characteristics and present special problems. A third of these are in Brazil – the Amazon Basin – and another one-quarter are in other Latin American countries; some are in west Africa, while others are in Asia and the islands of the East Indies. There are, of course, some biological differences among these areas, and there are many significant economic and political differences among the countries in which these forests lie.

Until recent decades, information about the rate of tropical deforesta-tion has been relatively scant. What information we did have came more from anecdotal evidence – provided by isolated investigations at single times and places – than from systematic studies conducted over large areas and lengths of time. As a result, there was little agreement as to the reliability and appropriateness of much of the data relating to these forests.

Table 33.2 presents the estimates of the world's forests assembled by Allen and Barnes combined with an earlier estimate (Zon and Sparhawk,

Table 33.2 World forested area: past and current estimates

Source and year	Million ha	Source and year	Million ha
Zon and Sparhawk, 1923[f]	3,031	Whittaker and Likens, 1973[b]	4,850
Weck, 1961[b]	2,477	Bruning, 1974[b]	4,150
WFI, 1963[d]	4,126	Windhorst, 1974, 1976[b]	2,393
Olson, 1970[a]	4,800	Olson, 1975[b]	4,800
Bazilevich et al., 1971[b]	5,290	Eckholm, mid-1970s	2,657
Bruning, 1971[b]	3,590	Eyre, 1978[e]	6,050
Whittaker and Wordwell, 1971[b]	5,000	Smith, 1978[e]	2,563
FAO, 1972[a]	3,800	Steele, 1979	3,799
Lieth, 1972[a]	5,700	FAO, n.d.	4,300
Lieth, 1972, 1975[b]	5,000	Openshaw, n.d.	3,712
Persson, 1973[c]	4,030	FAO/ECE, 1980	4,320
		Persson, 1985	4,147

Note: n.d. = no date
[a] Taken from Earl, 1975.
[b] Taken from Lieth, 1979.
[c] Taken from World Bank, 1978.
[d] Taken from Persson, 1974.
[e] Estimate of closed forests only.
[f] Zon and Sparhawk (1923). *Forest Resources of the World*, vol. 1. New York: McGraw-Hill.
Source: N. I. Bazilevich, L. F. Rodin and N. N. Rozov, "Geographical aspects of biological productivity," *Soviet Geography Review Translation*, 12, 293–317 (1971); E. F. Bruning *Forstliche Produktionslehre* (Europaische Hochschulschriften XXV/1, Bern, Frankfurt, 1971); E. F. Bruning, "Okosysteme in den Tropen," *Umschau*, 74, 405–10 (1974); D. E. Earl, *Forest Energy and Economic Development* (Oxford, Clarendon Press, 1975), 43; Eric Eckholm, *Planting for the Future: Forestry for Human Needs* (Washington, DC, Worldwatch Institute, 1979), 11; S. R. Eyre, *The Real Wealth of Nations* (New York, St. Martins Press, 1978); Helmut Lieth, "Uber die Primerproduktion der Pflanzendecke der Erde," *Angew Botanik*, 46, 1–37 (1972); Helmut Lieth, *Primary Productivity of the Biosphere* (New York, 1975); Helmut Lieth, "Forest uses in global and regional (USA) perspectives," in Stephen G. Boyce (ed.), *Biological and Sociological Basis for a Rational Use of Forest Resources for Energy and Organics*, proceedings of an international workshop sponsored by The Man and the Biosphere Committees of Canada, Mexico, and the United States held at Michigan State University, May 6–11, 1979, 70; J. S. Olson, *World Ecosystems* (Washington, DC, Seattle Symposium, 1975); Keith Openshaw, "Woodfuel – A time for Re-Assessment," *Natural Resources Forum*, 3, (1) (1978), 35–51; Reidar Persson, *World Forest Resources: Review of the World's Forest Resources in the Early 1970s* (Stockholm, Skogshogskolan, Royal College of Forestry 1974) 222; Nigel Smith, *Wood: An Ancient Fuel with a new Future* (Washington, DC, Worldwatch Institute, 1981) p. 134; R. C. Steele, "Some social and economic consequences and constraints to the use of forests for energy and organics in Great Britain," in Stephen G. Boyce (ed.), *Biological and Sociological Basis for a Rational Use of Forest Resources for Energy and Organics*, proceedings of an international workshop sponsored by The Man and the Biosphere Committees of Canada, Mexico, and the United States

Table 33.2 (*Continued*)

held at Michigan State University, May 6–11, 1979, 31; UN Food and Agriculture Organization, *Forestry for Rural Communities* (FAO Forestry Department, n.d.), 8; J. Weck and C. Wiebecke, Weltfurstwirtschaft und Deutschlands Forst- und Holzwirtschaft (Munchen, 1961); R. H. Whittaker and G. E. Likens, "Primary production: the biosphere and man," *Human Ecology*, 1, 357–69 (1973); H. W. Windhorst, "Das Ertragspotential der Walder der Erde," in *Studien zur Waldwirtschaffgeographic Beihefte zur Geographischen Zeitschrift H. 39* (Wiesbaden, 1974); H. W. Windhorst, "The forests of the world and their potential productivity," *Plant Research and Development*, 3, 40–9 (1976); The World Bank, *Forestry Sector Policy Paper* (Washington, DC, World Bank, 1978) 26; *World Forest Inventory*, 1963 (Rome, Food and Agriculture Organization, 1963); UN Food and Agriculture Organization (FAO) / Economic Commission for Europe (ECE), Forest Resources 1980 (FAO/ECE, Rome, 1985); R. Persson, unpublished report to Swedish International Development Authority (1985), Tables 1, 2, and 4.

1923) and two more recent estimates by Persson (1985) and the FAO/ECE (1985). As Allen and Barnes note, these estimates, coming from different sources and using different forest definitions, "are probably not very reliable as indicators of deforestation."

In table 33.3, Allen and Barnes go on to compare three sources of somewhat inconsistent data that were available in the early 1980s to estimate trends in forest area on a country-by-country basis over a period of four years or more. These sources are the FAO Production Yearbook (US FAO 1979), the recent Lanly-directed FAO study (1982), and a report to the National Academy of Sciences (Myers, 1980). Of these three studies, the 1982 FAO/UNEP study has come to be widely accepted as the most definitive. The 1982 FAO/UNEP study estimated the rate of deforestation of the closed tropical broadleaf forests at 7.1 million hectares per annum. The rate of deforestation is almost identical for Latin America, Africa, and Asia. When open and coniferous forest types in the tropics are included, the total area deforested annually is estimated to be 11.3 million hectares, with the percent decline in all tropical forests falling slightly to 0.6 percent per annum due to the larger total forest land areas involved.

The FAO has recently updated and expanded the 1982 study to assess the current situation and changes in the rate of tropical deforestation that may have occurred since that time. Estimates of tropical forest area reductions during the 1980s for 62 countries in the tropical regions, including almost all of the moist forest zone, indicate the rate of tropical deforestation for the period 1981–90 is 15.4 million hectares per annum, or about 0.8 percent of the total tropical forest area (table 33.1). This rate is roughly a 40 percent increase over the 0.6 percent reported in the 1983 report. As seen in table 33.1, Asia has the highest annual rate of deforestation of any continent over the past decade at 1.1 percent, while Latin American and Africa have each experienced an annual deforestation rate of 0.7 percent, according to FAO estimates.

TROPICAL DEFORESTATION AND
ENVIRONMENTAL CONCERNS

Although the conversion of a tract of tropical moist forest undoubtedly results in the destruction of most forms of vegetation and wildlife on that particular tract, the connection between the disappearance of a given tract and the loss of biodiversity is surely more complicated. Clearing one site need not mean that the same species and broad ecosystem may not continue on adjoining sites. Recent developments in the theory of island biogeography suggest a relationship between the area of habitat and the amount of biodiversity. However, this relationship is almost surely not linear, and the nature of this relationship is an empirical question subject to scientific inquiry.

Two measures are available to conserve the world's biological resources. *In situ* approaches leave species where they are found in nature, preserved in more or less natural settings. *Ex situ* approaches include removal of species from their natural habitat for preservation in permanent collections such as zoos and botanic gardens, and the preservation of seed and other genetic materials in a controlled environment (Harrington, 1982). Progress is being made on both fronts.

Today numerous activities are under way to collect and maintain germplasm. These involve not only private collection but also, for example, the National Seed Storage Laboratory operated by the US Department of Agriculture and the National Plant Germplasm System. In addition, about 40 germplasm banks around the world are now part of the International Board of Plant Genetic Resources and store over one million varieties of crop plants (Griggs, 1982). Recently, at the initiative of private timber interests and affected national governments, the Central America and Mexico Coniferous Resources Cooperative was established to collect and preserve endangered seed of the tropical pines resource of Central America. One activity will be to establish tropical pine orchards elsewhere to insure permanent genetic preservation.

Significant strides have been made in recent years to protect unique land and habitats. The world total of protected land doubled between 1970 and 1980 (WRI, 1986) and increased another 50 percent between 1980 and 1985. By the mid-1980s there were over 400 million hectares of protected areas, up from about 100 million in 1960 (IUCN, 1985). Nevertheless, as the efforts to preserve habitat reveal, the problem of the protection of the world's genetic resources remains a serious concern. It is estimated that 25 to 50 percent of the world's species reside in tropical forests. While documentation of actual tropical species extinction is rare, the common property characteristics of many of these forests and the wild genetic resources residing in them suggest susceptibility to overexploitation.

Table 33.3 Deforestation by country according to three different sources (from Allen and Barnes, 1982)

Country	(1) FAO Yearbook		(2) FAO/UNEP		(3) Myers	
	Rank	Annual change (%)[a]	Rank	Annual change (%)[b]	Rank	Annual change (%)[c]
*Ivory Coast	1	−3.7	1	−5.4	1	−5.3
Haiti	2	−3.1	4	−3.2	−	−
*Togo	3	−2.4	27	−0.64	−	−
*Philippines	4	−2.2	22	−1.02	8.5	−1.25[d]
Swaziland	5	−1.9	−	−	−	−
*Sri Lanka	6	−1.8	18	−1.42	5	−1.8[e]
*Liberia	7	−1.66	14	−1.89	6	−1.41[f]
Niger	8	−1.61	−	−	−	−
*Upper Volta	9	−1.56	−	−	−	−
*Rwanda	10	−1.46	10	−2.13	−	−
Brunei	11	−1.4	12	−1.99	−	−
*Ecuador	12	−1.14	13	−1.94	−	−
*Somalia	13	−0.94	32	−0.225	−	−
*Congo	14	−0.90	36	−0.103	−	−
*Mexico	15	−0.78	20	−1.09	−	−
*Maylasia	16	−0.75	21	−1.05	7	−1.28[g]
*Thailand	17	−0.54	6	−3.15	2	−4.3[h]
Afghanistan	18	−0.50	−	−	−	−
Vietnam	19	−0.46	25	−0.71	−	−
*Bolivia	20	−0.44	34	−0.146	−	−
Yemen PDR	21	−0.39	−	−	−	−
*Costa Rica	22	−0.38	5	−3.19	−	−
Sierra Leone	23	−0.36	24	−0.75	−	−
Belize	24	−0.344	26	−0.647	−	−
Surinam	25	−0.338	37	−0.0168	−	−
*Brazil	26	−0.246	29	−0.406	11	−0.342[i]
*Zaire	27	−0.232	33	−0.156	−	−
*Indonesia	28	−0.113	28	−0.47	8.5	−1.25[j]
*Nepal	29	−0.111	2	−3.68	−	−
*Argentina	30	−0.093	−	−	−	−
*Zambia	31	−0.087	19	−1.26	−	−
*Paraguay	32	−0.082	3	−3.39	−	−
*South Korea	33	−0.079	−	−	−	−
*Bangladesh	34	−0.067	23	−0.83	10	−0.83[k]
*Jamaica	35	−0.016	11	−2.05	−	−
*Madagascar	36	−0.001	17	−1.50	4	−2.21[l]
*Benin	37	0.004	7.5	2.83	−	−
Botswana	38	0.04	−	−	−	−
Kenya	39	0.05	16	−1.62	3	−2.6[m]
Uruguay	40	0.56	−	−	−	−
*Ethiopia	41	0.68	31	−0.227	−	−

Table 33.3 (Continued)

*India	43	0.4	30	−0.28	–	–
Guinea	44	0.5	15	−1.64	–	–
China	45	0.6	–	–	–	–
Burundi	46	1.6	9	−2.4	–	–
Cuba	47	2.83	35	−0.136	–	–

Notes: Dash indicates data not available. * indicates countries which are included in later regression analyses. Turkey, Morocco, Uganda, Burma, Dominican Republic, Guatemala, Nicaragua, and Panama are not listed in the table but are included in the later analyses.

[a] Average annual percent change in forests and woodland over the period 1968–78, using 1968 as a base year. Only complete removal of forest cover is measured. Forest and woodland refers to land under natural or planted stands of trees, whether productive or not, and includes land from which forests have been cleared but that will be reforested in the foreseeable future. Forests and woodland are not defined consistently in actual measurement practice by all countries nor over all years.

[b] Average annual percent change in natural woody vegetation in the form of closed broadleaf, coniferous, or bamboo forest over the period 1976–80. Excludes forest plantations. Only complete removal of forest cover is measured. Includes tropical countries only.

[c] Average annual percent change in tropical moist forest over various periods. This measure includes all types of forest conversion, including logging. Tropical moist forest includes evergreen or partly evergreen forests, with or without some deciduous trees, but never completely leafless, forest savanna mosaic where forests are not confined to streamsides, and coastal savanna mosaic. According to the FAO Committee on Forest Development in the Tropics, interpretation of the term "tropical moist forest" should be left to individual countries (Sommers, 1976, 6). Conversion of tropical moist forest can range from marginal modification to fundamental transformation (Myers, 1980, 9). Where Myers has no reliable figure for total forest area FAO/UNEP estimates of closed forest or natural woody vegetation was used as a base to calculate deforestation rates. See explanatory notes.

[d] "Forestlands . . . disappearing each year," 1971, 98.

[e] Forest "degraded" each year by shifting cultivation, 1950–80, 102.

[f] "Primary forest converted to degraded forest . . . or bushland by shifting cultivation," annually, 1980, 162.

[g] Actual rate of clearing in proposed agricultural lands, 1975–80, 83.

[h] "Loss of forest cover," 1961–78, 110.

[i] "Forests eliminated" 1966–75, according to Brazilian development office (SUDA), 128.

[j] Estimate of forest "eliminated," by shifting cultivation annually, 1980, 71.

[k] "Forest in the central and northern parts . . . declining, primarily through shifting cultivation . . . ," 65.

[l] Moist forest "disrupted and impoverished through shifting cultivation" annually 1980, 162.

[m] Net rate in Nandi Forest only, 1966–76, 159.

Although the extent of loss of biodiversity is difficult to assess, the recent research in island biogeography, which finds that viable populations of species are positively related to the area of habitat area, suggests that large losses in forest habitat could impact negatively on the number of plant and animal species.

Sources: (1) UN Food and Agriculture Organization, *1979 Production Yearbook* (Rome, UN Food and Agriculture Organization, 1980), 45–57; (2) Jean-Paul Lanly, tables compiled for the Forest Resources Division, UN Food and Agriculture Organization, 1981; and (3) Norman Myers, *Conversion of Tropical Moist Forests*, a report prepared for the Committee on Research Priorities in Tropical Biology of the National Research Council (Washington, DC, National Academy of Sciences, 1980) 62–167.

Finally, over the past decade, concern has also been expressed repeatedly by some observers about the dire effects on world climate resulting from increased CO_2 in the atmosphere, the warming effect this might have on the world's climate, and the role forest clearing may play in this process. Although deforestation may play some role in the buildup of atmospheric carbon, the major source is clearly fossil fuel burning (Detwiler and Hall, 1988). As long ago as 1982 the authoritative Carbon Dioxide Review 1982 (Clark, et al., 1982) stated flatly, "No one any longer suggests land-use changes will produce a significant fraction of man's total future releases of CO_2. If there is a carbon dioxide problem in the future, it will be due to the burning of fossil fuels; *not the burning of forests*" [emphasis added].

COMMON PROPERTY PROBLEMS AND SOME RECENT INNOVATIONS

For the transition to managed forests to occur successfully, a number of changes must occur that make forest management economically feasible. These include higher market prices and decreasing costs of management as more cost-efficient systems are developed. Also, institutional arrangements must be in place that allow the common property type of resources – such as wildlife, natural forests, foraging foods, genetic resources, and environmental services – to accrue predictably to individuals or groups who are willing to make the necessary investment in establishing and maintaining these resources or in preserving habitats necessary to maintain them. Although numerous such institutions exist, including the institution of private property rights as well as other institutions for public and collective action, these institutions are not fully functioning on all forest land for all outputs.

In some cases the common property problem relates to timber directly, as when tenure rights are poorly defined or governments lack enforcement power. In other cases the common property problem relates not to the timber values of the forest but to many of the nontimber values. A number of innovative attempts to protect tropical forests currently underway involve efforts to internalize some aspect of the common property problem. In the past decade a number of innovative efforts have been and are being undertaken to protect tropical forests. An early effort featured the so-called debt-for-nature-swaps whereby industrial countries provided some degree of debt relief in return for the establishment of a protected area. This approach allowed for the retirement of some of a developing country's debt, often at a discounted rate. In return the tropical country used local currency to finance the establishment of a new park or protected area. This procedure has been particularly attractive for many debt-ridden tropical countries.

A second innovative approach, still in its formative stages, provides for the commercialization of biodiversity. The concept is that tropical forests often have important values as habitat for unique genetic resources that may have valuable uses in drug development but whose returns are difficult for the owner of the habitat to capture. By allowing the forest land owner (perhaps the government) to capture financial returns from biodiversity, incentives for allowing the land to remain in native forest are increased, thereby promoting long-term conservation. In this approach, tropical countries would agree to provide access to unique indigenous genetic resources to pharmaceutical firms in return for financial payments. In some cases these payments may be as prices for access to the genetic resources directly; in other cases they may be in the form of royalties tied to the receipts generated from the commercial development of the genetic resource. In the now famous Merck-INBio arrangement in Costa Rica, the agreement had both payment features (Sedjo, 1992).

A third innovative approach is found in recent attempts to provide a "certification" of wood products as coming from sustainably managed forests. The sustainability of the management is verified by an independent group using well defined standards of certification. This effort is based partly on the notion that "green" values might result in some groups being willing to pay a premium for goods produced from sustainably managed forests. Should such a premium develop in the market, financial incentives would provide incentive for "certified" sustainable management.

CONCLUSION

The trends observed a decade ago, both positive and negative, have continued. Rates of tropical deforestation have increased significantly over the past decade, from about 0.6 to 0.8 percent annually. By contrast, the temperate region continues to experience an expansion of forest volumes and areas. There is little evidence of serious long-term lack of timber, and the role of plantation forestry continues to expand. However, environmental concerns related to forests persist, and indeed have increased, with a major focus now related to biodiversity.

REFERENCES

Allen, J. C., and Douglas F. Barnes (1982): "Deforestation, Wood Energy and Development." Washington, DC: Resources for the Future, unpublished paper.

Biswanger, Hans P. (1989): "Government Policies that Encourage Deforestation in the Amazon." Environmental Working Paper 16, World Bank, Apr.

Clark, William C., ed. (1982): *Carbon Dioxide Review 1982*. New York: Oxford University Press.

Clark, William C., and Kerry H. Cook, Gregg Marland, Alvin M. Weinberg, Ralph M. Rotty, P. R. Bell, and Chester L. Cooper (1982): "The carbon dioxide question: perspectives for 1982." In W. C. Clark, ed, *Carbon Dioxide Review 1982*, 4.

Clawson, Marion (1979): "America's forests in the long sweep of history." *Science*, 204(4398), 1168–74.

Detwiler, R. P., and C. A. S. Hall (1988): "Tropical Forests and the Global Carbon Cycle." *Science*, 239, 43–7.

Economic Commission for Europe/Food and Agricultural Organization of the United Nations (ECE/FAO) (1986): *European Timber Trends and Prospects to the Year 2000 and Beyond*. New York: United Nations.

———— (1989): "Outlook for the forest and forest products sector of the USSR." ECE/TIM/48, 75. New York: United Nations.

———— (1990): "Timber Trends and Prospects for North America." ECE/TIM/53, 68. New York: United Nations.

Griggs, Tim (1982): "FAO Acts to Strengthen Seed Conservation." *International Agricultural Development* (June), 14.

Harrington, Winston (1982): "Endangered species—a global threat." In *Resources*. Washington: Resources for the Future, October.

Holowacz, J. (1985): "Forest of the USSR." *Forestry Chronicle*, 61(5), 366–73. Oct.

Honer, T. G., W. R. Clark, and S. L. Gray (1990): "Determining Canada's Forest Area and Wood Volume Balance 1977–1986." Paper presented at the Conference on Canada's Timber Resources, June 3–6, 1990, Victoria, British Columbia.

Inter-American Development Bank (IDB) (1982): "Forest industries development strategy and investment requirements in Latin America: Technical Report No. 1." Prepared for IDB conference on Financing Forest-Based Development in Latin America, 22–25 June, Washington, DC.

International Union of Conservation Organizations (1985): "The United Nations List of National Parks and Protected Areas." Cambridge, Mass.

Kauppi, P. E., K. Mielikainen, and K. Kuusela (1992): "Biomass and Carbon Budget of European Forests, 1971 to 1990." *Science*, 256, 70–4.

Lanly, J. P. and J. Clement (1979): "Present and future natural forest and plantation areas in the tropics." *Unasylva*, 31(123), 12–20.

Manthy, Robert (1978): *Natural Resource Commodities – A Century of Statistics*. Baltimore: Johns Hopkins Press.

Persson, Reidar (1974): *World Forest Resources; Review of the World's Forest Resources in the Early 1970s*. Research Notes No. 17, Department of Forest Survey, Royal College of Forestry, Stockholm.

———— (1985): Unpublished report to the Swedish International Development Authority, presented in *World Resources 1986*, 62.

Postel, Sandra, and Lori Heise (1988): "Reforesting the Earth." *Worldwatch*. Washington, DC: Worldwatch Institute.

Potter, Neal, and Francis T. Christy (1962): *Trends in Natural Resource Commodities.* Baltimore: Johns Hopkins University Press.

Repetto, Robert, and Malcolm Gillis (1989): *Public Policies and the Misuse of Forest Resources.* New York: Cambridge University Press.

Sedjo, Roger A. (1983): *The Comparative Economics of Plantation Forests: A Global Assessment.* Baltimore: Johns Hopkins University Press/Resources for the Future.

———— (1983): "The Potential of U.S. Forest Lands in the World Context." In *Government Intervention, Social Needs and Forest Management*, ed. Roger A. Sedjo. Washington: Resources for the Future.

———— (1991): "Forest Resources: Resilient and Serviceable." In *America's Renewable Resources*, eds K. Frederick and Roger A. Sedjo. Washington: Resources for the Future.

———— (1992): "Property Rights, Genetic Resources, and Biotechnological Change." *The Journal of Law and Economics*, 35(1), 199–213.

———— and Marion Clawson (1984): "Global Forests." In *The Resourceful Earth*, eds Julian L. Simon and Herman Kahn. New York: Basil Blackwell.

Sivonen, S. (1971): "Havusahapuun Rantohinnan paa Sunntrainen Rehitya Suomessa vuosina 1920–67" (Report of the Committee on the Costs of Forest Planting and Seeding, Annex 5, 116–20). *Folia Forestalia*, 109. Helsinki, Finland.

Ulrich, Alice H. (1988): "U.S. Timber Production, Consumption, and Price Statistics, 1950–1986." US Department of Agriculture, Forest Service, Misc. Publication no. 1460. Washington, DC: USDA.

UN Food and Agriculture Organization (1980): *1979 Yearbook of Forest Products, 1967–1979*, 2–50. Rome: Food and Agriculture Organization.

UN Food and Agricultural Organization (FAO)/Economic Commission for Europe (ECE) (1985): *Forest Resources 1980.* Rome: FAO/ECE.

UNFAO/UNEP (1982): *Tropical Forest Resources*, by J. P. Lanly. FAO Forestry Paper No. 30, Rome: UNFAO/UNEP.

———— (1982): "The People's Republic of China: Environmental Aspects of Economic Development," 4–8. Rome: UNFAO/UNEP Office of Environmental Affairs, Projects Advisory Staff.

World Resources Institute and International Institute for Environment and Development (1986): *World Resources 1986.* New York: Basic Books.

World Wood (1981): *1981 World Wood Review*, 22(8).

Zon, R., and W. N. Sparhawk (1923): *Forest Resources of the World*, 2 vols. New York: McGraw-Hill.

34

Species Loss Revisited

Julian L. Simon and Aaron Wildavsky

Note to Reader: Since completing the first draft of this article, a book edited by Whitmore and Sayer (1992) and sponsored by the International Union for the Conservation of Nature and Natural Resources (IUCN), has confirmed our central assertion here – from the pens of the very conservation biologists who have been most alarmed by the threat of species die-offs. Lengthy quotations have been included below documenting the consensus that there is no evidence of massive or increasing rates of species extinction. That book was originally conceived as a response to the questions raised by our predecessor article and by others since then.

INTRODUCTION

Species extinction is a key issue for the environmental movement. It is the subject of magazine stories with titles like "Playing Dice with Mega-death" with a subhead "The odds are good that we will exterminate half the world's species within the next century" (Diamond, 1990, p. 55.) Species "loss" also is a focal point for fundraising from the public. And the Congress is asked again and again for large sums of public money to be used directly and indirectly for programs to protect species and for "debt for Nature" swaps.

The central assertion is that species are dying off at a rate that is unprecedently high, and dangerous to humanity. The World Wildlife Fund, which publicizes this issue widely, frames the proposition as follows: "Without firing a shot, we may kill one-fifth of all species of life on this planet in the next 10 years."

The issue came to scientific prominence in 1979 with Myers's *The Sinking Ark*, and then was brought to an international public and onto the US policy agenda by the 1980 *Global 2000 Report to the President* (referred to hereafter as "GTR"). These still are the canonical texts.

GTR forecast extraordinary losses of species between 1980 and 2000. "Extinctions of plant and animal species will increase dramatically.

Hundreds of thousands of species – perhaps as many as 20 percent of all species on earth – will be irretrievably lost as their habitats vanish, especially in tropical forests" (US, 1980, I, 3).

In 1984 we reviewed the data on the observed rates of species extinction. We found that the scientific evidence was wildly at variance with the by-then-conventional wisdom and did not provide support for the various policies suggested to deal with the purported dangers. We also reminded readers that recent scientific and technical advances – especially seed banks and genetic engineering, and perhaps electronic mass-testing of new drugs – had rendered much less crucial the maintenance of a particular species of plant life in its natural habitat than would have been the case in earlier years. But the bandwagon of the species extinction issue continues to roll with ever-increasing speed.

Now we revise our presentation of the empirical and theoretical situations in light of the literature that has appeared in the 1980s. We find that our earlier conclusions remain sound and may be considered strengthened by the absence of new countervailing material coming to light since then.

These are the key questions: Are species defined with sufficient clarity so that different people can arrive at satisfactorily similar estimates? What is the history of species extinction until now? What are the most reasonable forecasts of future extinction? What will be the effects of extinctions (including resulting new additions) on species diversity? What will be the economic and non-economic impacts of the expected course of species diversity?

Society properly is concerned about possible dangers to species. Individual species, and perhaps all species taken together, constitute a valuable endowment, and we should guard their survival just as we guard our other physical and social assets. But we should strive for as clear and unbiased an understanding as possible in order to make the best possible judgments about how much time and money to spend in guarding them, in a world in which this valuable activity must compete with other valuable activities, including the guarding of valuable aspects of civilization and of human life.

The importance of the topic is clear from the far-reaching extent of the policies suggested. Edward O. Wilson and Paul Ehrlich actually ask that governments act "to reduce the scale of human activities." More specifically, they want us "to cease 'developing' any more relatively undisturbed land," because "Every new shopping center built in the California chaparral, every swamp converted into a rice paddy or shrimp farm means less biodiversity" (1991, p. 761).

DEFINING SPECIES

Before discussing rates of extinction, we must touch on an issue that complicates such estimates – the definition of a species. Referring to the "never-ending arguments about the definition of the species category," Ernst Mayr infers that "those who do not work with species but with cells or molecules may think that the species is an arbitrary and insignificant concept in biology." He argues otherwise. "What, then," Mayr asks, "is biological classification? Unhappily," he concludes, "No agreement on the answer to this question exists yet among biologists."

The *Encyclopaedia Britannica* defines species as "groups of individuals that resemble one another more than they resemble any others." But what constitutes "resemblance"? If we adopt the definition of "who mates with whom," we will reduce the number of species; if we give the more general definition that "Species are groups of organisms sharing many traits, or characteristics in common," the classifiers have a lot of room to raise or lower the number. If we use common DNA as a criterion, we guess the number of species would greatly diminish.

SPECIES LOSS ESTIMATES

The basic forecast for loss of species comes from Lovejoy:

> What then is a reasonable estimate of global extinctions by 2000? Given the amount of tropical forest already lost (which is important but often ignored), the extinctions can be estimated. . . . In the low deforestation case, approximately 15 percent of the planet's species can be expected to be lost. In the high deforestation case, perhaps as much as 20 percent will be lost. This means that of the 3–10 million species now present on the earth, at least 500,000–600,000 will be extinguished during the next two decades. (US, 1980, II, p. 331.)

This extract summarizes a table that shows a range between 437,000 and 1,875,000 extinctions out of a present estimated total of 3–10 million species. The table in turn is based on a linear relationship running from zero percent species extinguished at zero percent tropical forest cleared, to about 95 percent extinguished at 100 percent tropical forest clearing. The main source of differences in the range of estimated losses is the range of 3–10 million species in the overall estimate.

The basis of any useful projection must be some body of experience collected under some range of conditions that encompass the expected conditions, or that can reasonably be extrapolated to the expected conditions. But none of Lovejoy's references contain any scientifically impressive body of experience. The only published source given for his key table (US, 1980, Table 13–30, p. 331) is Myers's *The Sinking Ark* (1979).

Myers's summary may be taken as the basic source:

> As a primitive hunter, man probably proved himself capable of eliminating species, albeit as a relatively rare occurrence. From the year A.D. 1600, however, he became able, through advancing technology, to over-hunt ani-

mals to extinction in just a few years, and to disrupt extensive environments just as rapidly. Between the years 1600 and 1900, man eliminated around seventy-five known species, almost all of them mammals and birds – virtually nothing has been established about how many reptiles, amphibians, fishes, invertebrates and plants disappeared. Since 1900 man has eliminated around another seventy-five known species – again, almost all of them mammals and birds, with hardly anything known about how many other creatures have faded from the scene. The rate from the year 1600 to 1900, roughly one species every 4 years, and the rate during most of the present century, about one species per year, are to be compared with a rate of possibly one per 1000 years during the "great dying" of the dinosaurs.

Since 1960, however, when growth in human numbers and human aspirations began to exert greater impact on natural environments, vast territories in several major regions of the world have become so modified as to be cleared of much of their main wildlife. The result is that the extinction rate has certainly soared, though the details mostly remain undocumented. In 1974 a gathering of scientists concerned with the problem hazarded a guess that the overall extinction rate among all species, whether known to science or not, could now have reached 100 species per year. [Here Myers refers to *Science*, 1974, pp. 646–7]

Yet even this figure seems low. A single ecological zone, the tropical moist forests, is believed to contain between 2 and 5 million species. If present patterns of exploitations persist in tropical moist forests, much virgin forest is likely to have disappeared by the end of the century, and much of the remainder will have been severely degraded. This will cause huge numbers of species to be wiped out. . . .

Let us suppose that, as a consequence of this man-handling of natural environments, the final one-quarter of this century witnesses the elimination of 1 million species – a far from unlikely prospect. This would work out, during the course of 25 years, at an average extinction rate of 40,000 species per year, or rather over 100 species per day. The greatest exploitation pressures will not be directed at tropical forests and other species-rich biomes until towards the end of the period. That is to say, the 1990s could see many more species accounted for than the previous several decades. But already the disruptive processes are well underway, and it is not unrealistic to suppose that, right now, at least one species is disappearing each day. By the late 1980s we could be facing a situation where one species becomes extinct each hour (1979, pp. 4–5).

We may extract these key points from the above summary quotation:

1 The estimated extinction rate of known species is about one every four years between the years 1600 and 1900.
2 The estimated rate is about one a year from 1900 to the present.

No sources are given for these two estimates, either on the page from which the quote is taken or on pages 30–1 of Myers's book, where these estimates are again discussed.

3 Some scientists (in Myers's words) have "hazarded a guess" that the extinction rate "could now have reached" 100 species per year. That is, the estimate is simply conjecture and is not even a point estimate but rather an upper bound. The source given for the "some scientists" statement is a staff-written news report (Holden, 1974). It should be noted, however, that the subject of this guess is different than the subject of the estimates in (1) and (2), because the former includes mainly or exclusively birds or mammals whereas the latter includes all species. While this difference implies that (1) and (2) may be too low a basis for estimating the present extinction rate of all species, it also implies that there is even less statistical basis for estimating the extinction rate for species other than birds and mammals than it might otherwise seem.

4 This guessed upper limit in (3) is then increased and used by Myers, and then by Lovejoy, as the basis for the "projections" quoted above. In *Global 2000* the language has become "are likely to lead" to the extinction of between 14 percent and 20 percent of all species before the year 2000 (US, 1980, II, p. 328). So an upper limit for the

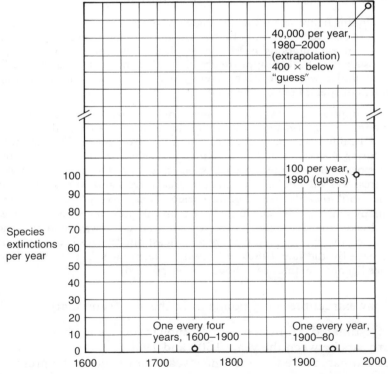

Figure 34.1 Myers–Lovejoy estimates of species extinction and their extrapolations to the year 2000

Source: N. Myers, *The Sinking Ark* (New York: Pergamon, 1979), and US CEQ and Dept of State (1980), The Global 2000 Report to the President, vol. 2 (Washington: GPO).

present that is pure guesswork has become the basis of a forecast for the future which has been published in newspapers to be read by tens or hundreds of millions of people and understood as a scientific statement.

The two historical rates stated by Myers, together with the yearly rates implied by Lovejoy's estimates, are plotted together in figure 34.1. It is clear that without explicitly bringing into consideration some additional force, one could extrapolate almost any rate one chooses for the year 2000, and the Lovejoy extrapolation would have no better claim to belief than a rate, say, one hundredth as large. Looking at the two historical points alone, many forecasters would be likely to project a rate much closer to the past than to Lovejoy's, on the basis of the common wisdom that in the absence of additional information, the best first approxima- tion for a variable tomorrow is its value today, and the best second approximation is that the variable will change at the same rate in the future that it has in the past. And the uncertainty about the definition of species adds to the confusion.

Projected change in the amount of tropical forests implicitly underlies the differences between past and projected species-loss rates in Love- joy's diagram. But to connect this element logically, there must be systematic evidence relating an amount of tropical forest removed to a rate of species reduction. We have found no reports of such empirical evidence. A recent survey document (Reid and Miller, 1989) says that "A useful rule of thumb is that if a habitat is reduced by 90 percent in area, roughly one-half of its species will be lost" (p. 35), and refers to a figure and an appendix, but no empirical studies are referred to, only speculation. The only empirical observation we found is by Lugo for Puerto Rico, where "human activity reduced the area of primary forests by 99 percent but, because of coffee shade and secondary forests, forest cover was never below 10 to 15 percent. This massive forest conversion did not lead to a correspondingly massive species extinction, certainly nowhere near the 50 percent alluded to by Myers" (1989, p. 28).

All this implies that there is no basis to choose between (a) Lovejoy's huge projected rates of extinction, and (b) modest rates continuing about the same as in the past – and this is the difference between the basis for recommending various national policies and not making any recommendations at all. (Again, this is not to say that no protection policies should be undertaken. Rather, it implies that other sorts of data to estimate extinction rates are needed as the basis for policy decisions.)

The discussion so far describes how matters stood when we wrote our 1984 article on the subject. In response to the questions we and others have raised, the "official" IUCN commissioned a book edited by Whit- more and Sayer (1992) to inquire into the extent of extinctions. The results of that project must be considered amazing. All the authors continue to be concerned about the rate of extinction. Nevertheless, they agree that the rate of *known* extinctions has been and continues to be very low. This is a sampling of quotations (with emphasis supplied), first on the subject of the estimated rates:

> ... *60 birds and mammals are known to have become extinct between 1900 and 1950 (Reid, 1992, p. 55)*

It is a commonplace that forests of the eastern United States were reduced over two centuries to fragments totalling 1–2% of their original extent, and that during this destruction, only three forest birds went extinct – the Carolina parakeet (*Conuropsis carolinensis*), the ivory-billed woodpecker (*Campephilus principalis principalis*), and the passenger pigeon (*Ectopistes migratorius*). Although deforestation certainly contributed to the decline of all three species, it was probably not critical for the pigeon or the parakeet (Greenway, 1967). *Why, then, would one predict massive extinction from similar destruction of tropical forest?* (Simberloff, 1992, p. 85)

IUCN, together with the World Conservation Monitoring Centre, has amassed large volumes of data from specialists around the world relating to species decline, and it would seem sensible to compare these more empirical data with the global extinction estimates. In fact, these and other data indicate that *the number of recorded extinctions for both plants and animals is very small.* . . . (Heywood and Stuart, 1992, p. 93)

Known extinction rates are very low. Reasonably good data exist only for mammals and birds, and the current rate of extinction is about one species per year (Reid and Miller, 1989). If other taxa were to exhibit the same liability to extinction as mammals and birds (as some authors suggest, although others would dispute this), then, if the total number of species in the world is, say, 30 million, the annual rate of extinction would be some 2300 species per year. This is a very significant and disturbing number, but it is much less than most estimates given over the last decade. (Heywood and Stuart, p. 94)

. . . if we assume that today's tropical forests occupy only about 80% of the area they did in the 1830s, *it must be assumed that during this contraction, very large numbers of species have been lost in some areas. Yet surprisingly there is no clear-cut evidence for this.* . . . Despite extensive enquiries we have been unable to obtain conclusive evidence to support the suggestion that massive extinctions have taken place in recent times as Myers and others have suggested. On the contrary, work on projects such as Flora Meso-Americana has, at least in some cases, revealed an increase in abundance in many species (Blackmore, pers. comm. 1991). An exceptional and much quoted situation is described by Gentry (1986) who reports the quite dramatic level of evolution in situ in the Centinela ridge in the foothills of the Ecuadorian Andes where he found that at least 38 and probably as many as 90 species (10% of the total flora of the ridge) were endemic to the "unprepossessing ridge". However, the last patches of forest were cleared subsequent to his last visit and "its prospective 90 new species have already passed into botanical history', or so it was assumed. Subsequently, Dodson and Gentry (1991) modified this to say that an undetermined number of species at Centinela are apparently extinct, following brief visits to other areas such as Lita where *up to 11 of the species previously considered extinct were refound,* and at Poza Honda near La Mana where six were rediscovered. (Heywood and Stuart, 1992, p. 96)

. . . actual extinctions remain low. . . . As Greuter (1991) aptly comments, *"Many endangered species appear to have either an almost miraculous capacity for survival,* or a guardian angel is watching over their destiny! This means that it is not too late to attempt to protect the Mediterranean flora as a whole, while still identifying appropriate priorities with regard to the goals and means of conservation." (Heywood and Stuart, p. 102)

. . . the group of zoologists could not find a single known animal species which could be properly declared as extinct, in spite of the massive reduction in area and fragmentation of their habitats in the past decades and centuries of intensive human activity. A second list of over 120 lesser-known animal species, some of which may later be included as threatened, show no species considered extinct; and the older Brazilian list of threatened plants, presently under revision, also indicated no species as extinct (Cavalcanti, 1981). (Brown and Brown, 1992, p. 127.)

Closer examination of the existing data on both well- and little-known groups, however, *supports the affirmation that little or no species extinction has yet occurred* (though some may be in very fragile persistence) in the Atlantic forests. Indeed, an appreciable number of species considered extinct 20 years ago, including several birds and six butterflies, have been rediscovered more recently. (Brown and Brown, 1992, p. 128)

And here are some comments from that volume on the lack of any solid basis for estimation:

. . . How large is the loss of species likely to be? *Although the loss of species may rank among the most significant environmental problems of our time, relatively few attempts have been made to rigorously assess its likely magnitude.* (Reid, 1992, p. 55)

It is impossible to estimate even approximately how many unrecorded species may have become extinct. (Heywood and Stuart, p. 95)

While better knowledge of extinction rates can clearly improve the design of public policies, it is equally apparent that *estimates of global extinction rates are fraught with imprecision. We do not yet know how many species exist, even to within an order of magnitude.* (Reid, 1992, p. 56)

. . . the literature addressing this phenomenon is relatively small. *. . .* Efforts to clarify the magnitude of the extinction crisis and the steps that can be taken to defuse the crisis could considerably expand the financial and political support for actions to confront what is indisputably the most serious issue that the field of ecology faces, and arguably the most serious issue faced by humankind today. (Reid, 1992, p. 57)

The best tool available to estimate species extinction rates is the use of species-area curves. *. . .* This approach has formed the basis for almost all current estimates of species extinction rates. (Reid, 1992, p. 57)

There are many reasons why recorded extinctions do not match the predictions and extrapolations that are frequently published. . . . (Heywood and Stuart, 1992, p. 93)

In the case of species extinction, as with many other public issues, there is a tendency – in both technical discussion and the press – to focus only upon the bad effects, and to exclude from consideration possible good effects of human activities. For example, Lugo notes that "Because humans have facilitated immigration [of species] and created new environments, exotic species have successfully become established in the Caribbean islands. This has resulted in a general increase in the total inventories of bird and tree species" (1989, p. 30). In tropical Puerto Rico where "human activity reduced the area of primary forests by 99%," as great a reduction as could be imagined, "seven bird species . . . became extinct after 500 years of human pressure . . . and . . . exotic [newly resident] species enlarged the species pool. More land birds have been present on the Island in the 1980s (97 species) than were present in pre-Colombian time (60 species)" (pp. 28 and 29).

Perhaps conservation biologists make mention of the extinctions but not of the newly-resident species because "there is a clear aversion to exotic [newly resident] species by preservationists and biologists (in cases such as predatory mammals and pests, with good reason!)" (Lugo, 1989, p. 30). This aversion to new species may involve the idea that humankind is somehow artificial and not "natural." Consider the language of Myers, who has played as important a role as any person in raising the alarm about species extinction: "[W]hereas past extinctions have occurred by virtue of natural processes, today the virtually exclusive cause is man" (1989, p. 42). If, however, it is species diversity that is at issue rather than only preserving species as they are today, then new species should count for as much as old ones.

Maintaining the Amazon and other areas in a state of stability might even have counterproductive results for species diversity, according to a recent body of research. Natural disturbances, as long as they are not catastrophic, may lead to discontinuity in environments and to consequent isolation of species that may "facilitate ever-increasing divergence" (Colinvaux, 1989, p. 103). Colinvaux goes on to suggest that "the highest species richness will be found not where the climate is stable but rather where environmental disturbance is frequent but not excessive." The same line of thought leads to possible benefits from interventions by humankind.

During the 1980s there has been increasing recognition that the rate of species loss really is not known. Myers now writes, "Regrettably we have no way of knowing the actual current rate of extinction in tropical forests, nor can we even make an accurate guess." And Colinvaux refers to the extinctions as "incalculable" (1989, p. 102). One would think that this state of affairs would make anyone leery about estimating future extinctions. Nevertheless Myers continues, "But we can make substantive assessments by looking at species numbers before deforestation and then applying the analytical techniques of biogeography. . . . According to the theory of island biogeography, we can realistically reckon that when a habitat has lost 90 percent of its extent, it has lost half of its species" (1989, p. 43). But this is mere speculation. And Lugo finds that in Puerto Rico, the "massive forest conversion did not lead to a correspondingly massive species extinction, certainly nowhere near the 50% alluded to by Myers" (1989, p. 28).[1]

Confirmation of the absence of scientific evidence for rapid species extinction is implicit in the nature of the "evidence" cited by, for example, Edward O. Wilson. He says that "the extinction problem" is "absolutely undeniable." But all he cites are "literally hundreds of anecdotal reports" (Charles C. Mann, "Extinction: Are Ecologists Crying Wolf?", *Science*, 253 (Aug. 1991), 736–8). The very reason for the scientific method in estimating rates is that anecdotal reports are of little value and often mislead the public and policymakers; that's why expensive censuses and other data-gathering instruments are mounted.

Some conservationists have become frustrated at their inability to document a rapid loss of species extinction that would justify calls for government regulation and funding; they also are annoyed at our writing about the actual state of the evidence. As a prominent conservationist responded to our article, "documenting degree of threat is often difficult, and economists and others who wish to downplay the risk of an extinction crisis can easily dispute this case or that case, casting doubt even on the claim that 5 percent of the world's birds are threatened" (Diamond, 1989, p. 41). Diamond therefore has suggested looking at the evidential issue in an entirely different fashion, one quite out of keeping with ordinary scientific practice. Normally, he writes,

> species are to be considered extant until proven extinct. . . . [But] For most species of the tropics or other remote regions – that is, for most of the world's species – a more appropriate assumption would be "extinct unless proven extant." We biologists should not bear the burden of proof to convince economists advocating unlimited human growth [an inaccurate description of one of the authors of this chapter] that the extinction crisis is real. Instead, it should be left to those economists to fund research in the jungles that would positively support their implausible claim of a healthy biological world. (p. 41)

This is an interesting twist, a "reversal of proof burden," as Western (1989, p. 33) puts it. It implies that it is enough for a warning to be sounded, a charge to be made, for the community to proceed as if the case has been proven. This intellectual strategy suggests that the biologists now despair of making their case with the usual tools of scientific inquiry and ask instead for support on the basis of non-evidential faith.

To go one step further: The conservationists premise their forecast of rapid species extinction on there being, now and in the future, a rapid rate of deforestation. We repeat that even if the rate of deforestation were indeed rapid, there would still be little or no basis for inferring a rate of species extinction of Lovejoy's projected magnitude. But their line of argument is rendered even less believable by the fact that the historical evidence does not support their projections of deforestation. (See chapter 33 by Sedjo and Clawson)

SOME OTHER ISSUES

1. Perhaps we should look backwards and wonder: Which species were extinguished when the settlers clear-cut the Middle West of the United States? Are we the poorer now for their loss? Obviously we do not know the answers. But can we even *imagine* that we would be enormously better off with the persistence of *any hypothetical* species? It does not seem likely. This casts some doubt on the economic value of species that might be lost elsewhere.

2. It is difficult to have a reasoned argument with biologists on species extinction. One reason is that they require an almost religious test of fealty and credentials before they will consider a person's testimony as relevant. The mention of a person's original training comes up again and again. In the recent survey volume that he co-edited, Western (1989, p. 33) writes: "The implications of an extinction spasm are also debatable, among both biologists and non-biologists," and Diamond (p. 37) says: "Our current concern with extinction is sometimes 'pooh-poohed' by nonbiologists with the one-liner 'Extinction is the natural fate of species'." In our view, the understanding of data is not the private province of any discipline, and the background of the analyst is not a test of the validity of the analysis. But as long as it is a criterion for biologists, the issue cannot be debated rationally.

Another difficulty is that conservation biologists' goals with respect to species diversity are not easy to understand. Sometimes they emphasize the supposed economic benefits of species diversity. For example, in its widely distributed 1990 fundraising letter (four letters received by the household of one of the present authors) The World Wildlife Fund asks, "Why should you care about the fate of these forests thousands of miles away?" and answers, "Because not only do they provide food and shelter to at least half the world's species of wildlife, these tropical forests are also the world's largest pharmaceutical factory – the sole source of lifesaving medicines like quinine, man's most potent weapon against malaria. Hundreds of thousands of people owe their lives today to these precious plants, shrubs, and trees. What would we do without them?" Diamond (1990, p. 59) answers similarly: "We need them to produce the oxygen we breathe, absorb the carbon dioxide we exhale, decompose our sewage, provide our food, and maintain the fertility of our soil."

But Quinn and Hastings say that "maximizing total species diversity is rarely if ever the principal objective of conservation strategies. Other aesthetic, resource preservation, and recreational values are often more important" (1987, p. 199). And Lovejoy says, most inclusively:

> What I'm talking about is rather the elusive goal of defining the minimum size [of habitat] needed to maintain the characteristic diversity of an ecosystem over time. In other words, I think the goals of conservation

aren't simply to protect the full array of plant and animal species on the planet, but rather also to protect them in their natural associations so that the relationships between species are preserved and the evolutionary and ecological processes are protected. (quoted in Iker, 1982, p. 29)

This vagueness of goals makes it very difficult to compare the worth of a species-saving activity against another value. What are the relative worths of maintaining the habitat on Mount Graham, Arizona, for about 150 red squirrels who could be kept alive as a species elsewhere, for example, versus using 24 acres for an observatory that would be at the forefront of astronomical science (*New York Times*, March 8, 1990, p. A1)? There is much less basis here for a reasoned judgment in terms of costs and benefits than there is even with such thorny issues as electricity from nuclear power versus from coal, or decisions about supporting additional research on cancer versus using the funds for higher Social Security payments or for defense or even for lower taxes.

Policymaking is also made difficult by conservationists asserting on the one hand that the purpose of conserving is that it is good for human existence, and on the other hand that human existence must be limited or reduced because it is bad for the other species. "There are many realistic ways we can avoid extinctions, such as by preserving natural habitats and limiting human population growth" (Diamond, 1990, p. 59) is a typical statement of that sort – by the same writer who urges that humans should preserve the species because humans need them for existence!

Still another difficulty in conducting reasoned discussion of the subject with biologists is their attitude toward economists, whose trade it is to assess the economics of supporting public programs. One of the most noted of conservation biologists, Peter Raven, views economists as follows: "Perhaps the most serious single academic problem in the world is the training of economists" (1988, p. 229). We have plenty of complaints ourselves about the training of economists, but our complaint is that the fundamental truths about the subject get lost in technical escalations. The complaint of Raven and other biologists is more far-ranging: They believe that the fundamental structure of thought of economics is perverted because it leads to unsound social choices by omitting considerations the biologists consider crucial. But the conservationists do not render those considerations into a form that a calculus of choice can deal with. Herein lies a major problem for the issue at hand.

3. The view that the interests of humans and of other species are opposed leads to humankind being seen in a rather ugly light. "[O]ur species has a knack for exterminating others, and we're becoming better killers all the time" (Diamond, 1990, p. 58). A recent article is entitled "Extinction on Islands: Man as a Catastrophe" (Olson, 1989, p. 5).

4. It is quite clear that species are seen by many as having value quite apart from any role they play in human life, a value that is seen as

competitive with the value of human life. Raven writes, "Although human beings are biologically only one of the millions of species that exist on Earth, we control a highly disproportionate share of the world's resources" (1988, p. 212), suggesting that it is unfair that we "control" more resources than do eagles, mosquitoes, or the AIDS virus.

These beliefs lead to policy recommendations toward the human race that hinge upon values about the worth of humans versus the worth of other species. "[P]opulation buildup" is seen as part of a "dismal scenario."

5. It is not the case, as some have asserted, that we "wish to downplay the risk of an extinction crisis" (Diamond, 1989, p. 41). Rather, we want to make as clear as possible how great the risk is. We want to separate the available facts from the guesswork and the purposeful misstatements, in order to improve the public decision-making process. And we want to comment upon how society may reasonably take into account the economic and non-economic worths of species, in light of our values for human and non-human aspects of nature and other aspects of life on earth. More generally, we would like to move the discussion in the direction of thinking as well as we can about this problem that is indeed difficult to think about sensibly, though it is probably much easier to think about than the greenhouse issue, which is much less subject to experimentation and observational comparison because there is only one atmosphere, whereas there are many separated areas whose diversity can be studied.

SUMMARY AND CONCLUSION

The scare about species extinction has been manufactured in complete contradiction to the scientific data.

The highest proven observed rate of extinction until now is only *one* species per year. Yet the "official" forecast has been *40,000* species dying out per year in this century, a million in all.

It is truth that is becoming extinct, not species.

The argument that because we do not know how many species are being extinguished, we should therefore take steps to protect them, is logically indistinguishable from the argument that because we do not know at what rate the angels dancing on the head of a pin are dying off, we should undertake vast programs to preserve them. And it smacks of the condemnation to death of witches in Salem on the basis of "spectral evidence" by "afflicted" young girls, charges that the accused could not rebut with any conceivable material evidence.

If something is unknowable at present but *knowable in principle*, then the appropriate thing is to find out. This does not necessarily mean finding out by direct observation only. A solid chain of empirical evidence can lead to a reasonable conclusion. But there must be some reasonable chain of evidence and reasoning.

If something is *unknowable in principle,* at least with contemporary techniques, then there is no warrant for any public actions whatsoever. To assert otherwise is to open the door to public actions and expenditures on behalf of anyone who can generate an exciting and frightening hypothetical scenario.

Some say the numbers do not matter scientifically. The policy implications would be the same, they say, even if the numbers were different even by several orders of magnitude. But if so, why mention any numbers at all? The answer, quite clearly, is that these numbers do matter in one important way: they have the power to frighten the public in a fashion that smaller numbers would not. We can find no scientific justification for such use of numbers.

Some have said: But was not Rachel Carson's *Silent Spring* an important force for good even though it exaggerated? Maybe so. But the account is not yet closed on the indirect and long-run consequences of ill-founded concerns about environmental dangers. And it seems to us that, without some very special justification, there is a strong presumption in favor of stating the facts as best we know them, especially in a scientific context, rather than in any manipulation of them.

At a time when there appear frequent reports on the extraordinary possibilities of genetic engineering (for example, "Animals Altered to Produce Medicine in Milk . . . Scientists Say Rare Drugs Could be Manufactured with Relative Ease" [*The Washington Post,* August 27, 1991, p. 1]), it is beginning to seem ludicrous to justify extraordinary expense to protect an animal like the grey squirrel – which may not even be genetically distinct – on the grounds that its gene pool will be valuable for human life in the future.

Still, the question exists: How should decisions be made, and sound policies formulated, with respect to the danger of species extinction? We do not offer a full answer. One cannot simply propose saving all species at any cost, any more than one can propose a policy of saving all human lives at any cost.

Then we must also try to get more reliable information about the number of species that might be lost with various forest changes. This is, of course, a very tough task, too, one that might exercise the best faculties of a statistician and designer of experiments. One suggestion: if the population sizes of selected species could be measured in a series of periods along with experimental or non-experimental changes in habitats, extrapolation might teach something about conditions that would cause species to approach or reach extinction.

Lastly, policy analysis concerning species loss must explicitly evaluate the total cost of the protection, for example, cessation of foresting in an area. And such a total cost estimate must include the long-run indirect costs of reduction in economic growth to a community's health, as well as the short-run costs of foregone wood or agricultural sales (see Wildavsky, 1988, and Keeney, 1990). To ignore such indirect costs because they are hard to estimate would be no more reasonable than ignoring the loss of species that we have not as yet identified.

We summarize the situation as follows: There is now no prima facie case for any expensive species-safeguarding policy without more extensive analysis than has been done heretofore. But the question deserves deeper thought, and more careful and wide ranging analysis, than has been done until now. As children say, just saying so does not make it so.

NOTE

Additional criticism of the "bio-geography" theory of extinction rates has recently been reported in *Science* (Charles C. Mann, "Extinction: Are Ecologists Crying Wolf?", *Science*, 253 (Aug. 1991), 736–8), but further discussion is beyond the scope of this paper.

REFERENCES

Atkinson, Ian (1989): "Introduced Animals and Extinctions." In *Conservation for the Twenty-first Century*, eds. Western and Pearl.

Brown, K. S., and G. G. Brown (1992): "Habitat alteration and species loss in Brazilian forests." In *Tropical Deforestation* and *Species Extinction*, eds. Whitmore and Sayer, 119–42.

Colinvaux, Paul A. (1989): "The Past and Future Amazon." In *Scientific American*, May: 102–8.

De Beijer, J. R. (1979): "Brazil." *World Wood Review*, July, 38–9.

Diamond, Jared (1989): "Overview of Recent Extinctions." In *Conservation for the Twenty-first Century*, eds. Western and Pearl.

——— (1990): "Playing Dice With Megadeath." In *Discover*, April, 55–9.

Ehrlich, Paul and Anne Ehrlich (1981): *Extinction*, New York: Random House.

——— and Edward O. Wilson (1991): "Biodiversity Studies: Science and Policy." *Science*, 253, August 16, 758–62.

Fyfe, W. S. (1981): "The Environmental Crisis: Quantifying Geosphere Interactions." *Science*, 213 (3) July, 105–10.

Heywood, V. H., and S. N. Stuart (1992): "Species extinctions in tropical forests." In *Tropical Deforestation and Species Extinction*, eds Whitmore and Sayer, 91–118.

Holden, Constance (1974): "Scientists Talk of the Need for Conservation and an Ethic of Biotic Diversity to Slow Species Extinction." *Science*, May, 647–8.

Iker, Sam (1982): "Islands of Life In a Forest Sea." In *Mosaic*, September/October, 24–9.

Instituto Brasiliero Desinuol (1980): "Brazil," *World Wood Review*, July, 56–7.

Keeney, Ralph L. (1990): "Mortality Risks Induced by Economic Expenditures." *Risk Analysis*, 10, December, 147–8, 155.

Lugo, Ariel E., ed. "Diversity of Tropical Species." In *Biology International*, Special Issue–19.

Muthoo, M. K. (1978): "Brazil," *World Wood Review*, May, 51–3.

Myers, Norman (1979): *The Sinking Ark*. New York: Pergamon.

——— (1989): "A Major Extinction Spasm: Predictable and Inevitable?" In *Conservation for the Twenty-first Century*, eds Western and Pearl.

Olson, Storrs L. (1989): "Extinction on Islands: Man as a Catastrophe." In *Conservation for the Twenty-first Century*, eds Western and Pearl.

Persson, R. (1974): *World Forest Resources*. Stockholm: Royal College.

Quinn, James F., and Alan Hastings (1987): "Extinction in Subdivided Habitats." In *Conservation Biology*, 1 (3) 198–208.

Raven, Peter H. (1988): "The Cause and Impact of Deforestation." In *Earth '88 – Changing Geographic Perspectives*, 212–29.

Reid, W. V. (1992): "How many species will there be?" in *Tropical Deforestation and Species Extinction*, eds Whitmore and Sayer, 55–74.

Reid, Walter V., and Kenton R. Miller (1989): *Keeping Options Alive: The Scientific Basis for Conserving Biodiversity*. Washington: World Resources Institute, October.

Rohter, Larry (1979): "Amazon Basin's Forests Going Up in Smoke," *The Washington Post*, January 5, A14.

Roush, G. Jon (1989): "The Disintegrating Web: The Causes and Consequences of Extinction." *The Nature Conservancy Magazine*, Nov./Dec., 4–15.

Sedjo, Roger A. (1980): "Forest Plantations in Brazil and Their Possible Effects on World Pulp Markets." *Journal of Forestry*, 78, November, 702–4.

Simberloff, D. (1992): "Do species-area curves predict extinction in fragmented forest?" In *Tropical Deforestation and Species Extinction*, eds Whitmore and Sayer, 75–90.

Sommer, Adrian (1976): "Attempt at an Assessment of the World's Tropical Mist Forests." *Unasylva*, 28, 5–27.

US CEQ and Department of State (1980): *The Global 2000 Report to the President*, vol. 2. Washington: GPO.

US CEQ and Department of State (1981): *Global Future: Time to Act*. Washington: GPO.

Waldrop, M. Mitchell (1981): "Wood: Fuel of the Future?" *Science*, Feb. 27, 914.

Western, David (1989): "Conservation Biology." In *Conservation for the Twenty-first Century*, eds Western and Pearl.

——— and Mary C. Pearl (1989): *Conservation for the Twenty-first Century*. New York, Oxford: Oxford University Press.

Whitmore, T. C., and J. A. Sayer, eds (1992): *Tropical Deforestation and Species Extinction*. New York: Chapman and Hall.

Whittaker, R. H., and G. E. Likens (1975): "The Biosphere and Man." In H. Lieth and R. H. Whittaker eds, *Primary Productivity of the Biosphere*. New York: Springer.

Wildavsky, Aaron (1988): *Searching for Safety*. New Brunswick: Transaction Press.

PART IV

Agriculture, Food, Land, and Water

35

Agricultural Productivity Before the Green Revolution

George W. Grantham

Crop yields measure man's success in bending nature's balance to his needs and desires. While natural environments commonly supply water, light, and nutrients in amounts sufficient to support dense populations of plants and animals, biological equilibria do not ordinarily occasion high sustained yields of useful species. The adventitiousness of productivity in natural conditions makes it necessary to create artificial environments favorable to the growth of the desired species. Before the second half of the nineteenth century this goal was usually achieved by mechanically eliminating competing plants through weeding and tillage. Nutrient supply was maintained by leaving land uncropped long enough to allow the natural cover to accumulate new reserves of mineralized nitrogen. Where conditions favored livestock husbandry the dung of domesticated farm animals accelerated this natural cycle.[1] In the elaborate forms of mixed husbandry developed by European farmers between 1400 and 1850, farmers planted nitrogen-fixing leguminous fodder crops, which they fed to farm animals returning most of the captured nitrates to the land in animal dung. This augmented the supply of mineralized nitrogen by as much as two thirds and raised average crop yields more than two-fold.[2] Between 1840 and 1875, new supplies of nitrates, potassium salts and phosphoric acid were discovered outside the agricultural sector and made commercially available to farmers, creating novel underlying conditions for agricultural productivity growth, although realization of their full potential required further advances in plant breeding and more perfect mechanization.

The history of crop yields prior to the development of concentrated chemical fertilizers in the 1840s and 1850s can be read in terms of variable responses to the opportunities latent in organic agricultural systems where the primary constraint on plant growth was the supply of available nitrogen. The upper bound to yields based on organic fertilizing is about 3,000 kilograms of cereal grain per hectare, because crops that are over-stimulated by nitrogen tend to lodge and are vulnerable to parasitic infection. This is true of all nitrate fertilizers and it is noteworthy that at the beginning of the twentieth century, when concentrated commercial fertilizers were already widely applied in northern Europe, the only countries where national average yields approached this limit were Denmark (2,850), Belgium (2,370), Ireland (2,320), the Netherlands (2,260), and Great Britain (2,110).[3] The 3,000 kilogram ceiling persisted through the 1930s and was lifted only with the development of new varieties capable of standing up to heavy dosings of fertilizer. The rise in average European yields from 964 kilograms to 1,410 between 1900 and 1935 was due to a narrowing of the gap between average and best practice.[4] By comparison, the average yield of wheat in all of Europe in 1989 was 3,600 kilograms and yields in northern Europe exceeded 6,000. This is a result of the development after 1950 of new agricultural technologies based on massive applications of chemical herbicides, pesticides, fertilizers, and accelerated plant breeding and developed primarily by means of sustained scientific experimentation.[5] This chapter is concerned with mechanisms of agricultural change before this Green Revolution.

THE ORIGINS AND EARLY DEVELOPMENT OF CEREAL CULTIVATION

Experiments in harvesting natural stands of wild wheat in the Black Mountain district of southeastern Turkey indicate that a man wielding a primitive flint-bladed sickle can harvest approximately 2.5 kilograms of grain per hour.[6] If we assume that it took 15 ten-hour days to reap a hectare of standing grain, it follows that under optimal conditions yields in natural stands of grain could not have exceeded 375 kilograms per hectare.[7] Effective yields were probably no more than half this amount owing to uneven ripeness of the stand, which caused much grain to be lost from spillage and inedibility of over- and under-ripe grain. The earliest farmers probably reaped no more than what was available in natural stands. Even so, cultivated stands were probably an improvement over wild varieties, as domesticated grain ripens more uniformly and is more strongly attached to the stalk, giving the plant more resistance to shattering during the harvest. These properties, which emerged ten thousand years ago in the Near East, most likely resulted from evolutionary adaptation caused by natural selection of wild cereals to make them attractive to human predators who were the unintentional agents of seed dispersal.[8] The subsequent diffusion of agriculture through central and into northwest Europe did not greatly improve on early yield levels, although a widening range of plants doubtless raised total food supplies. To judge from the size of settlements in Bronze and early Iron-Age Europe, which rarely exceeded one to three dozen persons, yields in northern Europe probably averaged around 200 to 450 kilograms through the early medieval age.[9]

A new level of performance was achieved in classical times on intensively cultivated estates. By the early Christian era, yields in the Mediterranean basin and on well-organized properties in Roman-controlled northern Europe attained 600 to 800 kilograms, and Cicero's estates near Mount Etna yielded 1,500 kilograms on fertile volcanic soils.[10] Yields as high as 2,500 kilograms per hectare can be inferred from surviving lease and tax records of Roman Egypt, a performance made possible only because the Nile annually deposited the equivalent of 20 tons of fertilizer per hectare during its annual flooding.[11] In the 1920s Egyptian grain yields averaged 1,700 kilograms, which probably represents the average for irrigated agriculture.[12]

It follows from this meager compilation that yields in pre- and early-medieval times fell into three broad classes: 225 to 450 kilograms obtained by slash and burn agriculture, shallow cultivation with primitive implements, and sparse sowing; a system employing animals to plow and teams of male and female slaves to hoe and weed that produced yields ranging from 500 to 750 kilograms; and irrigated desert agricultures which yielded upwards of 1,200 kilograms per hectare. Two obstacles barred the way to higher output per hectare. The first was the labor requirements of high-yield farming, which necessitated a rigidly authoritarian organization of labor to produce significant surpluses above the consumption requirements of the agricultural population; the second was the deficiency of farm animals and nitrogen-fixing plants to produce plant nutrients.

THE MEDIEVAL REVOLUTION

Inventories of Charlemagne's agricultural estates near Paris imply that around 800 AD yields on the main cultivated fields were about 300 kilograms per hectare. This miserable performance was due to defective tillage and extremely light sowing rates that gave gross yields less than twice the seed.[13] By 1100 cereal yields were ordinarily between 600 and 900 kilograms. The improvement is attributable to two changes. The first was the development of a heavy moldboard plough that by turning instead of stirring the soil permitted farmers to till more deeply and to extend the fields into potentially more fertile territory. The second was an increase in all inputs and an intensification of agriculture, possibly due in part to extra energy from improved nutrition, but ultimately owing to the ability of the recently evolved seigniorial system to extract more work from the dependent peasantry. By the year 1000, market towns of one to two thousand persons were sprouting throughout western Europe to provide, in weekly markets, incentives for peasants to intensify productive efforts on their own account.[14] By the late thirteenth century, this process of market-induced intensification had reached the point where farmers in the Low Countries and southeastern England had eliminated fallow and were sowing leguminous pulses in high-yield rotations. Around 1300 they attained yields of 1,500 kilograms per hectare, a level which would have been considered outstanding as late as the early nineteenth century.[15]

The subsequent history of European crop yields to the middle of the nineteenth century traces the advances, retreats, and elaboration of a system of farming the fundamental components of which were thus in place by the early fourteenth century. Leguminous fodder crops – clover, alfalfa, and sainfoin – added to the stock of nitrogen-producing crops, but only made more productive without fundamentally altering this system of farm husbandry that the eighteenth century mistakenly called "New," because in some places it had disappeared. The retreat and recovery of intensive mixed husbandry was mainly a function of its economic profitability.

DYNAMICS: MARKETS AND PRODUCTIVITY

What can be said about the long-term dynamics of agricultural change? Only a brief discussion is possible here. Traditional explanations of the course of agricultural productivity have tended to adopt Malthus's conjecture that the most powerful long-run forces in economic and agrarian history are demographic.[16] Recent research in agricultural and demographic history, however, indicates that most closed populations kept fertility well below the maximum permitted by human biology, which means that pre-modern populations were demographically homeostatic rather than uncontrolled. If so, the driving force for agricultural change could not have been unrestrained population growth. Scholars also find that outside the seasons of peak demand most peasant societies possess considerable reserves of labor that can be mobilized to increase the output of foodstuffs when it is to their advantage to do so. Medieval evidence does not indicate that farmers adopted advanced methods because of population pressure, and recent analysis of the archeological record indicates that the response of the earliest agricultural societies to pressure on sedentary foraging communities, one of which led to agriculture, produced divergent responses, including a regression to non-sedentary gathering.[17] Population pressure is therefore neither sufficient nor necessary for agricultural progress.

Recent ventilation of the full corpus of extant medieval records in England, however, does not support the gloomy alternative explanation that yields decline as population growth pushes farming into inferior lands.[18] Thousands of manorial records in England from the late twelfth to the early fourteenth century, along with French tithe records from the early fourteenth to early eighteenth century, make it abundantly clear that yields did not decline because overpopulation forced cultivation into inferior ground. Some regions with fertile soils had relatively modest yields; others whose soils were mediocre had high ones. The spatial incidence of yields, where it can be determined with accuracy, fails to conform to the Malthusian hypothesis.[19]

Average grain yields in England and northern France in the late thirteenth century were about 900 kilograms per hectare, with the best regular performances ranging from 1,200 to 1,500 kilograms. These yields were not generally maintained through the late medieval depression.[20] They declined in the fourteenth and fifteenth centuries and followed an uneven course well into the seventeenth. Neither hypothesis noted above adequately explains this history because both ignore, by their high level of aggregation, the spatial effects that led farmers exposed to strong grain markets by their physical proximity to invest in high-yielding systems of cultivation. Intensive husbandry was expensive, and farmers adopted it only where markets were sufficiently attractive to cover the extra costs.[21] When these markets slackened, farmers converted their farms to pasture and even to forests. Population growth on its own was not enough to counter this tendency, because a growing population of poor people did not automatically create economic incentives to produce grain for markets.

The dynamics of agricultural change, then, are largely connected to the history of European food markets, which are in turn connected to the more general economic history of European regions. The geographical patterns of cropping and productivity demonstrate that the intensity of cultivation before the transport revolution was inversely related to distance from major metropolitan markets for food. These geographical patterns can now be traced in England to the early thirteenth century, where recent reinterpretation of the archeological record has substantially raised previous estimates of the size of London and other major English towns.[22] The best known example of high yields induced by urbanization occurred in the Low Countries in the fifteenth and sixteenth centuries. Precocious urbanization and prosperity based on expanding foreign trade gave rise to an animal- and dung-intensive husbandry that came increasingly to depend on leguminous forage plants and the exchange of agricultural produce for urban night soil. As this "New Husbandry" diffused from its point of origin to other parts of Europe, it created "islands" of agricultural progress around cities. This pattern of diffusion allowed for a response to market opportunities, which was sufficiently elastic to feed Europe's growing population through the first crucial phase of industrialization.[23]

Institutional impediments caused by the small sizes of farms and collective regulation of agricultural practices are often cited as causes of agricultural backwardness, but their effect has been greatly exaggerated. Most eroded rapidly in the presence of active food markets. Where legal and political costs prohibited massive expropriation and rewriting of property rights, the necessary adjustments in farm size and compactness were carried out through rental markets. Close analysis therefore does not confirm the hypothesis that imperfect property rights constituted major barriers to agricultural progress.[24] When profit beckons, men find ways to bend law and custom to their will.

In sum, with the emergence of food, land, and labor markets on a broad scale after 1100, most of the institutional apparatus for exploiting the productivity inherent in an agriculture based on tillage and manuring was in place. Variations in agricultural performance then became a question of differential economic advantage. The major proximate cause of variations in agricultural productivity since the later Middle Ages is the uneven history of commercial and industrial development as between northern, southern, and eastern Europe.[25]

ESTIMATES

Meaningful average yields are impossible to construct from extant data prior to the nineteenth century because local experience was so diverse. The post-1800 experience is shown in figure 35.1.[26] National average yields and *a fortiori* European or world averages before the nineteenth century must remain conjectural. At present the most secure documentation comes from the English county of Norfolk, situated some 60 miles northeast of London. Figure 35.2 reconstructs yields in this county from

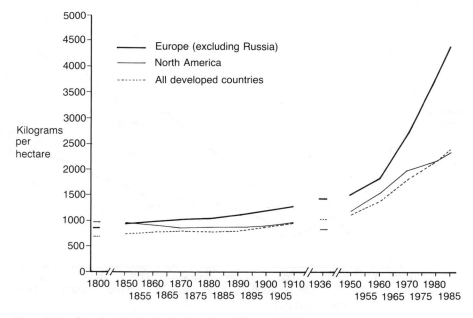

Figure 35.1   Cereal yields in the developed world since 1800
Source: Bairoch, "Les trois revolutions agricoles," 319.

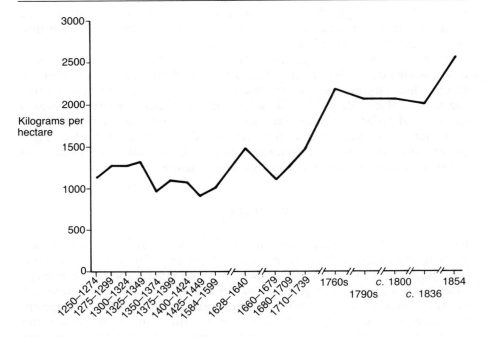

Figure 35.2   Yields in the county of Norfolk, 1250–1850
Source: Campbell, "English seignorial agriculture," 180.

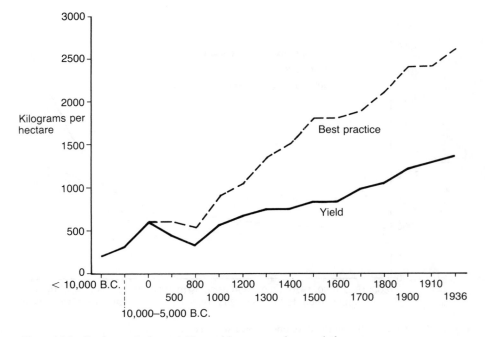

Figure 35.3   Conjectural wheat yields, prehistory to early twentieth century

the early thirteenth to the mid-nineteenth century.[27] It shows fluctuating yields between the middle of the thirteenth and the middle of the eighteenth century with a sharp rise after 1740 as the pace of England's economic growth accelerated. It should be recalled that the yields reported down to 1836 are no higher than what had been achieved on the most productive estates by the beginning of the fourteenth century. It was only around 1850 that the effects of concentrated fertilizers can be seen to mark the beginning of a new age of agriculture based on the inputs of scientific origin. For much longer stretches of time, only conjectures are possible – such may be seen for wheat in figure 35.3.

NOTES

1   Robert S. Shiel, "Improving soil productivity in the pre-fertilizer era." In Bruce M. S. Campbell and Mark Overton eds, *Land, Labor and Livestock, Historical Studies in European Agricultural Productivity*. Manchester: Manchester University Press, 1991.

2   G. P. H. Chorley, "The Agricultural Revolution in Northern Europe, 1750–1880: Nitrogen, Legumes, and Crop Productivty," *The Economic History Review*, 2nd series, vol. 34 (February, 1981), 71–93.

3   Institut International d'Agriculture. Service de Statistique Agricole, *Annuaire International de Statistique Agricole, 1911 et 1912*, Rome, 1914.

4   FAO *Yearbook of Food and Agricultural Statistics* I, (1948), for the years 1934–38. North American yields, based on extensive cultivation were below a thousand kilograms at this time and have therefore been omitted from this discussion, which is concerned primarily with technological possibilities.

5   FAO *Yearbook of Agricultural Statistcis* 43 (1989).

6   Jack R. Harlan, "A Wild Wheat Harvest in Turkey," *Archaeology* 20 (1967), 197–201.

7   The reaping labor input is based on observed rates in the most backward districts of eighteenth-century France, reported in G. Grantham, "The Growth of labor productivity in the production of wheat in the *Cinq Grosses Fermes* of France, 1750–1929," in Campbell and Overton, *Land, Labor and Livestock*, 340–63.

8   David Rindos, "Darwinian Evolution and Cultural Change: The Case of Agriculture," in Linda Manzanilla (ed.) *Studies in the Neolithic and Urban Revolutions*. Oxford, 1987, 97.

9   Peter S. Wells, *Farms, villages and cities. Commerce and Urban Origins in Prehistoric Europe* (Ithaca, 1984), pp. 42–3. Higher densities have been proposed for settlements in southern Poland, where yields are reckoned not to have exceeded 200 kilograms. Janusz Gatoja-Zagorski, "Demographic and Economic Changes in the Hallstatt Period of the Lusatian Culture," in D. Blair Gibson and Michael N. Geselovitz, *Tribe and polity in late prehistoric Europe, Demography, production, and exchange in the evolution of complex social systems* (New York and London, 1988), 119–35.

10   Keith Branagan, *Gatscombe, The Excavation and study of a Romano-British Villa Estate, 1967–1976*, British Archeological Reports 44 (Oxford, 1977), 193–6; M. M. Postan, ed. *The Cambridge Economic History of Europe I: The Agrarian Life of the Middle Ages*, 2nd edn (Cambridge, 1966), 104.

11 Napthali Lewis, *Life in Egypt Under Roman Rule* (Oxford, 1983), 109, 121–3.

12 Institut International d'Agriculture, *Annuaire international de statistique agricole, 1931–32* (Rome, 1932).

13 Georges Duby, *Guerriers et paysans, viie–xiie siècle premier essor de l'économie européenne* (Paris, 1974), 38.

14 Mavis Mate, "Medieval Agrarian Practices: The Determining Factors?" *Agricultural History Review* 33 (1985), 22–31.

15 Bruce M. S. Campbell, "Land, labour, livestock, and productivity trends in English seigniorial agriculture, 1208–1450." In Bruce Campbell and Mark Overton eds, *Land, Labour and Livestock* (Manchester, 1991), 145–81.

16 Ester Boserup, *The Conditions of Agricultural Growth: The Economics of Agrarian Change Under Population Pressure* (Chicago, 1965); Yujiro Hayami and Vernon W. Ruttan, *Agricultural Development: An International Perspective* rev. edn (Baltimore, 1985).

17 D. O. Henry, *From Foraging to Agriculture. The Levant at the End of the Ice Age* (Philadelphia, 1989).

18 Michael M. Postan, *The Medieval Economy and Society* (London: Penguin Books, 1972).

19 Bruce M. S. Campbell, "Land, labor, livestock, and productivity trends in English seigniorial agriculture," and Chris Thornton, "Determinants of Land Productivity on the Bishop of Winchester's Demesne of Rimpton, 1208/09 to 1402/03," in the same volume. The shortcomings of Postan's model are exhaustively reviewed by Kathy Biddick, in "Malthus in a Straitjacket? Analyzing Agrarian Change in Medieval England," *Journal of Interdisciplinary History* 20 (Spring, 1990), 623–35. For France see Hugues Neveux, *Les grains du Cambrésis, fin du xive – début du xviie siècles. Vie et déclin d'une structure économique* (Paris, 1980).

20 Campbell, "English Seignorial agriculture."

21 For a demonstration of profit-dependent diffusion in a somewhat later period, see G. Grantham, "The Diffusion of the New Husbandry in Northern France, 1814–1840," *Journal of Economic History* 38 (June, 1978), 311–37.

22 Bruce Campbell, "Towards an Agricultural Geography of Medieval England," *Agricultural History Review* 36 (1988), 87–98.

23 See G. Grantham, "Agricultural Supply in the Industrial Revolution: French Evidence and European Implications," *Journal of Economic History* 49 (March, 1989), 43–72.

24 For Britain see Robert Allen, "The Efficiency and Distributional Consequences of Eighteenth-Century Enclosures," *Economic Journal* 92 (1982), 937–53. The position that enclosures did matter is argued in Donald McCloskey, "The Economics of Enclosure: A Market Analysis." In William N. Parker and Eric Jones, eds, *European Peasants and their Markets* (Princeton, 1975), 123–76. French experience is documented in G. Grantham, "The Persistence of Open-Field Farming in Nineteenth-Century France," *Journal of Economic History* 40 (September, 1940), 515–32, and Philip Hoffman, "Institutions and Agriculture in Old Regime France," *Politics and Society* 16 (1988), 241–64.

25 This is the conclusion of J. L. Van Zanden, "The First Green Revolution: The growth of production and productivity in European agriculture, 1870–1914," *Economic History Review* XLIV:2 (May, 1991), 215–39.

26 Post-1800 yields have been constructed for the developed countries in Paul Bairoch, "Les trois révolutions agricoles du monde développé: rendements et

productivité de 1800 à 1985," *Annales, économies, sociétés, civilisations* (1989), 317–53. These yields are reproduced in figure 35.1.

27   Campbell op. cit.

EDITOR'S NOTE (JULIAN L. SIMON)

See chapter 36 by Avery for recent data on agricultural productivity.

Figures 35.4 and 35.5 show data on corn yields in the United States since 1492 and rice yields in Asian countries since 600. Figure 35.6 shows changes in productivity in three crops in the United States since 1800.

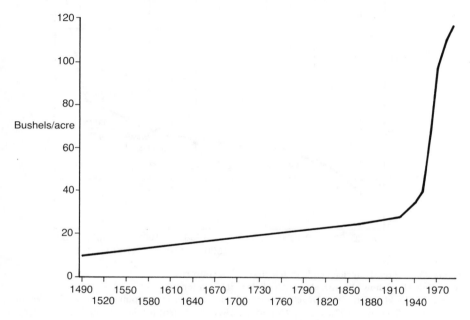

Figure 35.4   Estimated North American corn yields, 1490–1990
Source: Adapted from "Corn: America's Crop," *Farm Journal* (mid-March, 1990), AC-4A.

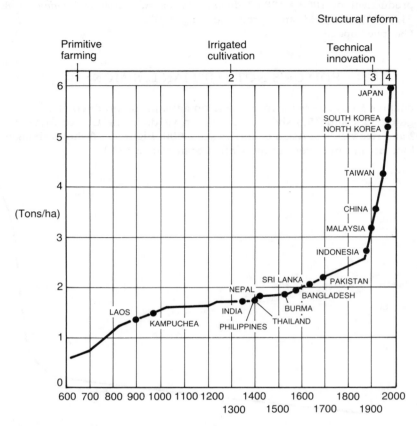

Figure 35.5 Trends in rice yields in selected Asian countries
Note: Solid line shows historical growth of rice yields in Japan.
Source: W. D. Hopper, "The Development of Agriculture in Developing Countries," *Scientific American* 235 (Sept., 1976), 200. Copyright 1976 by Scientific American, Inc. All rights reserved. Reprinted by permission.

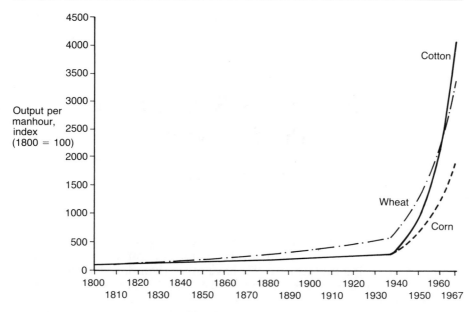

Figure 35.6 Index of US farm labor productivity in corn, wheat, and cotton, 1800–1967
Note: Trend lines calculated from average annual percent gains 1800–1937 and 1937–67.
Source: Calculated from data in I. Welfeld, *Where We Live* (New York: Simon and Schuster, 1988).

36

The World's Rising Food Productivity

Dennis Avery

The world's food strategy now stands revealed as one of our great successes. The secret is rising productivity, generated by agricultural science and capital investment. We have more than doubled world food output in the past 30 years. We have raised food supplies per person by 25 percent in the populous Third World. And the quality of those calories has been sharply upgraded, including far more of such resource-expensive foods as cooking oil, meat, fruits, and vegetables. Meanwhile, the real cost of food continues to decline.

Remarkably, this larger and more nutritious food supply is coming from about the same land base as in 1960.[1] With the exception of Africa, agriculture has not had to expand onto more fragile lands, deforest continents, or push more wildlife out of its habitat to meet food needs.

The most important productivity factor has been higher yields of crops per acre of land and per acre-foot of available water.[2]

RISING WORLD CROP YIELDS

The world's food production gains since 1950 have been truly remarkable, as exemplified in figure 36.1, "World Cereal Production, 1950–1990." The increased food production has permitted a sharp increase in per capita food supplies for most of the poorest countries, despite rising population. For most of the world's countries, the gains are shown most effectively in cereal production. (See figure 36.2, "Per Capita Grain Production, 1950–1990.")

Most of the world's food gains have come from rapidly-rising crop yields. (See figure 36.3, "World Cereal Yields, 1950–1990.") Some observers have mistakenly attributed the lack of expansion in world crop plantings to a lack of additional good farmland. On the contrary, the world still has more than a billion acres of unplanted or underused arable land. Large chunks of this land are in the United States, Argentina, Brazil, Zaire, Sudan, and Turkey.

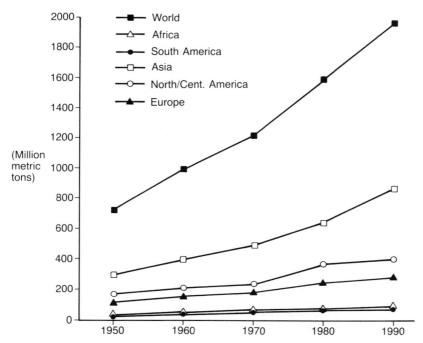

Figure 36.1a World cereal production, 1950–90
Source: FAO.

But higher-yielding seeds, modern fertilizers, and improved pest control have made it cheaper to increase food production on existing land than to forge out onto frontiers. It takes heavy capital investment in roads, schools, and other infrastructure to bring new land into commercial farming.

World cereal yields have risen 144 percent since 1950. In the years before 1960, the yield trend was led by North America. Since 1960, the Green Revolution has truly been extended to nearly every region of the world. Cereal yields have more than doubled in such diverse countries as France, India, and Indonesia, and nearly doubled in such countries as Poland, Brazil, and Kenya. Some of the most dramatic progress has been made in China, where grain production has increased 180 percent (figure 36.4) and grain yields have increased 264 percent since 1960 (figure 36.5)

Yields of other important crops such as potatoes, cassava, coffee, cotton, oilseeds, and oil palm have also increased radically.[3]

CAN THE FOOD PROGRESS CONTINUE?

There is no indication that the potential yield gains have been "used up." In the 1980s:

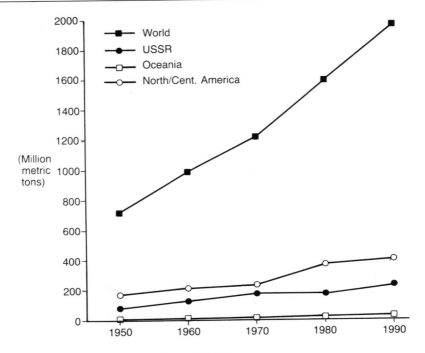

Figure 36.1b World cereal production, 1950–90
Source: FAO.

- Wheat production (non-United States) has been rising at 2.0 percent annually, with yields up by 3.2 percent per year and declining acreage needed.[4]
- Rice production has been rising at 3.5 percent annually, although yields have been rising faster (at 2.1 percent) than population growth.[5]
- Demand for coarse grains has risen slowly, due to slow economic growth and big US corn surpluses. Non-United States corn production has risen by only 1.6 percent per year.[6] However, yields on much of the corn planted in Asia, Latin America, and Africa could be tripled – with seed varieties currently on the shelf – if consumer incomes were strong enough to buy more meat and thus support the cost of improved seeds.[7]
- Oilseed production has been rising most rapidly of all crops (4.8 percent) because of the demand for higher quality diets. But oilseeds are a luxury from the standpoint of hunger. Most of the 360 million acres growing oilseeds in 1990 could have produced twice as many calories if planted in cereals.

In the five years 1985–1990, farm output in the developing countries – where more and better food is still urgently desired – rose 14.4 percent

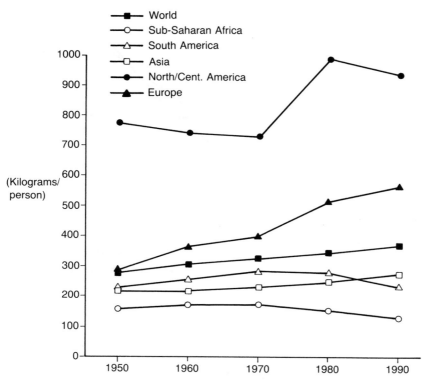

Figure 36.2a Per capita grain production, 1950–90
Note: More than half of Africa's food calories come from root crops such as cassava and yams.
USSR grain is reported bunker weight rather than cleaned and dried as in other countries.
Source: FAO.

(2.9 percent annually). (See figure 35.6, Index of World Agricultural Production, 1984–89.) The world's total production rose far less – 8.6 percent – because farm product demand is already saturated in the affluent countries, where farm production is rising just 0.3 percent annually. Overall, world crop yields have stayed comfortably abreast of, and generally slightly ahead of, world population growth, with the only apparent constraint being the costliness of producing surpluses unconsumed by the population. (See table 1, World Crop Yields, 1980/81–1990/91.)

Table 36.1 World crop yields, 1980/81–1990/91 (tons/hectare)

	1980/81	*1990/91*	*Average annual percent gain*
Wheat	1.87	2.54	2.70
Rice	2.80	3.50	1.80
Coarse grain	2.10	2.61	1.73
World population			1.75

Source: USDA/FAS

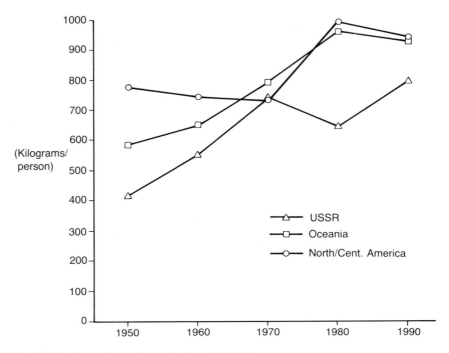

Figure 36.2b Per capita grain production, 1950–90
Note: More than half of Africa's food calories come from root crops such as cassava and yams.
USSR grain is reported bunker weight rather than cleaned and dried as in other countries.
Source: FAO.

The world has more scientists, working in more countries, with better tools (like the new electron scanning microscope) and a stronger base of knowledge (now including the blueprint for all biological heredity, the DNA spiral). The world's stock of capital (our labor pool, machinery, manufacturing plants, etc.) is larger and growing faster.

United States crop yields continue to trend strongly upward, even though the trend has been running longer in the United States than in most countries. (US Index of Crop Production Per Acre, 1950–89). The recent dips in production were due to external forces: In 1980, drought and a new race of corn blight which forced an emergency re-breeding program; in 1983 a sharp expansion of government setaside combined with drought; and in 1988 severe drought. None of these factors has changed the long-term trends.

People have always thought of food as some sort of gift from the heavens, conveyed through sunlight, rainfall, and soil resources. Now, investments in agricultural research, in relatively higher prices to farmers, and in expanded infrastructure (roads, storage bins, processing plants, etc.) raise food output substantially, reliably and sustainably.

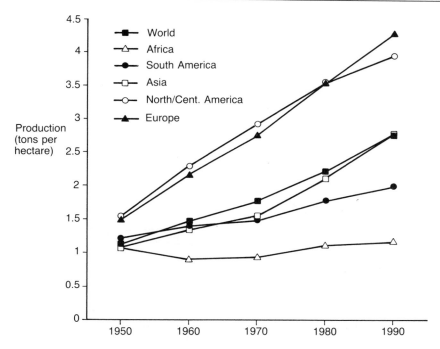

Figure 36.3a World cereal yields, 1950–90
Source: FAO.

FARM TECHNOLOGY IS PROTECTING THE ENVIRONMENT

The trend toward higher yields on existing cropland has also been an important protection for the environment. Without it, rising population would mean more deforestation, soil erosion and loss of wildlife habitat as primitive agriculture was extended onto more fragile land.

Science is making even more direct gains for the environment:

- A new farming system can cut the environmental impact of slash-and-burn farming in the Amazon by 90 percent. One crop of a legume named kudzu can restore soil fertility as effectively as 14 years of the traditional bush fallow.[8]
- Conservation tillage leaves a heavy crop residue in the upper layer of the soil, cutting erosion by half. It is being used on millions of acres in Western Europe and the United States.[9]

THE MAJOR SCIENTIFIC CONTRIBUTIONS TO HIGHER YIELDS

We are living in the Golden Age of plant breeding. It took 100 years for Gregor Mendel's study of plant heredity (done in the 1860s) to develop

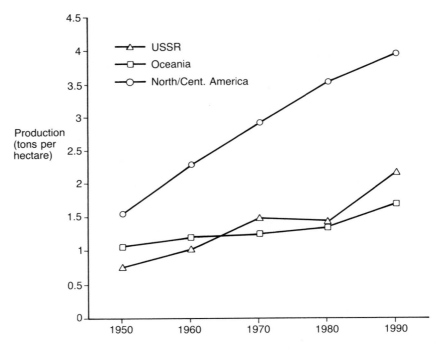

Figure 36.3b World cereal yields, 1950–90
Source: FAO.

worldwide impact. Since 1960, however, the world has seen an explosion of higher-yielding, shorter-season, drought-tolerant, pest-resistant crop plants. A worldwide fraternity of plant breeders now communicates by satellite and computer, drawing on an expanding international network of gene banks. At least some higher-yielding plants are now available for perhaps 80 percent of the world's arable land, with the range increasing almost daily.

- Shorter-season corn hybrids have widened the world's Corn Belt by 500 miles clear around the circumference of the earth, benefiting Canada, the USSR, Poland, China, and Argentina.[10]
- The erucic acid and glucosinates have been bred out of rapeseed, making it a more useful and healthful oilseed as well as higher yielding.
- Winter wheat and barley now tolerate colder weather, giving temperate-zone grain crops a flying start in the spring.
- Hybrid sunflower is expanding vegetable oil supplies in dozens of warm-climate countries.
- The first efforts to breed pest-resistant cassava have tripled yields of this vital food staple in Africa and Southeast Asia.[11]
- The first African sorghum hybrid (for Sudan) is far more drought tolerant, with triple the yield potential.[12]

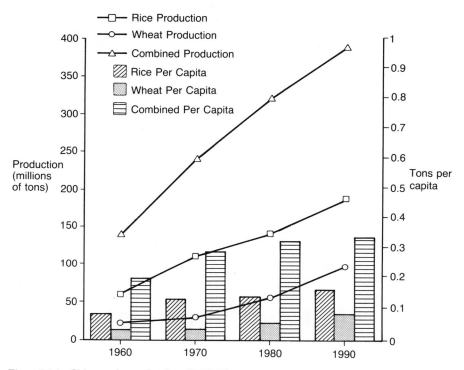

Figure 36.4 China grain production, 1960–90
Sources: Production, FAO. Population, 1960, UNESCO Statistical Yearbook 1965; 1970–90, World Bank.

- Acid tolerant corn, rice and forage crops have recently been bred for a billion acres of now-barren acid-soil savannahs in Latin America, southern Africa, and southeast Asia.[13]
- Cloning and tissue culture have shortened the tree breeding cycle from decades to months, with fourfold gains already achieved in yields of palm oil, coconut, and rubber.[14]

New farming systems

- Asia has learned to grow another 100 million tons of dry-season wheat between its rice crops.
- A redesigned ox-drawn plow makes possible two good crops instead of one poor one on huge tracts of "cracking clay" vertisol soils in India and (potentially) Ethiopia.[15]
- African rice research indicates that high-yielding rice could be grown on 500 million acres of inland wetlands, thanks to improved seeds and improved human disease control; only about 12 million acres of this land are currently farmed.[16]

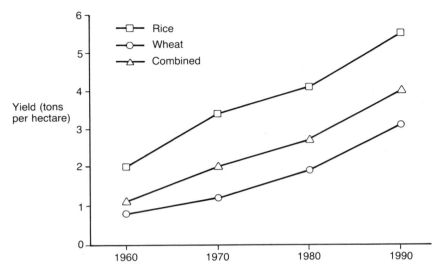

Figure 36.5 China grain yield, 1960–90
Source: FAO.

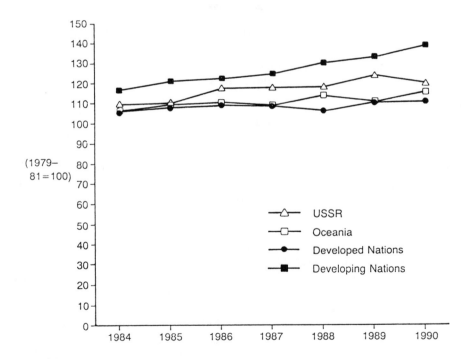

Figure 36.6a Index of world agricultural production

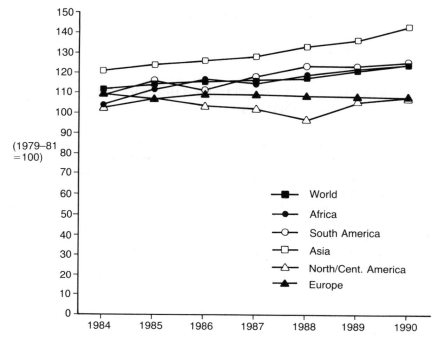

Figure 36.6b Index of world agricultural production

- Alley cropping is the first West African farming system which requires neither shifting cultivation nor chemical fertilizer.[17] Rows of trees interplanted with rows of crops provide shade and mulch, cut erosion, and add plant nutrients. Alley cropping should permit up to four times as many families to support themselves with subsistence farming.

Chemical fertilizers amplify the impact of the new seeds and farming systems. Fertilizer is absolutely necessary for high-yield farming, and global applications have increased about fivefold. Used carefully to replace the plant nutrients removed by cropping (and with conservation systems to minimize runoff and leaching), the fertilizer has little impact on humans or the natural environment.

Pesticides. The bad news is that mankind cannot yet effectively protect crops and livestock from insects, weeds, and diseases without chemical pesticides. The good news is that the modern pesticides are far less persistent, much narrower in toxicity, and applied as ounces per acre instead of pounds. (One of the latest weed killers is no more toxic than table salt, and an aspirin-sized tablet will treat a whole acre.) Science is gaining some successes with biological pest control; the biggest is the predator pests which are controlling the cassava mealybug and green spider mite in Africa's huge cassava belt.[18] Researchers also have some modest experimental successes with genetic engineering, and more are expected. The best news is that in the future, researchers hope to include far more pest resistance in the seeds themselves, eliminating much of the current need for chemicals.

MAJOR CONTRIBUTIONS OF CAPITAL TO FOOD PRODUCTION

Irrigation. Agriculture already accounts for 70 percent of the world's water use, but with an efficiency rate of only about 40 percent.[19] Sprinkler systems can triple water use efficiency over simple flood systems. Trailing tube systems which put metered amounts of water right at the root zones of the plants, are even more efficient. Better drainage and more realistic (higher) prices for water can prevent most of the problems with salinity and waterlogging of irrigated soils. India is tapping underground water flows with tubewells in its eastern states.[20] Turkey is building a set of major dams in the Upper Euphrates Valley – which will create a virtual replica of California's famously productive San Joaquin Valley.[21]

Farm-to-market transport. A study in Bangladesh found that new roads typically raised crop production about one-third.[22] Roads cut the cost of improved seeds and fertilizer and the cost of moving farm products to urban consumers. Brazil projects that railroads into the brushy Cerrado Plateau would cut the cost of moving grain and oilseeds to coastal ports by three-fourths.

Storage facilities and processing plants. At least one-fourth of the Third World's annual farm production is lost to birds, rodents, insects, and spoilage. Storage and processing facilities cut these losses – and spread seasonal surpluses into year-round supplies.

AFRICA AND FARMING TECHNOLOGY

Africa has became synonymous with the world's hunger problems – but Africa is a relatively small remainder of the hunger problem which threatened the whole Third World in 1960. Africa has only 12 percent of the world population. Quite correctly, the first hunger efforts focused on Asia, where food supplies for 60 percent of the world's population were much more precariously balanced.

Figure 36.1 confirms that Africa's grain production has been increasing at roughly the same rate as world grain output in recent years. However, that is not yet enough progress to provide real food security for all its fast-growing population. Nor has Africa stored grain surpluses for its drought years.

Africa has severe agricultural constraints, but it also has substantial farm potential which is only beginning to be developed. A great deal of Africa's current food problem is due to a late start in agricultural research, and to national policies which discouraged farmers in most African countries.

Grain production in sub-Saharan Africa during 1988–90 was 22 percent higher than during the 1981–3 period. This reversed a decline in per capita food production which had prevailed since the beginnings of African independence in the 1960s. Nevertheless, the record in Africa is erratic, strongly affected by changing government policies. Yet there are encouraging signs:

- Recent droughts and food shortfalls have forced African governments to put food policy higher in their priorities. They are offering farmers better prices, more realistic foreign exchange rates, and lower taxes.
- Africa is also beginning to get some gains from agricultural research, after a very late start.[23] Until recent decades, Africa had plenty of land for its inexpensive bush-fallow farming system. It also proved difficult to transfer much technology to Africa from other regions. Most of Africa's gains to date have come from the internationally sponsored research units. The African nations' own research units are still woefully underfunded. But the recent grain production increases in such countries as Tanzania, Mauritania, Zimbabwe, and Ghana are particularly notable, coming as they have from improvements in both policy and technology.
- Zimbabwe's corn production reflects the continent's best corn breeding program and extension of hybrid corn beyond white commercial farmers to the big traditional farming sector. (That's why yields have gone down and production up.)
- Tanzania discouraged its farmers with low prices and decrepit transportation in the 1970s, but production has soared in the 1980s as the constraints have been removed.[24] Improved rice varieties are boosting irrigated yields.
- Mauritania shifted land on the northern banks of the Senegal River from communal to fee-simple ownership – and the new owners dug their own irrigation ditches and bought small diesel pumps to expand rice production.[25]
- Ghana shifted from socialist to market-oriented farm policies in the late 1970s. The country is also beginning to plant new high-yield corn and cassava varieties from the international research center in Nigeria. Corn and cassava production has increased sharply as a result.

Africa's food security is still precariously balanced. The next major continental drought could produce almost as much suffering as the last one in 1983–4.

CHINA'S FOOD PROGRESS

China has developed one of the Third World's best agricultural research systems, but its recent food progress has been dominated by (1) a major shift in its agricultural policy, and (2) investments in chemical fertilizer production.

China broke up its big, inefficient communal farms in 1979–80, and leased most of the land back to family farmers. Prices were raised about 25 percent. Stimulated by individual price incentives and fertilizer expansion, total farm output rose more than a third in the next six years.

In the 1980s, China developed the world's first successful hybrid rice, with yield gains of about 30 percent. China also extended its use of semi-dwarf wheat and hybrid corn.

Despite China's continuing technical progress, its recent crop production patterns reflect primarily the importance of national economic policies. Farm output surged when the government offered profits to its family farmers. Production plateaued again in the late 1980s, as Chinese officials attempted to reestablish the old command system over farmers in lieu of profits. (China's recent grain production may also be underreported, since the incentive for farmers to "hide" grain and market it as meat has increased significantly since 1985.)

FEEDING ANOTHER BILLION PEOPLE

The world could readily feed another billion people, right now, without stressing any fragile acres or putting on heavy doses of farm chemicals. Here's how:

In the United States, at least 30 million acres of good farmland are being diverted from production in most years under federal farm programs. On the average, economists say this land would yield about 80 percent of the United States average crop yield – which has recently been about 1.5 tons per acre. Multiplying 30 million acres times 80 percent of 1.5 tons would total about 45 million tons of grain.

Argentina has recently been pasturing cattle on about 75 million acres of the world's finest farmland, in the rich, well-watered pampas. If the world needed more grain, this land could readily be planted to wheat, with the cattle shifted to less valuable land. Moreover, Argentine farmers have been using virtually no fertilizer or pesticides due to the heavy import taxes on such inputs and the heavy export taxes on the crops they produce. Observers say that better price incentives to farmers could lead to Argentina's doubling or even tripling its total farm output. We can assume that a sudden need to feed another billion people would offer some such price incentives.

If 75 million additional acres of Argentine land were planted to wheat and fertilized, another 200 million tons of grain could be produced annually and sustainably.

Thus the United States and Argentina together could produce an additional 245 million tons of grain per year on land that is currently unused or underused.

Dividing this grain total by the current per capita grain consumption in India (370 pounds per capita) we find that an additional 1.4 billion people could be supplied with enough grain to more than meet their caloric needs.

If the extra 100 million acres of land were planted primarily to the new high-protein corn varieties developed recently at the International Maize and Wheat Improvement Center (CIMMYT) in Mexico, we could expect even higher grain yields and a higher level of protein in the grain. (We could, of course, insist that conservation tillage techniques be used to minimize soil erosion.)

Thus, just by fully utilizing the spare land in the United States and Argentina, with the latest sustainable farming technology, we could offer another 1.4 billion people a better diet than the average citizen of India currently receives.

The United States and Argentina are offered in this example because the farmland in both countries is already served by relatively well-developed infrastructure. (The US infrastructure, admittedly, is in far better shape than Argentina's.) That means the land could be brought into crops much more readily than more remote stretches of unplanted arable land in places like the upper Nile Valleys or Brazil's interior Cerrado Plateau.

NOTES

1   World Land Use, 1960–88, from Table 1, *FAO Production Yearbook*, 1976 and 1989 editions; see also Total Cereals, Area, *FAO Production Yearbook* (table 9 in the 1976 edition and table 15 in the 1989 edition).

2   Total Cereals, Area, Yield and Production, Table 9, *FAO Production Yearbook 1976*, and table 15, *FAO Production Yearbook 1989*.

3   Total Cereals, Area, Yield and Production, *FAO Production Yearbook*, editions cited.

4   Wheat Area, Yield and Production Tables, *World Crop Production*, USDA/FAS, WCP 5–87, May, 1987, and *World Agricultural Production*, WAP-1-91, January, 1991.

5   Rice Area, Yield and Production Tables, *World Crop Production*, USDA/FAS, WCP 5–87, May, 1987, and *World Agricultural Production*, WAP-1-91, January, 1991.

6   Coarse Grains Area, Yield and Production Tables, *World Crop Production*, USDA/FAS, WCP 5–87, May, 1987, and *World Agricultural Production*, WAP-1-91, January, 1991.

7   *Indonesia Grain and Feed Annual Report, 1989*, USDA/FAS, Jakarta, April, 1989. See also *Philippines Grain and Feed Annual Report*, USDA/FAS, Manila,

Feb., 1990; *Chile Grain and Feed Annual*, USDA/FAS Santiago, February, 1990; and *Argentina Grain and Feed Annual*, USDA/FAS, Buenos Aires, May, 1989.

8 Personal communication, Dr Pedro Sanchez, Coordinator, TROPSOILS, North Carolina State University, Raleigh, NC, see also Pedro Sanchez, "Low-Input Cropping for Acid Soils of the Humid Tropics," *Science*, vol. 238, December 1987, 1521–7, and Pedro Sanchez, Testimony before the US House of Representatives Committee on Science, Space and Technology, February 23, 1989.

9 Conservation Tillage Information Center, West Lafayette, IN.

10 Personal communication, Dr J. L. Geadelmann, corn breeder, Department of Agronomy, University of Minnesota.

11 *IITA Strategic Plan 1989–2000* (Ibadan, Nigeria: International Institute of Tropical Agriculture, 1988), 59–60, and *International Institute of Tropical Agriculture Annual Report 1989/90* (Ibadan, Nigeria: IITA, 1990), 12–13. See also *Indonesia Annual Agricultural Situation Report 1988*, USDA/FAS, Jakarta, March 1988.

12 Ejeta, Gebisa, "Development and Spread of Hageen Durra 1, the First Commercial Sorghum in the Sudan," *Journal of Applied Agricultural Research*, vol. 3, no. 1, 1988.

13 Personal communication, Dr Pedro Sanchez, Coordinator, TROPSOILS, North Carolina State University, Raleigh, NC. See also Pedro Sanchez, "Low-Input Cropping for Acid Soils of the Humid Tropics," *Science*, vol. 238, December 1987, 1521–7, and Pedro Sanchez, Testimony before the US House of Representatives Committee on Science, Space and Technology, February 23, 1989.

14 Oilseeds: personal conversation, Alan Holz, Oilseeds Division, USDA Foreign Agricultural Service, 1991. Cocoa: *Malaysia Cocoa Annual Report 1990*, USDA/FAS, Kuala Lumpur, September 1990. Rubber: *Thailand Agricultural Situation Report 1989*, USDA/FAS, Bangkok, April 1989.

15 "Animal Drawn Tiller/Planter/Weeder Tested," *ILCA Newsletter*, vol. 8, no. 3, International Livestock Centre for Africa, Addis Ababa, Ethiopia, July, 1989. See also *ICRISAT Annual Report 1975/76*, 190, and *ICRISAT Annual Report 1987*, 318–20, International Center for Research in the Semi-Arid Tropics, Patancheru, Andhra Pradesh, India.

16 IITA Annual Report 1989/90, 66–8.

17 B. T. Kange, G. F. Wilson, and T. L. Lawson, *Alley Cropping, A Stable Alternative to Shifting Cultivation* (Ibadan, Nigeria: International Institute for Tropical Agriculture, 1984); see also *IITA Annual Report 1989/90*, IITA, Ibadan, Nigeria, 1990, 14; see also *Highlights from Centers Week*, CGIAR Secretariat document, November 15, 1989.

18 "IITA/CIAT Research in Biological Control Wins the 1990 King Baudouin Award," *International Institute of Tropical Agriculture Annual Report*, 1989/90, 17–19.

19 Sandra Postel, "Saving Water for Agriculture," *State of the World 1990* (Washington, DC: Worldwatch Institute), 1990.

20 *India, Structural Change and Development Perspectives*, World Bank Report No. 5593-IN, vol. 1 (Washington, DC: World Bank, April 1985), xv, 92–3.

21 "Turkey Diverts Euphrates River," *Washington Post*, January 14, 1990, A22; see also, "The Southeastern Anatolia Project," *Turkey 2000*, Turkish government publication, 114–16.

22 *Development Impact of Rural Infrastructure in Bangladesh*, IFPRI Research Report No. 83, International Food Policy Research Institute, October 1990.

23 *Agricultural Potential of Mid-Africa: A Technological Assessment*, Winrock International Institute for Agricultural Development, Morrilton AR, November, 1989.

24 *Tanzania Agricultural Situation Report 1990*, USDA/FAS, Nairobi, April 1990, and *Tanzania Grain and Feed Report 1989*, USDA/FAS, Nairobi, December 1989.

25 Mauritania, Rice Area, Yield and Production, table 17, *FAO Production Yearbook 1989* (Rome: UN Food and Agricultural Organization), 118.

EDITOR'S APPENDIX (E. CALVIN BEISNER)

The story of rising agricultural productivity can be summarized by saying that we are getting more food by employing fewer resources (land and livestock) and employing more resources with less labor. As a result, we are able to feed (and clothe) more people at less cost in both labor and resources. Figures 36.7 and 36.8, illustrating increasing productivity in the chief crop and livestock divisions of American agriculture, make the story clear. In crops, acreage employed per manhour rose from the early 1960s to the mid-1980s by anywhere from a low of 15 percent (wheat) to a high of 852 percent (cotton); yield per acre rose by anywhere from a low of 22 percent (cotton) to a high of 76 percent (corn); and yield per manhour rose by anywhere from a low of 72 percent (wheat) to a high of 900 percent (cotton). In livestock, yield per manhour rose by anywhere from a low of 187 percent (laying chickens) to a high of 1,099 percent (turkeys); we're even getting more milk per cow (69 percent) and tending more cows per manhour (282 percent), giving us five times more milk per manhour.

EDITOR'S NOTE (JULIAN L. SIMON)

See chapter 35 by George W. Grantham for long-run data on agricultural productivity.

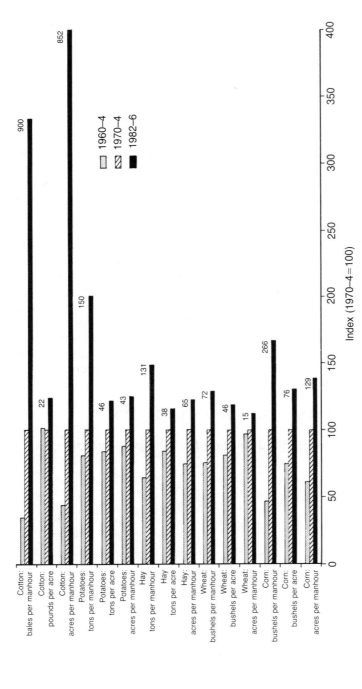

Figure 36.7 Index of crop productivity, USA, 1960–4 to 1982–6

Note: The number above the 1982–6 bar for each series represents the percent gain for the 22-year period.

Source: Statistical Abstract of the United States, 1988, table 1090; after E. C. Beisner, *Prospects for Growth: A biblical View of Population, Resources, and the Future* (Wheaton, Ill.: Crossway Books, 1990), 127.

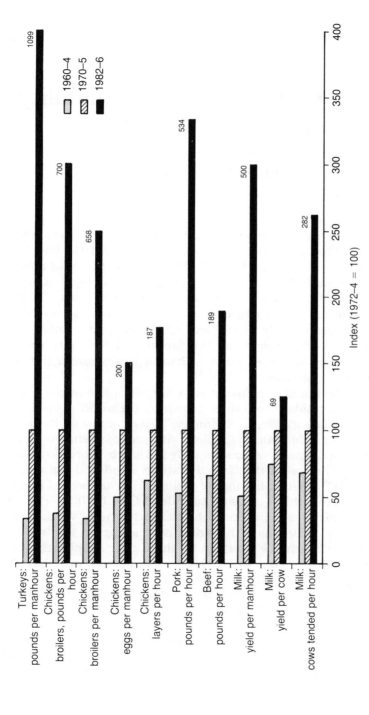

Figure 36.8 Index of livestock productivity, USA, 1960–4 to 1982–6

Note: The number above the 1982–6 bar for each series represents the percent gain for the 22-year period.

Source: *Statistical Abstract of the United States, 1988*, table 1090; after E. C. Beisner, *Prospects for Growth: A Biblical View of Population, Resources, and the Future* (Wheaton, Ill.: Crossway Books, 1990), 127.

37

Recent Trends in Food Availability and Nutritional Well-Being

Thomas T. Poleman

Until recently evidence on changes in nutritional well-being was largely anthropometric. That the average American soldier in 1941 had been better fed than his counterparts in 1917, 1898, and 1861 could reasonably be inferred from the larger uniforms the Army was obliged to supply him; and that few of us could squeeze into the medieval suits of armor on display in museums is strong evidence that things have improved since the Middle Ages. Only in the last half century have other data come into existence that might lay the foundation for more precise quantification. These, however, must be handled with care if they are not to be seriously misleading.

Ask, for example, the man in the street what life is like in the poorer countries – wherein resides about three-fourths of the world's population – and he will confidently reply, "Hungry." His confusion is understandable; the popular literature is replete with pronouncements to this effect. We were, for instance, assured as early as 1950 by Lord Boyd-Orr, the first Director General of the United Nations Food and Agriculture Organization (FAO), that "a lifetime of malnutrition and actual hunger is the lot of at least two-thirds of mankind" (Boyd-Orr, 1950, 11); by one of the many alarmist articles induced by the misnamed "World Food Crisis" of the early 1970s that "the world is teetering on the brink of mass starvation" (Ehrlich, 1975, 152); by the Carter Administration's Presidential Commission on World Hunger a decade ago that the "world hunger problem is getting worse rather than better" (*RPCWH*, 1980, 182); and in 1989 by the UN's World Food Council that "hunger and malnutrition are growing and will continue to grow" (World Food Council, 1989, iv).

All these statements are false. They depend for statistical support on a small number of "hunger quantification" studies conducted by FAO (1946, 1952, 1963, 1977, 1985) and the World Bank (Reutlinger and Pellekaan, 1986; Reutlinger and Selowsky, 1976). It is easy to demonstrate that these studies are riddled with errors and grossly overstate the extent of hunger (Poleman, 1981, 1983). Less easy is to quantify with certainty the actual amount of nutritional deprivation existing today and how it has evolved over time.

PROBLEMS OF QUANTIFICATION

In order to confidently plot trends in global nutrition, one would like evidence, country by country, on changes in food availability and information about how access to these supplies varies among consumers with differing incomes. Estimates of national availabilities are derived by constructing a food balance sheet, incorporating on the supply side measurements of production, trade, and stocks changes, and on the utilization side such items as seed and feed use and losses in storage. Availabilities for human consumption are derived as a residual and thus reflect the totality of error.

For the developed countries of Europe and North America the residual error is and has for some time been negligible. Food balance sheets were first constructed during World War I, when both the German and Allied governments found them useful tools in the management of food supplies, and are now quite accurate.

In the developing countries, however, the collection of agricultural statistics is relatively recent and the findings of the balance sheet calculation can be very misleading. The general tendency is for newly created statistical services to understate production. Wheat production in the United States, for instance, is recognized to have been 30 to 40 percent above that officially reported during the first decade (1866–75) of USDA's statistical efforts (Working, 1926, 260). In Mexico, the extent of understatement for maize during 1925–34, the *Direccion General de Economia Rural*'s first decade, was over 50 percent (Poleman, 1977, 16, 19).

To generalize about the extent to which food supplies in the developing countries have been and are now understated is not possible. A reasonable assumption for most countries is that the accuracy of production estimates has improved with time – and thus that some of the apparent gains in food availabilities are spurious – but the opposite may well have taken place in sub-Saharan Africa. There, independence has frequently been accompanied by a deterioration in the reporting systems established by colonial administrators. When perfection may be anticipated is anybody's guess – it was not until 1902, 36 years after the effort began, that USDA began reporting wheat output with an acceptable margin of error, and not until the mid-1950s, with 30 years of experience in hand, was Mexico able to confidently measure its maize harvest – and it is for this reason that FAO's country-by-country estimates of trends in average per capita calorie availabilities are untrustworthy guides to what has happened in many poorer countries and are not reproduced here.

It is also the reason why the indices of total and per capita food production shown in figure 37.1 are best seen as general, not precise indicators. Still, the impression they convey is valid: over the past four decades the less-developed countries have expanded production rather more rapidly than the developed ones. Population growth, to be sure, has absorbed most of the gains, but with the exception of sub-Saharan Africa modest per capita improvements have occurred.

This record of steady agricultural progress does not, however, mean that everyone in the developing world is now better fed than at mid-century. Since the early 1970s it has been a commonplace in serious pronouncements on the world food situation that, equitably distributed, global supplies are sufficient to feed all. The problem is that all within a country do not have equal access to existing supplies. Access to food is a function of income. Those with adequately paying jobs are easily able to afford an acceptable diet; their less fortunate neighbors sometimes cannot.

Our insights into the effect of income on food habits comes from household budget and consumption surveys and, again, while the evidence for most industrialized countries is acceptable, carefully conducted surveys of broadly representative samples are still few for the developing countries.

PLOTTING TRENDS IN NUTRITIONAL WELL-BEING

Despite these problems of measurement, it is evident that at no time in history has the world been as well fed as it is today. To argue otherwise would be to deny a basis for the increase – from about 40 to over 60 years – that has occurred since mid-century in life expectancy in the poorer countries. In the rare instances where outright famine has occurred recently – in Ethiopia, Somalia, and Sudan, for instance – the problem has been localized and attributable to political conflict.

Several other points regarding postwar trends in food availability and nutritional well-being can also be made with confidence:

- There is a global sufficiency of food, and despite poverty most people in developing countries behave as if they manage to eat adequate, if not especially tasty, diets. The "quality" of such diets as measured by the starchy staple ratio has improved since at least 1965 and most probably throughout the entire postwar period.
- Nutritional deprivation, where it exists, is likely to be most acute among the young, and the incidence of infant malnourishment is declining.
- Where this decline is least evident – in sub-Saharan Africa and among the very poorest in South Asia – corrective action lies in political reform (Africa) and more and better jobs (Asia).

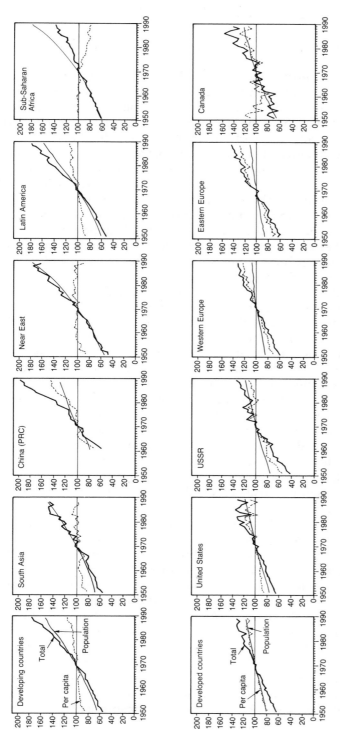

Figure 37.1 Indices of total and per capita food production, 1951–89 (1969–71 = 100)

Source: Data from US Dept of Agriculture, Econ. Res. Ser., *World Indices of Agricultural and Food Production* (various issues); ibid., *World Agricultural Trends and Indicators* (various issues); FAO, *Quarterly Bulletin of Statistics* (various issues); and ibid., *Production Yearbook* (various issues).

Behavioral Evidence of Perceived Dietary Adequacy

The statistically most reliable, although not widely used, method for measuring hunger is to identify behavior associated with perceived dietary adequacy or inadequacy. Essential to monitoring such behavior is understanding how poor people over the centuries have contrived to feed themselves and the sorts of changes their diets undergo as they become more wealthy. This was the subject of some truly pioneering research carried out during the 1930s and 1940s by Merrill K. Bennett (Bennett, 1954). He noted that the very poor everywhere seek to maximize the nutritional return per outlay for food by building their diets around foods composed principally of starch: wheat, rice, potatoes, cassava, and the like. This is so because of the cheapness of these starchy staples, whether expressed as market price or production cost. Far less land and far less labor are needed to produce a thousand calories of energy value in the form of the starchy staples than in the form of any other foodstuff. Meat producers by comparison are inefficient converters; an animal must be fed between two and ten pounds of grain for it to produce a pound of meat. But most people enjoy meat, and they turn away from the starchy staples as they become wealthier.

A simple way to rank diets is according to the percentage of total calories supplied by the starchy staples, and an easy way to record change is to monitor shifts in this starchy staple ratio. In the United States the ratio stood at 55 percent 125 years ago, when our great-great-grandparents consumed large amounts of bread and potatoes. Today our diets are dominated by meat, fats and oils, sugar, vegetables, and dairy products; and the starchy staple ratio has dropped to 21 percent. We pay more for such a diet and presumably enjoy it more. But it does not follow that it is a better diet. Indeed, current thinking among nutritionists is that because of its high sugar and fat content it may well be a poorer one.

A few years ago Neville Edirisinghe analyzed five budget/consumption surveys – for Sri Lanka, Indonesia, Bangladesh, Peru, and Brazil – for evidence regarding the extent to which households in the developing countries behave as if adequately nourished. This was taken to be wherever they chose to purchase "quality" rather than quantity as their income rose (Edirisinghe and Poleman, 1983). Figure 37.2, based on data collected in Brazil in 1974/75, is typical of his findings. It is apparent that the diet at the lower end of the income range is that of poor people: the four starchy staples (cassava, maize, rice, and wheat) supply over half of total energy availability. But it would not appear to be the diet of people who perceive themselves threatened with hunger. Additional calories are not purchased as income increases. Instead, consumption of cassava, the least preferred staple, falls off sharply, its place taken by rice and wheat bread.

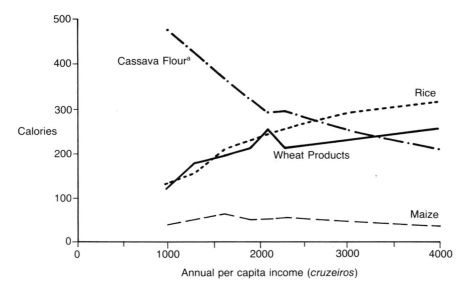

Figure 37.2 Northeast Brazil: apparent per capita daily consumption of major starchy staples among low-income classes, 1974–5 (calories)
Note: [a]Plus other roots and tubers (5 percent of total).
Source: Data from N. Edirisinghe and T. T. Poleman, *Behavioral Thresholds as Indicators of Perceived Dietary Adequacy or Inadequacy* (Cornell/Int'l. Agric. Econ. Study 18, Ithaca, NY, 1983), 28.

This type of behavior, which is also evident in the data for the other countries examined, is less suggestive of widespread hunger in the developing countries than of the ability of the people there to shrewdly allocate their limited resources so as to get by on what is, by the standards of the industrialized world, very little. And lest it be thought that the poor were excluded from the surveys, be assured that they were not. The Brazilian data are for the north-eastern part of the country, Brazil's poorest region, and the survey was suppressed by the government because of the social inequality it reveals.

Although reliable budget survey material is still available for only a handful of countries, other less sensitive evidence suggests that the behavior they reveal obtains throughout the developing world, and further that the quality of the average diet as monitored by the starchy staple ratio has improved with time. Figure 37.3 indicates that between 1964–6 and 1984–6, the earliest and most recent periods for which FAO has published food balance sheets for a large number of countries, even in the poorest countries the starchy staple ratio dropped.[1] The sole exceptions are the Philippines, where economic and dietary stagnation have gone hand in hand, and sub-Saharan Africa. The quality of African data is such that for most countries meaningful starchy staple ratios cannot be calculated.

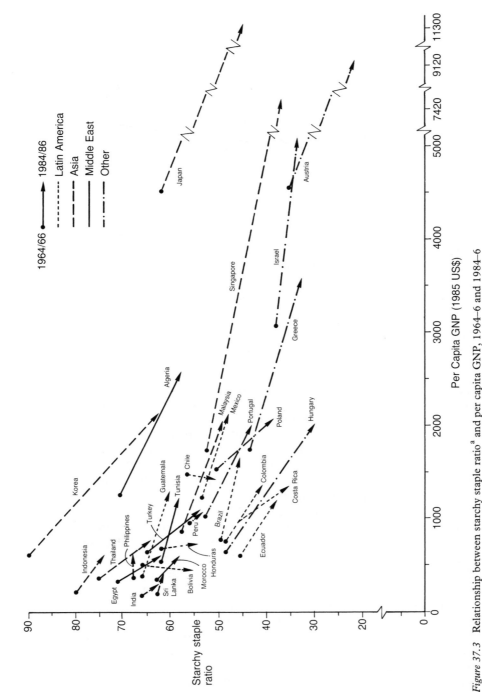

Figure 37.3 Relationship between starchy staple ratio[a] and per capita GNP, 1964–6 and 1984–6

Note: [a]Percent of total calories derived from cereals and roots and tubers.

Source: Data from World Bank, *World Development Report, 1987* (Oxford University Press, 1987); FAO, *Food Balance Sheets, 1964–6 Average* (Rome, 1971); and FAO, *Food Balance Sheets, 1984–6 Average* (Rome, 1991).

The Nutritionally Vulnerable Within the Household

If we do not know how many among the poor in the developing countries suffer nutritional deprivation, there is agreement that the preschool child and the pregnant and lactating mother are those most likely to be adversely affected. There are several reasons for this. The early growth and reproduction phases are nutritionally the most demanding in the life cycle. Yet it is precisely the mother and young child whose needs can be reflected least in the choice of foods purchased by the household and who may be the residual claimants on that which has been prepared for all to eat.

Estimating the extent of malnutrition among young children and their mothers is not easy. It involves the new science of nutritional anthropometry, and a body of reliable evidence is yet to be collected. Further, a debate attends the standards it should employ for healthy children, the measurements – whether weight for age, height for age, or weight for height – it should involve, and where the cut-off criteria should be established. (For two very different opinions see Mason (1984) and Sukhatme (1982).) If the number of those at risk is to be estimated, therefore, we have no alternative but to do so rather arbitrarily.

One approach is to use the infant mortality rate as an indicator of maternal and child nutrition. Since low birth weights and feeding problems are associated with higher infant mortality rates, this is a reasonable linkage. But again, debate attends the conditions that should be equated with particular rates. Is, for instance, anything below 50/1,000 indicative, as some would argue, of a population in which most children are reasonably well-nourished?[2] Or would some other figure be more appropriate?

It is clear from figure 37.4, which includes countries containing over 75 percent of the population of the developing world, that infant mortality has fallen dramatically during the past quarter century. Rising income levels and improved access to food have presumably contributed to this, but it is impossible to say whether this has been of greater or lesser importance than the spread of education or enhanced public health and medical facilities (United Nations, 1985; Hobcraft, et al., 1984). (China and Sri Lanka are oft-cited examples of the role that noneconomic factors can play in reducing infant mortality.) Whatever the case, much of Latin America is now in comparatively good shape, as are the more prosperous countries of Asia, plus – impressively – China. The principal exceptions are India, Pakistan, and Bangladesh, and virtually all of sub-Saharan Africa. Here, despite declines, infant mortality remains above 100/1,000.

The Special Problems of Africa and the Impoverished of South Asia

To the generalizations that overall diet quality and infant nutrition have improved during the past half century, two groups must be excepted:

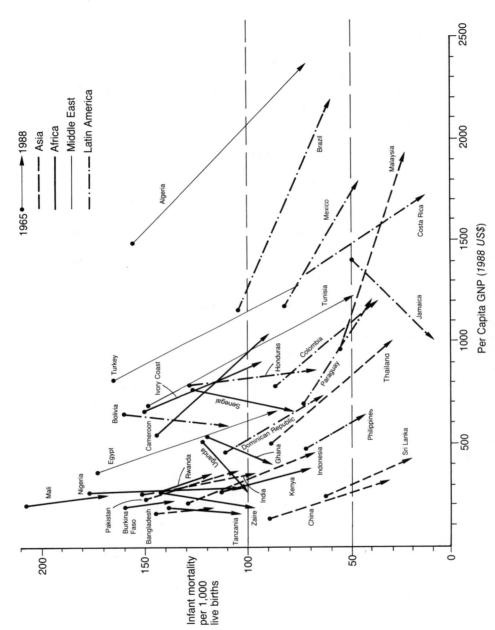

Figure 37.4 Relationship between infant mortality rate and per capita GNP, 1965 and 1988
Source: Data from World Bank, *World Development Report, 1990* (Oxford University Press, 1990).

those living in that portion of Africa south of the Sahara and north of South Africa, and the impoverished of the three countries of the Indian sub-continent.

The problem in India, Pakistan, and Bangladesh is one of poverty and insufficient effective demand. The sub-continent is one of the homes of the Green Revolution, and food production in all three countries during the past several decades has increased rather faster than population growth. Were it equitably distributed, the supply of food would be adequate to feed all. But none of the countries has been able to implement an effective program to provide food to the poor, and continued high population growth has frustrated the incorporation of all within the process of economic growth. The outlook for those unfortunate enough to be left out is not encouraging.

Even less promising is the situation in Africa. Population growth continues there unabated, and it is the only region for which the data in figure 37.1 suggest per capita food production has not increased in recent years, but actually declined. The extent to which this fall-off – of roughly 20 percent, if the data are to be believed – reflects reality and not merely the deterioration of the crop reporting system defies measurement. Sub-Saharan Africa is a statistician's nightmare. The reason data for it are not included in figure 37.3 is that the food balance sheets for many countries show declines in both energy availabilities and starchy staple ratios – a course of dietary evolution inconsistent with any logical explanation.

The failure of food production to progress in the newly independent states of Africa reflects many things, all of them correctable, but none easily. These include political instability, counterproductive food pricing policies, and an absence of many of the ingredients that made the Green Revolution possible elsewhere. Whether, as some suggest, the continuation of rapid population growth and agricultural regression will lead over the next several decades to a rise in the death rate and even local fulfillment of the Malthusian nightmare in parts of Africa is an open question. Certainly without an improvement in the political scene it should not be ruled out. It is a blot on a picture that, almost everywhere else, is colored with accomplishment.

NOTES

1 Although the starchy staple ratio is everywhere related inversely to income, in countries where rice is the dominant staple the ratios at a given income level are higher than elsewhere. Thus the seemingly anomalous behavior of Japan, Korea, and Singapore in figure 37.3.

2 Infant mortality in the United States is slightly below 10/1,000, down from 25/1,000 in 1965. Among African-American children the current figure is just under 19/1,000.

REFERENCES

Bennett, M. K. (1954): *The World's Food*. New York: Harper.

Boyd-Orr, Lord John (1950): "The Food Problem," *Scientific American*, vol. 183, no. 2, August.

Edirisinghe, Neville and T. T. Poleman (1983): *Behavioral Thresholds as Indicators of Perceived Dietary Adequacy or Inadequacy*. Cornell/Int'l. Agric. Econ. Study 18, Ithaca, NY.

Ehrlich, Anne, and Paul Ehrlich (1975): "Starvation: 1975," *Penthouse*, July.

FAO (1946): *World Food Survey*. Washington, DC.

——— (1952): *Second World Food Survey*. Rome.

——— (1963): *Third World Food Survey*, Freedom from Hunger Basic Study 11. Rome.

——— (1977): *The Fourth World Food Survey*, Statistics Series 11. Rome.

——— (1985): *The Fifth World Food Survey*. Rome.

Hobcraft, J. N., J. W. McDonald, and S. O. Rutstein (1984): "Socio-economic Factors in Infant and Child Mortality: A Cross-national Comparison." *Population Studies*, XXXVIII, 2.

Mason, J. B., et al. (1984): *Nutritional Surveillance*. Geneva: World Health Organization, 1984.

Poleman, T. T. (1977): "Mexican Agricultural Production, 1896–1953: An Appraisal of Official Statistics." Cornell Agric. Econ. Staff Paper No. 77–30, Ithaca, NY.

——— (1981): "Quantifying the Nutrition Situation in Developing Countries." *Food Research Institute Studies*, vol. XVIII, no. 1.

——— (1983): "World Hunger: Extent, Causes, and Cures." In D. G. Johnson and G. E. Schuh, eds, *The Role of Markets in the World Food Economy*. Boulder, CO: Westview Press.

Report of the Presidential Commission on World Hunger (1980): *Overcoming World Hunger: The Challenge Ahead*. Washington, DC, March.

Reutlinger, Shlomo, and J. V. Pellekan (1986): *Poverty and Hunger*. Washington, DC: World Bank.

——— and Marcelo Selowsky (1976): *Malnutrition and Poverty*. Washington, DC: World Bank Staff Occasional Paper 23.

Sukhatme, P. V., ed. (1982): *Newer Concepts in Nutrition and Their Implications for Policy*. Pune, India.

United Nations, Dept. Int'l. Econ. and Social Affairs (1985): *Socio-Economic Differentials in Child Mortality in Developing Nations*. New York: United Nations.

Working, Holbrook (1926): "Wheat Acreage and Production in the United States Since 1866: A Revision of the Official Estimates." *Wheat Studies of the Food Research Institute*, II, 7, June.

World Food Council (1989): *Ending Hunger: The Cypress Initiative*. 15th Ministerial Session, Cairo, 22–5 May.

38

Trends in Grain Stocks

William J. Hudson

The single measure of "annual grain stocks" has become a popular way of judging the trend in world food abundance. There is both power and hazard, however, in this kind of single-index simplification. When the index dips for a few years, due to weather or policy changes, some analysts see the downturn as evidence for the approach of food shortage or general famine. Attempts to disaggregate the figure to explain the overall trend more justly are ignored by the press and labeled pure optimism by those most worried about the earth's resources.

In figure 38.1, the "total grains" line shows the usual way of computing and displaying a single index for "world annual grain stocks." In 1989 and 1990, this index dipped to its lowest level for nearly 15 years,

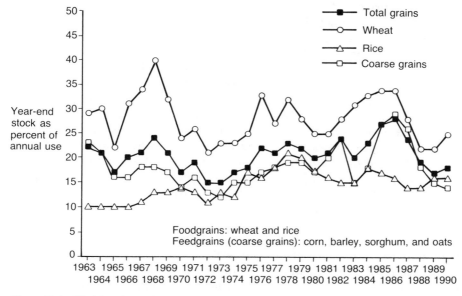

Figure 38.1 World grain stocks, 1963–90

and observers such as Lester Brown of Worldwatch Institute sounded alarms similar to those heard in 1974, that the world would soon be unable to produce all the food it needs.

The concept in figure 38.1 is one of "inventory adequacy," in which stocks are stated as a percentage of annual usage. The line for total grains combines all types of grain together into a single "bin," as though the world's different grains were all harvested at the same time (in different latitudes of the southern hemisphere as well as the northern) and used for the same purpose (food, as opposed to seed, livestock feed, industrial chemicals, and other uses). The method leaves out the extraordinary role of *price* in determining the stocks-to-use ratio: When harvests are smaller than usual, implying that the next year's inventory ("ending stocks") will be small, then price rises and usage is cut, particularly for livestock feeding. With higher prices, the signal goes out to produce more grain in succeeding years by means of greater acreage or higher yield achieved by increasing inputs of seeds, fertilizer, and management, or both. The interaction of supply, demand, and price is ignored by observers such as Brown, whose model of the world is "resource-based" rather than "market-based" or "species-based." Brown's view is that grain inventories must surely decline because resources are limited; the alternative view is that demand has much more impact on grain and food availability than natural resources by themselves.

Figure 38.2 shows the relationship between "world annual grain stocks" and "world grain price," as represented by the average export price of wheat and corn, cost including freight, at Rotterdam. The figure shows that price did not rise as high in 1989 and 1990 as it did in the mid-1970s, or as might have been expected by the 15-year low in grain stocks. In other words, unlike Brown, participants in world grain markets were not unduly alarmed about the level of grain stocks. Price certainly reacted to the severe American drought of 1988, but no lasting signal was sent to the world's farmers for a leap in production.

To understand the relatively weak market price for grain in 1989, 1990, and 1991 (in contrast to the rather low inventories of grain), we must disaggregate the stocks calculation into more meaningful categories. Figure 38.1 also shows world *foodgrain* stocks (wheat and rice, shown separately) and *feedgrain* stocks (corn, barley, sorghum, and oats, shown together as coarse grains). In both cases, stocks-to-use ratios were quite low in 1989 and 1990. Why then did the market price not rise more?

Part of the answer is provided by figures 38.3 and 38.4, which indicate what stocks could have been without the US government's policy of annual Acreage Reduction Programs (ARPs), had world demand been sufficient to entice farmers to harvest the land they instead set aside under the ARPs. American farmers must participate in ARPs in order to receive price support payments for wheat and feedgrains. Basically, the United States (with only 5 percent of the world's population but 25

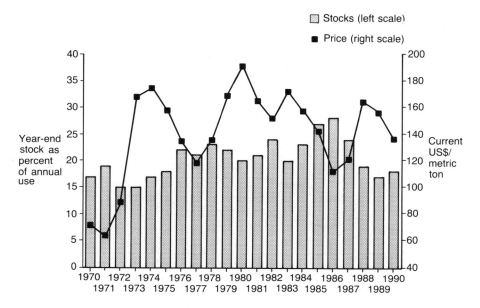

Figure 38.2 World total grain stocks vs. price
Note: Price is average for wheat and corn, cost including freight, Rotterdam. Note that periods
of low stocks are associated with higher prices, and vice versa.

percent of its food production capacity) is faced with chronic *surpluses* of
grain, cheap prices, and low farm income. Since the 1930s, American
farm legislation has sought to raise farm incomes by means of price

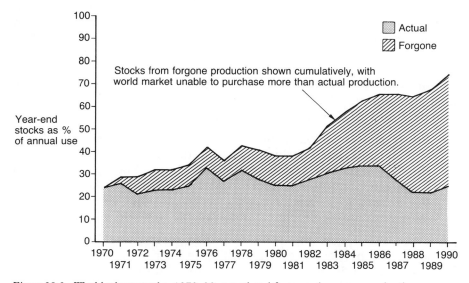

Figure 38.3 World wheat stocks, 1970–90, actual and foregone by acreage reduction

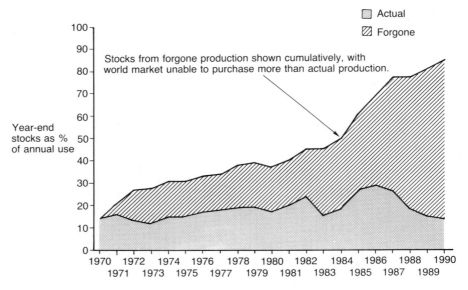

Figure 38.4 World coarse grain stocks, 1970–90, actual and foregone by acreage reduction

supports, one form of which is acreage reduction ("set-aside"). In the 1980s, the average annual production foregone due to ARPs (and lately to the Conservation Reserve Program (CRP) as well) was about two billion bushels per year, which is equivalent to two bushels for each of the world's one billion hungriest people, people whose diet is only 400 to 500 pounds of grain per year.

Grain markets are well aware of the US policy to remove substantial acreage from production when price is low (i.e., when stocks are high), and markets know that if price makes a clear upward signal, then this surplus acreage will only be added back the following year, resulting once again in over-production.

Besides the separation of stocks into food grains and feedgrains, we must also consider three regional "bins," as suggested by Sharples and Krutzfeldt of USDA-ERS and as shown in figure 38.5:

- *Bin 1, with 35 percent of world's stocks*: China, India, and the Soviet Union, stocks used exclusively within the country.
- *Bin 2, with 50 percent of world's stocks*: United States, Canada, and European Community, stocks available to world market.
- *Bin 3, with 15 percent of world's stocks*: all other countries, stocks mostly "domestic pipeline," not available to world market.

Each of the major importing countries of the world (Bin 1) has problems of a particular kind regarding grain production and stocks. China's grain production is very large, but its transportation system is weak, resulting in the isolation of grain stocks from urban areas – coastal

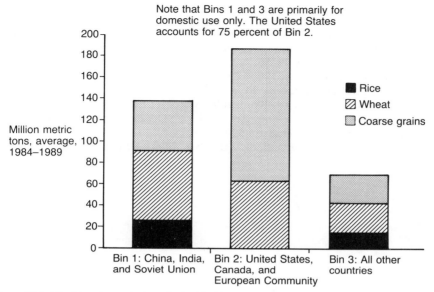

Note that Bins 1 and 3 are primarily for domestic use only. The United States accounts for 75 percent of Bin 2.

Million metric tons, average, 1984–1989

- Rice
- Wheat
- Coarse grains

Bin 1: China, India, and Soviet Union

Bin 2: United States, Canada, and European Community

Bin 3: All other countries

Figure 38.5 World grain stocks by principal holders, major importers and exporters, and others

cities, which find it easier to import supplies from world markets. India's grain production is also large, but not on a per capita basis, and it is subject to extremely volatile, Monsoon climate – leading to a government policy of high domestic grain reserves. The former Soviet Union is plagued by a volatile, northerly climate, and (as could also be said of China and India) the absence of a market economy. The role of the market, in the United States for instance, is to offer farmers higher prices for later delivery of grain than at harvest time. This spread between near and deferred futures contracts provides profit potential to anyone who wants to store grain and has led to an efficient system of grain storage and distribution in the United States. In the former Soviet Union, which has no futures market, just the opposite is true; during the 1980s, average "dockage and waste" every year was nearly one billion bushels – which is, once again, enough grain for one bushel for each of the world's one billion hungriest people.

As should be clear from this brief discussion, the single index of "world annual grain stocks" probably does more to hide than to illuminate the truth about world food abundance. Analysis of the index into its components shows a chronic condition of grain surplus in such regions as North America and Europe that is much more than enough to eradicate hunger and malnutrition in other parts of the world. The challenge is not so much limited resources as enabling poorer countries to produce enough wealth from their own store of human capital to exert enough demand to draw some set-aside lands back into production.

REFERENCES

Joy L. Harwood and C. Edwin Young (October, 1989): "US Wheat; Background for 1990 Farm Legislation," United States Department of Agriculture Economic Research Service, AGES 89–56.

Linwood Hoffman, ed. (May, 1990): "US Feed Grains; Background for 1990 Farm Legislation," USDA Economic Research Service, AIB 604.

Jerry A. Sharples and Janette Krutzfeldt (March, 1990): "World Grain Stocks; Where They Are and How They Are Used," USDA Economic Research Service, Agricultural Information Bulletin 594.

"World Grain Situation and Outlook" (October, 1990): USDA Foreign Agricultural Service, FG10-90.

39

Trends in Food from the Sea

John P. Wise

The *Global 2000 Report to the President* said: "[T]he world harvest of fish . . . is expected to rise little, if at all, by the year 2000. . . ." That was in 1980, when world fish landings were about 70–75 million tons annually. They are now around 100 million tons. (See figures 39.1.)

The overall rate of increase in world fish landings for the ten years ending in 1990 was well over 3 percent per year, well above the 1.7–2.2 percent annual growth in the world's human population estimated for 1965–2000. World fish landings per capita have been increasing for at least the last 40 years.

The fact that increases in fish harvests have at least kept pace with increases in human population is confirmed by the history of fish prices. Prices increase when commodities become scarce. But current prices for most edible fish products, after adjustment for inflation, are comparable to those of 30 years ago, at least in the United States.

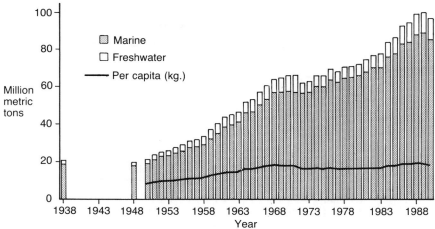

Figure 39.1 World fisheries landings, 1938–90
Note: Freshwater is groups 11–13. Marine is all other. Per capita is total.
Source: From *FAO Yearbook*, vol. 70 and preceding vols.

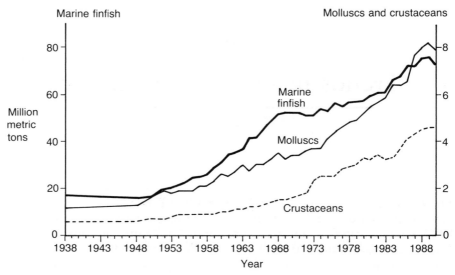

Figure 39.2   Selected world fisheries landings, 1938–90
Source: See fig. 39.1

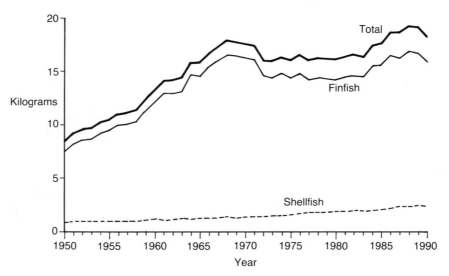

Figure 39.3   World fish landings per capita, 1950–90
Source: See fig. 39.1.

Eight species make up nearly a third of world marine landings. (See table 39.1.)

Table 39.1   World marine landings (1990)

Species	Million metric tons
Alaska pollock	5.8
Japanese sardine (=Japanese pilchard)	4.7
Chilean sardine (=South American pilchard)	4.3
Chilean jack mackerel	3.8
Peruvian anchovy (=Anchoveta)	3.8
European sardine (=Pilchard)	1.5
Atlantic herring	1.5
Atlantic cod	1.5
Total for these species	26.9
World marine landings	85.2

It might appear that the growth of the world marine catch is due only to the five species that currently make up a quarter of the total – Alaska pollock, Japanese sardine, Chilean sardine, Chilean jack mackerel, and Peruvian anchovy. But even when these species are left out of the calculation, marine landings increased over 2 percent annually during 1981–90.

Alaska pollock, Japanese sardine, and Chilean sardine show annual rates of increase of 3 to 5 percent for the last ten years. Chilean jack mackerel catches have increased even faster. These rates obviously cannot be maintained indefinitely. Catches of Alaska pollock, Japanese sardine, and Chilean sardine in fact *decreased* from 1988 to 1990. (Recent figures are subject to revision.) And substantial fluctuations do occur. Enormous variations in landings of Peruvian anchovy are well documented, and there have been long-term variations over several orders of magnitude in landings of the Japanese sardine.

Total landings in 1990 decreased just over three million tons from 1989 figures. It would be premature to say that these figures represent a permanent downturn – Peruvian anchovy and Japanese sardine catches taken together decreased about two million tons, not unusual for these extremely variable species. The USSR's catches decreased nearly a million tons, probably because of political problems.

Many species show large growth potential. It has been estimated that the world catch of squids and octopuses, at present about 2.2 million tons, could surpass ten million metric tons per year (Gulland, 1971). The South Pacific Commission has estimated the potential catch of skipjack tuna in the Central Pacific alone at several million tons annually, compared with a present world catch of about one million tons. The largest unknown factor is the Antarctic krill. Annual production of this shrimp-like creature has been estimated from 75 million to 1.35 *billion* tons per year, but recent landings have been less than half a million tons annually.

There have been many attempts to evaluate the overall potential of world marine fisheries. Estimates have been as high as 2,000 million metric tons per year; the most generally accepted by fisheries experts have been around 100 million tons per year of "conventional species," based on the 1971 work by J. A. Gulland and his collaborators. If ways are found to fish the Antarctic krill, total production could more than double. Other unconventional species may also make large contributions.

About 25–30 percent of the world catch is used principally for fish meal and oil – some 27 million tons in 1990. The proportion has held remarkably constant since the mid-1960s. Most fish meal is used in livestock feedstuffs. It represents a very large potential increase for direct use as human food. The increase would be something less than 25–30 percent, since an efficiency of 10–20 percent in conversion of fish meal to meat and eggs must be discounted. Elimination or at least substantial reduction of discards at sea of edible fish, combined with more efficient processing with less spoilage and wastage could lead to further substantial increases in food production from the sea without major increases in catches.

The bottom line is that tripling or even quadrupling production of human food from the sea remains a reasonable possibility. Achievement of maximum sustained yield will depend on appropriate management of the resources and their fisheries, a goal that so far has proven difficult to achieve.

World aquaculture production, included in preceding figures, is currently estimated at about 11 million metric tons per year. Sixty-six percent by weight is finfish, 28 percent is molluscs, 6 percent is crustaceans. (Some three million tons of acquatic plants are not included.) About 85 percent of all aquaculture takes place in Asia. Primary constraints to expansion of aquaculture, at least in the United States, are political, administrative and economic.

A few years ago I wrote:

> The predicted disastrous consequences for fisheries of oil spills have not materialized. Their effects, if any, on important fisheries have been below the measurable level. Even their influence on local small-scale fisheries [has] usually been transitory. See, for example the discussion in *Science*, 8 July, 1983, of the effects of a spill of nearly a quarter of a million tons of oil off the Brittany coast.

Now, evidence is beginning to mount that even the effects of the deliberate dumping of large amounts of oil in the Persian Gulf during the 1991 war have been far smaller than originally forecast. Articles are appearing in both the popular and the technical press suggesting that oil spills may be much less harmful than previously believed. There is no doubt that the spills are dramatic and messy. They make for spectacular television coverage. They kill large numbers of birds and occasionally other animals. But their overall effects are transient, and there is no evidence that they have any measurable effects on large marine fisheries.

REFERENCES

Anonymous (1982): *The Global 2000 Report to the President.* New York: Penguin Books. (Originally published in 1980 by the US Council on Environmental Quality and US Department of State.)

———— (1990): "Spilled Oil Looks Worse on TV." *Science* 250 (4979), 371.

———— (1991): "Oil Pollution. Local Troubles." *The Economist,* March 30, 39.

Bernton H. (1990): "In Fouled Alaska Sound, Record Salmon Harvest." *The Washington Post,* August 25. Committee on Assessment of Technology and Opportunities for Marine Aquaculture in the United States, and Marine Board, Commission on Engineering and Technical Systems, National Research Council (1992). *Marine Aquaculture: Opportunities for Growth.* Washington: National Academy Press.

FAO (1991): *Aquaculture production (1986–1989).* Fisheries Circular 815, Revision 3. Rome: Food and Agriculture Organization of the United Nations.

FAO (1992): *Yearbook: Fishery Statistics – Catches and Landings 1990,* vol. 70. Rome: Food and Agriculture Organization of the United Nations.

Gulland, J. A., ed. (1971): *The Fish Resources of the Ocean.* Surrey, England: Fishing News (Books) Ltd.

Joint Subcommittee on Aquaculture (1983): *National Aquaculture Development Plan.* Washington: Joint Subcommittee on Aquaculture of the Federal Coordinating Council on Science, Engineering, and Technology, established within the US Office of Science and Technology Policy.

Leal, D., and M. Copeland (1991): "A Second Look at Oil Spills." *PERC Viewpoints* No. 10. Bozeman, Montana: Political Economy Research Center.

Marine Resources Service (1990): *Review of the State of World Fishery Resources.* Fisheries Circular No. 710, Revision 7. Rome: Food and Agriculture Organization of the United Nations.

Mielke, J. E. (1990): *Oil in the Ocean: The Short- and Long-term Impacts of a Spill.* 90-356 SPR. Congressional Research Service. Washington: The Library of Congress.

Nicol, S., and W. de la Mare (1993): "Ecosystem Management and the Antarctic Krill." *American Scientist* 81 (1), 36–47.

Readman, J. W., S. W. Fowler, J.-P. Villeneuve, C. Cattini, B. Oregioni, and L. D. Mee (1992): "Oil and Combustion-product Contamination of the Gulf Marine Environment Following the War." *Nature* 358, 662–5.

Wise, J. P. (1984): "The Future of Food from the Sea." In *The Resourceful Earth,* eds. Julian L. Simon and Herman Kahn. Oxford and New York: Basil Blackwell.

Wise, J. P. (1991): *Federal Conservation and Management of Marine Fisheries in the United States.* Washington: Center for Marine Conservation.

40

Trends in Soil Erosion and Farmland Quality

Bruce L. Gardner and Theodore W. Schultz

Soil erosion is a universal phenomenon, but it only becomes a social problem in particular circumstances. This chapter considers the extent of soil erosion in the United States today, the threat of erosion becoming a problem in the future, and the rationale for and accomplishments of US policies aimed at reducing soil erosion.[1]

EXTENT OF EROSION

The 1987 National Resources Inventory (NRI) is our best source of data on land characteristics and soil erosion. It is based on soils and related resource data collected at 300,000 sample sites by the Soil Conservation Service of the US Department of Agriculture. One of the major findings of the survey is an estimate that 123 million acres (29.1 percent) of US cropland is highly erodible (USDA, 1989, p. 36). The previous NRI, conducted in 1982, also estimated 123 million acres in this category.

For cropland to be classified as "highly erodible," it must have an erodibility index of eight or more. The erodibility index (EI) is calculated as the maximum of

$$EI_1 = \frac{R \cdot K \cdot LS}{T}$$

or

$$EI_2 = \frac{C \cdot I}{T}$$

where R is a measure of the erosive force of rainfall, incorporating intensity and duration of precipitation; K is an index of the erodibility of the soil, based on its texture and adherence; LS is a topographic index incorporating the length and steepness of slope at the location; T is a soil-loss tolerance measure, indicating the maximum rate at which natural regeneration of soils permits erosion to occur without impairing productivity; C is a climatic factor incorporating the strength and direction of wind along with an evapotranspiration index to indicate dryness; and *I* is an index of the surface texture of the soil, indicating susceptibility to wind erosion.

Each of the components of the erosion indexes is difficult to measure, and perhaps even more difficult is their proper aggregation to obtain an overall index of erodibility. The specification used in the equations has been derived from observations on experimental plots, and while its generalizability and accuracy for any particular plot is open to question, the approach does not appear to be either too optimistic or too pessimistic about erodibility. The erodibility index can be criticized, however, as being uninformative about the extent of actual erosion.

The 1987 survey is the latest in a series of efforts going back to 1934 to assess the state of US farmland. It has been obvious from the beginning that the crops grown and practices used in growing those crops were as important as the characteristics of soils in determining the extent of erosion that actually occurs. USDA uses a universal soil-loss equation to estimate the tons of soil lost. The essential step is to multiply the numerator of the erodibility index as specified above by a factor that represents crops and practices used.

The soil-loss equation involves difficulties even beyond those of the erodibility index in producing accurate data on soil loss. But it too is not seen by the community of agricultural professionals as generating biased estimates of erosion. Applying the soil-loss equation to the NRI sample and extrapolating to the United States as a whole, the USDA obtained an estimate of an average of 3.8 tons of soil erosion per acre of cropland annually caused by rainfall (USDA, 1989, p. 23). The trend over time in this estimate is not available for long-run comparisons. But the 1982 survey, with which the 1987 survey was constructed so as to be comparable, indicated a rate of soil loss of 4.3 tons per acre. Thus, there appears to be notable progress over this five-year period. With respect to wind erosion, however, USDA estimates 3.3 tons of soil loss per acre of cropland in 1987, up from 3.1 tons in 1982.

The rates of erosion vary substantially from state to state. Water erosion rates are least in western states such as Arizona, Nevada, and Wyoming, where wind erosion is highest (and the acreage of cropland is very small). The sum of wind and water erosion rates ranges from a low of 1.8 tons per acre in Florida to 13.1 tons per acre in Texas.

Although data directly comparable to the 1987 survey are available only from 1982, earlier surveys do permit some assessment of current conditions as compared to prior decades. Bills and Heimlich (1984) discuss data from the 1977 National Resource Inventory but develop their own "new taxonomy" of cropland erosion which is not comparable to the assessments quoted from 1982 and 1987. Their thoughtful discussion indicates that erosion problems are likely overstated by the earlier estimate that 29 percent of US cropland was highly erodible as of 1989.

Nonquantified but pervasive evidence of dust storms and land degradation during the 1930s is so extensive as to indicate a significantly more serious problem at that time. Droughts in 1974, 1980, and 1987 gave rise to some wind erosion problems, but by no account as widespread and severe as in the 1930s.

Lowdermilk (1953) estimated that three fourths of the roughly 400 million acres of US cropland in 1953 was eroding at rates faster than the rate at which the soil was regenerating. In the terminology used in 1987, this means soil loss at a rate greater than T in the soil-loss equation. USDA estimates that 50 million acres had average sheet and rill erosion at a rate greater than T in 1987, and 37 million acres had average wind erosion at that level.[2] This adds up to 87 million acres, or 21 percent of the 423 million acres of cropland in 1987. Even allowing that several million acres probably have erosion rates greater than T for the sum of wind and water erosion, though not for each source separately, the 1987 (and 1982) incidence of erosion is far less than the 75 percent estimated in 1953.

Swanson and Heady (1984) review the main assessments made since the 1930s and find all of them concluding that substantial improvements have occurred over time in reducing the rate of soil loss.

Improvements notwithstanding, the 1987 survey estimates an impressive tonnage of soil loss from US cropland. With 7.1 tons lost per acre from both water and wind erosion, the 421 million acres of cropland will send an average of almost three billion tons of topsoil to another location each year. To give a sense of how much soil this is, if it were all shipped to Louisiana and spread out evenly over the whole state, the pile would be two thirds of an inch deep.[3] While 7.1 tons lost from an acre of cropland sounds like a lot, it is only about 1/20 of an inch of topsoil. And since natural regeneration of soils averages about 5 tons per acre annually, the net loss of topsoil is considerably less – about 2.2 tons, or 1/65 of an inch of topsoil.

EROSION AS A SOCIAL PROBLEM

What's wrong with soil erosion? The main cause of alarm in the past, and of concern persisting today, is the threat to our food production capacity. We have seen fields ruined by gullies and find the sight distressing. However, loss of productivity of our soil does not seem to have been a serious problem in the past and is surely not today on a national scale. National average crop yields per acre have continued to increase. Moreover, holding the quantity of inputs constant, including fertilizers and other chemicals, total factor productivity in US agriculture has continued to increase, and the rate of growth has not slowed in recent years (see figure 40.1).

A more serious social problem is the effect of soil runoff on the places that receive the soil. In the 1930s, the events that arguably caused the

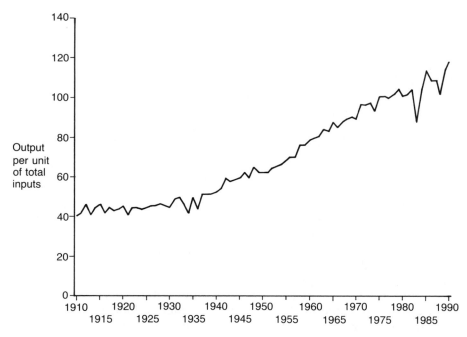

Figure 40.1 US agricultural productivity
Source: US Dept of Agriculture.

greatest distress and cost involved the blowing and settling soil, not the
state of the fields after the eroded soil had gone. Today's water erosion
likewise imposes costs on users of streams or lakes, which become
silt-laden or into which nitrogen or pesticides may be carried by soil
particles. Swanson and Heady (1984, p. 211) summarize the sketchy
evidence available as of the early 1980s. It indicates that off-site damage
from soil erosion is substantially greater than the cost of on-site productivity loss. More recently Ribaudo (1989) estimated off-site damages at
over $7 billion annually. Clearly, finding ways to charge these external
costs back to farmers would increase their incentive to reduce erosion –
and thus both on-site and off-site damage – further.

While it is in farmers' interest, other things being equal, to preserve
their soil, some argue that much erodible acreage is being depleted
through farmers' ignorance, short-sightedness, or negligence. But beyond the lack of aggregate evidence of yield declines, even in states with
the most highly erodible soils, the field-level connection between erodibility and productivity is yet to be made. In the most careful and detailed
study available of the data, Heimlich (1989) finds soils classified as
highly erodible to be practically as productive as less erodible soils. He
concludes: "The conventional wisdom regarding the current productivity of highly erodible soils is incorrect."

As so often when environmental concerns exist, public policy has moved ahead of our knowledge of the benefits versus costs of soil erosion control measures. The Food Security Act of 1985 introduced a Conservation Reserve Program (CRP) to remove 40 to 45 million acres of cropland (roughly 10 percent of the US total) from crop production. It also introduced "sodbuster" and "conservation compliance" provisions according to which farmers would not be eligible for price support benefits if they brought erodible land into crop production after 1985, and if they had not filed a conservation plan for any highly erodible land on their farms by the end of 1989. The Food, Agriculture, Conservation and Trade Act of 1990 extended and focused the regulatory details of these programs.

Through 1990, 34 million acres of cropland had been enrolled in the CRP under 10-year rental contracts. The average rental rate paid by the government to enrolled farmers was $50 per acre, indicating an annual government cost of the program of about $1.7 billion. (In addition, federal cost-sharing of practices such as permanent grass cover or tree planting adds about $300 million annually.)

Congress made eligible for the program all cropland that had an erodibility index of eight or more (the same criteria used in the right-hand column of table 40.1). The average eligible land is estimated to have an erosion rate of 19 tons per acre, which placing the land in grass reduces to 1.5 tons per acre. Thus, with average eligible land in the program, soil erosion would be reduced by 34 million × 17.5 = 595 million tons annually. USDA estimates that actual reductions have been 655 million tons, because the enrolled land is slightly more erodible than the average eligible land. This amounts to about one fifth of the estimated US annual erosion of three billion tons.

What are the benefits of this reduction in erosion? Ribaudo (1989) estimates the off-site benefits to be $10 per acre. The on-site soil productivity benefits are sure to be lower, and indeed on Heimlich's evidence would be essentially zero.

For an overall benefit/cost assessment of the CRP, it is necessary to consider not only the $50 paid to landowners and the $10 of off-site benefits per acre but also other gains and losses that the program generates. The $50 payment is a cost to the government, but it is a receipt to the landowner. Thus, except for costs of tax collection and program administration, it is a matter of redistribution rather than resource costs. A net resource cost occurs because land in the conservation reserve no longer grows crops, and idling a valuable resource reduces national income at the margin. The best estimate of the value of the crops lost by idling an acre of land is the market rental value of that land.

Congressional intent was clearly for the government to pay farmers the rental value of the land placed in the program and not more. Nonetheless, the rentals paid exceeded the market rental rate, especially in the western plains where most of the participation occurred (US General

Table 40.1 Soil loss per cropland acre and erodibility, 1987, selected states

State	Water (sheet and rill) erosion (tons per acre)	Wind erosion (tons per acre)	Highly erodible cropland (% of total cropland)
Northeast:			
New York	2.7	0.0	33.4
Pennsylvania	5.6	0.0	68.0
Maryland	5.1	0.2	35.0
South:			
Virginia	5.1	0.1	45.2
North Carolina	6.0	0.3	28.0
Georgia	5.9	0.0	15.0
Florida	1.5	0.3	4.1
Alabama	5.6	0.0	26.5
Mississippi	7.3	0.0	23.0
Tennessee	8.8	0.0	46.2
Kentucky	8.5	0.0	53.0
Midwest:			
Ohio	3.5	0.3	19.0
Indiana	4.2	0.8	14.4
Illinois	5.2	0.0	17.0
Minnesota	2.5	0.8	7.4
Iowa	6.5	1.9	32.0
Missouri	7.0	0.8	40.4
Plains:			
North Dakota	1.9	3.9	11.5
South Dakota	2.3	2.4	9.0
Nebraska	4.7	1.9	35.0
Kansas	2.6	3.7	53.0
Oklahoma	2.3	4.3	42.0
Texas	2.5	11.6	40.1
West:			
Montana	1.9	7.7	58.5
Idaho	4.2	3.9	35.0
Colorado	2.2	10.0	75.4
Washington	5.6	3.4	38.0
Oregon	2.9	1.7	26.0
California	1.2	5.3	75.4
Arizona	0.4	9.7	70.0
United States	3.8	3.3	29.0

Accounting Office, 1989). The market value of land in the program is not known with much precision. The General Accounting Office estimates the market rental rates for land in the program at $30 per acre. On this accounting, the net annual loss to society would be $20 per acre enrolled in the CRP ($30 in crop value given up minus the $10 in off-site benefits saved). To this must be added the one-time establishment costs of conservation practices (mostly planting grasses), both the costs incurred by farmers and by the federal cost-share. At an average establishment cost of $15 per acre, annualized at $1.50 per year over the 10-year contract, the social cost of the program is $21.50 per acre annually.

This accounting is not the end of the story, however, because some of the land placed in the reserve would have been idled under the annual acreage reduction programs if it had not been placed in the CRP. USDA estimates that about 11 million acres of wheat land is enrolled in the program. At the same time, another five to 10 million acres of wheat land has been enrolled in annual acreage reduction programs during the CRP. Had it not been for the CRP, most and perhaps the large majority of the CRP land would have been enrolled in a larger annual program. A larger annual acreage reduction would have been necessary since the percentage idled is chosen so as to achieve a given price and carryover stock level, and this would have required more acreage had the CRP land not been idled. Assuming that half of the CRP land would have been idled anyway, the overall cost of the CRP program to the United States as a whole would be reduced from $21.50 to $11.50 per acre. With 34 million acres in the program we obtain an estimate of $390 million per year as the net social cost, sometimes called deadweight loss, caused by the CRP program.

The Food, Agriculture, Conservation and Trade Act of 1990 extends and broadens the CRP program. It requires 40 million acres to be enrolled by 1995 under a set of criteria that includes targeting for enrollment of wetlands in permanent easements, for new programs that will permit the land to be farmed but in such a way as to generate less runoff of sediments while being cropped for placing trees on marginal pastureland in the CRP.

The stated aims of many of these changes focus on water quality, that is, on the off-site benefits to be obtained by changes in farming practices. What policymakers principally have in mind is not just sedimentation of streams but the nitrogen and pesticide pollution that have gained much prominence in environmental lobbying groups' political positions. The new program will add six million acres of land, about a 15 percent increase over the land already enrolled to the Conservation Reserve. The magnitude of social benefits and net social costs is likely to be much smaller than in the pre-1990 CRP. Indeed, because of the focus on smaller-scale practices that are tuned to an individual farm situation, it is conceivable that these programs could generate a net social gain. But this remains to be seen.

CONCLUSION

Where Ted Schultz was born and reared in South Dakota, the landscape has been tamed. Gone are the awesome prairie fires. They no longer take the lives of animals and birds, nor do they destroy people's homes.

Cyclone cellars have been replaced by better protection against tornadoes. Belts of trees now serve as windbreaks, sheltering pheasants and deer, improving the attractiveness of the countryside, and slowing down the wind a bit. Farmers have adopted other and better ways to reduce wind erosion.

Soil erosion is a social problem. It is a problem because it causes damage to people other than those who cause it to occur. But soil erosion is not threatening our agricultural productivity. And, the potential for off-site damages because of water quality degradation or other consequences of soil runoff are less severe now than they have been in both the recent past and in the more distant past of the 1930s. Ribaudo (1989) estimates the total off-site damages caused by soil erosion to be in the range of $5 to $18 billion annually (Ribaudo 1989, p. 12). While these are substantial losses, the policies undertaken in attempts to remedy soil erosion problems have so far resulted in costs greater than the benefits generated. Soil erosion is not an easily cured problem. Hard as it is to make this choice, the evidence indicates that the best approach is to take soil erosion as we take the common cold: a problem that is irritating but not seriously threatening, and one for which the available cures are likely to cause as many problems as they solve.

NOTES

1   This chapter incorporates three paragraphs previously published in Schultz (1983).
2   "Average" is USDA terminology that might better be understood as "potential." It does not mean that 87 million acres actually eroded at a rate greater than T, because that would depend on rainfall and drought patterns. The EI index and soil-loss estimates are based on conditions expected in a typical year.
3   Based on 31.1 million acres in the state and an acre-inch of soil weighing 150 tons = 3,000,000,000 tons/31,100,000 acres/150 ton/inch.

REFERENCES

Bills, Nelson L., and Heimlich, Ralph E. (1984): "Assessing Erosion on U.S. Cropland." US Department of Agriculture, Economic Research Service, Agricultural Economic Report No. 513, July.

Heimlich, Ralph E. (1989): "Productivity and Erodibility of U.S. Cropland." USDA, Economic Research Service, Agricultural Economic Report, No. 604, January.

Lowdermilk, W. C. (1953): "Conquest of the Land Through Seven Thousand Years." US Department of Agriculture, Soil Conservation Service, Agricultural Information Bulletin, No. 99.

Ribaudo, Marc O. (1989): "Water Quality Benefits from the Conservation Reserve Program." US Department of Agriculture, Economic Research Service, Agricultural Economic Report, No. 606, February.

Schultz, Theodore W. (1983): "The Dynamics of Soil Erosion in the United States." In John Baden, ed., *The Vanishing Farmland Crisis*. 45–57. Lawrence: University of Kansas Press.

Swanson, Earl R., and Heady, Earl O. (1987): "Soil Erosion in the United States." In Simon and Kahnr, eds, *The Resourceful Earth*. 202–23. Oxford: Basil Blackwell.

US Department of Agriculture (1989): "Summary Report: 1987 National Resource Inventory." Soil Conservation Service, Statistical Report No. 790, December.

—— (1991): "Production and Efficiency Statistics." Economic Research Service, ECIFS 9-4, April.

US General Accounting Office. (1989): "Conservation Reserve Program Could Be Less Costly and More Effective." Washington, DC: GAO/RCED-90-13, November.

—— (1989): "The Second RCA Appraisal." Washington, DC, June.

41

Water, Water Everywhere
But Not a Drop to Sell

Terry L. Anderson

DOOMSDAY PREDICTIONS

Environmental concerns about water quantity and quality are increasing. According to the Global 2000 Report (1980, 137) published by President Carter's Council on Environmental Quality in 1977, "of all the substances found on the earth, those most fundamental to the existence of man, or to the existence of life itself, are unquestionably water and air." The report's water projections find that "because of the regional and temporal nature of the water resource, water shortages even before 2000 will probably be more frequent and more severe than those experienced today. By the year 2000 population growth alone . . . will cause at least a doubling in the demand for water in nearly half the countries of the world" (158). On Earth Day 1970, Paul Ehrlich, one of the leading doomsayers, predicted that America would have water rationing by 1974 and that the oceans could be as dead as Lake Erie by 1979. These statements may be extreme, but they typify concerns about water quantity and quality.

Such dire conclusions are the inevitable result of water planning models that assume fixed supplies of water but exponential growth in consumption. For example, the National Water Summary (USGS, 1990, 125) projects water supplies on the basis of "water budgets" where hydrological factors such as evapotranspiration, precipitation, surface-water outflow to the oceans and neighboring countries, and stocks of groundwater determine water availability. Demand, on the other hand, is often based on an extrapolation of past trends (see USGS, 1990, 118). Figure 41.1 shows total water withdrawals in the United States increasing at a rapid compound annual rate of 2.33 percent between 1960 and 1985. At this rate consumption will double in 20 years. Combining this with fixed water budgets, it is easy and not

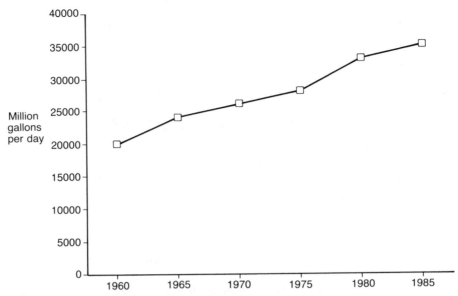

Figure 41.1 US water withdrawal, 1960–85, by public supply systems
Source: US Geological Survey, *National Water Survey, 1987*, Washington, DC: GPO, 1990, 72.

unreasonable to conclude that we will encounter severe shortages in the
future unless something changes.

POLITICAL CONTROL OF SUPPLY AND DEMAND

Two very different solutions are typically offered to counter the shortages: (1) augmenting supply, and (2) curtailing demand through command and control. After considering each, we will turn to a third
alternative – increased use of water marketing. More reliance on the
latter leads to a very different vision of the future.

Augmenting Supply

Supply or structural solutions dominated water policy in the United
States from the passage of the Newlands Reclamation Act in 1902 until
the 1970s. As shown in figure 41.2, storage capacity of reservoirs increased dramatically between 1910 and 1970. Since the late 1970s,
however, additions to reservoir capacity waned. "At least three factors
contributed to this slowing trend: A paucity of remaining good reservoir
sites; the recent change in Federal water policy, which requires increased non-Federal contributions to the funding of Federal water projects; and the increased concerns about the environmental cost of
additional damming of free-flowing streams" (USGS, 1990, 139).

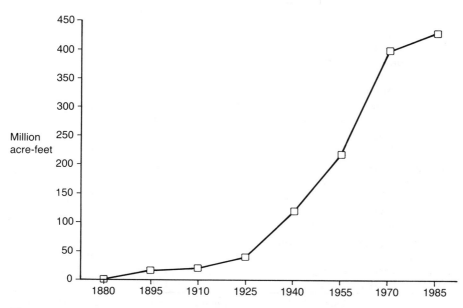

Figure 41.2   Reservoir storage in the USA, 1880–1985
Source: US Geological Survey, *National Water Survey, 1987*, Washington, DC: GPO, 1990, 139.

Indeed, these constraints are not new, but they are only just beginning to take effect in the political arena. When the Bureau of Reclamation was formed in 1902, subsidies were necessary; "while private enterprise had managed to bring under successful irrigation an impressive and substantial acreage of land, a point had been reached where further development would need stronger support by the Federal and state governments" (Golze, 1961, 12). Comparing the value of water to farmers to the actual cost of storing and delivering water (see table 41.1) illustrates why the "stronger support" from government was necessary. Column 1 is the value of water in agriculture (based on the difference between irrigated and nonirrigated land prices) divided by the amount farmers actually repay the Federal government for project construction. Since all ratios are greater than 1, the irrigation projects are a good deal for farmers. Column 2, however, tells a different story. It shows the value of water to farmers (again measured by the difference between irrigated and nonirrigated land prices) divided by the full cost of storing and delivering water. As indicated by ratios less than one, 60 percent of the projects fail the benefit–cost test.[1]

In addition to the physical and fiscal limits on augmenting supply, environmental constraints have become binding. There is probably no other form of subsidized environmental destruction that compares to

Table 41.1   Value of water in agriculture compared to price of water

	Column 1[a]	Column 2[b]
Glenn-Colusa, California	51.1	4.6
Moon Lake, Utah	10.1	4.4
Elephant Butte, New Mexico	5.9	2.2
Truckee Carson, Nevada	8.5	1.4
Goshen, Wyoming	5.4	1.4
Cachuma, California	7.3	1.4
Lower Yellowstone No. 1, Montana	3.9	1.1
Grand Valley, Colorado	5.6	0.8
Wellton-Mohawk, Arizona	7.1	0.8
Imperial, California	2.9	0.8
San Luis Unit, California	4.7	0.7
Coachella, California	2.3	0.7
Lugert-Altus, Oklahoma	6.6	0.7
Black Canyon No. 2, Idaho	5.3	0.6
East Columbia Basin, Washington	12.8	0.4
Malta, Montana	5.0	0.4
Orovill-Tonasket, Washington	1.5	0.3
Farwell, Nebraska	1.9	0.1

Source: Richard W. Wahl, *Markets for Federal Water: Subsidies, Property Rights, and the Bureau of Reclamation.* Washington, DC: Resources for the Future, 1989, 40.
[a] Column 1 is the value of water to farmers (measured by the difference between irrigated and nonirrigated land prices) divided by the cost to farmers.
[b] Column 2 is the value of water to farmers (measured by the difference between irrigated and nonirrigated land prices) divided by the total cost of storing and delivery of the water.

water storage and delivery. Throughout the western United States, canyons have been flooded and massive flows have been diverted to "make the desert bloom like a rose."

On the international scene, the environmental disruption associated with larger water projects is well illustrated by the problems on the Aral Sea in Central Asia (see Elliot, 1991). In the early 1950s, the Aral Sea, the fourth-largest inland sea in the world, was one of the most productive fisheries in the region. This changed later in the decade when planners decided that the Soviet Union should become self-sufficient in cotton, and two rivers feeding the sea were diverted to irrigate cotton fields. As a result, the Aral Sea today is approximately 60 percent of its original size and shrinking. Salt deposits from newly exposed shoreline are blown to surrounding land, reducing its productivity. The fishery has all but disappeared as salinity in the sea has increased. Elliot (1991, 29) concludes that the Aral Sea water project "was probably the greatest man-made environmental catastrophe of the modern world, Chernobyl notwithstanding." Such environmental disasters in the Former Soviet

Union and around the world are making environmentalists more skeptical of centralized water management.

Limiting Demand Through Command and Control

Lacking the potential to increase supplies, governmental agencies have turned to micromanagement of demand. California water allocation typifies this approach. Water authorities have threatened that homes exceeding their water allocation will have water flow restrictors installed, the number of new hookups to municipal water systems has been limited, and rationing has been implemented. Although some cities are having their supplies curtailed, others like Sacramento (coincidentally the state capital) do not even have water meters. The political allocation of water has forced cities like Santa Barbara to consider building desalinization plants to produce fresh water at a cost of $2,000 per acre foot, while farmers in the Central Valley pay as little as $9 per acre foot. This policy of micromanagement "turns a fundamentally economic issue – how best to use a natural resource – into a political issue" (Bay Area Economic Forum, 1983, p. 1).

THE MARKET ALTERNATIVE

Like every resource that exhibits shortages, the problem with water allocation is caused by the lack of the proper information and incentives generated by markets. Water is "cheaper than dirt" and consequently treated that way. As long as consumers are supplied their water at low prices, they have little or no incentive to conserve. Demand will continue to grow exponentially while supply remains constant. This has been especially true for agriculture in the western United States. Consuming approximately 85 percent of supplies, farmers apply water to relatively low-valued crops as long as the incremental value of output covers the minimal price they pay to federal or state authorities. In most cases they face the rule of "use it, or lose it," meaning that if they reduce consumption through conservation measures, the water is available to other users. If they could sell their water to higher valued users, the opportunity cost of using the water would rise and they would have an incentive to conserve.

The main argument against a market-based approach to allocation is that water is a necessity, making it impossible for consumers to respond. Whereas it is true that an individual could not live without some water each day, many conservation possibilities certainly exist, given that average household consumption per capita in the United States is 183 gallons per day, the highest in the world. If water becomes more expensive, car washing might be less frequent, lawns and gardens might be planted to include less water-intensive plants, and leaky faucets might be fixed. Table 41.2 and figure 41.3 suggest the magnitude of residential

Table 41.2 Comparison of metered versus flat-rate use rates in the United States

	Metered areas	Flat-rate areas
Annual average	Gallons per day	per dwelling unit
Leakage and waste	25	35
Household	247	236
Sprinkling	186	420
Total	458	691
Maximum day	979	2354
Peak hour	2481	5171
Annual	Inches of water	
Actual lawn sprinkling	14.1	39.4
Potential requirements	22.5	14.8
Summer		
Actual lawn sprinkling	7.4	25.5
Potential requirements	11.5	10.3
Maximum day		
Actual lawn sprinkling	0.15	0.51
Potential requirements	0.25	0.29

Source: F. P. Linaweaver, John C. Geyer, Jerome B. Wolff, "A Study of Residential Water Use." Washington DC: Department of Housing and Urban Development, 1967, 50.

response as a result of metering water and thus raising the price for additional units of consumption.

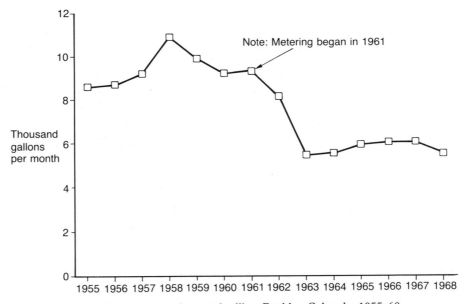

Note: Metering began in 1961

Figure 41.3 Total water consumption per dwelling, Boulder, Colorado, 1955–68
Source: S. H. Hanke, "Demand for Water Under Dynamic Conditions," *Water Resources Research*, vol. 6 (Oct., 1970), 5, 1260.

Table 41.3 Variance in industrial unit water withdrawals around the world, 1960

Product or user	Unit	Draft (gallons)		
		Maximum	Mean	Minimum
Steam-electric power	kWh	170	80	1.32
Refining	gallons of oil	44.5	18.3	1.73
Steel	ton	65,000	40,000	1,400
Soaps	pounds	7.5	—	1.57
Carbon	pounds	14	4	0.25
Natural rubber	pounds	6	—	2.54
Butadine	pounds	305	160	13
Glass containers	ton	667	—	118
Automobiles	per car	16,000	—	12,000
Trucks	per unit	20,000	—	15,000

Source: Jack Hirshleifer, *Water Supply: Economics, Technology, and Policy*. Chicago: University of Chicago Press, 1960, 27.

The empirical evidence for demand being price responsive is convincing. Table 41.3 shows the maximum, minimum, and mean quantities necessary for producing various commodities in the 1960s. Though no more recent study is available, it is reasonable to assume that advances in technology since then would expand the alternatives available if the price incentive is there. The main findings from studies of price responsiveness or elasticity of demand for water are summarized by Patrick Mann (1981, iii): "[A]ggregate municipal demand is relatively price inelastic; however, price elasticity appears to vary positively with water price levels; i.e., there is more usage-price sensitivity with higher rates than with lower rates. . . . [S]easonal demand generates higher price elasticities than non-seasonal demand. Finally, commercial and industrial demands appear to be more sensitive to price changes than residential demand." Given that 57 percent of US water consumption is in industrial uses and 34 percent is in agricultural uses, these conclusions suggest that higher prices can make a difference in total water demand.

To have price determined in a market, it is necessary to have clearly specified water rights that can be traded. Prior appropriation water law in the American West offers the best example of how this can work. This doctrine establishes rights to water on the basis of "first-in-time, first-in-right," meaning that individuals who used water first established rights with priority over later users (see Anderson, 1983). Under this system, quantities which are clearly specified and recorded can be traded as long as other water rights holders are not impaired. There is growing evidence that such a system can greatly enhance water-use efficiency.

Even the environment can benefit under a market for water flows. Consider, for example, the pollution of the Kesterson National Wildlife Refuge in California (see Wahl, 1989, 197–219). In 1983, the Fish and Wildlife Service discovered that waterfowl at the Kesterson National Wildlife Refuge were being hatched with grotesque deformities. The cause was traced to high selenium concentrations being carried to the refuge in agricultural drainage water. The water in question was being delivered at highly subsidized rates to lands "marginal in their suitability for irrigated agriculture . . . because of highly saline, slowly permeable soils with anticipated or present drainage problems" (US Department of Interior, 1978, 163). At prices that cover only 15 percent of the project costs (Wahl, 1989, 203), farmers had an incentive to use large quantities of water for irrigation even though the water was leaching salts, selenium, and other trace elements from the soil and depositing them in Kesterson.

To solve the Kesterson problem, the Federal government proposed reducing deliveries to the Westlands Water District and halted use of the drain that led to Kesterson. Needless to say, water users, farm communities, and banks with loans based on the irrigated value of land opposed this solution.

A program of water marketing would have eliminated this problem. In the first place, elimination of Federal subsidies for agricultural water would substantially reduce the amount of water used and hence the amount of selenium leached from the soil and delivered to Kesterson. But even with the subsidies in place, allowing the farmers to sell their water to other users would provide substantial incentives to reduce irrigation and to clean up drainage water. Wahl (1989, 208) estimates that voluntary water sales by farmers in the Westlands Water District to municipalities could net them as much as $3,151 per acre above the value of the water in agriculture! Given the potential for switching to less water-intensive crops and for desalting the water after it is used, this potential profit from water marketing could have a tremendous positive impact on the environment.

CONCLUSION

In the absence of market incentives for both supply and demand, gloom and doom predictions regarding water will likely come true. Without price rationing, demand will continue to grow as long as there is a supply. But, since the quantity that people use cannot exceed the quantity supplied, eventually the price will rise. The only question is what form the price increase will take. If markets are allowed to ration, signals will evolve to encourage conservation and stimulate supply. If the rationing takes the form of political command and control, Mark Twain's prophetic words will apply: "Whiskey is for drinkin', water is for fightin'."

NOTE

1   See Rucker and Fishback, 1983, p. 62, for further evidence of uneconomic Federal projects.

REFERENCES

Anderson, Terry L. (1983): *Water Crisis: Ending the Policy Drought.* Washington, DC: CATO Institute.

Bay Area Economic Forum (1991): *Using Water Better: A Market-Based Approach to California's Water Crisis.* San Francisco: Bay Area Economic Forum, September.

Elliot, Michael (1991): "The Global Politics of Water." *The American Enterprise,* September/October.

Golze, Alfred R. (1961): *Reclamation in the United States.* Caldwell, ID: The Caxton Printers.

Mann, Patrick C. (1981): *Water Service: Regulation and Rate Reform.* Columbus, OH: The National Regulatory Research Institute.

President's Council on Environmental Quality (1980): *Global 2000 Report,* Vol. 2. Washington, DC: GPO.

Rucker, Randal R. and Price V. Fishback (1983): "The Federal Reclamation Program: An Analysis of Rent-Seeking Behavior." In: Terry L. Anderson ed. *Water Rights: Scarce Resource Allocation, Bureaucracy, and the Environment.* San Francisco: Pacific Research Institute.

US Department of Interior, Bureau of Reclamation (1978): *Special Task Force Report on San Luis Unit, Central Valley Project, California.* Washington, DC: GPO.

US Geological Survey (USGS) (1990): *National Water Summary,* 1987. Washington, DC: GPO.

Wahl, Richard W. (1989): *Markets for Federal Water: Subsidies, Property Rights, and the Bureau of Reclamation.* Washington, DC: Resources for the Future.

42

Trends in Land Use in the United States

H. Thomas Frey

The landscape of the present United States (48 contiguous states only) has changed greatly since the early English colonies were established on the Atlantic seaboard. From the standpoint of agriculture, the interval of nearly four centuries can be divided arbitrarily into (1) the period of expansion (1607–1930) and (2) the period of relative stability (1930–present). The former was characterized by the spread of settlement, agriculture, and related activities across the continent. The latter is characterized by internal adjustments and development within the context of surplus production capacity.

THE PERIOD OF EXPANSION

Agricultural expansion, as documented by early censuses of agriculture, has been summarized by Wooten and Anderson (1957). Improved land in farms, which totaled 113 million acres in 1850, increased to 163 million acres in 1860 and 189 million in 1870. Cropland, a less inclusive category, totaled 188 million acres in 1880 and more than doubled to 402 million acres in 1920. The historical increase slowed thereafter and peaked at 413 million acres in 1930. Meanwhile, the acreage of grassland pasture and range, estimated at more than 900 million acres in 1880, decreased correspondingly (figure 42.1).

Forestland, the source of most cropland in the eastern and southern states, probably contributed little on net to the rapid expansion of cropland after 1880. Although substantial acreages were cleared for crops during this period, they were largely offset by reversion of cropland to forest.

Urban and other special uses of land became increasingly important in the early decades of the twentieth century. Wooten et al. (1957) estimated that acreages in the most intensive uses, i.e., urban areas, highway, road, and railroad rights-of-way, increased from 29 million acres in

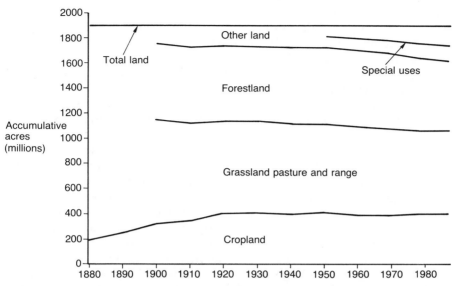

Figure 42.1 Trends in major uses of land, 48 contiguous states, 1880–1987
Notes: Cropland excludes cropland used only for pasture. Grassland pasture and range includes grassland and other nonforested pasture and range including cropland used only for pasture. Forestland excludes forestland in parks and other special uses of land. Special use areas include rural transportation areas (highway, road, and railroad rights-of-way, plus airports), national and state parks (areas administered primarily for recreational purposes; includes wilderness areas), federal and state wildlife areas (areas administered primarily for fish and wildlife protection and propagation), and national defense areas (areas administered by the Dept of Defense for military purposes). Other land is unclassified land including urban areas, various special-purpose uses, desert, marshes, etc.
Sources: A. B. Daugherty (1991): *Major Uses of Land in the United States: 1987*, AER-643, Washington: US Dept Agr., Econ. Res. Serv.; H. T. Frey (1982): *Major Uses of Land in the United States: 1978*, AER-487, Washington: US Dept Agr., Econ. Res. Serv.; id. (1973): *Major Uses of Land in the United States: Summary for 1969*, AER-247, Washington: US Dept Agr., Econ. Res. Serv.; id. and R. Hexem (1985): *Major Uses of Land in the United States: 1982*, AER-535, Washington: US Dept Agr., Econ. Res. Serv.; H. H. Wooton (1953): *Supplement to Major Uses of Land in the United States*, TB-1082, US Dept Agr., Bur. Agr. Econ.; id. and J. R. Anderson (1957): *Major Uses of Land in the United States: Summary for 1954*, AIB-168, US Dept Agr., Agr. Res. Serv.; id., K. Gertel, and W. C. Pendleton (1962): *Major Uses of Land and Water in the United States: Summary for 1959*, AER-13, US Dept Agr., Econ. Res. Serv.

1910 to 35 million in 1930. However, these estimates are not comparable with those of recent decades.

THE PERIOD OF RELATIVE STABILITY

Agricultural and Forestry Uses

Trends in land uses since cropland peaked in 1930 have, with some qualification, been characterized by relative stability at the national level (figure 42.1). The total acreage of cropland has fluctuated near the 400 million acre level, ranging from the historic high of 413 million acres in 1930 to the modern low of 384 million in 1969.

Grassland pasture and range trended downward throughout the 1930–87 period. This category is a source of land for all other major categories. Part of the decrease represented shifts to urban, recreational, park, wildlife and forest uses; part served to maintain the cropland base; and a substantial acreage of arid rangeland was reclassified as unsuitable for grazing.

Estimates of forestland, exclusive of acres reserved for parks and other special uses, increased somewhat during 1930–59 and decreased sharply thereafter. However, much of the decrease was attributable to reclassification rather than to actual changes in vegetative cover. Approximately 18 million acres, one-third of the indicated decrease, were reserved for park and wildlife purposes. A substantial but undetermined acreage of woody vegetation was reclassified from forestland to open rangeland.

This extended period of relative stability, especially in the cropland base, is attributable to large increases in productivity per acre. During the 1949–88 period, crop production per acre almost doubled while the acreage used in crop production decreased about 10 percent (figure 42.2). From 1981 to 1989, total crop production was sufficient to permit the export of products from the average equivalent of 107 million acres annually while, at the same time, an average of more than 50 million acres were diverted from crop production (US Dept of Agriculture, 1990b).

Much of the increase in productivity per acre is due to improvements in technology, but part is attributable to regional shifts resulting in an increase in the quality of the average acre in crop production. For example, large acreages in the East and South have reverted to pasture and forest due to hilly terrain, infertile soils, small and irregularly shaped fields, and other reasons (see map in Healy, 1984). Additional acreages of uneconomic cropland in parts of the Great Plains have reverted to pasture and range (Frey, 1983a). Concurrently, limited acreages of new land with high productivity potential have been added to the cropland base in the North Central States, lower Mississippi Valley, and other favored regions.

Cotton, which peaked at 45 million acres in 1926 (US Dept of Commerce, 1975) and recently has ranged between 8 and 12 million acres annually (US Dept of Agriculture, 1991), is a classic example of the shift of crop production to land of higher quality as well as the decline in acreage of a major crop. In 1919 the States of Alabama, Georgia, North Carolina, and South Carolina had one-third of the 34 million acres of cotton harvested nationally; in 1990 these States had only 9 percent of a much smaller 12 million acre total. Arkansas, Louisiana, and Mississippi increased their share from 20 to 24 percent during this period, but the smaller 1990 acreage was concentrated on

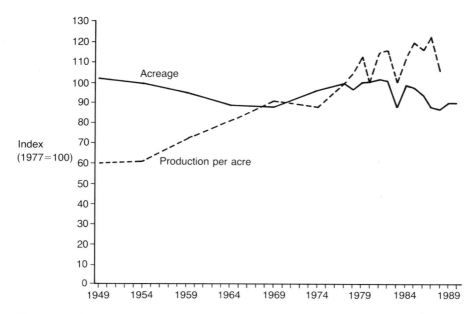

Figure 42.2 Cropland used for crops, 1949–90, index of land use and productivity
Note: The index depicts the sum of cropland harvested, crop failure, and cultivated summer fallow and is computed from unrounded data.
Sources: A. B. Daugherty (1991): *Major Uses of Land in the United States: 1987*, AER-643, Washington: US Dept Agr., Econ. Res. Serv.; US Dept of Agriculture, Economic Research Service (1990); *Economic Indicators of the Farm Sector: Production and Efficiency Statistics, 1988*, ECIFS 8–5; US Dept of Agriculture, Economic Research Service (1976): *Changes in Farm Production and Efficiency: A Summary Report*, Statistical Bulletin 561.

fertile alluvium in the Mississippi Valley, much of which was still forested and undrained in 1919. A similar internal shift occurred in Texas, the dominant cotton state, where acreages declined on small farms in the east and expanded on large, mechanized farms in the west, mainly on irrigated land. In 1990 more than four-fifths of the total cotton acreage was concentrated in Arkansas, Louisiana, Mississippi, Texas, Oklahoma, Arizona, and California, compared with less than three-fifths in 1900 (US Dept of Commerce, 1962).

Special Uses

A number of nonagricultural uses, including urban areas, transportation areas, federal and state parks and wildlife areas, and military bases absorb agricultural and forestland to a varying extent. Urbanization is of special interest because of its growth and competitive strength relative to agricultural uses of land. Measures of the area in urban areas that are generally comparable in terms of definitions and methods are available for 1960, 1970, 1980 and 1990 from the Bureau of the Census.

Figure 42.3 illustrates the trends in rural and urban land use as percent of total US land.

The 56 million acres classified as urban in 1990 compare with the totals of 47 million in 1980, 34 million in 1970 and 25 million in 1960. These totals yield average annual urbanization increments ranging between 0.9 and 1.3 million acres between 1960 and 1990. The 1990 total accounted for 2.9 percent of total land area. However, several factors combine to give these acreages inflated values.

First, Census urban area data are developed primarily for the purpose of computing population densities rather than as measures of urban area per se. The definitions and methods, while generally comparable, resulted in the inclusion of open land without offsetting omissions of developed land. Second, as areas designated by the Bureau of Census as Urbanized Areas enlarge in area, the low-density or vacant land needed to eliminate enclaves, close indentations in boundaries, and link outlying areas tends to increase more rapidly then actual development. Third, since 1960 there has been an increasing trend toward the extension of official incorporated place boundaries to include territory essentially rural in character. Thus, Census urban totals substantially overstate the areal extent of urbanization. This overstatement is offset somewhat by the exclusion of small population concentrations (fewer than 2,500 inhabitants) outside Urbanized Areas (Frey, 1983).

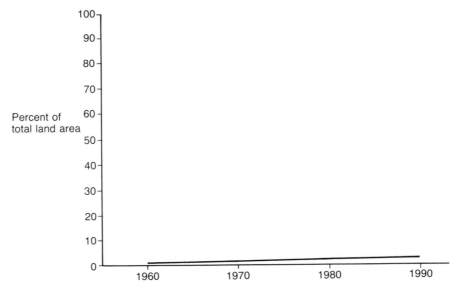

Figure 42.3 Urban land use, 48 contiguous states, 1960–90

Notes: Urban area is defined by the Bureau of the Census as urbanized areas and places of 2,500 or more population outside urbanized areas. Urban acreage data for 1960, 1970, and 1980 are as developed by the Geography Division, Bureau of the Census, summarized in (Frey, 1990b). Acreage for 1990 from (US Dept of Commerce, 1992).

Researchers in the Economic Research Service, USDA, have analyzed the results of several urbanization studies based on aerial photo interpretation and sampling procedures. Based on a synthesis of data from these studies plus indirect measures from the Census of Population, they currently estimate urbanization at 1.1 million acres annually. Approximately 34 percent of this total is from cropland and high-quality pasture (Heimlich, 1991).

Several other nonagricultural uses of land involve substantial acreages (figure 42.4). However, none of these uses competes strongly with cropland. Currently, the mileage added to the highway and road system annually is small. Because of its physical and locational characteristics, very little of the land in parks and wildlife areas is of cropland quality, but in some cases rangeland and forestland of commercial quality are involved. Several million acres used for defense purposes were once used for crops and pasture, but land in this category is not increasing.

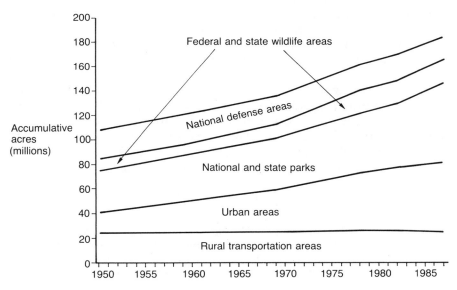

Figure 42.4 Trends in special land uses, 48 contiguous states, 1950–87
Note: The 1950 urban area data point was derived by extrapolation of 1960 and 1970 data.
Sources: A. B. Daugherty (1991): *Major Uses of Land in the United States: 1987*, AER-643, Washington: US Dept Agr., Econ. Res. Serv.; H. T. Frey (1982): *Major Uses of Land in the United States: 1978*, AER-487, Washington: US Dept Agr., Econ. Res. Serv.; id. (1973): *Major Uses of Land in the United States: Summary for 1969*, AER-247, Washington: US Dept Agr., Econ. Res. Serv.; id. and R. Hexem (1985): *Major Uses of Land in the United States: 1982*, AER-535, Washington: US Dept Agr., Econ. Res. Serv.; H. H. Wooton (1953): *Major Uses of Land in the United States*, TB-1082, US Dept Agr., Bur. Agr. Econ.; id., K. Gertel, and W. C. Pendleton (1962): *Major Uses of Land and Water in the United States: Summary for 1959*, AER-13, US Dept Agr., Econ. Res. Serv.

Estimates of the area in artificial reservoirs have not been developed regularly, as water areas are excluded from total land area. However, examination of unpublished data indicates that as late as the 1950–60 decade, the water area of large (40 acres or more) artificial reservoirs increased an average of 210,000 acres annually. More recently, construction of artificial reservoirs has not been a major activity due to a scarcity of good sites and environmental considerations.

DEFINITIONS AND EXPLANATIONS OF THE DATA

The estimates of major land uses presented herein are from an ongoing series of land use inventories conducted by the Economic Research Service (ERS), USDA and its predecessor agencies for years in which a census of agriculture is conducted. The author was primarily responsible for developing and analyzing the estimates for the 1964–85 period.

Data for these inventories are obtained from several Federal agencies, including USDA and the Departments of Commerce, Defense, Interior, Transportation, and others. State agencies are also important sources of data because several states own relatively large acreages of agricultural and forested land. Most states also maintain park systems and wildlife refuges, and all administer land in institutional uses and for other purposes.

The estimates, with some exceptions, were synthesized or otherwise adapted, modified, or adjusted from available data rather than used exactly as developed by source agencies. This process is necessary because land use data typically are obtained in surveys that differ in scope, methods, definitions, and other characteristics. Individual sources account for only one or a few uses and for only a limited part of the total land area.

REFERENCES

Daugherty, Arthur B. (1991): *Major Uses of Land in the United States: 1987 AER–643*. Washington: US Department of Agriculture, Economic Research Service.

Frey, H. Thomas (1983a): *Acreage Formerly Cropped in the Great Plains*. ERS Staff Report AGES830404. Washington: US Department of Agriculture, Economic Research Service.

———— (1983b): *Expansion of Urban Area in the United States: 1960–1980*. ERS Staff Report AGES830615. Washington: US Department of Agriculture, Economic Research Service.

———— (1982): *Major Uses of Land in the United States: 1978*. AER–487. Washington: US Department of Agriculture, Economic Research Service.

———— (1973): *Major Uses of Land in the United States: Summary for 1969*. AER–247. Washington: US Department of Agriculture, Economic Research Service.

———— and Roger Hexem (1985): *Major Uses of Land in the United States: 1982*. Aer–535. Washington: US Department of Agriculture, Economic Research Service.

Healy, Robert G. (1984): *Competition for Land in the American South: Agriculture, Human Settlement, and the Environment.* Washington: The Conservation Foundation.

Heimlich, Ralph E. Personal communication, March 26, 1991.

US Department of Agriculture, Economic Research Service (1976): *Changes in Farm Production and Efficiency: A Summary Report.* Statistical Bulletin 561.

——— Economic Research Service (1990a): *Economic Indicators of the Farm Sector: Production and Efficiency Statistics, 1988.* ECIFS 8–5.

——— Economic Research Service (1990b): *Agricultural Resources: Cropland, Water and Conservation.* AR 19.

US Department of Agriculture, National Agricultural Statistics Service (1991): *Crop Production: 1990 Summary.* CrPr 2–1 (91).

US Department of Commerce, Bureau of the Census (1962): *A Graphic Summary of Land Utilization.* Special Reports, vol. V, part 6, chapter 1.

——— Bureau of the Census (1975): *Historical Statistics of the United States, Colonial Times to 1970.* Bicentennial Edition, Part 1.

——— (1992): *Statistical Abstract of the United States: 1992.* Washington: Government Printing Office.

Wooten, H. H. (1953): *Major Uses of Land in the United States.* TB–1082. US Department of Agriculture, Bureau of Agricultural Economics.

——— (1953): *Supplement to Major Uses of Land in the United States.* TB–1082. US Department of Agriculture, Bureau of Agricultural Economics.

——— and J. R. Anderson (1957): *Major Uses of Land in the United States: Summary for 1954.* AIB–168. US Department of Agriculture, Agricultural Research Service.

———, Karl Gertel, and William C. Pendleton (1962): *Major Uses of Land and Water in the United States: Summary for 1959.* AER–13. US Department of Agriculture, Economic Research Service.

PART V

Pollution and the Environment

43

Long-Run Trends in Environmental Quality

William J. Baumol and Wallace E. Oates

When we undertook this study several years ago, we had definite preconceptions about the general trends in environmental quality.* Because of growth in population and industrial activity, we were convinced that virtually all forms of environmental damage were increasing and that, in the absence of powerful countermeasures, they would continue to accelerate more or less steadily. A preliminary study of available data seemed to support this view.[1] However, a more careful and extensive reexamination of the evidence has led us to revise this simplistic view of the course of environmental decay. We have found on closer study that the trends in environmental quality run the gamut from steady deterioration to spectacular improvement. This chapter presents our accumulated data on environmental trends.

COLLECTING EVIDENCE ABOUT THE ENVIRONMENT

Because widespread and systematic concern about environmental issues is relatively recent, it was not surprising that we found it difficult to obtain reliable evidence on environmental trends extending back more than a few years. Since the experience of a short period of time often can be deceiving and heavily colored by transient, irrelevant influences, we could not rely on easily accessible, short-term information.[2] The process of tracking down longer-term evidence took us beyond libraries to repositories of dusty records, to the Swedish Fisheries Bureau in

* This chapter was written at a much earlier date for another volume. The authors are happy to have it reappear here, but note that they should not be assumed to subscribe to all the conclusions reached in this article. As environmentalists, it is their conviction that public policy must play an active role in the protection of our natural resources. [The editors, and probably all the other contributors also, are not in principle against government intervention in any particular situation requiring environmental protection. – *Ed.*]

Gothenberg, to the offices of St Paul's Cathedral in London, and to the conservationists' offices in the Louvre. We want to emphasize that our survey in this chapter presents *all* the long-term data that we have been able to discover; systematic evidence over past centuries (or even decades) is, indeed, very scarce. However, what we have found is, in some instances, quite intriguing. Our searches often led to dead ends, but in other cases turned up evidence that revealed that broad statements reported in the popular press were often either misleadingly simplistic or completely untrue.[3] This has forced us to revise our earlier, naive view that environmental deterioration has been a universal, accelerating process whose source is modern industrialization and population growth.

Environmental deterioration caused by natural processes. Some environmental damage, whose source at first appears to be industrial pollution, is in fact primarily the result of natural forces. An instructive illustration is the quality of the deeper waters of the central Baltic Sea. The data in figure 43.1 indicate that, in the central waters of the Baltic roughly midway between Stockholm and Helsinki, the oxygen content has been falling steadily from about 300 ml/l at the beginning of the century to virtually zero today. The data depicted in the diagram give the figures for only one of the sampling stations, but nearby stations show very similar trends. From such figures, some observers have concluded that the Baltic is becoming a "dead sea;" the inference has been that the pollution from its shores is destroying the Baltic.[4]

Yet it is by no means clear how closely the trends in oxygen content are related to pollution. For example, the northern part of the Bothnian

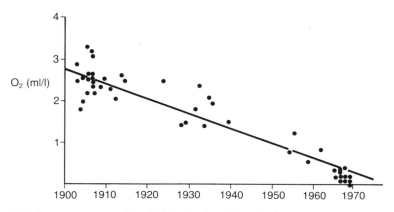

Figure 43.1 Oxygen content of the Baltic Sea (in milliters per liter), Station F 74 (depth of approximately 150 meters)

Sources: Compiled from data from: (a) Conseil Permanent International pour l'Exploration de la Mer (ICES), *Bulletin des Résultats acquis pendant les Courses Périodiques*, Copenhagen (up to 1959); (b) *ICES Oceanographic Data Lists*, Copenhagen (1959–62); and (c) *Hydrographical Data*, Harsfiskelaboratorret, Götenborg (1963–69).

Gulf of the Baltic, which is quite shallow, receives a considerable influx of pollutants from nearby paper plants[5] and is consequently one of the portions of the Baltic most heavily subjected to oxygen-demanding effluents. Yet the data on the oxygen content of this portion of the Baltic (see figure 43.2) seem to exhibit no trend such as that displayed in the previous diagram. According to Stig H. Fonselius, the primary reason for the decrease in oxygen content in the central Baltic has been an increase in salinity.[6] Moreover, he attributes this rise in salinity to meteorological factors.[7] Specifically, "The main reason[s] . . are changes in the atmospheric circulation which have been observed over a long period . . . [causing] a decrease in precipitation resulting in a corresponding decrease in runoff [that is, a diminution in the inflow from rivers feeding into the Baltic]."[8] Examining the data on three rivers in the eastern Baltic region going back to the nineteenth century, he concludes, "There seems to be a general decreasing trend of the runoff in all three rivers from the beginning of the 20th century. If this is a true trend and it does continue there is not much hope for improved oxygen conditions in the Baltic deep water."[9]

Another rather curious illustration of environmental deterioration not attributable to recent human abuse is the case of "Cleopatra's Needle," the obelisk now standing in New York City's Central Park. Three sides of this monument are badly eroded, and the damage is often attributed to air pollution and continuous vibration from nearby traffic. We learned, however, that the obelisk, which originally stood at Heliopolis

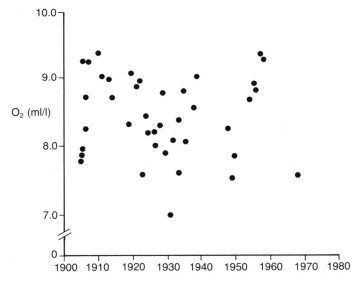

Figure 43.2 Oxygen content (in milliters per liter) at 100 meters depth at Station F 12 in the Bothnian Bay of the Baltic Sea between 1900 and 1968
Source: S. H. Fonselius, *Hydrography of the Baltic Deep Basins, III*, Fishery Board of Sweden Series Hydrography Report no. 23 (Lund, Sweden: Carl Bloms Boktryckeri, 1969), 46.

on the east bank of the Nile River, was tipped over by Persian invaders and remained on its side for some five and a half centuries until the Roman emperor, Augustus, had it moved and re-erected at Alexandria. From there it was brought to New York in 1880. According to E. M. Winkler, "The present east (undamaged) face probably faced downward during the monument's prostrate position between 500 B.C. and about 43 B.C." During that period there was "capillary migration from the iron-rich ground water on the flood plain silt of the Nile River."[10] That is, the stone absorbed Nile water and accumulated salt, which through normal capillary action was stored in the portions of the stone farthest away from the point of entry (that is, close to the other three sides). As a result "more than a few hundred pounds of granite flakings off the obelisk were cleaned up after a few years of exposure to the moisture-loaded atmosphere of New York City, which caused the hydration and expansion of the salts entrapped in the capillaries."[11] Much of the remaining damage to the monument is probably attributable to "the strong abrasive action of drifting sand" while the monument stood in Heliopolis and Alexandria. Thus, Winkler concludes that "the disastrous disintegration of granites in city atmospheres, as exemplified by Cleopatra's Needle, is a myth. It is therefore hoped that the obelisk will be eliminated from textbooks of physical geology as 'a good example of weathering in cities.' "[12]

Not all damage to our environment can be traced to economic growth and industrialization.

Environmental deterioration as an historical phenomenon. Another thing history makes very clear is that pollution is not a modern invention; technological developments have not always been an unmixed curse upon the environment. When the automobile began to replace the horse and rid the streets of odorous dungheaps, it was hailed as a major contributor to public health and sanitation. Certainly the modern city, whatever its state of cleanliness, is an improvement on the relatively tiny, but incredibly filthy, streets and waterways of medieval and Renaissance cities. It is reported that in about 1300 under Edward I, a Londoner was executed for burning sea coal in contravention of an Act designed to reduce smoke. During the reign of Edward I's grandson, some seventy years later, we find the following proclamation, one of many designed to protect the cleanliness of late medieval rivers and streets, all apparently equally unsuccessful (the reader will note especially the remarkable, if not wholly credible, assertion which we have italicized):

Edward, by the grace of God etc., to our well-beloved, the Mayor, Sheriffs, and Aldermen, of our City of London, greeting, *Forasmuch as we are for certain informed that rushes, dung, refuse, and other filth and harmful things, from our City of London, and the suburbs thereof, have been for a long time past, and are daily, thrown into the water of Thames, so that the water aforesaid, and the hythes thereof, are so greatly obstructed, and the course of the said water so greatly narrowed, that great ships and vessels are not able, as of old they were wont, any longer to come up to the same city, but are impeded therein*; to the most grievous damage as well of ourselves as of the city aforesaid, and of all the nobles and others of our people to the same city resorting; we, wishing to provide a fitting remedy in this behalf, do command you, on the fealty and allegiance in which unto us you are bound, strictly enjoining that, with all the speed that you may, you will cause orders to be given that such throwing of rushes, dung, refuse, and other filth and harmful things, into the bed of the river aforesaid, shall no longer be allowed, but that the same shall be removed and wholly taken away therefrom; to the amendment of the same bed of the river, and the enlarging of the watercourse aforesaid; so behaving yourselves in this behalf, that we shall have no reason for severely taking you to task in respect hereof. And this, as we do trust in you, and as you would avoid our heavy indignation, and the punishment which, as regards ourselves, you may incur, you are in no wise to omit. Witness myself, at Prestone, the 20th day of August, in the 46th year of our reign in England, and in France the 33rd.[13]

A description of the quality of the atmosphere in London in 1700 reminds us of its state two and a half centuries later:

. . . the glorious Fabrick of St. Paul's now in building, so Stately and Beautiful as it is, will after an Age or Two, look old and discolour'd before 'tis finish'd, and may suffer perhaps as much damage by the Smoak, as the former Temple did by the Fire.[14]

The author goes on to point out:

By reason likewise of this Smoak it is, that the Air of the City, especially in the Winter time, is rendered very unwholesome: For in case there be no Wind, and especially in Frosty Weather, the City is cover'd with a thick *Brovillard* or Cloud, which the force of the Winter-Sun is not able to scatter; so that the Inhabitants thereby suffer under a dead benumming Cold, being in a manner totally depriv'd of the warmths and comforts of the Day . . . when yet to them who are but a Mile out of Town, the Air is sharp, clear, and healthy, and the Sun most comfortable and reviving.[15]

It is thus important to recognize that modern, industrialized society has no monopoly on pollution and environmental damage from either human sources or natural forces. Evidence of significant environmental damage does not necessarily mean that the damage is growing, and evidence that deterioration is growing may not mean that the source of the problem is human activity. We do not intend here to deny the seriousness of environmental decay, but rather to point out the complexities besetting an understanding and interpretation of levels and trends in environmental quality. With this in mind, we turn to the presentation of our accumulated data on these trends.

CASES OF MIXED OR IMPROVING
ENVIRONMENTAL TRENDS

Our examination of the available facts suggests that the trends in environmental damage are far less uniform than we had initially expected. This section presents some examples in which environmental quality is actually improving or in which a varied pattern of deterioration and/or improvement is apparent.

The Great Lakes. The deteriorating water quality of two of the most vulnerable of the Great Lakes (in terms of population and water volume), Lake Erie and Lake Ontario, has received widespread attention. Figures 43.3–43.5 show trends for these two lakes in the concentrations of three substances closely tied to industrial and municipal pollution: dissolved solids, sulfates, and calcium.[16] The figures show very clearly that, at least since early in the twentieth century, these concentrations have been increasing dramatically. Figures on concentrations of other substances in these lakes (chloride, sodium, and potassium, also associated with industrial wastes) tell much the same story (see figure 43.6). But the water quality of Lakes Erie and Ontario differ sharply from that of the other Great Lakes. Although Lakes Michigan and Huron have

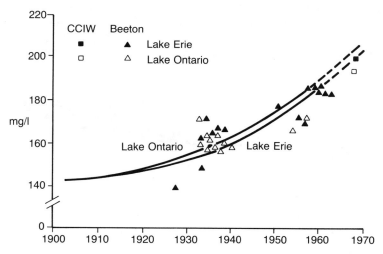

Figure 43.3 Changes in the concentrations of total dissolved solids in Lakes Erie and Ontario
Sources: Proceedings of the Conference on Changes in the Chemistry of Lakes Erie and Ontario, 5–6 Nov., 1970, *Bulletin of the Buffalo Society of Natural Sciences* 25(2), (1971). In the key CCIW refers to Canada Centre for Inland Waters: Beeton refers to A. M. Beeton, "Eutrophication of the St Lawrence Great Lakes," *Limnology and Oceanography* 19 (1965): 240–54; and Kramer refers to J. R. Kramer, "Theoretical Model for the Chemical Composition of Fresh Water with Application to the Great Lakes," in *Great Lakes Research Division* (Ann Arbor, Michigan: University of Michigan, 1964), 147–60.

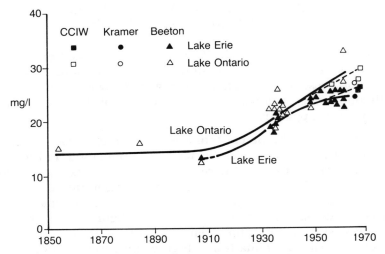

Figure 43.4 Changes in the concentrations of sulfate in Lakes Erie and Ontario
Source: See figure 43.3.

suffered from an increase in dissolved solids (see figure 43.7), this growth has been far slower than in Lakes Erie or Ontario. Moreover, Lake Michigan has suffered no growth in calcium content, and its concentrations of sodium and potassium leveled off soon after the turn of the century. Lake Superior's purity seems to have been increasing rather steadily or holding constant, at least in terms of the dissolved solids for which we have data. Figure 43.6 shows the great diversity in

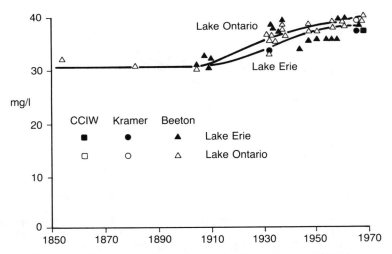

Figure 43.5 Changes in the concentrations of calcium in Lakes Erie and Ontario
Source: See figure 43.3.

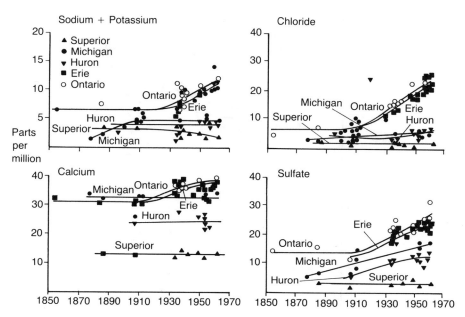

Figure 43.6 Changes in the concentrations of sodium-plus-potassium, chloride, calcium, and sulfate in each of the Great Lakes
Source: Beeton, "Eutrophication of the St Lawrence Great Lakes," 248.

calcium, chloride, sodium, and potassium concentrations over time in the Great Lakes, with these chemicals apparently presenting an increasing problem only in Lakes Erie and Ontario. Sulfates, on the other hand, have increased in concentration in every one of the lakes except Lake Superior, the largest, deepest, and most isolated of the lakes.[17] A reporter could paint the most dismal picture of the state of the Great Lakes by singling out Lakes Erie and Ontario, while the investigator who chooses to study only Lake Superior could easily conclude that there is no cause for alarm.[18]

Water quality in the New York City Harbor. The quality of the waters surrounding New York City has a mixed history. Figure 43.8 shows the trend in dissolved oxygen concentrations in five waterways surrounding the city for the years 1910–70.[19] (Any biodegradable emissions such as human wastes, food products, or waste paper are gradually transformed and assimilated by natural processes that use up oxygen. A waterway heavily polluted by biodegradable emissions will, therefore, tend to have a relatively low dissolved oxygen content.)[20] The decline indicated by the graph for the period prior to 1920 suggests that oxygen-using pollution in the city's waterways increased rapidly during the first years of the twentieth century. In about 1920, however, the process of deterioration suddenly halted. For the next three decades, the dissolved-oxygen level

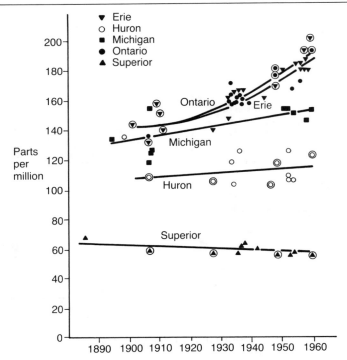

Figure 43.7 Concentrations of total dissolved solids in the Great Lakes (circled points are averages of 12 or more determinations)
Source: Beeton, "Eutrophication of the St Lawrence Great Lakes," 246.

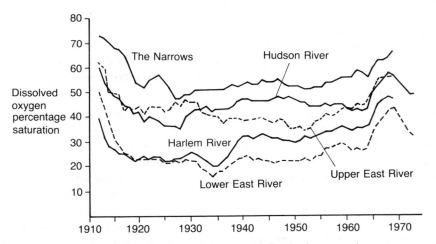

Figure 43.8 Five-year moving averages of the annual dissolved oxygen (as percentage of saturation) for the main branches of New York Harbor
Source: *New York Harbor Water Survey, 1970,* provided by City of New York, Environmental Protection Administration, Dept of Water Resources, Bureau of Water Pollution Control.

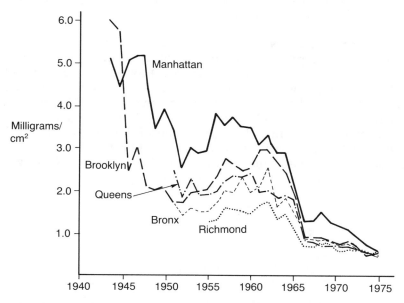

Figure 43.9 Settleable particulate matter for five New York City boroughs
Source: Dept of Air Resources, New York.

remained relatively stable, or even improved slightly. It has been suggested that the decrease in dissolved oxygen up to 1917 was the result, not of increased industrial activity in the area, but of rapid population growth. This population growth increased the amount of sewage dumped virtually untreated into the city's waterways. According to this view, it was the restriction of immigration from abroad that accounted for the stabilizing of water quality in New York's rivers. Unfortunately this explanation is only partly tenable at best, since the population of the city did continue to grow at least until 1930.[21]

In any event, by the early 1950s a major expansion in waste treatment facilities was begun, first in the East River and later in the Hudson. By the middle of the 1960s, dissolved oxygen concentrations had increased sharply in all five of the waterways described in the graph. Indeed, there was observable improvement throughout the Hudson River:

> almost all fishermen, marine biologists, environmentalists and government officials [are] agreeing that the river is cleaner now than it has been in recent years.
> By almost every measure available – amount of money spent, number of sewage treatment plants constructed, number of crabs returning, number and size of fish, visibility of sewage, number of people swimming – the 155-mile-long main stem of the Hudson River between New York City and Troy is improving.
> In the fishing season just ending, fishermen took in more and bigger

blue-claw crabs, hauled in bigger weakfish, caught large numbers of "lafayettes" for the first time in 30 years. . . .[22]

US air quality trends. New York City's air quality has shown similar improvement in terms of certain pollutants. Figure 43.9 shows the trends in settleable particulate matter (soot) for all five boroughs of the city during the postwar period. The change is certainly startling. In Brooklyn the figure has fallen to about one-sixth of its 1945 level, and in Manhattan it has declined by more than two-thirds. Similarly, there has been a decline in the sulfur dioxide content of the atmosphere, not only in New York but in other major cities as well. The data depicted in figure 43.10 show a dramatic improvement in New York and Chicago and

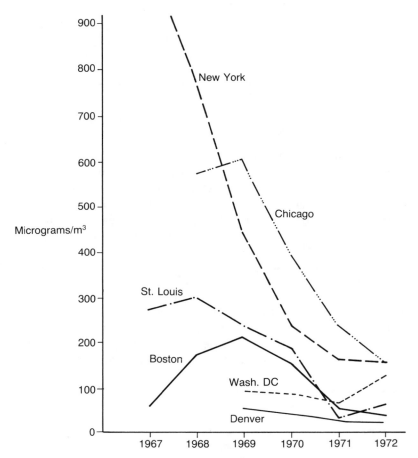

Figure 43.10 SO$_2$ air quality data for six US cities (micrograms of SO$_2$ per cubic meter) *Source*: Derived from *Environmental Quality, The Fourth Annual Report of the Council on Environmental Quality* (Washington, DC: US Government Printing Office, 1973), 273. They cite as their source: EPA data from the National Air Sampling Network.

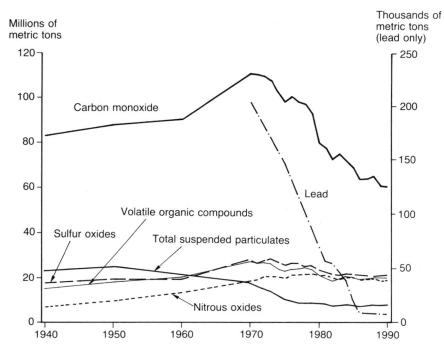

Figure 43.11 National air pollutant emissions
Source: Statistical Abstract of the United States, 1992.

more modest gains in the other cities such as Boston and St Louis (although the latter two cities exhibited something of a reversal between 1971 and 1972). This sort of evidence suggests that environmental policy can be effective and can produce results that are both rapid and substantial.[23] Figure 43.11 depicts national emission trends for six major air pollutants for the period 1940–90. The data indicate that between 1940 and 1970 emissions of five of these pollutants were generally increasing. Between 1970 and 1990, however, particulate emission levels dropped by about 59 percent, sulfur dioxide declined 25 percent, carbon monoxide dropped 41 percent, lead dropped 97 percent, and volatile organic compounds dropped 25 per cent, while only one pollutant rose – nitrous oxides, by 6 percent. (And nitrous oxide emissions fell 6 percent from 1980 to 1990.)

London: air and water quality trends. Air quality in London displays striking parallels to the New York experience. London's long history of air pollution was largely the result of the heavy use of coal fires for both industrial and domestic heating. One of the consequences was the prevalence of filthy fogs, sometimes in garish colors produced by the chemical content of the air; such fogs became as much a symbol of London as Westminster Abbey and Big Ben. Only after the disaster of December 1952, during which in a two-week period abnormally high concentrations of sulfur dioxide and smoke resulted in an estimated 4,000 excess deaths,[24] did Parliament adopt more stringent regulations for the protection of the atmosphere. The Clean Air Act of 1956 and subsequent legislation established smoke-control zones in which strict codes governed allowable smoke emissions.[25] The results have been impressive: pea-soup fogs have disappeared, and the number of hours of sunshine in London has climbed significantly. [For detailed discussion of air pollution trends in the UK, see the chapter in this volume by Derek Elsom.]

The English have also made considerable progress in the cleanup of some rivers and estuaries. We noted earlier the marked increase in recent years in the dissolved-oxygen content of New York City's rivers. Figure 43.12 depicts similar trends in levels of dissolved oxygen in the Thames near London. In contrast to the period of deterioration from the 1930s until 1954, when the dissolved oxygen content of the Thames just downstream from the City was close to zero, the curve for 1969 indicates a marked improvement. The Royal Commission on Environmental Pollution reported:

> The oxygen content of the Thames for some 10 miles above and 30 miles below London Bridge had been diminishing for decades, and the consequences were beginning to be very serious. In 1949 the Water Pollution Research Laboratory began an investigation into the causes of the deterioration. When these were diagnosed, the Port of London Authority

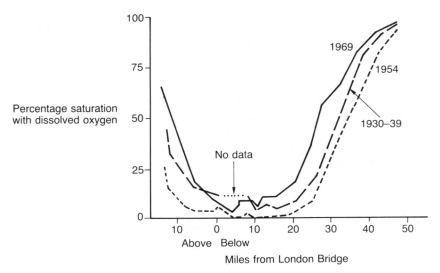

Figure 43.12 Analysis of water of river Thames: percentage saturation with dissolved oxygen at high water (average July–Sept.)

Source: Royal Commission on Environmental Pollution, *First Report* (London: Her Majesty's Stationery Office, Feb. 1971), 20.

launched a programme to improve the quality of the water. The success of this programme has been shown by the return of many kinds of fish. In 1957–58 a survey showed no fish between Richmond (15 miles above London Bridge) and Gravesend (25 miles below). By 1967–68 some 42 species were present and migratory forms were able once again to move through the polluted zone.[26]

These cases support the view that environmental policy can work. It does not follow, however, that the battle against pollution is essentially won or that improvement is universal and unambiguous. There are important and documented instances of progressive decline in environmental quality. In the next section, we will discuss some evidence of growing human abuse of the environment.

CASES OF DETERIORATING ENVIRONMENTAL QUALITY

Another look at urban air quality.   In the previous section we described the decreasing levels of atmospheric sulfur dioxide and particulate matter in New York and other cities as an example of improving air quality.[27] The trends in other measures of air quality are not so satisfying. Figure 43.13, for example, shows that concentrations of carbon monoxide in New York City, apart from short-term fluctuations, remained constant from 1958 to 1976. This is also largely true for most European cities. Similarly, figure 43.14, which depicts trends in the levels of suspended particulates in five major cities from 1967–72, shows that other cities have not done as well as New York in this respect. Indeed, the air in Denver has apparently been growing steadily more polluted in terms of particulate matter.[28]

Trends in atmospheric lead pollution.   A striking case of increasing environmental deterioration is the steady growth of lead concentrations in the earth's atmosphere. Analyses of ice layers in the Arctic and Antarctic

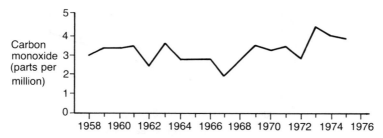

Figure 43.13 Carbon monoxide concentrations in New York City, 1958–75 (parts per million)
Source: Dept of Air Resources, New York. Figures represent average annual measurements taken at Station Laboratory 121 (located at 121st St) fifteen feet above street level. This station was chosen because it is the oldest in the city and its data are more complete. Most of the air pollutant measuring stations in New York City began operation after 1969.

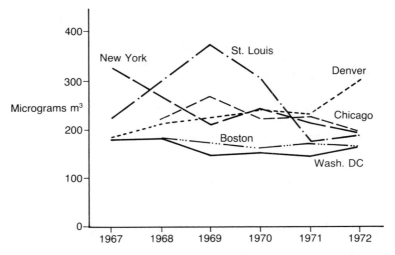

Figure 43.14 Total suspended particulates for six cities, 1967–72 (micrograms per cubic meter)
Source: *Environmental Quality, The Fourth Annual Report of the Council on Environmental Quality* (1972), 273.

regions have produced estimates of long-term trends in lead concentrations (and other pollutants). Figure 43.15 shows the findings of one of the most recent, and apparently most systematic, of these studies.[29] The graph, which covers over twenty-five centuries, is certainly startling. It

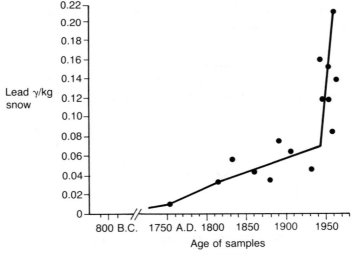

Figure 43.15 Industrial lead pollution at Camp Century, Greenland, since 800 BC, micrograms per kilogram
Source: M. Murozumi, T. J. Chow, and C. Patterson, "Chemical Concentrations of Pollutant Lead Aerosols, Terrestrial Dusts and Sea Salts in Greenland and Antarctic Snow Strata," *Geochimica et Cosmochimica Acta* 33, 10 (Oct. 1969), 1285.

shows that, even in the remote regions of northern Greenland, there has been a persistent, accelerating rise in lead pollution; today lead concentrations at Camp Century, Greenland, are well over 500 times "natural" levels.[30] The study's measurement of other "impurities" in the ice samples showed no discernible trends; that is, of the seven items whose concentrations were estimated as far back as 800 BC, it was lead alone that exhibited any long-term growth pattern. It should be noted that even the carefully gathered evidence of the Greenland study has been criticized on two grounds: first, similar results were not obtained in the Antarctic samples, and, second and perhaps more serious, the Greenland figures may reflect a very special source of lead contamination: "a major US military base was established in Thule, Greenland, during World War II and a large camp was set up between 1959 and 1960 at Camp Century. This is only 80 [kilometers] from the Virgin Trench site where samples of snow, dated 1952–65, were collected. The base was supplied by aircraft using leaded fuel . . . [thus] snow at the Trench site may well have been contaminated by these activities" (A. L. Mills, "Lead in the environment," *Chemistry in Britain*, 7 [April 1971], 161). It may be relevant to note that Mr Mills is chairman of the Institute of Petroleum's advisory committee on health. Both charges have evoked responses. The difference in the results for the Northern and Southern studies has been attributed to "barriers to north–south tropospheric mixing . . . which hinder the migration of aerosol pollutants from the northern hemisphere to the Antarctic" (Murozumi, Chow, and Patterson, "Chemical concentrations," p. 1247). As for the second charge, an authoritative defender of the Greenland study has replied, "Firstly, the greatest amounts of lead were found to be deposited in the winter months when precipitation was heaviest and air traffic lightest. Secondly, the lead levels from sites between the Virgin Trench and the bases showed no elevation attributable to significant contamination from the bases. Thirdly, the Virgin Trench site was predominantly upwind from the bases. It is also worth noting that the major increase in ice-lead levels began about 20 years before the closer base was established" (D. Bryce-Smith, "Lead pollution from petrol," *Chemistry in Britain*, 7 [July 1971], 285).

The figures constitute grounds for genuine concern. Lead has been described as

one of the most insidiously toxic of the heavy metals to which we are exposed, particularly in its ability to accumulate in the body and to damage the central nervous system including the brain. . . . It inhibits enzyme systems necessary for the formation of haemoglobin . . . and has been said to interfere with practically any life-process one chooses to study. Children and young people appear specially liable to suffer more or less permanent brain damage, leading [among other things] . . . to mental retardation, irritability and bizarre behavior patterns. . . . More serious occupational exposure can lead to insanity and death. . . .[31]

Further evidence of long-term increases in lead concentrations is provided by a recent study in Peru comparing the lead content of six-century-old human bones with more modern samples. Lead concentrations in the modern bones were, on the average, over ten times as high as those in the earlier ones.[32] However, a Polish study showed ". . . that the levels of lead in modern Polish bones do not differ significantly from those found in bones from the third century, although levels in the Middle Ages were often very high."[33] While there is strong evidence that the prevalence of this poison has increased markedly over the centuries, it is still impossible to reach completely unqualified conclusions. However, at least in the United States, high emissions and concentrations of airborne lead pollution in earlier decades of this century have given way, with recognition of possible health hazards, to sharply downward trends. [See editors' appendix.]

The accumulation of solid waste.   We turn next to a source of indisputable environmental deterioration: the burgeoning *quantity* of solid wastes that society produces. Growth in population and in output per capita can be expected to increase the amount of solid waste. While time series for long-term trends in this area are not easy to obtain, the evidence available indicates that this has, indeed, been true. It is reported, for example, that in the 1960s the flow of solid wastes in New York City was increasing 4 percent per year.[34] We have collected some data for Cincinnati that go back more than 40 years. Figure 43.16 indicates that the amount of solid waste collected in Cincinnati has grown at an average rate of about 4.5 percent per year.[35] If records were obtained for other cities, they would no doubt show very similar results. There is no

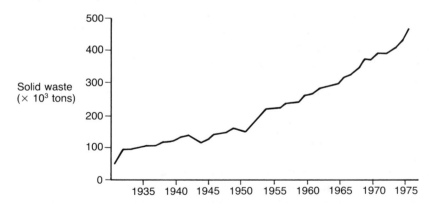

Figure 43.16   Total tonnage of solid waste received at city disposal sites, Cincinnati, Ohio, 1931–76
Source: City of Cincinnati, Dept of Public Works, Division of Waste Collection, Letters of 9 Nov. 1973, and 17 Feb. 1977, from R. D. Behrman, Administrative Assistant, and A. H. Schuck, Acting Superintendent.

question that this trend poses ever-increasing problems for society. Already many cities are having trouble disposing of the mounting heaps of trash. Neighboring areas are reluctant to serve as the cities' dumps, and locations near cities that are suitable for landfill operations are getting scarce. Other methods of waste disposal are now recognized to create problems of their own. Burning garbage pollutes the air, while treatment of liquid wastes leaves a sludge which must be disposed of. Moreover, we are learning that dumping wastes into the ocean nearby is not costless to society; sludge dumped into the sea can kill or contaminate marine life and pollute nearby waters and beaches. The changing composition of solid wastes also adds to the problem of disposal. For example, plastics (which are nondegradable and often have harmful combustion properties) make up an increasing percentage of solid wastes. All in all, the problem of solid waste disposal can hardly be viewed with equanimity; it surely represents a major environmental problem that is likely to grow worse.

CONCLUSIONS

The trends in environmental deterioration are varied and uneven. While the evidence presented in this chapter may undermine some of the more rash and unqualified predictions of imminent ecological disaster, there is no justification for complacency and inaction. Unquestionably, various types of environmental damage directly associated with human activity have grown rapidly and without interruption for a very long period of time. Some of them, with little doubt, produce serious consequences. Certain forms of pollution, besides making existence uglier and far less pleasant, have almost certainly increased the frequency of illness and added significantly to death rates. Some forms of pollution may even pose serious hazards for human survival. In recent decades facts have often caught up with and surpassed the inventions of science fiction; some of the more bizarre horrors threatened by environmental abuse cannot be ruled out with any high degree of confidence. But scare tactics are not necessary to make a case for strong environmental policy. The demonstrable ill effects of pollution on health and longevity, despite the untidy diversity of trends that accompany them, surely justify the adoption of effective countermeasures.

APPENDIX A: A LESSON FROM VERY LONG TIME SERIES: THE CASE OF THE NILE

This chapter has presented a number of time series for environmental data, some of considerable duration. Yet from a historical point of view they are relatively brief; few of them extend more than 75 years.

In one area, the study of climate, time-series data are available over extraordinarily long periods of time.[36] From the evidence of tree rings, glaciers, and other sources, experts have accumulated figures spanning hundreds of years. One of the most remarkable of these series provides the annual figures on the height of the Nile River. They seem not to have been assembled by scientists, but rather by tax authorities who based their tax levies on these data as an indicator of the agricultural prosperity of the Nile valley. Before presenting a graph that summarizes a substantial portion of the data (which extend more than seven centuries), let us consider briefly the data in figure 43.17, which gives, in five-year averages, the behavior of the annual low-water mark of the Nile River. The pronounced and steady downward trend over the period of more than 30 years should be clear enough. It may suggest the onset of a period of drought. Surely, the consistency of its decline presages unpleasant things for the succeeding years. An observer looking at this trend might well project dire consequences for the future of the river valley and its inhabitants.

Yet the companion figure 43.18, which gives the trend of the *high-water* levels over the same period, already provides grounds for doubt. There is still something of a downward trend, but it is not nearly so pronounced or persistent as the annual minima. Moreover, figures 43.19 and 43.20 tell quite a different story; they repeat the data of figures 43.17 and 43.18 along with their sequels. The time paths can hardly be described as steady downward trends.

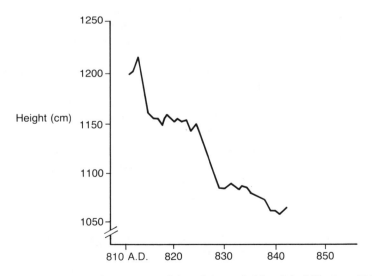

Figure 43.17 Five-year moving averages of the minimum height of the Nile river, 810–50
Source: Prince Omar Tousson, "Mémoire sur l'histoire du Nil," Mémoires présentés, à l'Institut d'Egypt et Publiés sous les Auspices de sa Majesté Fouad 1er, Roi d'Egypt, 10 (Cairo: Institut Français d'Archéologie Orientale, 1925), 361–411.

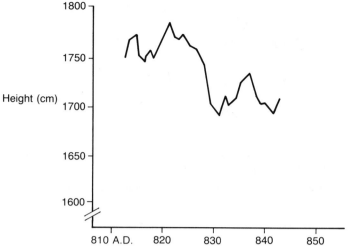

Figure 43.18 Five-year moving averages of the maximum height of the Nile river, 810–50
Source: See figure 43.17.

The full history, however, is revealed by our last graph, figure 43.21, which shows the variations in the height of the Nile from 641 through 1451. It is certainly not easy to discern any sharp trends in the data. The moral should be clear: it is dangerous to extrapolate from a consistent trend in a data series, even one persisting over decades.

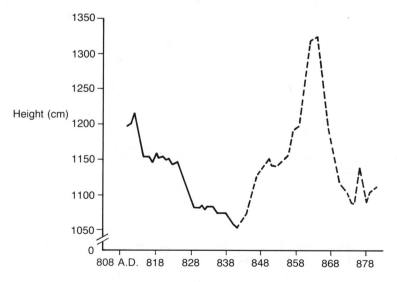

Figure 43.19 Five-year moving averages of the minimum height of the Nile river, 808–78
Source: See figure 43.17.

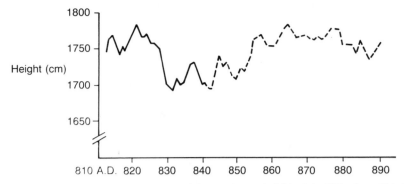

Figure 43.20 Five-year moving averages of the maximum height of the Nile river, 810–90
Source: See figure 43.17.

NOTES

1 See William J. Baumol, "Environmental protection, international spillovers and trade," *The Wicksell Lectures for 1971* (Uppsala, Sweden: Almquist & Wiksells Boktryckeri Ab, 1971), pp. 15–16. These pages also represented the view at that time of Oates, who had read and commented extensively on the manuscript during its preparation.
2 For an illustration of the dangers in drawing conclusions about trends from data spanning even many decades, see Appendix A which examines water-flow figures for the Nile River that cover many centuries.
3 A case in point is the matter of the progressive deterioration of works of art that are located out of doors. We had assumed from numerous studies in the press that there had been a marked acceleration in decay in recent years coincident with growing pollution. However, interviews with some of the world's leading authorities soon made our naiveté apparent. Where there is deterioration of these works of art, the causes are not completely understood and in some cases are clearly attributable in good part to natural phenomena. Even where visible deterioration has increased sharply in recent years, it is not safe to assume that the cause is recent. For example, a piece of stonework may have been decaying beneath its surface for centuries; when the weakened structure finally collapses, it hardly can be blamed on twentieth-century abuse. The experts did *surmise* that chemicals emitted into the atmosphere do increase the incidence of "stone sickness" which leads to the crumbling of buildings and sculptures, but repeatedly emphasize the absence of conclusive evidence confirming this plausible conjecture.
   There are, of course, a few noteworthy exceptions – cases in which the evidence of accelerating deterioration is persuasive. For example, there are the casts of some portions of the Parthenon Frieze made by Lord Elgin in the early nineteenth century. A comparison of those casts with the originals in Athens confirms that, in the sixteen decades since Elgin was in Greece, the marbles have indeed suffered enormous visible deterioration relative to what they underwent in the more than twenty centuries before. Details that were sharp and clear in 1800 are virtually unrecognizable today (see the photographs in H. J. Plenderleith, *The Conservation of Antiquities and Works of Art*, 2nd edn

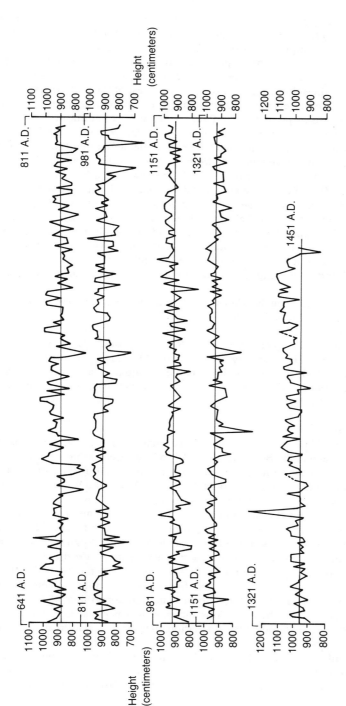

Figure 43.21 Variations in height of Nile Flood. The sloping line indicates the secular raising of the bed of the Nile by the deposition of silt
Source: C. E. P. Brooks, "Periodicities in the Nile Floods," *Memoirs of the Royal Meteorological Society*, 2(12) (Jan. 1928), 10.

(London: Oxford University Press, 1971), plates 40A and B, p. 316). However, except for catastrophes such as the Florentine flood, there are few other documented examples of recent acceleration in deterioration.

Despite all that has been printed and said about the threat of modern industry to our artistic heritage, we were able to find very little *conclusive* evidence on the subject one way or another. Most of the authorities with whom we spoke were not even willing to conjecture that matters were getting worse. A notable case in point was St Paul's Cathedral, London, at the end of the seventeenth century when St Paul's was under construction, could hardly boast about the purity of its atmosphere. For example, a German tourist reported in 1710 that in order to see the vista from the tower of the cathedral, he had to get there very early in the morning "in order that we might have a prospect of the town from above before the air was full of coal smoke" (*London in 1710, From the Travels of Zacharias Conrad von Uffenbach*, trans. ed. W. H. Quarrell and Margaret Mare (London: Faber and Faber, 1934), pp. 31–2). It is not surprising, therefore, that the new cathedral had to be cleaned within three decades after construction was started and before it was even finished. How, then, can one conclude with any degree of confidence that architectural stonework in London is being damaged more heavily today than it was three centuries ago?

4 See, for example, Stig H. Fonselius, "Stagnant Sea," *Environment*, 12 (July–August 1970), 2–3, 6, 28, 40.

5 Ibid., p. 42.

6 Stig H. Fonselius, *Hydrography of the Baltic Deep Basins, III*, Fishery Board of Sweden Series Hydrography Report No. 23 (Lund, Sweden: Carl Bloms Boktryckeri, 1969), p. 91. He lists, as the secondary reason for the increase, a rise in phosphorus concentrations which may well be related to the influx of sewage (pp. 90–1).

7 Ibid., p. 90.

8 Ibid., p. 56.

9 Ibid., p. 62. There is also some evidence suggesting that low-oxygen problems may have plagued the Baltic in the middle of the nineteenth century, well before the onset of extensive industrial activity. For the Vuoksi River in Finland, the uninterrupted data on runoff go back to 1847. While the Vuoksi does not itself flow into the Baltic, its runoff figures since 1900 follow a time pattern very similar to those for the two other rivers for which data are available only since 1900, a relationship that is hardly surprising since they are all presumably replenished from the same regional sources. This is significant, because it suggests that the behavior of the other rivers feeding into the Baltic could be expected to have paralleled that of the Vuoksi for the second half of the nineteenth century. Moreover, the figures suggest that, during the period 1847–1900, the runoff of the Vuoksi exhibited something of a rising trend. Indeed, in the 1850s the runoff was almost as low as it has been in recent times; this raises the likelihood that stagnation may have occurred in the Baltic in earlier periods too.

10 E. M. Winkler, "Decay of Stone," *International Institute for Conservation, 1970 New York Conference on Conservation of Stone and Wooden Objects, I* (London: The International Institute for Conservation of Historic and Artistic Works, 1971), p. 6.

11 Ibid.

12 E. M. Winkler, "Weathering rates as exemplified by Cleopatra's Needle in New
    York City." *Journal of Geological Education*, 13 (2) (1965), 50–2.

13 "Royal Proclamation against the Pollution of the Thames," *Memorials of London
    and London Life in the XIIIth, XIVth, and XVth Centuries, Being a Series of
    Extracts, Local, Social, and Political, from the Early Archives of the City of London,
    A.D. 1276–1419*, select., trans. and ed. Henry Thomas Riley (London: Long-
    mans, Green and Co., 1868), pp. 367–8. A reviewer comments, "I assume that
    what seems incredible about the quote is that the refuse could block navigation.
    I think this was probably a legally necessary assertion even if not an objective
    truth. If my offhand memory of legal history serves me well, the King's power
    over the condition of the waterways stemmed from a navigational servitude in
    favor of the crown which gave the crown such powers over the waterways as
    were necessary to preserve the crown's right of navigation thereon" (letter from
    Professor Marcia Gelpe, University of Minnesota Law School, 3 February
    1977).

14 Timothy Nourse, *Campania Felix* (London, 1700), p. 352.

15 Ibid. Perhaps the quotation should be taken with a grain of salt since Nourse
    was a violent critic of cities, but the evidence from other sources certainly
    suggests that his description was based on fact.

16 For a table listing the major air and water pollutants with their sources,
    characteristics, and effects on human health, see W. Baumol and W. Oates,
    Economics, Environmental Policy and the Quality of Life, ch. 3.

17 The table below represents the latest figures on total dissolved solids, sulfates,
    calcium, chloride, and potassium available from the Great Lakes Basin Com-
    mission, Ann Arbor, Michigan, May 1977 (parts per million).

	Superior	*Huron*	*Michigan*	*Erie*	*Ontario*
Total dissolved solids	52	118	150	198	194
Sulfates	3.0	15	16	26	29
Calcium	13	25	32	37	40
Chloride	1.2	5.4	6	25	28
Potassium	0.5	0.8	1.0	1.0	1.0

18 Environmental programs have apparently produced some recent improvements
    in the purity of some of the Great Lakes. See *New York Times*, 23 May 1974,
    p. 1 and 9 June 1974, Sect. 4, p. 2. The quality of Lake Superior, however, has
    been threatened by discharges of the Reserve Mining Company, which pours
    some 67,000 tons of taconite tailings into the lake every day. Besides just
    dirtying the lake and its shores, it has been alleged by some experts that "the
    asbestos-like fibers emptied into the water – which is used in its pure form as
    drinking water by Duluth and other communities – are a health hazard, since
    they have been known to cause diseases such as cancer among asbestos workers
    in other parts of the country, though sometimes not until 20 years after
    inhalation" (*New York Times*, 9 June 1974, Sect. 4, p. 2).

19 To eliminate from the data the confusing and irrelevant year-to-year fluctua-
    tions that are heavily influenced by fortuitous meteorological conditions, the
    graphs represent five-year averages rather than the raw annual data from which
    the averages are derived. That is, the figure shown for 1912 actually represents

an average of the data for 1910–14, the figure for 1913 is an average of the data for 1911–15, etc.

20 We should emphasize that the level of dissolved oxygen is only one determinant of water quality; as we saw in the case of the Great Lakes, there are other important elements affecting the quality of a body of water. More on this will be said later.

21 The population of the city as given by US Census figures from 1900 to 1950 was: 1900, 3.4 million; 1910, 4.8 million; 1920, 5.6 million; 1930, 7.3 million; 1940, 7.5 million; 1950, 8.0 million (Ira Rosenwaike, *Population History of New York City* (Syracuse: Syracuse University Press, 1972), p. 133).

22 *New York Times*, 29 September 1973, p. 1. However, the reader will note in figure 16.8 that the most recent data for two of the rivers, the Hudson and the Lower East Rivers, indicate a reversal of this trend with dissolved-oxygen levels declining in the early 1970s.

23 Allen Kneese, in a letter to us, attributes much of the improvement in air quality, not to environmental policy, but to economic considerations which led to the substitution of oil and natural gas for coal "first in home heating (which was a terrible low level source of harmful and damaging substances) and later in industrial and electrical power generation. . . . This raises some interesting questions about what will happen when large scale reconversion to coal occurs."

24 Lester B. Lave and Eugene P. Seskin, *Air Pollution and Human Health* (Baltimore: Published for Resources for the Future by The Johns Hopkins University Press, 1977), chap. 9, p. 188.

25 Albert Parker, "Air Pollution Research and Control in Great Britain," *American Journal of Public Health* 47 (May 1957): 569.

26 Royal Commission on Environmental Pollution, *First Report* (London: Her Majesty's Stationery Office, February 1971), p. 23.

27 It now seems evident that in fact, sulfur dioxide pollution is not as serious a threat to human health as are sulfate concentrations. Though SO_2 pollution in the cities has decreased markedly in the last decade (apparently largely because of the relocation to less populated areas of the main sources of urban SO_2, the municipal power plants), national ambient sulfate levels have remained fairly stable. This difference in the trends in sulfate and sulfur dioxide concentrations is apparently rather mysterious, since sulfates are a product of SO_2.

28 The evaluation of the overall pollution content of the atmosphere of a particular city or of a waterway is a complex matter, and data on any single pollutant or group of pollutants can easily be misleading. This point was made forcefully in a letter to us from Thomas McMullen of the Monitoring and Reporting Branch of the Environmental Protection Agency. Commenting on the data underlying figure 43.9, he remarked, "This measurement can in no way be interpreted as an index of general air quality because the complex character of air pollution has been changing over recent decades. Dustfall levels have been diminishing as restrictions on use of soft coal, conversions of home heating systems to gas, and changes in industrial practices have reduced the quantity of larger particles emitted. However, concurrent growth in vehicular traffic, expansion of urbanization, and the burgeoning diversity of industrialization has increased the volume of other pollutant emissions and multiplied the variety of trace pollutants. I think it might be difficult to specify an index, implying reference to a base year or to some common denominator, applicable to the evolving nature of air pollution over the last several decades."

29 See M. Murozumi, T. J. Chow, and C. Patterson, "Chemical Concentrations of Pollutant Lead Aerosols, Terrestrial Dusts and Sea Salts in Greenland and Antarctic Snow Strata," *Geochimica et Cosmochimica Acta* 33, No. 10 (October 1969): 1247–94.

30 Ibid., p. 1285.

31 D. Bryce-Smith, "Lead pollution – a growing hazard to public health," *Chemistry in Britain*, 7 (February 1971), 54. In a later note Bryce-Smith adds the significant point that "no other toxic chemical pollutant appears to have accumulated in man to average levels so close to the threshold for potential clinical poisoning" ("Lead pollution from petrol," p. 286).

32 See "Lead in ancient and modern bones," *Scientist and Citizen*, 10 (3) (April 1968), 89.

33 Mills, "Lead pollution," p. 161, citing a study by Z. Jaworoski.

34 *New York Times*, 27 March 1970, p. 49.

35 The figure of 4.5 percent was computed from data supplied by R. D. Behrman, Administrative Assistant, and A. H. Schuck, Acting Superintendent, Department of Public Works, Division of Waste Collection, City of Cincinnati, in letters of 9 November 1973, and 17 February 1977.

36 A good source on this subject is Emmanuel Le Roy Ladurie, *Histoire du Climat Depuis l'An Mil* (Paris: Flammarion, 1967).

EDITORS' APPENDIX (E. CALVIN BEISNER AND JULIAN L. SIMON)

Since the mid-1970s, pollution levels in the Great Lakes have fallen dramatically (figures 43.22–43.25). Drinking water quality has also improved significantly in the United States (figure 43.26) – Julian Simon.

Recent atmospheric lead levels measured within the United States (figure 43.27) show ambient lead concentrations falling by 94.5 percent from 1976 through 1990. The figure also shows significant drops in concentrations of all five other important air pollutants.

Municipal solid waste generation in the United States rose 105 percent between 1960 and 1988, but during the same period the *rate* of recycling of municipal solid waste rose 96 percent, of incineration for energy recovery rose 6,700 percent, and of incineration without energy recovery fell 98 percent (figure 43.28), indicating trends toward increasingly efficient disposal of waste. The combined tonnage recycled and burned for energy recovery rose over 800 percent, while the tonnage burned without energy recovery fell about 96 percent. The tonnage landfilled reached a peak in 1986 and then fell 4 percent by 1988. Nearly 71 percent of municipal wastes generated in 1988 were biodegradable (paper and paperboard 40 percent, textiles 2.1 percent, wood 3.6 percent, food wastes 7.4 percent, yard wastes 17.6 percent). (Franklin Associates, Ltd., *Characterization of Municipal Solid Waste in the United States: 1990 Update*, prepared for US Environmental Protection Agency; reported in *Statistical Abstract of the United States, 1992*, 216.)

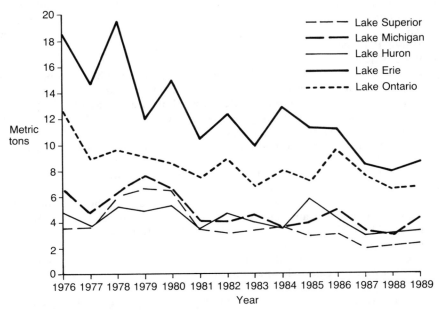

Figure 43.22 Estimated phosphorus loadings to the Great Lakes
Source: Council on Environmental Quality, *Environmental Quality, 22nd Annual Report*, March 1992.

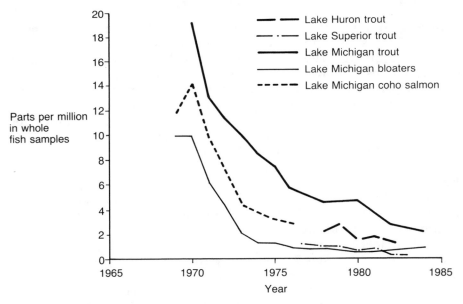

Figure 43.23 DDT levels in the Great Lakes
Source: Council on Environmental Quality, *Environmental Quality, 17th Annual Report*, 1986.

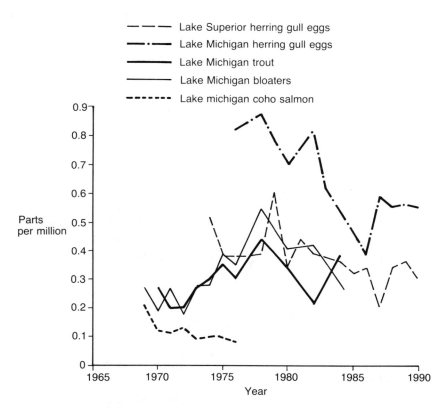

Figure 43.24 Dieldrin levels in the Great Lakes
Source: Council on Environmental Quality, *Environmental Quality, 17th Annual Report*, 1986, C-45, 263, and *22nd Annual Report*, 1992.

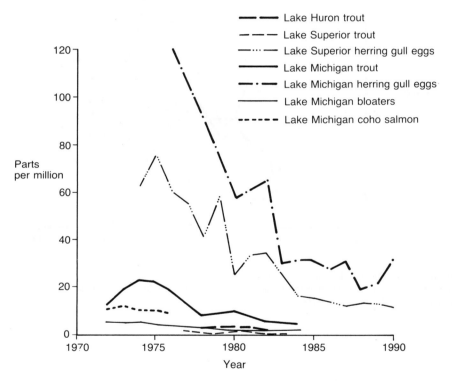

Figure 43.25   PCB levels in the Great Lakes
Source: Council on Environmental Quality, *Environmental Quality, 17th Annual Report*, 1986,
C-44, C-45, 263, and *22nd Annual Report*, 1992.

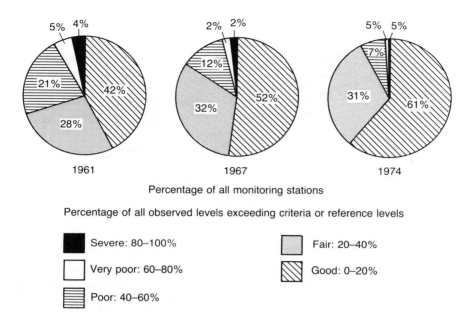

Percentage of all monitoring stations

Percentage of all observed levels exceeding criteria or reference levels

Severe: 80–100%

Very poor: 60–80%

Poor: 40–60%

Fair: 20–40%

Good: 0–20%

Figure 43.26 Trends in the quality of drinking water in the USA
Source: US Council on Environmental Quality, Annual Report, 1975, 352.

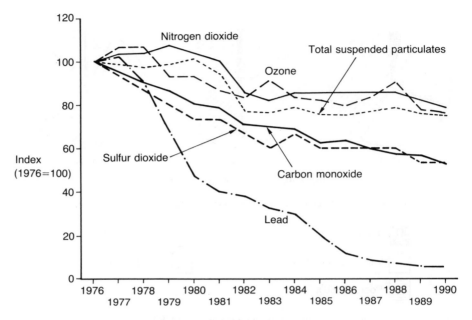

Figure 43.27 Ambient air pollutant concentrations, USA, 1976–90
Source: *Statistical Abstract of the United States*, 1992.

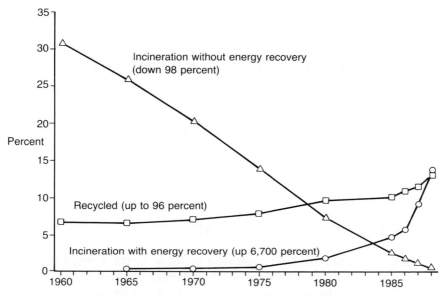

Figure 43.28 Municipal waste recovery and incineration, US, 1960–90
Note: Excludes mining, agricultural, and industrial processing, demolition and construction wastes, sewage sludge, and junked autos and obsolete equipment wastes.
Source: *Statistical Abstract of the United States*, 1992.

44

Atmospheric Pollution Trends in the United Kingdom

Derek M. Elsom

INTRODUCTION: EARLY HISTORICAL TRENDS

The earliest documented air pollution incident in the United Kingdom occurred in AD 1257 when the wife of Henry III visited Nottingham Castle and found the air so full of the stench of smoke from coal burning that she left for fear of her health (Brimblecombe, 1987). Generally, the thirteenth century marked the beginning of urban air pollution problems in the United Kingdom as cheap coal, containing 3 percent sulfur, began replacing increasingly expensive wood and charcoal as the major industrial fuel. In London, large quantities of coal began to be imported by sea from Newcastle in northeast England for use in lime burning and smelting. By 1307, so many wealthy people and visiting nobles objected to the offensive smells produced by the burning of coal in London that Edward I issued a Royal Proclamation prohibiting its use in open furnaces.

Emissions arising from burning coal in furnaces and lime kilns gave rise to frequent complaints in the following centuries, but it was not until the seventeenth century, when coal was adopted widely as a domestic fuel, that pollution problems worsened markedly. Coal use increased at this time because extensive deforestation around major cities had created shortages in wood-based fuels which, in turn, raised their prices. Between 1580 and 1680, coal imports into London from Newcastle increased twentyfold (Brimblecombe, 1987). In 1661, John Evelyn, a London pamphleteer, described the appalling air quality that characterized the city. He wrote: "Most Londoners breathe nothing but an impure and thick mist, accompanied by a fuliginous and filthy vapour, corrupting the lungs, so that catarrhs, coughs and consumptions rage more in this one city, than in the whole Earth." In a booklet, entitled "Fumifugium or The Inconvenience of the Aer and the Smoake of

London Dissipated," he proposed that air quality in the city could be improved by moving polluting industries to the outskirts of the city, by the use of cleaner fuels, and by planting trees and shrubs. Unfortunately, his proposals were ignored.

The rapid deterioration in air quality during the seventeenth century is indicated by estimates of annual concentrations of smoke and sulfur dioxide derived for London by Brimblecombe (1977). Brimblecombe calculated London's historical trend in air quality for the period 1585–1940 using a model based simply on the amount of coal consumption, the diameter of the built-up area, and the sulfur content of the coals burned. Figure 44.1 shows that sulfur dioxide rose rapidly from 1585 to around 1700, steadied for a century, and then increased slightly to 180 μg/m³ (micrograms/cubic meter) around 1850. Smoke concentrations increased steadily from 1585 to a peak of 125 μg/m³ in the 1890s. Both pollutants declined rapidly after the 1890s. These pollution levels are averages for the whole urban area of London, so central London pollution levels were likely to have been much worse.

The relatively high smoke and sulfur dioxide concentrations during the early nineteenth century caused a marked increase in the incidence of bronchitis (known initially as the "British disease") in such sooty, sulfur-smelling cities as Bradford, London, Manchester, Wolverhampton, and Sheffield. This prompted the government to take its first faltering

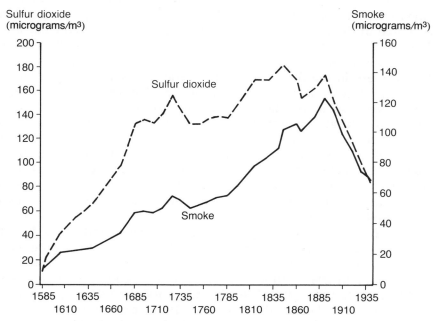

Figure 44.1 Smoke and sulfur dioxide levels, London, 1585–1940
Note: Estimated decadal mean smoke and sulfur dioxide concentrations.
Source: Brimblecombe (1977).

step toward addressing air pollution problems. In 1819, a parliamentary committee was appointed to consider "how far it may be practicable to compel persons using steam engines and furnaces in their different works to erect them in a manner less prejudicial to public health and public comfort." However, no effective legislation resulted. Further committees were established during the 1840s and 1850s, but although there was increasing recognition of the unhealthy effects of smoke, a well-organized industrial lobby blocked several attempts to pass smoke abatement laws. Eventually, with a growing acceptance among owners of furnaces that smoke emissions represented wasted fuel, a Smoke Nuisance Abatement (Metropolis) Act was passed in 1853 (amended in 1856). This was later followed by the Alkali Act of 1863, which is regarded as providing the genesis of the modern approach to air pollution control in Britain (Elsom, 1992).

THE TWENTIETH-CENTURY DECLINE IN SMOKE AND SULFUR DIOXIDE

No reliable air quality measurements are available to highlight air quality improvements from the nineteenth into the twentieth century, but, because smoke particles encourage fog formation and persistence, the record of fog frequency provides a surrogate measure of the amount of suspended particulates in the air. The "Great stinking fogs" of London, first noted in the late seventeenth century, increased in frequency after 1750, reached a peak in the 1890s, and then declined rapidly through this century (Brimblecombe, 1982; Brimblecombe and Rodhe, 1988). This decrease in fog frequency (e.g., average annual number of foggy days in London fell from 30 in the 1880s to 11 in the 1900s), together with an increase in the hours of bright sunshine (increasing by 70 percent in central London from the 1880s to the 1900s), has been attributed to reductions in smoke emissions, which were a consequence of fuel changes (e.g., gas fires replacing coal), stricter controls on industry, the decline in importance of coal, and the greater dilution of emissions which was a function of the spreading out of the city emission sources due to an improved transport network (Ashby, 1975; Bernstein, 1975; Brimblecombe, 1987).

Although average annual fog frequency and smoke and sulfur dioxide concentrations in London decreased after the 1890s, lethal smogs continued to occur. London experienced a sequence of "killer smogs," including that in 1948 (300 excess deaths attributed to the smog), 1952 (3,800 deaths), 1956 (480 deaths), 1957 (300–800 deaths) and 1962 (340–700 deaths). During the December 1952 "peasouper" smog, peak daily smoke and sulfur dioxide concentrations reached 4,500 and 3,700 $\mu g/m^3$ respectively. Other cities experienced similar peak concentrations, including Manchester, with maximum daily smoke concentrations exceeding 5,000 µg/m³ and sulfur dioxide around 3,500 µg/m³ during smogs in December 1961, January 1962, and December 1962. However, by the 1970s, peak daily smoke and sulfur dioxide concentrations during similar meteorological episodes of calm conditions, subfreezing temperatures, and a strong low-level inversion reached 750–1,200 µg/m³, and, by the 1980s, peak daily concentrations seldom exceeded 250–500 µg/m³ in such extreme meteorological conditions.

The dramatic improvement in urban smoke and sulfur dioxide levels in this century is most marked in the late 1950s or early 1960s. Some researchers link this reduction with the passing of the Clean Air Act in 1956. This act was introduced following intense public and media pressure to prevent the recurrence of a 1952-type smog. It introduced controls on smoke emissions from large industrial plants and on chimney heights for new industrial plants (the tall stacks policy), and permitted the designation of urban smoke control zones in which only smokeless fuels could be burned in domestic premises. Undoubtedly, although this act contributed to the decrease in both pollutants, urban air quality records (figure 44.2) indicate that improvements in some locations had begun much earlier than 1956 (as also suggested by the Brimblecombe model for London's air quality). Similarly, indirect indicators of smoke levels such as the mean number of hours of bright sunshine showed an increase well before the mid-1950s (figure 44.3).

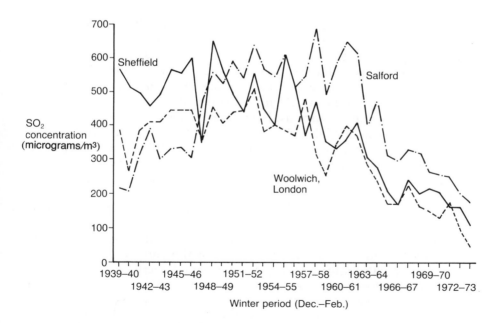

Figure 44.2 Mean winter SO₂ concentrations, selected sites in the United Kingdom
Source: Martin and Barber (1988).

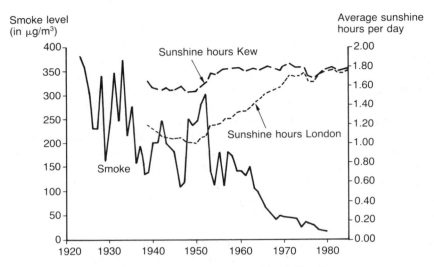

Figure 44.3 Smoke level and mean hours of winter sunshine in London
Source: J. L. Simon adaptation of Elsom figure and data plus Brimblecombe and Rohde (1988) data.

Consequently, although the Clean Air Act of 1956 contributed to reductions in smoke and sulfur dioxide concentrations (probably more so in the less prosperous areas of the country such as northern England), technological, social, and economic factors were already causing levels to fall (Auliciems and Burton, 1973; Elsom, 1992). For example, traditional domestic open coal fires were being replaced by less-polluting central heating systems as affluence followed the postwar depression. Slum clearance programs replaced dense terraced housing characterized by multiple low-level inefficient chimneys with multi-story dwellings having efficient central heating systems. Economic changes included the continued decline in use of coal and its replacement with gas, electricity, and later, oil. Electricity generation began to be consolidated into very large (2,000 MW) power station plants which were located away from city centers and which employed tall (200 m) stacks to disperse emissions.

Mean annual sulfur dioxide levels in London were generally around 300–400 µg/m³ from the late 1930s to the early 1960s after which there was a steady, uninterrupted decline through to the 1980s (Laxen and Thompson, 1987). A national network of sites that monitored smoke and sulfur dioxide levels was established in the early 1960s, and data from this network indicate clearly the continued decrease in smoke and sulfur dioxide levels through the 1970s and 1980s (figure 44.4). In Manchester, as evident in other cities, one of the benefits of the decrease in smoke and sulfur dioxide concentrations was a marked reduction in the incidence of bronchitis (figure 44.5). Legislation, such as the strengthening of the Clean Air Act in 1968, continued to help reduce

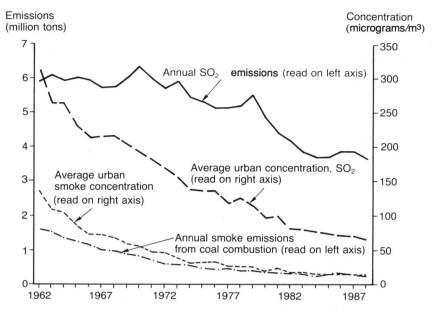

Figure 44.4 SO₂ and smoke emission and concentration, United Kingdom, 1962–88
Source: Compiled from data presented in Dept of Environment (1990).

emissions of both smoke and sulfur dioxide, but so did other changes. One major influence was the national economic recession, triggered by the OPEC oil price increase in 1973, since it not only reduced emissions through plant closures but also encouraged adoption of improved fuel efficiency as well.

In the 1970s, legislation was introduced specifically to tackle a potential sulfur dioxide pollution problem in city centers. The intensive development of office blocks heated by means of fuel oil with a high sulfur content (3.5 percent) was causing concern that sulfur dioxide levels might rise significantly. This prompted the City of London to introduce legislation limiting the sulphur content of fuel oil to 1 percent in new premises, beginning in 1972 (existing premises were given 15 years to comply). Other city authorities were given this option when it was included in the Control of Pollution Act of 1974. This provided the means to achieve compliance eventually with the European Community sulfur dioxide air quality standard introduced in 1983. European Community air quality standards (limit and guide values) were also introduced for airborne lead in 1982, smoke in 1983, and oxides of nitrogen in 1985.

ACID DEPOSITION, SULFATES, AND NITRATES

The marked decrease in urban sulfur dioxide concentrations in recent decades reflects the dramatic reduction in emissions from low-height

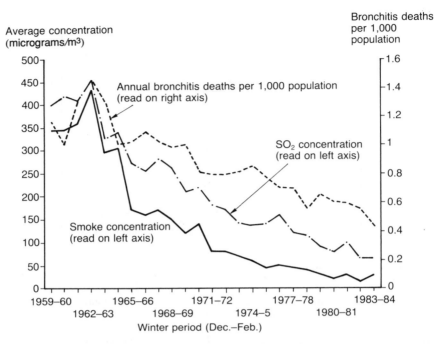

Figure 44.5 Air pollution and bronchitis deaths, Manchester, 1959–60 to 1983–4
Source: M. Eastwood, Director, Environmental Health and Consumer Protection Dept, Manchester City Council.

sources rather than changes in total national emissions of sulfur dioxide. For example, although total emissions fell by 38 percent between 1963 and 1988, average urban concentrations decreased by nearly 78 percent, reflecting the 75 percent fall in low-level emissions during the same period (Department of Environment, 1990).

Clearly, urban areas have benefited from the reduction in emissions from low-level sources, but it may have been partly at the expense of rural and distant areas receiving increased pollution, albeit at very low concentrations. This arises because industrial stacks discharge waste gases and particles at higher velocities, greater heights, and higher temperatures. This improves dilution of the pollutants before they reach ground level, but it also results in their wider dispersal, with pollutants sometimes being transported distances of hundreds, even thousands, of kilometers. Occasionally, weather conditions may favor long-distance transport of pollutants with only limited dilution, and this can result in areas experiencing episodes of unusually high pollution concentration such as the black acidic snowfalls that occurred in Scotland (Davies et al., 1984).

Wider dispersal of pollutants from urban-industrial areas has been taking place for more than a century, but its impact increased greatly as national emissions of sulfur dioxide increased from around 1.2 million tons in 1860 to 6.0 million tons by the 1960s. Europe's emissions increased from 4.0 million tons around 1940 to 24 million tons in 1950 and to over 50 million tons by the 1970s. Not surprisingly, this situation led to increased acid deposition in rural and remote areas throughout Europe. Reliable historical records of acid deposition are rare despite the pioneering work undertaken by Robert Angus Smith, the first Alkali Inspector for Manchester, who introduced the term "acid rain." Acid precipitation is commonly expressed in terms of pH values, where pH is a measure of the concentration of hydrogen ions present. The pH scale extends from 0 to 14, with unpolluted precipitation having a pH of around 5.0 to 5.6. Because the scale is logarithmic, each whole-number increment represents a tenfold change in acidity; thus, solutions with pH of 6, 5, 4, and 3 contain 1, 10, 100, and 1,000 microequivalents of hydrogen ions per liter, respectively. Reinterpreting the figures given by Smith (1872) for free acidity, rural values for various regions of the United Kingdom in 1870 were estimated to have been around pH 5.1, while urban and industrial regions were in the range pH 3.4–3.9 (UK Review Group on Acid Rain, 1983). In comparison, the highest precipitation-weighted mean annual pH in the period 1986–8 was pH 4.1–4.3, recorded in the north Midlands and northern England (UK Review Group on Acid Rain, 1990).

Data revealing long-term changes in atmospheric acidic deposition have become available recently through the analysis of fossil diatoms found in lake sediments. Diatoms are microscopic algae with silica skeletons which are relatively resistant to decay. When a diatom dies, it sinks to the bottom of the lake where its remains accumulate as sediment. Different diatom species have different pH preferences, so a core of lake sediment containing various diatoms provides a record of changing lake acidity which, in turn, reflects changing rates of atmospheric acidic deposition or land-use changes within the catchment, and sometimes both. The upper sediment layers can be dated using the concentration of naturally occurring isotope lead-210. Figure 44.6 shows the 700-year reconstructed pH record using fossil diatoms from Round Loch of Glenhead, an unforested granite catchment with mainly blanket peat soils, in southwest Scotland (Flower and Battarbee, 1984; Flower et a.l, 1987). It reveals that the pH of the lake remained at 5.6–5.8 from AD 1300 to around 1850, after which acidification increased steadily until it reached 4.8 in 1960, a value maintained to the core top (dated 1980). Such a change in acidity in this Scottish lake is the consequence of the massive increase in national and European emissions of sulfur dioxide (and oxides of nitrogen) which began in the mid-nineteenth century.

Results from the monitoring of precipitation at Pitlochry, Scotland, for the period 1973–84, indicate that mean annual precipitation acidity has lessened slightly in the United Kingdom since the late 1970s,

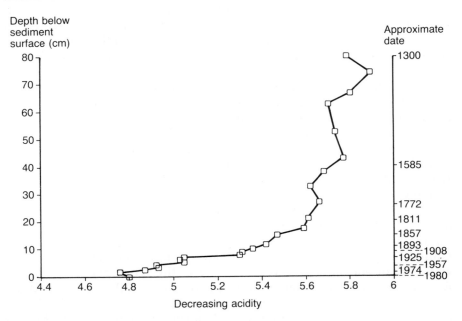

Figure 44.6   Acidity of Round Loch, Scotland, 1300–1980
Note: Acidity inferred from analysis of fossil diatoms in lake sediments in Round Lock of Glenhead, Galloway, Scotland.
Source: Modified from Flower and Battarbee (1983) with permission of R. J. Flower.

reflecting the recent reduction in national and European emissions of sulfur dioxide (Laxen and Schwar, 1985; UK Review Group on Acid Rain, 1987). However, the beneficial effects of a lessening of acidic deposition on aquatic and terrestrial ecological systems likely will be delayed for some years to come because acidity accumulated in soils and lakes for over a century will continue to affect terrestrial and aquatic systems. At the present time, upland ecology in some areas of the country shows significant adverse effects due to acidification (Woodin, 1989). Annual surveys of trees in the United Kingdom to assess signs of damage indicate a serious situation. In 1988, 64 percent of all trees revealed slight to severe defoliation (damage classes 1–4) and 27 percent showed moderate to severe defoliation (classes 2–4). Some of this damage may be due to atmospheric pollution, although it is difficult to assess how much (for example, see chapter 57 by Kenneth Mellanby, in this volume in which he urges caution in claiming air pollution is the cause of forest damage). Nevertheless, the degree of damage being experienced by trees in the United Kingdom is similar to that suffered in Czechoslovakia, Poland, and western parts of the former Soviet Union, where high levels of air pollution are being blamed as the primary cause (refer to chapter 46 by Mikhail Bernstam in this volume for a broader discussion of pollution in socialist countries).

In general, sulfuric acid, derived from emissions of sulfur dioxide, contributes 60–70 percent of UK precipitation acidity, while nitric acid, derived from emissions of oxides of nitrogen, contributes 30–40 percent. However, whereas sulfur dioxide emissions have fallen steadily since the 1970s, national emissions of oxides of nitrogen have not changed markedly since 1970. During the past two decades, they have remained in the range 2.2–2.6 million tons, with 40 percent being contributed by large combustion plants (Department of Environment, 1990). Consequently, nitric acid is increasing its relative contribution to acid deposition compared with sulfuric acid. This is confirmed in the record of changing levels of aerosol sulfate and nitrate for a rural site in central southern England (figure 44.7).

An important factor in the decline of sulfate concentrations beginning in the 1970s was the economic recession following the 1973 OPEC increase in oil prices. In contrast, nitrate levels continued to increase after the 1970s, albeit at a slower rate after 1973, and they have doubled since 1954. The reduction in sulfate concentrations during the 1970s is also confirmed indirectly by an improvement in mean summer long-range visibility recorded at southern England rural meteorological stations, since sulfates are the aerosols primarily responsible for light scattering (Elsom, 1992; Lee, 1984).

Sulfur dioxide emissions in the United Kingdom are expected to continue decreasing as a result of the European Community (EC)

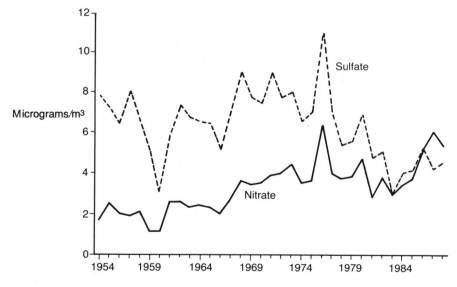

Figure 44.7 Sulfate and nitrate concentrations in atmospheric aerosols, Harwell, United Kingdom
Source: Adapted from figure 4 in Atkins et al. (1990) with permission of Harwell Laboratory, Oxfordshire.

Directives in 1988 which limit emissions from large combustion plants. Total sulfur dioxide emissions from plants exceeding 50 megawatts capacity (which account for 84 percent of all sulfur dioxide emissions in the United Kingdom) are required to be reduced by 60 percent by the year 2003, taking 1980 emissions as the baseline. Similarly, total emissions of oxides of nitrogen from large combustion plants are required to be reduced by 30 percent by 1998, taking 1980 emissions as the baseline.

PHOTOCHEMICAL POLLUTION AND VEHICLE EXHAUST EMISSIONS

Photochemical pollutants, of which low-level ozone is the most important, are secondary pollutants formed by a complex series of reactions between oxides of nitrogen and volatile organic compounds (includes hydrocarbons) in the presence of sunlight. Episodes of elevated levels of ozone throughout the United Kingdom have been observed and monitored since the 1970s. Annual ozone concentrations and the number of days or hours each year exceeding World Health Organization guidelines display great variability from year to year, reflecting variations in the meteorological conditions that favor photochemical activity (figure 44.8). Generally, the number of days and hours during which such guidelines are exceeded is significantly higher at suburban and

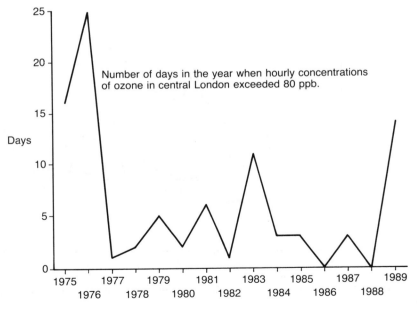

Figure 44.8 Ozone concentrations, central London
Source: London Scientific Services (1990a).

rural sites compared with urban sites (UK Photochemical Oxidants Review Group, 1987). This arises because urban areas produce large quantities of fresh emissions of nitric oxide that inhibit ozone formation but, as these emissions travel downwind of the urban area, the photo-chemical reactions in the atmosphere change to favor ozone formation. The London pollution plume may enhance ozone levels downwind by 40 ppb (0.04 ppm) or even 70 ppb (0.07 ppm) on some occasions (Varey et al., 1988).

The frequency of high ozone episodes increases in a direction from southeast to northwest Britain, suggesting that precursor emissions transported from continental Europe contribute to these episodes. Con-sequently, reductions in ozone levels will require reductions in emissions of VOCs (volatile organic compounds) and oxides of nitrogen at the national and European levels. In 1988, 22 million motor vehicles in the UK contributed 45 percent of the national emissions of oxides of ni-trogen and 30 percent of VOCs, so the recently agreed EC Directive on vehicle emissions, requiring the introduction of catalytic converters on new cars in 1993, ought to reduce vehicle emissions of both pollutants. However, with the government forecasting road traffic increases of 83–142 percent by the year 2025, increasing petrol consumption may offset the effects of this legislation. In London, the concentrations of both nitric oxide and nitrogen dioxide are increasing (figure 44.9).

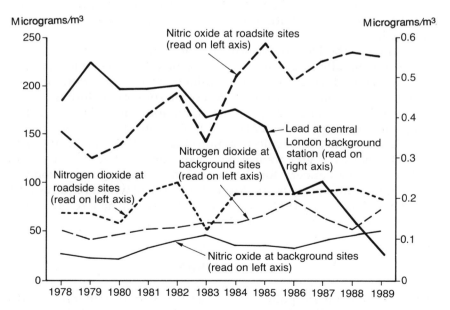

Figure 44.9   Nitrogen oxides and lead levels, central London, 1978–89
Source: London Scientific Services (1990b) and London Planning Advisory Committee (1990).

Central London's roadside station regularly exceeds the one-hour EC limit value of 200 µg/m³ (Holman, 1989).

Lead in the atmosphere comes mainly from lead in petrol. The maximum amount of lead permitted in petrol has been reduced successively from 0.84 g/l in 1972 to 0.40 g/l in 1981 and 0.15 g/l in 1985. As a result of the latter reduction, airborne lead levels fell by an average of 55 percent between 1985 and 1986 (Department of Environment, 1990). Unleaded petrol first became available in 1986, and by the beginning of 1990, 90 percent of petrol stations were selling unleaded petrol and unleaded sales had reached 30 percent. Emissions of lead from petrol-powered road vehicles are estimated to have fallen from 7,500 tons in 1980 to 2,600 tons in 1989 (Department of Environment, 1990).

GREENHOUSE GASES

The infrared absorbing gases of carbon dioxide, methane, chlorofluorocarbons (CFCs), nitrous oxide, low-level ozone, and water vapor are contributing to global warming through the "greenhouse effect." Currently, carbon dioxide contributes about half of the global greenhouse effect. Emissions of carbon dioxide in the United Kingdom, about 3 percent of global emissions, declined in the early 1980s but rose again between 1984 and 1988 (Department of Environment, 1990). In November 1990, the government committed itself to stabilizing emissions by the year 2005 but subsequently brought the target date forward to the year 2000 in line with other member nations of the European Community.

According to the Department of Environment (1990), methane emissions have generally fluctuated around or below the 1970 level and fell sharply in 1984 during the national miners' strike. In 1988, emissions reached 3.4 million tons from a variety of sources including animals (33 percent), coal mines (29 percent) and landfill sites (21 percent), gas leakage and venting (16 percent), and fuel combustion (1 percent).

In relative terms, molecule-per-molecule, CFC-11 is 12,400 times more effective as a greenhouse gas than carbon dioxide. Given this situation and the concern for the depletion of stratospheric ozone, the United Kingdom is party to the Montreal Protocol which requires CFCs to be phased out. By 1989, the United Kingdom had reduced CFC emissions from the 1986 baseline by 50 percent.

CONCLUSION

Trends in atmospheric pollutants in the UK vary considerably. Urban smoke and sulfur dioxide concentrations have decreased dramatically during the past few decades and, currently, there are few occasions when these pollutants exceed health-based air quality standards. During the 1970s, public concern for smoke and sulfur dioxide began to be replaced by concern over increasing urban concentrations of oxides of nitrogen, VOCs, airborne lead, and carbon monoxide as well as the transfrontier pollution problems of photochemical pollution and acidic deposition. During the 1980s, stratospheric ozone depletion by CFCs and global warming due to increasing greenhouse gases were added to the list of atmospheric pollution problems. With the exception of airborne lead, whose levels have fallen markedly since 1980 following a strong public and media campaign highlighting the possible effects of airborne lead on the health of young children, the other pollution problems have shown only limited, if any, improvement.

REFERENCES

Ashby, E. (1975): "Clean Air over London." *Clean Air*, 5, 25–30.

Atkins, D. H. F., D. V. Law, C. Healy, R. Sandalls, H. Jeffery, and J. Sandalls (1990): *Trends in the Concentration of Sulphate and Nitrate in the Ambient Aerosol at a Site in the United Kingdom (1954–1988)*. Report AERE R 13288. Harwell: AEA Environment & Energy.

Auliciems, A., and I. Burton (1973): "Trends in Smoke Concentrations before and after the Clean Air Act of 1956." *Atmospheric Environment*, 7, 1063–70.

Bernstein, H. T. (1975): "The Mysterious Disappearance of Edwardian London Fog." *The London Journal*, 1, 189–206.

Brimblecombe, P. (1977): "London Air Pollution, 1500–1900." *Atmospheric Environment*, 11, 1157–62.

—— (1982): "Long Term Trends in London Fog." *The Science of the Total Environment*, 22, 19–29.

—— (1987): *The Big Smoke*. London: Methuen.

—— and H. Rodhe (1988): "Air Pollution – Historical Trends." *Durability of Building Materials*, 5, 291–308.

Davies, T. D., P. W. Abrahams, M. Tranter, I. Blackwood, P. Brimblecome, and C. E. Vincent (1984): "Black Acidic Snow in the Remote Scottish Highlands." *Nature*, 312, 58–61.

Department of Environment (1986): *Digest of Environmental Protection and Water Statistics, 1985 (and Additional Tables)*. London: Her Majesty's Stationery Office.

—— (1990): *Digest of Environmental Protection and Water Statistics, 1989 (and Statistical Bulletin "Air Quality" Supplement)*. London: Her Majesty's Stationery Office.

Elsom, D. M. (1992): *Atmospheric Pollution*, 2nd edn. Cambridge, MA, and Oxford, UK: Blackwell.

Flower, R. J., and R. W. Battarbee (1984): "Diatom Evidence for Recent Acidification of Two Scottish Lochs." *Nature*, 305, 130–3.

—— and P. G. Appleby (1987): "The Recent Palaeolimnology of Acid Lakes in Galloway, Southwest Scotland: Diatom Analysis, pH Trends, and the Role of Afforestation." *Journal of Ecology*, 75, 797–824.

Holman, C. (1989): *Air Pollution and Health*. London: Friends of the Earth.

Laxen, D. P. H., and M. J. R. Schwar (1985): *Acid Rain and London*. London: Greater London Council.

Laxen, D. P. H., and M. A. Thompson (1987): "Sulphur Dioxide in Greater London, 1931–85." *Environmental Pollution*, 43, 103–14.

Lee, D. O. (1984): "Trends in Summer Visibility in London and Southern England 1962–1979." *Atmospheric Environment* 17, 151–9.

London Planning Advisory Committee (1990): *Air Pollution Associated with Transport in London*. London: LPAC.

London Scientific Services (1990a): *London-wide Ozone Monitoring Programme: Summary of Results for 1989*. Report LSS/LWMP/109 (1990). London: LSS.

—— (1990b): *London Air Pollution Monitoring Network: Fourth Report—1989*. Report LSS/LWMP/120 (1990). London: LSS.

Martin, A., and F. R. Barber (1988): "Two Long Term Air Pollution Surveys Around Power Stations." *Clean Air, U.K.*, 8, 61–73.

Smith, R. A. (1872): *Air and Rain. The Beginnings of a Chemical Climatology*. London: Longmans, Green and Co.

UK Photochemical Oxidants Review Group (1987): *Ozone in the United Kingdom. Interim Report*. London: Department of Environment.

UK Review Group on Acid Rain (1983): *Acid Deposition in the United Kingdom*. Stevenage: Warren Spring Laboratory.

—— (1987): *Acid Deposition in the United Kingdom 1981–1985*. Stevenage: Warren Spring Laboratory.

—— (1990): *Acid Deposition in the United Kingdom 1986–1988*. Stevenage: Warren Spring Laboratory.

Varey, R. H., D. J. Ball, A.-J. Crane, D. P. H. Laxen, and F. J. Sandalls (1988): "Ozone Formation in the London Plume." *Atmospheric Environment*, 22, 1335–46.

Woodin, S. J. (1989): "Environmental Effects of Air Pollution in Britain." *Journal of Applied Ecology*, 26, 749–61.

45

Trends in Air Pollution in the United States

Hugh W. Ellsaesser

The earliest air pollution concerns were aroused by the visual detection of smokes and dusts, both airborne and settled, and the olfactory detection of odors. While carbon monoxide was of early concern, this was primarily because of the indoor hazard from stoves and furnaces used for cooking and space heating (Eisenbud and Ehrlich, 1972). Outdoor levels, except in tunnels, etc., were generally too low to be either of concern or easily measurable with the instruments then available. Sulfur dioxide was the first invisible gas pollutant of concern probably because of both its low odor threshold and the relative ease with which it could be measured. Thus it is not surprising that sulfur dioxide and particles were the first air pollutants of concern and were given a large boost in public awareness by the so-called "Air Pollution Episodes" of Meuse Valley, Belgium, December 1–5, 1930 (Firket, 1936); Donora, Pennsylvania, October 27–31, 1948 (Schrenk et al., 1949), and London, England, December 5–9, 1952 (Ministry of Health, 1954).

These early recognized air pollutants were most apparent in dense urban areas and in the winter season during periods of damp cold with little wind ventilation and in the United States in the northeastern part of the country. In the mid-1940s, when people became aware of Los Angeles smog, they soon learned they faced something different. "After smoke and oxides of sulfur were controlled, the pollution problem (known now as photochemical smog) persisted, . . . the black industrial cloud . . . was replaced by a structureless, grayish, eye-irritating haze covering all Los Angeles and stretching far into the adjacent valleys of San Fernando and San Gabriel" (Haagen-Smit, 1972). This photochemical smog differed from the earlier recognized forms of air pollution "by occurring more in hot than in cold periods, being oxidizing rather than reducing, occurring more frequently during dry rather than wet weather, being related to petrol rather than coal combustion and being largely derived from motor vehicle exhaust rather than from household heating" (Goldsmith, 1969).

While until roughly 1960 the primary indicators of the severity of this smog were the number of public complaints of eye irritation, percentage of total plant leaf damage observed in special monitoring plant test boxes, and the degree of cracking of strips of raw rubber, these symptoms appear not to have been the primary driving force behind the Los Angeles clean air movement. In the words of Haagen-Smit (1972), "Cities – e.g., Pasadena which for a long time had been regarded as a haven for those who valued the quality of their mild climate and scenery – daily saw mountains disappearing in ugly haze. This, perhaps more than anything else, was responsible for a clean air movement by private citizens."

LONG-TERM DATA: PROBLEMS IN COMPARABILITY AND REPRESENTATION

Data on airborne pollutants in the United States are very skimpy prior to World War II. The parameters measured, the measuring instruments, and the number and location of observing stations can only be described as having been in a continuing state of flux, at least through 1979. According to Ludwig et al. (1970), "The routine monitoring for gases in a fashion amenable to trend analyses was started by the federal government in 1962 and then in only six urban areas, two of which were relocated in 1964–65." Accordingly, to attempt to gain insight on trends prior to about 1960, reliance must be placed on observations of dust fall (e.g., figure 45.1), visibility at a few isolated monitoring stations, and occasional records of isolated measurements. From those skimpy and not always comparable observational records, it appears that all measures of airborne pollutants in the United States have shown downward, if any, long-term trends since the beginning of records. The one exception that might be claimed is a record of suspended particulates at 20 *nonurban* sites.

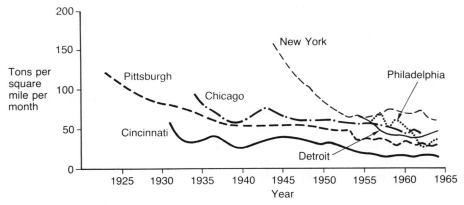

Figure 45.1   Trends in settleable dust in six cities
Source: Ludwig et al. (1970).

Because of the continuing evolution in parameters measured, measuring instruments, and mix of sites available, the most credible trend records for 1950 to 1970 are probably provided by figure 45.2 from the California Air Resources Board (ARB).

PROBLEMS OF SUBSTITUTING EMISSION FOR CONCENTRATION DATA

Because of the many problems in establishing long-term trend curves for airborne pollutants over extended periods, there has been a tendency to fill the void with *estimated* data on *emissions*. Ludwig et al. (1970) included two figures with such *estimates* for the future based on concepts such as their model of our ability to pollute given as a product of population times living standard raised to a power not less than one. The contrast between the rising curves in these figures and the downward curves based on their observational data illustrates a problem that has plagued modelers of air pollution from the beginning.

The problem is not simply that airborne concentrations refuse to rise with estimated emissions, even when the latter are based on good data, but that airborne concentrations have also, to a large extent, refused to

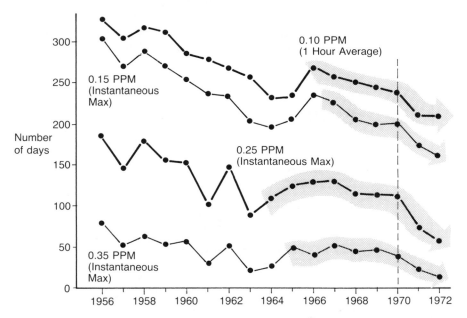

Figure 45.2 Number of days per year with highest concentration equal to or greater than levels shown

Note: Number of days per year that total oxidants reached one hour maxima of 0.10 ppm and instantaneous values of 0.15, 0.25, and 0.35 ppm in the Los Angeles Air Pollution Control District as compiled by California Air Resources Board (1974).

Source: Los Angeles County Air Pollution Control District.

fall as emissions have been reduced by control regulations or, in some cases, have fallen even more rapidly than the emission estimates. While there may be many reasons for this paradox, the two most common ones have usually been the failures to take into account natural sources of the substances treated as pollutants and the natural removal processes that are constantly operating to remove these same substances from the atmosphere. An excellent discussion of the difficulties in attempting to obtain observational confirmation of other presumptive expectations concerning the behavior and effects of pollution is given by Baumol and Oates (1984; partially reprinted in this volume).

THE GENERAL PICTURE: DECLINING AIR POLLUTANT CONCENTRATIONS

Since about 1972 there are reasonably good records for determining trends in airborne concentrations of most standard pollutants. However, there has also been a continuing growth in the number of reporting stations, and lumping new stations with different background levels of pollutants would distort any actual trends. For this reason, EPA and others preparing such data for trend analysis have usually presented different curves for each decade using a fixed set of stations over the decade.

In figure 45.3 trend curves for particulates, sulfur dioxide, carbon monoxide, ozone, and lead are presented, prepared by compositing decadal trend curves from EPA (1977, 1984, and 1990).

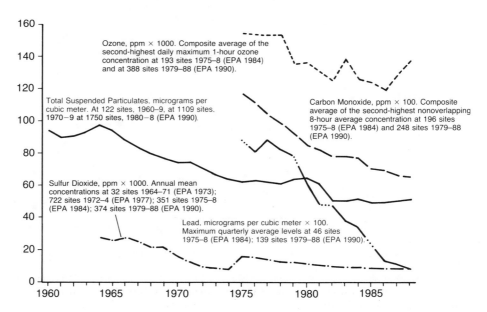

Figure 45.3 Air pollutant trends, 1960–88

The observations for total suspended particulates (TSP) go back farthest – to 1960. The composite curve indicates a more or less undulatory decrease in TSP with a leveling off in recent years. From the curves for settleable dust in Ludwig et al. (1970) shown in figure 45.1, it is a reasonable assumption that this is a valid characterization for airborne particles back at least to the 1920s. The abrupt decline circa 1980 to 1982 occurred with introduction of a new type of filter in 1982. EPA (1990) states, "Although the difference between 1979 and post-1981 is real, the pattern of the yearly change in TSP between 1979 and 1981 is difficult to assess and most of the large apparent decrease in pollutant concentrations between 1981 and 1982 can be attributed to a change in these filters."

A similar composite curve occurs for the annual mean concentration of sulfur dioxide from 1964 through 1988. When combined with data in Ludwig et al. (1970), this again indicates an undulatory decrease at least since 1962 with the rate of decrease diminishing with time. EPA (1973) shows annual mean SO_2 curves for three metropolitan areas with "historically severe SO_2 problems," New Jersey/New York/Connecticut, Philadelphia, and Chicago. These areas showed declines of over 80, 50, and 75 percent, respectively, through the relatively short period from 1967 to 1973. At that time, in association with the ill-fated CHESS program (EPA, 1974), the major health concern was attributed to the sulfate fraction of TSP. Accordingly, EPA (1973) reported, "In light of the recently assessed downward trends in total suspended particulate matter and SO in urban areas [refs], one might expect corresponding decline in the amount of sulfate in the particulate samples. The results of this study, based on data from 62 sites over the years 1964 through 1970, do not bear out this expectation."

In view of the formation of EPA in 1970, the following comments from EPA (1977) are of interest. "The entire sulfur dioxide picture has changed in the 1970s. The early 1970s saw dramatic decreases in ambient sulfur dioxide levels in the nation's urbanized areas [EPA, 1997]. . . . Less than 10 percent of these [722] sites had data in 1970–1971, when sulfur dioxide levels were rapidly reduced in many areas; therefore, graphs are presented only for the 1972–1976 time period." For the most recent decade, from 1979 to 1988, EPA (1990) reported a decrease of 30 percent, or about 4 percent per year, and a sulfur oxide emissions decrease of 17 percent. It stated, "The disparity between the 30 percent improvement in SO_2 air quality and the 17 percent decrease in SO_2 emissions can be attributed to several factors." The two most important given appeared to be that "the 200 highest SO_x emitters . . . account for 59 percent of all SO_x emissions" and that because of the use of tall stacks "measured ground level concentrations . . . may not reflect local emissions." It is difficult to avoid the conclusion that non-EPA mandated local controls were having more effect on airborne concentrations of SO_2 than were EPA-mandated controls.

The composite curves for the second highest nonoverlapping 8-hour average concentration of carbon monoxide from 1975 through 1988 again show a general decline slowing with time. Trends prior to 1975 can be inferred from Eisenbud and Ehrlich (1972), and the following quote from EPA (1973): "Overall carbon monoxide averages have decreased notably since 1968. Even after adjusting for the upward bias from 1 to 4 ppm (1.1 to 4.6 mg/m^3 [*sic*]) due to instrumental modification for sites in Los Angeles County prior to April 1968, there has still been a decline in CO concentrations in Los Angeles County."

The trend for ozone is not so clear. As reported by CEQ (1983), "Although the trend indicates an overall decrease in ozone concentrations of 18 percent between 1975 and 1982, the pattern shows fairly consistent levels from 1975 through 1978, followed by a drop between 1978 and 1979 that may be due to a change in monitor calibration procedures." This refers to the switch-over from measuring ozone by KI (potassium iodide) to UV (ultra violet absorption) as the reference method. The available data suggest significant declines in oxidant (or ozone) prior to about 1972, but because of the numerous changes in monitoring procedures and the essentially constant levels of this pollutant since the implementation of the most stringent control measures, it appears more likely that these apparent declines are due to measurement artifacts rather than any reduction in emissions of ozone precursors. It is significant that ARB (1983) presented maps with isolines of percent change in three-year averages of ozone over the five years from 1974–6 to 1979–81. These were far from indicative of long-term air quality trends. In the South Coast Air Basin ozone "decreased in the San Fernando and San Gabriel Valleys but increased in the coastal and eastern areas." In the Bay Area, "concentrations decreased in the northern and southern San Francisco Bay Area but increased in a band through Marin and Contra Costa Counties."

It is also noteworthy that there was an episode in the South Coast Air Basin September 7–19, 1979 described as "the worst Southern California smog siege in 25 years" (ARB, 1979b). For eight consecutive days, ozone reached or exceeded 35 pphm. After a comparison with the previous sieges of September 1971 (6 days), June 1973, July 1973, and June 1978 (all 4-day sieges), however, ARB (1979b) concluded that "because of some data base problems, it is probably not appropriate to quantitatively compare the 1971 siege to the 1979 siege" and the "degree of difference seen, between the 1978–79 days and the 1973–74 days, should not be accepted with confidence, however, because of uncertainties in the 1973–74 instrumental methods of measuring oxidant."

Controls on lead in leaded gasoline were begun in the early to mid-1970s. These were further tightened in 1972, along with the introduction of unleaded gasoline for cars equipped with catalytic converters. In October 1978, EPA promulgated a new air quality standard for lead itself. In addition, the lead content of leaded gasoline was further reduced from 1.0 gram/gallon to 0.5 gram/gallon July 1, 1985 and to 0.1 gram/gallon January 1, 1986. These actions account for the significant drop in airborne lead. ARB (1979a) contains graphs for airborne lead in individual air basins of California going back as far as 1970. In general these show little trend other than a tendency for a maximum circa 1976, except for the South Coast Air Basin, which shows a substantial decline (about 30 percent) from 1971 to 1973.

While the sources cited herein have also given data for nitrogen dioxide, the earlier data suffer from even greater measurement difficulties than do those for ozone. The data over the last decade suggest a 7 percent decline from 1979 to 1983 and are essentially trendless thereafter.

REFERENCES

ARB (Air Resources Board) (1979a): *Air Quality Trends in the South Coast Air Basin through 1979*. Sacramento: ARB Technical Services Division.

—— (1979b): *California Air Quality Data XI(3)*. Sacramento: ARB Technical Services Division.

—— (1983): *California Air Quality Data XV (4)*. Sacramento: ARB Aerometric Data Division.

CEQ (Council on Environmental Quality) (1983), *Environmental Quality 1983*, 14th annual report. Washington: Superintendent of Documents.

Eisenbud, M., and L. R. Ehrlich (1972): "Carbon Monoxide Concentration Trends in Urban Atmospheres." *Science* 176, 193–4.

EPA (Environmental Protection Agency) (1973): *Monitoring and Air Quality Trends Report, 1972*, PB-234 445. Research Triangle Park, NC.

—— (1974): *Health Consequences of Sulfur Oxides: A Report from CHESS, 1970–1971*, EPA-650/1-74-004. Research Triangle Park, NC.

—— (1977): *National Air Quality Emissions Trends Report, 1976*, EPA-450/1-77-002. Research Triangle Park, NC.

—— (1984): *National Air Quality and Emissions Trends Report, 1982*, EPA-450/4-84-002. Research Triangle Park, NC.

—— (1990): *National Air Quality and Emissions Trends Report, 1988*, EPA-450/4-90-002. Research Triangle Park, NC.

Firket, J. (1936): "Fog Along the Meuse Valley." *Transactions of the Faraday Society*, 32, 1192–7.

Goldsmith, J. R. (1969): "Los Angeles Smog." *Science Journal* (March) 44–9.

Haagen-Smit, A. J. (1972): "Abatement Strategy for Photochemical Smog." In *Photochemical Smog and Ozone Reactions, Advances in Chemistry Series 113*. Washington: American Chemical Society.

Ludwig, J. H., and G. B. Morgan and T. B. McMullen (1970): "Trends in Urban Air Quality." *EOS* 51(5), 468–75.

Ministry of Health (1954): "Mortality and Morbidity During the London Fog of December 1952." In *Reports on Public Health and Medical Subjects No. 95*, London: Her Majesty's Stationery Office.

Schrenk, H. H., H. Heimann, G. D. Clayton, W. M. Gafafer, and H. Wexler (1949): "Air Pollution in Donora, Pa." In *Public Health Bulletin No. 305*, Washington: Federal Security Agency.

EDITORS' APPENDIXES (CALVIN BEISNER AND JULIAN SIMON)

Despite difficulties in comparability of data over long periods described in this chapter, the general picture on air pollution in the United States is unquestionably improving.

In the 15 years from 1976 through 1990, concentrations of the six most significant forms of air pollution fell across the board, according to EPA data – by between 20 and 30 percent for ozone, total suspended particulates, and nitrogen dioxide, by over 40 percent for carbon monoxide and sulfur dioxide, and by over 90 percent for lead (figure 45.4). Even more significant than a simple index of concentrations is an index of those concentrations compared with the EPA's air quality standard for each of the pollutants. After all, knowing actual concentrations means little unless we also have some basis for evaluating the health implications of those concentrations. Figure 45.5 shows that even by 1976 four of the six pollutants' average levels were already well within the quality standard (sulfur dioxide, TSPs, nitrogen dioxide, and lead), and all four of these fell significantly farther afterward. A fifth (carbon monoxide) met the standard by 1980 and has fallen significantly since then – never again topping it. And the sixth (ozone) met the standard for the first time in 1986, rose again in 1987–8, and fell significantly below it in 1989–90. – E. Calvin Beisner

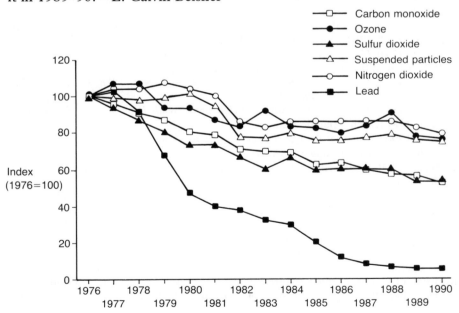

Figure 45.4 Ambient air pollutant concentrations, USA, 1976–90
Source: US Environmental Protection Agency, National Air Quality and Emissions Trends Report, annual.

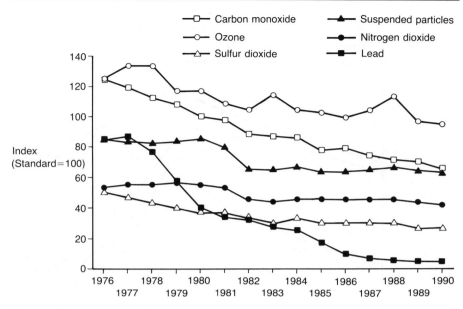

Figure 45.5 Ambient air pollutant concentrations, index of deviations from standards
Source: US Environmental Protection Agency, National Air Quality and Emissions Trends Report, annual.

Emissions of major air pollutants in the United States have fallen significantly in recent decades. Generally upward trends in the 1940s reversed in the 1950s for suspended particulates and in the late 1960s for all other pollutants (figure 45.6). Air quality trends in major cities have also been positive in recent decades (figure 45.7). Most cities have seen the annual number of days on which air quality failed to meet the PSI index level fall from between 125 and 225 to around 10; even Los Angeles, which has the greatest problems, has seen its record improve by some 40 percent. Air pollutant concentrations for which data are available have all fallen nationwide over the period 1960–90. – Julian L. Simon

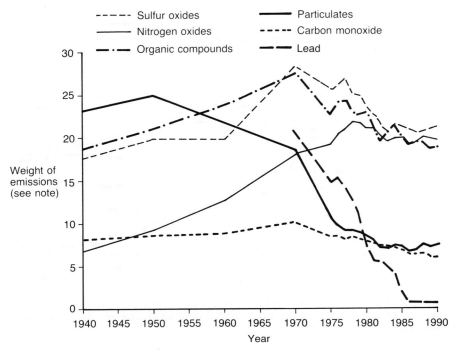

Figure 45.6 Emissions of major air pollutants in the USA
Note: In millions of metric tons per year, except lead in ten thousands of metric tons per year, and carbon monoxide in ten million metric tons per year.
Source: Council on Environmental Quality, Environmental Quality, 22nd Annual Report, 1992, 273. J. L. Simon, *The Ultimate Resource*, rev. edn. (Princeton: Princeton University Press, forthcoming).

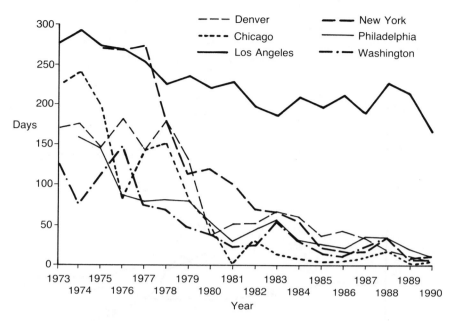

Figure 45.7 Air quality trends in major urban areas (number of days greater than the PSI index level)

Source: Council on Environmental Quality, Environmental Quality, 22nd Annual Report, 1992, 277. Council on Environmental Quality, Environmental Quality 1981, 12th Annual Report, 1981, 244.

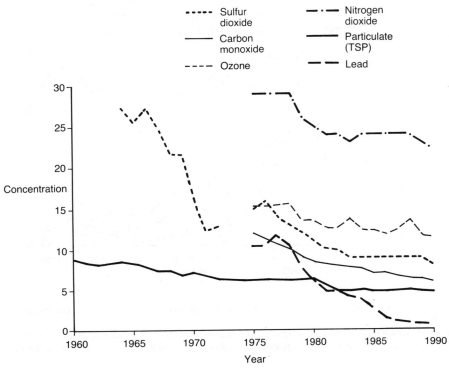

Figure 45.8 Pollutants in the air, USA, 1960–90
Source: Council on Environmental Quality, Environmental Quality, 22nd Annual Report, 1992,
276. Council on Environmental Quality, Environmental Quality 1981, 12th Annual Report,
1981, 243.

46

Comparative Trends in Resource Use and Pollution in Market and Socialist Economies

Mikhail S. Bernstam

If pollution grows in direct proportion to economies, then for the contemporary world with its $3,700 income per capita in 1988 to develop to the level of Western economies ($16,500 per capita in 1988) would mean multiplying global pollution per capita by 4.5 times.[1] This means a ninefold increase of global pollution when the world's population doubles, or ten times if Western economies continue to grow without reducing pollution. Can people of the world sustain a tenfold increase in environmental disruption?

If pollution must increase with economic growth, as virtually all the literature says,[2] long-term economic growth is self-defeating. Yet, the facts contradict the literature, and this chapter assembles data that offer a different perspective. The evidence shows divergent environmental trends in market and socialist economies. The data demonstrate that resource use and pollution actually decline in developed market economies as these economies grow. The data also indicate that pollution trends depend primarily on economic systems that determine the patterns of resource use, not on economic growth and industrial development per se.

DIVERGENT ENVIRONMENTAL TRENDS

The most significant development in modern environmental history has been the divergence in the trends in pollution and resource use in the last several decades along the lines of economic systems.

Pollution and Economic Growth in Western Market Economies

Pollution declined consistently in virtually all Western market economies in the 1970s and 1980s. Figure 46.1 presents the data on atmo-

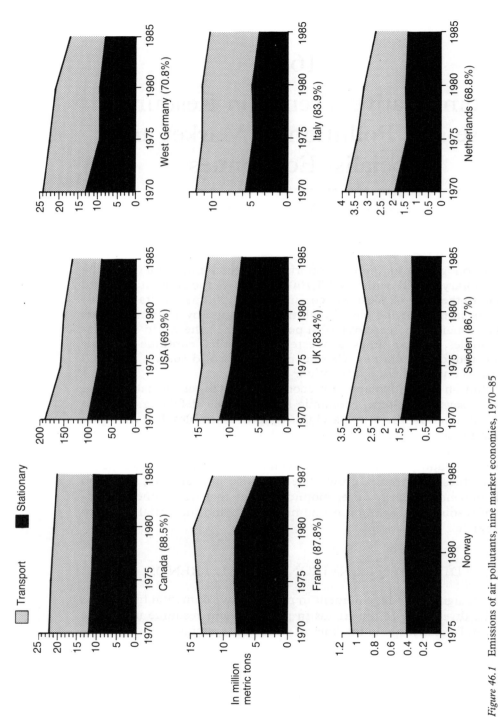

Figure 46.1 Emissions of air pollutants, nine market economies, 1970–85

Note: Percentages in parentheses show the total amounts of emissions in 1985 in a given country as a percent of those in 1970. The data for France refer to 1987 in lieu of 1985; no data for Norway for 1970 are available.

Sources: OECD, *OECD Environmental Data Compendium, 1989*, Paris: OECD, 1989, 21–9.

spheric pollution from both transportation and stationary sources (fuel combustion, industrial processes, etc.) in nine market economies from 1970 to 1985. Trends in Japan and other developed market economies (DMEs) are similar, but the data are sketchy.[3] Importantly, the most consistent and rapid declines have been in pollution from stationary sources. This primarily encompasses resource use in production sectors of market economies. Transportation, especially household use of vehicles, was less consistent and more recalcitrant in terms of economizing resources and reducing pollution. Overall trends are unambiguous nonetheless: emissions of atmospheric pollutants declined from 12 percent to 31 percent in the period 1970–85, depending on the country.

The downward trend encompasses the entire economic system. The United States was a leader in pollution reduction despite (or rather, because of) the fact that its economic growth was on par with Western Europe in the 1970s and twice as high in the first half of the 1980s.[4]

Figures 46.2–4, provide a detailed series of long-term data for the United States and reveal a major untold story of modern pollution. First, figure 46.2 shows that emissions of air pollutants in the USA were lower in 1988 than in 1940 by 12.5 percent (127.2 million metric tons [MMT] against 145.4 MMT, respectively). This is despite (or, because of) the fact that the US gross national product increased 5.21 times in constant 1982 dollars from $772.9 billion in 1940 to $4024.4 billion in 1988 (figure 46.4).

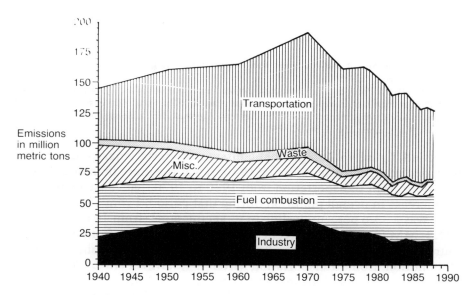

Figure 46.2 Emissions of air pollutants, by sources, USA, 1940–88
Source: US Environmental Protection Agency, *National Air Pollutant Emission Estimates, 1940–88*, Washington, DC: USGPO, 1990, 12–21.

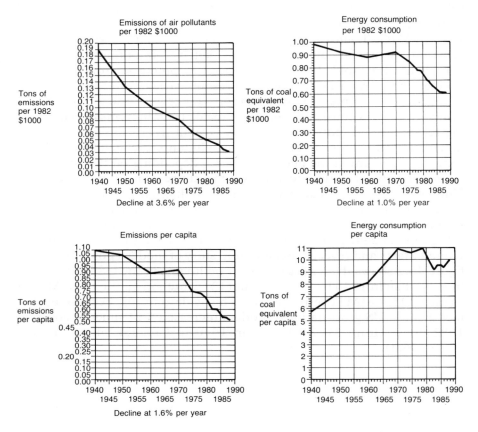

Figure 46.3 Indexes of the real gross national product, energy use, and emissions of air pollutants, USA, 1940–88
Sources: GNP: US President, *Economic Report of the President, 1990*, Washington, DC: USGPO, 1990, 296; energy: *UN Energy Yearbooks*, various years; *UN Statistical Yearbooks*, various years; emissions: US Environmental Protection Agency, *National Air Pollutant Emission Estimates, 1940–88*, Washington, DC: USGPO, 1990, 2.

Second, figure 46.2 shows that emissions from stationary sources have been declining consistently since 1940. It was the more than doubling of emissions from transportation from 1940 to 1970 that caused a temporary increase in overall pollution during that period. From 1970 to 1988, there was a 33.6 percent decline in total emissions (from 191.5 MMT to 127.2 MMT), a 28.5 percent decline in emissions from stationary sources (from 97.2 MMT to 69.5 MMT), and a 38.9 percent reduction of pollution from transportation (from 94.5 MMT to 57.7 MMT). Transportation emissions in 1988 were lower than in 1950 (57.7 MMT and 59.5 MMT, respectively). Emissions from both industrial processes and stationary fuel combustion were lower in 1988 than in 1940. Interestingly, the biggest decline in the early period 1940–60 was from the

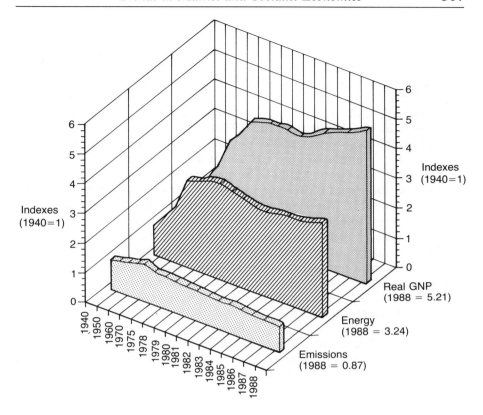

Figure 46.4 Emissions of air pollutants and energy use per capita and per 1982 $1,000, USA, 1940–88

Sources: GNP: US President, *Economic Report of the President, 1990*, Washington, DC: USGPO, 1990, 296; energy: *UN Energy Yearbooks*, various years; *UN Statistical Yearbooks*, various years; emissions: US Environmental Protection Agency, *National Air Pollutant Emission Estimates, 1940–88*, Washington, DC: USGPO, 1990, 2. Population: US Bureau of the Census, *Statistical Abstract of the United States, 1990*, Washington, DC: USGPO, 1990, 7.

so-called uncontrollable miscellaneous sources. Specifically, these were forest fires and home heating and cooking. In both cases, this was due to an unsung substitution of electricity, oil, and natural gas for wood and thus a great reduction in wood consumption.[5]

Figure 46.3 presents the key features of pollution decline in the USA since 1940. Atmospheric emissions per constant $1,000 of GNP have been consistently declining at 3.6 percent per year (from 188 kg in 1940 to 32 kg in 1988). Emissions per capita declined at 1.6 percent per year (from 1.1 tons to 0.5 tons). Energy consumption per constant $1,000 declined at 1.0 percent per year. Energy consumption per capita increased from 1940 to 1970 and then declined. One can notice that energy use (which constitutes 50–55 percent of the total mass of physical resource inputs in industrial countries) has more inertia and is more resistant to reductions over the course of technological progress than other resources. Reductions in the use of nonenergy resources as well as substitution toward less-polluting energy sources have been the main factors of the long-term decline in pollution per capita and per \$1,000 of GNP. Increase in energy efficiency per \$1,000 of GNP over the long run and per capita in the recent two decades was an additional factor. Interestingly, figure 46.3 implies that increasing efforts of environmental regulation in the 1970s and the 1980s did not make an identifiable impact on the long-term trend of declining atmospheric pollution.

Figure 46.4 summarizes the US trends in air pollution with respect to energy use and economic growth in the period 1940–88. Panel A shows actual data and Panel B presents indexes of emissions, energy consumption, and real GNP (the 1940 values equal unity). In short, the economy grew 5.2 times, total energy use increased only 3.2 times (an increase in the 1970s was small and there was no increase in the 11 years from 1978 to 1988), and pollution declined to 0.87 of its level in 1940. These trends, and not a parallel or other kind of joint increase, represent the true relationship between economic growth, energy use, and pollution.

Declining trends, similar to those in atmospheric emissions, can be observed in other forms of pollution and its human impacts in developed market economies. Figure 46.5 shows some improvements in water quality in the USA, the United Kingdom, and other Western countries (the Danube River of Hungary is presented for contrast). Figure 46.6 presents major declines in human exposure to pesticides in the USA, UK, and Japan over several decades.

Evaluating Pollution in Socialist Countries

Since comprehensive, comparable, and reliable data on air pollution in socialist countries are difficult to obtain, the stock measure of pollutant concentrations is preferable to the flow measure of emissions. The latter is subject to official assumptions of what proportion of polluting residuals of production and fuel combustion is abated by scrubbers and precipitators; the former measure is straightforward and the relevant data collection has much better coverage.[6] Figure 46.7 and 46.8 show comparable concentrations of three major air pollutants (SO_2, NO_2, and total suspended particulates) in market and socialist countries in 1985–87. Figure 46.7 presents absolute values and figure 46.8 contains "the Japanese indexes" (national levels normalized by Japanese levels of respective pollutants).

The figures are striking, for one would have expected similar levels of pollutant concentrations in market and socialist countries. Given similar population densities, if income and production levels per capita are two to three times higher in market economies and pollution per \$1,000 were two to three times higher in socialist economies, the two ratios should cancel each other and yield equal concentration levels. This

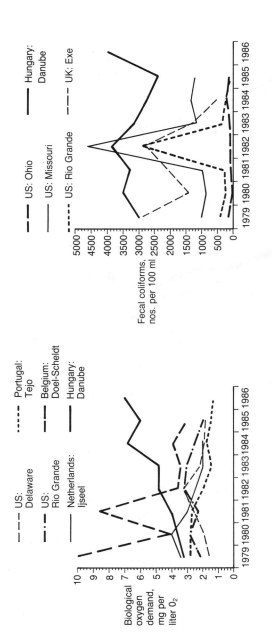

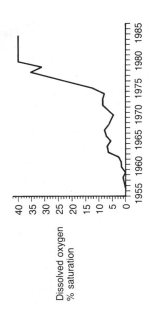

Improvements in dissolved oxygen, the Thames, 1955–85

Figure 46.5 Trends in water quality in developed countries, pollution indicators in rivers, USA and five European countries, 1979–86

Sources: USA and five European countries, 1979–86: UN Environment Programme, *Environmental Data Report*, London: Blackwell, 1989, 108–10, 117–19; the Thames: UN Environment Programme, *Environmental Data Report*, London: Blackwell, 1987, 52.

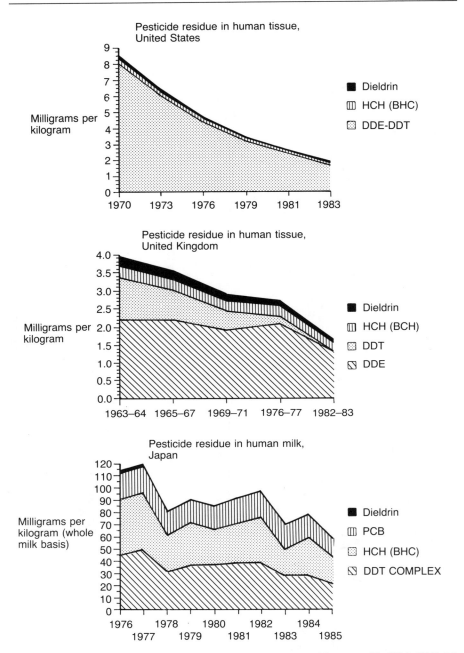

Figure 46.6 Pesticide residue levels in human adipose tissue and human milk, USA 1970–83, United Kingdom 1963–83, Japan 1976–85

Sources: USA and UK: UN Environment Programme, *Environmental Data Report*, London: Blackwell, 1987, 100–1; Japan: UN Environment Programme, *Environmental Data Report*, London: Blackwell, 1989, 196.

seems to be true only for Hungary, Poland, and Yugoslavia and also for the concentrations of nitrogen dioxide in the USSR. Low concentration of the latter, only 23 percent above the US level, is due to the fact that the bulk of NO_2 derives from transportation, and there are 9.5 times fewer vehicles per 1,000 persons in the USSR than in the USA,[7] while concentrations per vehicle are much higher. Unexpectedly high concentrations of sulfur dioxide and total suspended particulates (TSP) in the USSR, China, Czechoslovakia, and East Germany are higher than the levels in Japan and the USA by an order of magnitude of three to seven times.

By the law of conservation of mass, emissions of respective air pollutants per capita in these socialist countries must be roughly as high, relative to the USA, as the levels of pollution concentrations (given similar population densities). There are, however, conflicting data sources on the amounts of emissions of major air pollutants. The USSR State Committee on Statistics and similar agencies in other socialist countries employ idiosyncratic definitions and assumptions concerning estimates of emissions. Only estimates of emissions from transportation are reliable. They are based on the registered number of vehicles and the emissions per vehicle that are easy to estimate given a limited variety of vehicles and uniform qualities of fuels. Estimates of emissions from stationary sources include only pollution from fuel combustion and industrial processes and exclude emissions from uncontrollable miscellaneous sources and solid waste,[8] that is about 20–30 percent of pollution from stationary sources by Western standards.

More importantly, most official estimates of emissions from stationary sources in the former USSR and other developed European socialist countries are based on the assumed amounts of emissions from given technological processes and the assumed coefficients of emissions abatement by scrubbers and precipitators.[9] These coefficients are reported to have increased linearly in the USSR from 73 percent of all emissions produced in 1980 to 77 percent in 1988.[10] However, influential sources from the then Soviet Union and Czechoslovak, such as a leading environmental spokesman of the then newly elected USSR parliament, Professor Aleksei Yablokov, and the spokesman for the then newly created Czechoslovak Ministry of the Environment have insisted that these coefficients of pollution abatement represent pure bureaucratic fiction. Czechoslovakia used Soviet-made scrubbers and precipitators. Their actual capacity of abating emissions is about 30 percent according to the new Ministry of the Environment.[11] In the former USSR, only 50 percent of industrial enterprises do have working scrubbers and precipitators; they are largely ineffective, break often, and are often turned off during the night shifts. In Moscow, the actual coefficient of abatement of emissions from fuel combustion is 28 percent; it is about 18 percent in other major industrial cities.[12] In addition, emissions only from 46,000 industrial units are taken into account, while there are more than 80,000 registered industrial units that produce pollution.[13]

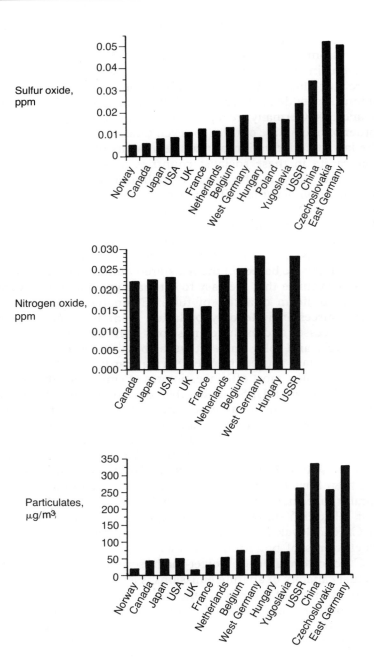

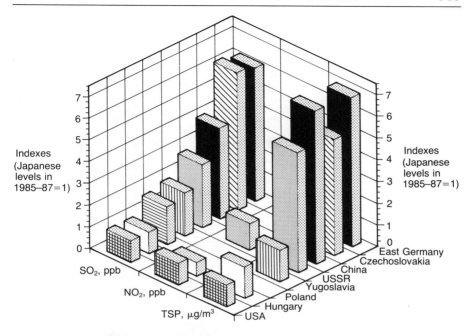

Figure 46.8 The Japanese indexes of concentrations of air pollutants, eight countries, 1980s, latest available year
Sources: Derived or calculated from the following sources by countries: Japan, UK, and Yugoslavia: OECD, *OECD Environmental Data Compendium, 1989*, Paris: OECD, 1989, 32–3, 35, 37. For other countries, see sources for figure 46.7.

In this situation, I chose to use two bounds of estimates of atmospheric pollution in developed European socialist countries (this procedure does not apply to China). The lower bound employs the data on emissions from stationary sources as reported by statistical agencies with

---

Figure 46.7 (see opposite)   Concentrations of three air pollutants, 16 countries, 1985–7
Sources: Derived or calculated from the following sources by countries: Western Europe, Canada, Japan, and Yugoslavia: OECD, *OECD Environmental Data Compendium, 1989*, Paris: OECD, 1989, 32–3, 35, 37; Yugoslavia, China, Poland: UN Environmental Programme, *Environmental Data Report*, London: Blackwell, 1989, 32–4; Yugoslavia: Yugoslavian Federal Statistical Office, *Statisticki Godisnjak Yugoslavije, 1989*, Beograd, 1989, 85–6; China: D. Elsom, *Atmospheric Pollution: Causes, Effects and Control Policies*, London: Blackwell, 1987, 235; Li Ming, *Zhongguo: Zhuang Zheng She Ji Liu Feng Jie*, Tainjin, 1989, 79. Hungary: Hungarian Central Statistical Office, *Statisztikai Kiado Vallalat, 1988*, Budapest, 1990, 339–40; Czechoslovakia: Jiri Pehe, "A Record of Catastrophic Environmental Damage," *Report on Eastern Europe*, vol. 1, no. 12 (1990), 5–6; GDR: H. F. French, *Green Revolutions: Environmental Reconstruction in Eastern Europe and the Soviet Union*, Washington, DC: Worldwatch Institute, 1990, 11; USSR: USSR State Committee on Statistics, *Okhrana Okruzhaiushchey Sredy i Ratsionalnoe ispolzoyanie Prirodnykh Resursov y SSSR Statsticheskii Sbornik*, Moscow: Finansy i Statistika, 1989, 28–31; US: US Bureau of the Census, *Statistical Abstract of the United States, 1990*, Washington, DC: USGPO, 1990, 203; US Environmental Protection Agency, *National Air Quality and Emissions Trends Report, 1988*, Washington, DC: USGPO, 1990, passim.

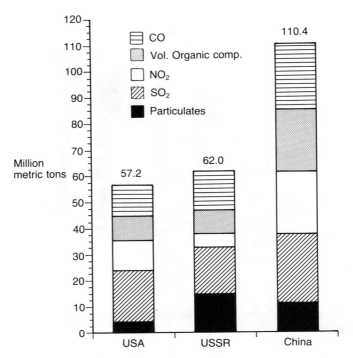

Figure 46.9 Emissions of air pollutants from fuel combustion and industrial processes, by types, USA, USSR, and China, 1988
Sources: US: US Environmental Protection Agency, *National Air Pollutant Emission Estimates, 1940–88*, Washington DC: USGPO, 1990, 17–21. USSR: USSR State Committee on Statistics, *Narodnoe Khoziaistvo SSSR y 1988 Godu: Statisticheskii Ezhegodnik*, Moscow: Finansy i Statistika, 1989, 249. (Note: the data is based on the assumption that 77 percent of emissions are abated while in reality about 30 percent may be abated. If the latter is the case, the total amount of these emissions was 180 MMT.) China: derived and calculated from State Statistical Bureau of People's Republic of China, *China: Statistical Yearbook, 1989*, Beijing, 1990, 683.

their assumed 73 percent (for 1980) and 77 percent (for 1988) levels of abatement. The higher bound employs the same data recalculated with the assumption that the actual abatement proportion is 30 percent of emissions from stationary sources. The data on pollution from transportation are taken as given. I give more confidence to the higher bound of estimates. It is more consistent with the data on both actual pollution concentrations and the trends in Soviet use of energy and other natural resources relative to the USA.

As figure 46.9 shows, even at the lower bound of estimates, atmospheric emissions from fuel combustion and industrial processes are higher in the USSR than in the USA in 1988 (62.0 MMT and 57.2 MMT, respectively). The amount of emissions from fuel combustion and industrial processes in China in 1988 is calculated to be 110.4

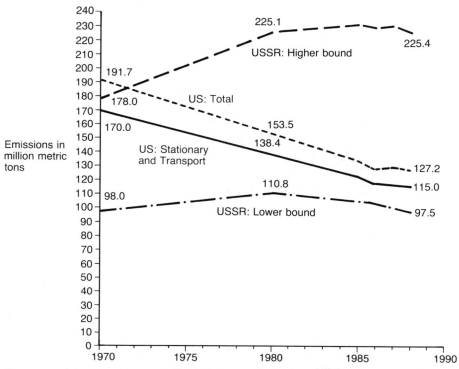

Figure 46.10 Emissions of air pollutants, USA and USSR, 1970–88 (selected years), different types of measurement

Sources: US: US Environmental Protection Agency, *National Air Pollutant Emission Estimates, 1940–88*, Washington DC: USGPO, 1990, 2, 12–21. USSR: R. McIntyre and J. Thornton (1970): "Environmental Divergence: Air Pollution in the USSR," *Journal of Environmental Economics and Management*, vol. 1 (Aug. 1974), 119. (Note: it is assumed, in accordance with the data for the 1980s, that 60 million tons of emissions in 1970 refer to stationary sources only. It is also assumed, in the absence of other data, that 38 million tons of emissions from transport in 1980 can be applied to 1970.) USSR, 1980–8: USSR State Committee on Statistics, *Okhrana Okruzhaiushchey Sredy i Ratsionalnoe ispolzoyanie Prirodnykh Resursov y SSSR Statsticheskii Sbornik*, Moscow: Finansy i Statistika, 1989, 7, 83. (Note: the lower bound of emissions in the USSR refers to official estimates whereas emissions from stationary sources assume 73 percent to 77 percent abatement ratios. The higher bound estimates assume that emissions from stationary sources are abated by 30 percent; official estimates of emissions from transportation remain unchanged. Since the data on Soviet emissions include those from transportation and stationary sources and exclude emissions from solid waste, forest burning, and other miscellaneous sources, the US data are presented in a comparable series of emissions from transportation and stationary sources. The total US emissions series is also presented.)

MMT. This is almost twice as high as in the USA, while Chinese GNP constituted only 10.4 percent of that in the USA.[14]

Figure 46.10 shows comparative trends in total amounts of emissions of atmospheric pollutants in the USA and the USSR in 1970–88. At the higher bound of Soviet estimates, which are more reliable, the trends clearly diverge. Comparably defined and estimated emissions start at

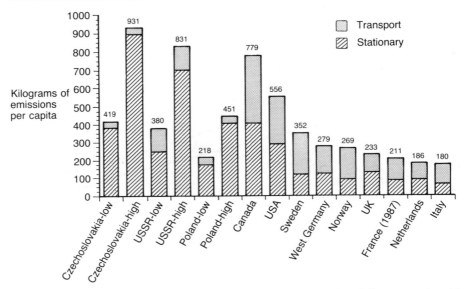

Figure 46.11 Emissions of air pollutants per capita, market and socialist economies, 12 countries, 1985

Sources: OECD countries: *OECD Environmental Data Compendium, 1989*, Paris: OECD, 1989, 21–9, 305. Socialist countries: USSR State Committee on Statistics, *Okhrana Okruzhaiushchey Sredy i Ratsionalnoe ispolzoyanie Prirodnykh Resursov y SSSR Statsticheskii Sbornik*, Moscow: Finansy i Statistika, 1989, 7, 84–5; population data from national statistical yearbooks. (Note: the lower bound of estimates of emissions in socialist countries assumes that about 76 percent of emissions from stationary sources is abated; the higher bound of estimates assumes that 30 percent of emissions from stationary sources is abated.)

approximately similar levels in 1970 – 170 MMT in the USA and 178 MMT in the USSR. Emissions decline rapidly in the USA to 115 MMT and increase as rapidly in the USSR to 225 MMT in the 1970s and the early 1980s. They continued to decline in the USA in the second half of the 1980s. There was also a slight decline in pollution in the USSR in the second half of the 1980s due to political reasons. A number of the most polluting plants were closed down due to popular pressures in the new democratic situation.[15] The best estimates suggest that, comparably evaluated, Soviet emissions were almost twice as high as those in the USA in 1988.

Figures 46.11 and 46.12 compile the data on emissions of atmospheric pollutants per capita and per $1,000 in nine developed market countries and three developed socialist countries, the USSR, Czechoslovakia, and Poland, in 1985. Both the lower and the higher bounds of estimates are presented for socialist countries. Even the lower bound estimates in socialist countries were two or more times as high per $1,000 as in market economies (with the exception of Canada). The higher bound estimates increase the ratio to four to six times. On a per capita basis, the lower bound estimates in socialist countries were lower

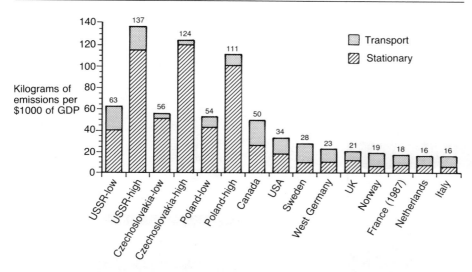

Figure 46.12   Emissions of air pollutants per $1,000 of GDP, market and socialist economies, 12 countries, 1985
Sources: Pollution: market countries: OECD, *OECD Environmental Data Compendium, 1989*, Paris: OECD, 1989, 21–9, 305. Socialist countries: USSR State Committee on Statistics, *Okhrana Okruzhaiushchey Sredy i Ratsionalnoe ispolzoyanie Prirodnykh Resursov y SSSR Statsticheskii Sbornik*, Moscow: Finansy i Statistika, 1989, 7, 84–5. (Note: the lower bound of estimates of emissions in socialist countries assumes that about 76 percent of emissions from stationary sources is abated; the higher bound of estimates assumes that 30 percent of emissions from stationary sources is abated.) GDP: Market countries: US Bureau of the Census, *Statistical Abstract of the United States, 1990*, Washington, DC: USGPO, 1990, 841; Socialist countries: V. Marynov, "SSSR i SShA po Materialam Mezhdunarodnykh Sopostavlernii OON i SEV (Raschety Goskomstata SSSR)," USSR State Committee on Statistics, *Vestnik Statistik* 9 (1990), 11, 13, 15. (Note: estimates of GDP are given in purchasing power parities in 1985 dollars.)

than emissions in the USA and Canada but higher than in western European countries. The higher bound estimates per capita in the USSR and Czechoslovakia (but not in Poland) were about twice as high as in most market countries (Canada was an exception). Characteristically, the bulk of emissions in market economies derives from transportation, that is primarily from personal consumption, while in socialist countries most emissions derive from industrial production (fuel combustion and industrial processes).

Divergence between market and socialist economies also occurred in the last several decades in trends in water pollution, human exposure to pesticides, and other environmental aspects.[16] I showed improvements in water quality in western countries in figure 46.5. figure 46.13 compares water pollution discharges per capita in the USSR and ten market economies and presents the dynamics of Soviet water pollution in the 1980s. Soviet discharges are more than twice as high as in most market economies and more than seven times as high as those in West Germany

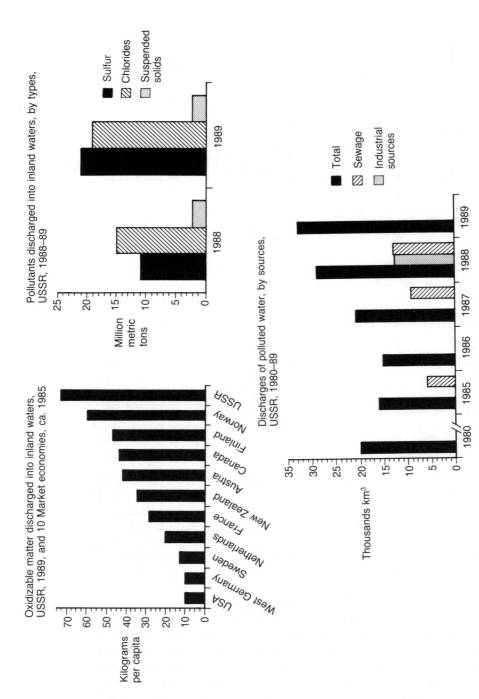

Figure 46.13 Inland water pollution, USSR, 1980–9, and ten market economies, ca 1985
Sources: Market economies: OECD, *OECD Environmental Data Compendium, 1989*, Paris: OECD, 1989, 59. USSR: USSR State Committee on Statistics, *Sotsialnoe Razvitie v SSSR: Statisticheskii Sbornik*, Moscow, 1990, 351, 356; USSR State Committee on Statistics, *Okhrana Okruzhaiushchey Sredy i Ratsionalnoe ispolzoyanie Prirodnykh Resursov y SSSR Statsticheskii Sbornik*, Moscow: Finansy i Statistika, 1989, 79–80; USSR State Committee on Statistics, *Narodnoe Khoziaistvo SSSR y 1988 Godu: Statisticheskii Ezhegodnik*, Moscow: Finansy i Statistika, 1989, 569 (hereafter *NKh* and reference year); *NKh 1988*, Moscow, 1989, 247; *NKh 1989*, Moscow, 1990, 248.

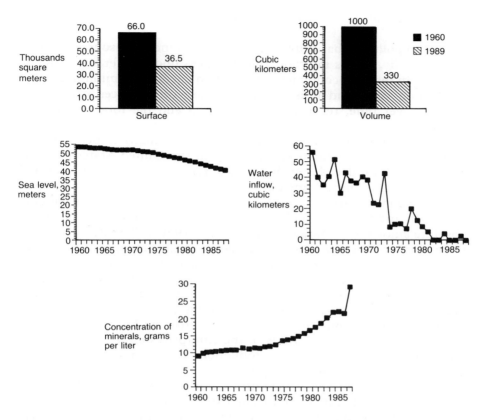

Figure 46.14   The contraction and pollution of the Aral Sea, 1960–89

Sources: E. D. Rakhimov, A. N. Kravchenko, et al., *Sotsialno-Ekonomischeskie Problemy Arala i Priaral'ia*, Tashken: fan, 1990, 8–9; S. Melnikov, "Aral," *Rabochaia Tribuna*, Moscow, Oct. 5, 1990, 3.

and the USA. Figure 46.13 also shows rapid growth in water pollution in the former Soviet Union. One can observe and estimate that discharges increased from both industrial sources and sewage.

Figure 46.14 illustrates the best known Soviet environmental case. This is the destruction of the Lake of Aral (the Aral Sea) in Central Asia due to the diversion of the inflow of rivers Syr Daria and Amu Daria to irrigation of cotton plantations and to the residual discharge of mineral fertilizers and pesticides into the sea. From 1960 to 1989, the inflow of rivers virtually ceased. Sea level declined from 53 meters to 40 meters, the surface of the sea shrank by almost one half, the volume of water declined to one third of the 1960 level. Most of the sea has virtually disappeared and the remaining water is heavily poisoned, to the detriment of the population, which has no alternative sources of drinking water.

A Note on Resource Use and the Underlying Systemic Forces

Trends in pollution basically derive from the trends in resource use and, more broadly, trends in production practices under different economic systems. In market economies, competition encourages minimization of production costs and thus reduces the use of resources per unit of output. Over time, resource use per capita and the total amounts of resource inputs also decline and this, in turn, reduces pollution. For example, figure 46.3 shows that energy consumption per constant $1,000 of the GNP has been declining in the United States over the course of several decades at the average rate of 1 percent per year and that the speed of this decline has been accelerating. This acceleration of the decline in resource use led to the decline of energy use per capita in the 1970s and the 1980s and, as figure 46.4 shows, to the decline in the absolute amounts of energy use in the early 1980s. Elsewhere, I discuss in detail a more general case of the decline in the amounts of consumption of various resources across developed market economies.[17]

By contrast, regulated state monopolies in socialist economies maximize the use of resources and other production costs. This is because under a regulated monopoly setting, prices are cost-based and profits are proportional to costs. Accordingly, the higher costs justify higher prices and higher profits of the enterprises *vis-à-vis* the government which controls the entire production. This high and ever-growing use of resources per unit of output explains the high extent of environmental disruption in socialist countries. For example, in the USSR, total energy consumption increased from 1,074 MMT of coal equivalent in 1970 to 1,473 MMT in 1980 to 1,954 MMT in 1988, that is, by 37.2 percent and by 32.7 percent respectively.[18] This means an acceleration of the growth of energy use from 3.2 percent per year to 3.6 percent per year. Thus, the recent environmental split between market and socialist economies merely reflects their systemic differences in resource use.

NOTES

1   Calculated from The World Bank, *World Development Report 1990*. New York: Oxford University Press, 1990, 178–9; and US Directorate of Intelligence, *Handbook of Economics Statistics, 1990*. Washington, DC: USGPO, 1990, 28, 31, 36, 50–1. The GNP for socialist countries is reduced according to V. Martynov, "SSSR i SShA po Materialam Mezhdunarodnykh Sopostavlenii OON i SEV. (Raschety Goskomstata SSSR)." USSR State Committee on Statistics, *Vestnik Statistiki*, 9 (1990), 10–15.

2   Robert U. Ayres and Allen V. Kneese, "Production, Consumption and Externalities." *The American Economic Review*, 59 (2) (May 1969), 282–97; Allen V. Kneese, Robert U. Ayres, and Ralph C. d'Arge, *Economics and the Environment: A Materials Balance Approach*. Baltimore, MD: The John Hopkins Press for Resources for the Future, Inc., 1970; US Commission on Population

Growth and the American Future, Ronald G. Ridker, ed. *Population, Resources, and the Environment,* vol. 3. Washington, DC: USGPO, 1972, 43–9; John P. Holdern and Paul R. Ehrlich, "Human Population and the Global Environment." *American Scientist,* 62 (1974), 282–92; Allen V. Kneese, *Economics and the Environment.* New York: Penguin Books, 1977, 64–71; Julian L. Simon, *The Ultimate Resource.* Princeton, NJ: Princeton University Press, 1981, 241; Margaret E. Slade, "Natural Resources, Population Growth and Economic Well-Being." In *Population Growth and Economic Development: Issues and Evidence,* D. Gale Johnson and Ronald D. Lee, eds. Madison: The University of Wisconsin Press, 1987, 339; and many others. The former Soviet President Mikhail S. Gorbachev observed that there was "the conflict between industry and nature," and "the conflict between consumer society and nature." He concluded: "It would be an environmental catastrophe if all the countries of the world tried to achieve the standard of living of the US" (Mikhail S. Gorbachev, "Interview to *Time* Magazine," *Time,* June 4, 1990, 28).

3 OECD, *OECD Environmental Data Compendium, 1989* Paris: OECD, 1989, 21–9.

4 US President, *Economic Report of the President, 1987.* Washington, DC: USGPO, 1987, 368.

5 F. Landis MacKellar and Daniel R. Vining, Jr, "Natural Resource Scarcity: A Global Survey." In *Population Growth and Economic Development: Issues and Evidence,* D. Gale Johnson and Ronald D. Lee, eds. Madison: University of Wisconsin Press, 1987, 312; US Bureau of the Census, *Statistical Abstract of the United States, 1990* Washington, DC: USGPO, 1990, 673 (hereinafter *SAUS* and the year of reference).

6 The data on concentrations of air pollutants are collected in almost 500 cities including virtually all major industrial sites. The data on emissions are collected from only 46,000 industrial units while the total number of polluting industrial units is more than 80,000 (E.Iu. Bezuglaia and A. S. Zaitsev, "Chem Dyshit Gorod." In *Ekologichskaia Alternativa,* M.Ia. Lemeshev, ed. Moscow: Progress, 1990, 174, 178.

7 *SAUS 1990,* 843 (applied to national population data).

8 This is obvious from the fact that the sum of the reported amounts of emissions from stationary sources that are claimed to be abated and that are claimed to be discharged is always equal to the amount abated divided by the assumed abatement proportion. See USSR State Committee on Statistics, *Okhrana Okruzhaiushchey Sredy i Ratsionalnoe Ispolzovanie Prirodnykh Resusrsov v SSSR. Statisticheskii Sbornik.* Moscow: Finansy i Statistika, 1989, 7, 83–5 (hereinafter *Okhrana*). See also State Statistical Bureau of People's Republic of China, *China Statistical Yearbook, 1989.* Beijing, 1990, 683. However, emissions from uncontrollable sources (forest fires, home heating and cooking, etc.) and solid waste are not subject to abatement by scrubbers and precipitators and cannot be estimated by this official procedure. Generally, these two sources are never mentioned or otherwise considered in official reports as sources of air pollution.

9 E.Iu. Bezuglaia and A. S. Zaitsev, op. cit., 178.

10 *Okhrana,* pp. 7, 83; USSR State Committee on Statistics, *Narodnoe Khoziaistvo SSSR v 1989 Godu. Statisticheskii Ezhegodnik.* Moscow, Finansy i Statistika, 1990, 251 (hereinafter *NarKhoz* and the year of reference).

11 Marlise Simons, "Pollution's Toll in Eastern Europe: Stumps Where Great Trees Once Grew." *The New York Times,* March 19, 1990, 9.

12   Aleksei Yablokov, "Probuzhdenie ot Ekologisheskoi Spiachki." *Rodina*, 4 (Moscow, 1990), 66.

13   E.Iu. Bezuglaia and A. S. Zaitsev, op. cit., 178.

14   *SAUS 1990*, p. 840. The GNP data are for 1987.

15   Mikhail S. Gorbachev, op. cit., 28.

16   Water pollution and exposure to residue of pesticides and mineral fertilizers made a major adverse impact on infant mortality in the areas of cotton plantations in Soviet Central Asia in the 1970s and the 1980s. See statistical analysis in Elwood Carlson and Mikhail S. Bernstam, "Population and Resources Under the Socialist Economic System." In *Resources, Environment, and Population: Current Knowledge, Future Options*, Supplement to *Population and Development Review*, Kingsley Davis and Mikhail S. Bernstam, eds, 16 (1990).

17   Mikhail S. Bernstam, "The Wealth of Nations and the Environment," ibid.

18   United Nations, *Statistical Yearbook 1971*. New York: United Nations, 1972, 339; United Nations, *World Energy Supplies, 1950–1974*. New York: United Nations, 1976, 111; United Nations, *1983 Energy Statistics Yearbook*. New York: United Nations, 1985, 31; *1988 Energy Statistics Yearbook*. New York: United Nations, 1990, 31.

47

Acid Rain

J. Laurence Kulp

In the 1970s, some scientists (e.g., Likens et al., 1979) speculated that increased acidity in rain from industrial activity was causing serious damage to crops, forests, human health, building materials, surface waters, and visibility. Further, they assumed that if additional controls were not instituted immediately the problem would get progressively worse, leading to ecological disaster. The media picked up this theme with enthusiasm, and today the average American is convinced that "acid rain" is a serious problem. Fortunately, this is not the case (NAPAP, 1987; Lefohn, 1988; and NAPAP, 1991).

After a decade of research by hundreds of scientists costing roughly $0.5 billion coordinated by the US National Acid Precipitation Assessment Program (NAPAP), the effects of present levels of acid rain were found to range from net positive (e.g., on crops) to modest negative (e.g., surface water). There is no evidence that any significant worsening of these effects will occur over the next half century even if the present levels of pollution were to continue. Actually, current regulations will produce a steadily decreasing acidity of rain in the USA, Europe, and Japan. Finally, new technology for control has emerged so that in the future new coal-burning plants can operate with negligible levels of acid-forming emissions at no greater cost than plants using the older technology (NAPAP, 1991).

CAUSES OF ACID RAIN

Virtually all rain in nature, prior to the Industrial Revolution, was acid. The relatively recent term *acid rain* refers to the *additional* acidity in rain coming from emissions of sulfur dioxide, primarily from coal-burning utility plants, and nitrogen oxides, primarily from vehicles. *Acid rain* is often used loosely in the press to include all air pollutants such as ozone, sulfur dioxide, and nitrogen dioxide gases, in addition to sulfuric and nitric acids of industrial origin, but in this chapter and in strictly scientific terms only the acids are included.

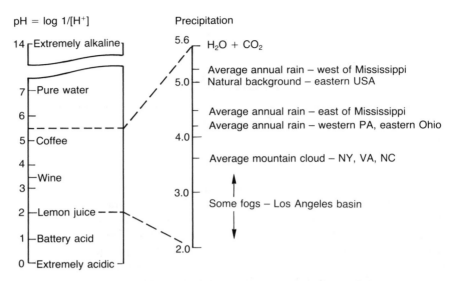

Figure 47.1   pH scale, acidity of common substances and current rain/clouds/fog

To define acidity, the chemist uses a pH scale in which each lower unit of pH represents a ten-fold increase in acid concentration. A pH of 7 means a solution is neutral, of 14, highly basic, and 1, highly acid. Figure 47.1 shows the relative acidity of various natural waters and other substances.

Pre-industrial (natural background) rain over forested areas in the temperate zone had an average annual pH of about 5.0, whereas the most polluted rain (annual weighted average) that falls in the eastern USA is about eight times more acid at a pH of 4.2. The pre-industrial rain contained sulfuric, nitric, and organic acids of natural origin. Note also that fogs and clouds carry higher concentrations of pollutants than does accompanying rain. Individual rains and parts of rains differ in their acidity, but the annual or seasonal average is generally the most important value insofar as effects are concerned (NAPAP, 1987).

The bulk of the acidity in rain comes from the reaction of sulfur dioxide (SO_2) with hydrogen peroxide in clouds, which produces sulfuric acid. This is the important mechanism in the summertime, when most acidic deposition occurs. The hydrogen peroxide is formed from photochemical reactions of volatile organic compounds derived from such divergent sources as trees and automobile exhaust. In the eastern USA in the summer, the hydrogen peroxide in clouds is commonly sufficient to convert virtually all of the SO_2 in the clouds to sulfuric acid. In the winter, however, the hydrogen peroxide concentration is considerably lower, so much of the emitted SO_2 goes out to the Atlantic Ocean without reacting.

Nitric acid is formed in air by the oxidation of nitrogen oxides. It contributes only a fraction of the acidity in acid rain and decreases more rapidly than sulfuric acid with distance from the source of emissions.

Since the emission of volatile organic compounds from natural sources, and therefore the consequent hydrogen peroxide, cannot be controlled, only reduction of SO_2 emissions from large coal-burning utility plants can significantly decrease the sulfuric acid component of the rain. Control of nitrogen dioxide emissions from these plants will make only a secondary contribution to the reduction of acidity in rain over most of the country.

DISTRIBUTION AND TRENDS

The distribution of the average annual acidity of rain over the USA is shown in figure 47.2 for 1990. (The pattern has not changed greatly over the last decade.) The area having the highest acidity in rain – i.e. western Pennsylvania, eastern Ohio, southwestern New York, and northern West Virginia – lies about one day's average travel time downwind of air masses from the Ohio River Valley area, which has the highest concentration of uncontrolled coal-burning plants. West of the Mississippi valley, north of Maine, and in southern Florida the acidity of the rain approaches that of the pre-industrial, unpolluted level (i.e., pH = ~5.0). In the western USA, aside from the heavily polluted Los Angeles basin, the average annual acidity in rain is less than the natural background of the eastern forested area (i.e., pH > 5.0) due to the alkaline dust in the air, which neutralizes the acid. Also in the west the SO_2 emissions are much less than in the eastern USA. This pattern also emphasizes that acid rain is not a global but rather a regional problem, with highest deposition occurring within several hundred kilometers of the source.

The trend in the emissions of SO_2 and NO_x in the USA since the turn of the century is shown in figure 47.3. It is evident that industrial acid rain has been with us throughout this entire period and, therefore, so have the effects.

The emissions of SO_2 today are about what they were in 1930. Prior to 1972, the SO_2 curve reflects coal use. Without emission control, these emissions reached a maximum in 1972, after which implementation of the 1970 Clean Air Act caused emissions to decline. Since the peak of emissions in 1972, however, the acidity of rain has not decreased as sharply as the SO_2 emissions. This is probably due to the concurrent control of particulate matter, required by the Clean Air Act, which is generally alkaline and thus became less effective in neutralizing the acids.

From 1972 to 1990, the SO_2 emissions have *decreased* by about 30 percent, while the total coal use has *increased* by about 80 percent. This contradicts the popular fallacy that more energy use means more pollu-

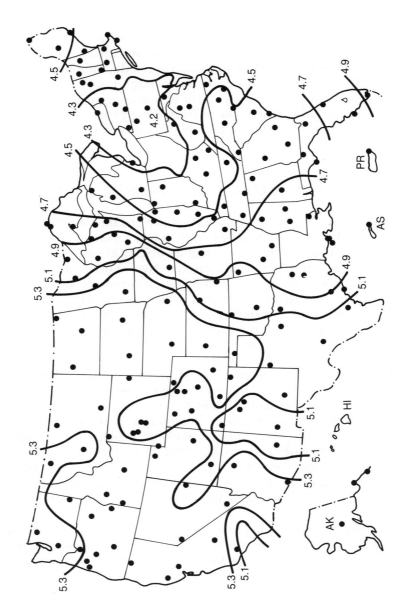

Figure 47.2 Distribution of acid rain over the USA in 1990 (NAPAP data)

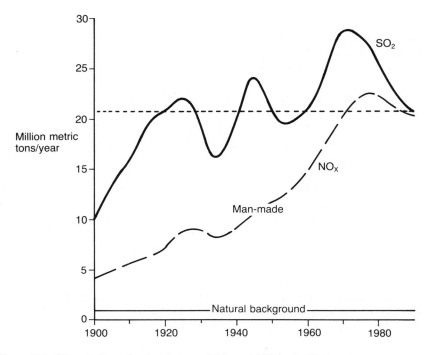

Figure 47.3 Historical trends of emissions of SO_2 and NO in the USA, 1900–90

tion. In fact, energy generation and pollution are independent parameters. Pollution related to energy production and use can be reduced to any desired level given resources and appropriate technology.

NO_x emissions began at a lower level than SO_2 emissions at the turn of the century, primarily tracked the growth of vehicles, reached a maximum in the later 1970s, and now will decline based on current laws and regulations. (See chapters 44 and 45 by Elsom and by Ellsaesser in this volume.)

EFFECTS OF ACID RAIN

The effects (positive and negative) of acid rain at current (ambient) concentrations are now rather well understood. It is important to emphasize the term *ambient*, for at sufficiently high concentrations of acidity almost anything will be damaged. What is relevant is the benefit or damage attributable to acid rain for a particular receptor at observed concentrations of acidity.

The six receptors that have been of concern in the acid rain issue are discussed below in order of increasing negative effects ("damage").

Crops (clearly beneficial)

Based on numerous controlled experiments in which many species have
been exposed over a growing season to simulated acid rain, it has been
established that there is no measurable retardation of growth at current
levels of acid rain observed in the agricultural areas of the US. Figure
47.4 shows the results of some representative experiments on several
major crops. There is no negative effect on yield even at ten times
greater acidity (i.e., pH = 3.2) than is encountered under current
conditions in areas of highest acid rain, e.g., eastern Ohio.

The average nitrate component (3 kg/ha/yr) and, to a lesser extent, the
average sulfate component (7 kg/ha/yr) of acid rain east of the Rocky
Mountains are actually beneficial to nearly all crops. This fertilization,
valued at several hundreds of millions of dollars per year, reduces the
amount the farmer must otherwise employ (Irving, 1987).

Forests (probably slight net benefit)

Controlled experiments exposing a wide range of tree species as seedlings to simulated acid rain over the ambient range in concentration have
shown no negative effects. The final NAPAP report (NAPAP, 1991,
chapter 16) concludes, "The vast majority of forests in the U.S. and

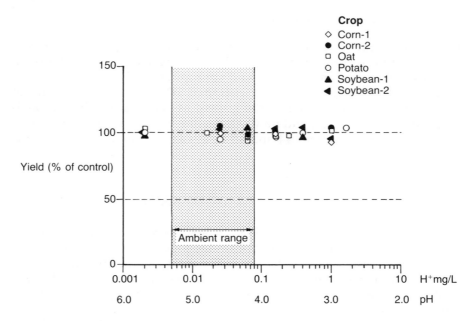

Figure 47.4 Effect of simulated acid rain over the growing season on the yield of several
important agricultural crops

Canada are not affected by decline. . . . Moreover there is no case of forest decline in which acidic deposition is known to be a predominant cause." Figure 47.5 shows the results of the first carefully controlled experiment (NAPAP, 1987) using seedlings from three very different tree species (grown in relatively poor soil to maximize any negative soil effects on nutrition from the simulated acid rain) over a period of three years. Like the results with agricultural crops, there is no negative effect on growth of these trees even at ten times the acidity of the average rain on the eastern US forests (Kulp, 1987).

Many other controlled experiments on the dose/response of simulated acid rain in the frequency and amount observed in the forested areas show similar results, including many genotypes of loblolly pine, eastern hardwoods, and red spruce. Further experiments were done with saplings so that the critical physiological age range of trees was covered. Again, no negative results were observed. It must be concluded that in the ambient range of acid rain in the United States there is no impairment of growth by foliar damage.

Theoretically there could be deleterious effects on the soil from continued acid deposition over the years due to a variety of mechanisms such as leaching of nutrients, impairment of the decomposition of the humus layer, injury to microorganisms required for plant growth, aluminum toxicity, etc. However, controlled experiments and field observations on a wide variety of soil types have not shown significant effects in the ambient range of acid rain. Theoretical models of soil leaching have been constructed that predict that some soils might show negative effects for tree nutrition after 50 years. However, acid rain at similar concentration to today's has been present for at least the last 60 years

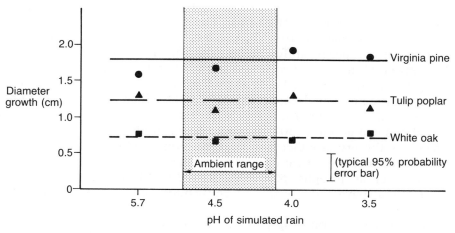

Figure 47.5 Diameter growth of 2- to 5-year-old seedlings exposed to different concentrations of simulated acid rain over 30 months

(figure 47.3), and no tree decline has been found in the USA that could be specifically attributed to decreased soil nutrition from acid rain.

Extensive surveys in natural forests and commercial plantations at low elevation, i.e., below average cloud base, over the eastern and north-western states, have failed to identify any regional decline that could not be attributed to natural causes (NAPAP, 1991). (These low elevation forests, i.e., below the cloud base of about 700 meters in the Appalachian mountains, represent more than 99.9 percent of the forested area east of the Mississippi.)

In the northern Appalachians at 800–1,000 meters, red spruce has experienced a decline in growth accompanied by increased mortality following a succession of severe winters in the early 1960s. Associated birch and balsam fir were not affected. The mature spruce trees that were heavily damaged in the 1960s are now mostly dead, but the younger and less damaged specimens seem to be recovering. There are also vigorous red spruce seedlings and saplings in the mountains, despite nearly constant average cloud acidity over these years. This red spruce decline of the 1960s has been attributed by most researchers to winter freezing and desiccation injury due to the severe winters of the late 1950s and early 1960s. Similar widespread mortality occurred over the Adirondack Mountains in New York and the Green and White Mountains in Vermont and New Hampshire respectively a century ago, coincident with another series of severe winters but long before there was any significant air pollution. Unusually severe drought during the mid-1960s may also have added to the stress. Therefore this decline can be entirely accounted for by natural causes without invoking acid rain.

However, red spruce is much more susceptible to freezing damage than is fir or birch, and, at a high enough acidity (pH < 3.0, mist applied to red spruce seedlings during the growing season has been shown to reduce their frost hardiness, but there is little effect at pH 3.6, the average pH of the ambient clouds. The fraction of the time that the ambient clouds have a pH < 3.0 is less than 3 percent. One other experiment comparing different branches of the same tree exposed to different combinations of mist and ozone at ambient and lower acidity showed slightly less frost hardiness when exposed during the growing season to the more acid mist (NAPAP, 1991). Since the acidity of rain, or the clouds, had no abrupt change during the early 1960s – or indeed probably since about 1930 – it would seem unlikely that such a mist effect was the primary factor in triggering this decline. Natural conditions (primarily the severe winters of the early 1960s followed by other stresses such as drought) probably initiated the decline. The acidity of the mist may have had a secondary contributing effect in decreasing the frost hardiness.

As with agricultural crops, most tree species grow in a forest floor environment that is nitrogen deficient. Thus in commercial plantations of Douglas fir in the northwest and loblolly pine in the southeast USA, nitrogen fertilization is practiced. It would appear, therefore, that the nitrogen component of acid rain would also be generally beneficial to natural forests over most of the nation.

The most recent scientific results (Blank et al., 1988; Roberts et al., 1989) on the forests of Germany are consistent with the notion that acid rain is not a primary factor in the various types of regional decline displayed by some species in different parts of that country.

Health (probably no significant effect)

The ambient sulfur dioxide and NO_x air quality standards are currently being met essentially everywhere in the USA. These are set so that there are no significant health effects.

Normal exercising adults have been exposed to acidic aerosols resulting from acid rain formation at concentrations of about 1,000 micrograms per cubic meter with no ill effects. Asthmatics show sensitivity (but no long-term or serious effects) above about 100 micrograms per cubic meter. The average rural regional concentrations of acidic aerosols lie in the range of only a few micrograms per cubic meter. If there are any effects, they would appear to be insignificant under ambient conditions (NAPAP, 1991).

Further, researchers have not demonstrated indirect health effects in drinking water from acid rain.

Building Materials and Monuments (small negative effects)

Three types of building materials have been examined for damage by acid rain: galvanized steel, carbonate stone, and surface coatings. All are affected by many agents in their outdoor exposures, including solar radiation, ozone, thermal cycling, moisture cycling, bacterial activity, natural organic acids, and carbonic acid in rain. The problem is to determine the incremental degree of erosion or corrosion of these materials due to ambient levels of acid rain. Work to date suggests that the acid rain effect is second order (NAPAP, 1991).

The primary issue is whether ambient acid rain contributes significantly to reducing the time to replacement or repair of the building material or monument. If, for example, the reason for repainting a house is to change its color, the corrosion agent or its corrosion rate is irrelevant. If a galvanized sheet or tube has rusted beyond acceptable limits for its application, the question is how much its life was shortened by exposure to ambient acid rain. If the increment in time to replacement is a small fraction of the total, the acid rain effect can be ignored. Unfortunately, to date, these factors have not been quantified, so economic loss cannot be calculated. It seems doubtful with the data at hand that acid rain will be shown to play a significant role in the time to replacement for any of these materials.

*Lakes and Streams (negative effects on a small percentage of
the surface waters in the northeastern USA)*

All rocks and minerals are alkaline and therefore neutralize acidic
ground water. The degree of neutralization depends on the time of
contact with the minerals and their individual reactivity. For example,
coarse feldspar, a major component of granite, reacts very slowly, whereas fine-grained calcium carbonate reacts rapidly. The acidity of a lake or
a headwater stream is therefore dependent on the natural acidity of the
soil and of the rainfall, the reactivity of the minerals along the flow paths
of the rainwater through the watershed (soil and rock fractures), and the
time the water is in contact with the neutralizing medium.

In preindustrial time, rain averaging pH = 5.0 could produce lake or
stream water of similar acidity if the water fell on a smooth granite
surface and ran directly into the lake or stream. If the same water moved
slowly through a soil rich in limestone fragments, the resulting stream or
lake water might have a pH > 7. Organic acids from humic layers in the
forest soil, with pH as low as 4 or even lower, could be diluted by
rainwater. Thus at present in the areas of highest acid deposition in the
eastern USA, where the acidity of the rain on average is about eight
times preindustrial rain, the lake and headstream acidity may range from
pH = 4.2 upward.

The principles for small headwater streams are the same as for lakes,
and similar examples can be cited. Small streams may show considerably
more variation in acidity during the year due to "acid episodes," which
may be brought on by spring thaw and runoff or by summer storms on
already saturated ground. In these cases the impact on organisms may
be locally severe for a short duration. The regional importance of these
episodes on fish populations is not yet understood but appears to be
minor based on the absence of commonly associated fish mortality.

Finally, as would be expected, large lakes with large watersheds and
rivers draining large areas are not significantly acidified by acid rain.

The national inventory of lakes taken over the *most sensitive* areas of
the USA, i.e., where there is little or no limestone, where there are only
thin soils in the watershed, or where the rain acidity is high, is shown in
table 47.1. For each region the median pH is given. Even in the area of
highest acid rain (the Adirondacks), the average lake has a pH of nearly
7, i.e., neutral. Also shown are the percentages of the lake area (excluding very large lakes like the Great Lakes or Lake Champlain) with
pH < 5.0 and < 6.0. When the acid neutralizing capacity of the lake
equals zero, the pH lies in the range of 5.0–5.5. In preindustrial time,
many lakes in these sensitive areas had pH less than 6 ("naturally
acidic"), since the rain was 5.0 and if it percolated through the humic
layer it would have been made more acid before it entered the stream or
lake.

Table 47.1 Summary of US National Lake Survey

Region		Percent of lake area by pH	
	Median pH	*<5.0*	*<6.0*
Adirondacks	6.9	2	10
Northeast	6.7	1	6
Upper Midwest	7.4	0	2
Southern Blue Ridge	7.0	0	0
Mountainous West	7.4	0	0

Source: NAPAP, 1991.

Table 47.2 Potential benefits (costs) of reducing acid rain by 50 percent (estimated from data in NAPAP, 1991)

Effect	Benefit of control (billion $/year)	Comment
Crops	– 0.3	without grasslands
Forests	– 0.1	partial fertilization gain
Lakes and streams	+ 0.1	saturated with fishermen
Materials	small	unquantifiable
Health	negligible	ambient aerosol too low
Visibility	+ 0.6	highly uncertain contingent valuation
Overall	≪1	

Based on field observations, most species of sports fish tolerate pH levels above 5.5, but relatively few species can sustain populations in water with pH below 5.0. The fraction of lake area with pH < 6 ranges from zero to small in these sensitive regions.

The most seriously impacted area in the USA is the Adirondack Park of New York, where 10 percent of the lake area has a pH < 6. This corresponds fairly well to the fraction of lake area that has substandard fish populations. About 25 percent of these acidic lakes were produced by natural causes; of those lakes acidified to a measurable extent by acidic deposition, most are very small in area.

The timing of the acidification is poorly known, but some may have been at similar pH for centuries, some have become progressively more acidic with the regrowth of a forest after a fire or logging, and some became acidic from atmospheric deposition. In the latter case, more of those in the glaciated areas probably reached current levels of acidity by 1930–40, when they equilibrated with the rain.

In summary, less than 2 percent of the lake area in the Adirondacks (the most highly impacted region) is uninhabitable for sports fish, and less than 10 percent may have some deleterious effect on some species. Similar percentages appear to apply to the most highly impacted areas of southeastern Canada.

Visibility (impacted from sulfate aerosols, but hard to quantify)

The sulfuric acid produced from the emissions of SO_2 reacts with alkaline compounds in the air to form fine particles of ammonium sulfate and calcium sulfate, among others. These fine particles produce haze. The effect is multiplied at high levels of humidity. It is particularly dramatic in the relatively clean air areas of the western states.

In many of the National Park areas of the west, the sulfate creates 50 to 60 percent of the degradation in visibility. In the east, it is closer to 70 percent of a much lower natural visibility. The effect over short periods is difficult to quantify because many factors, including humidity and carbonaceous and soil particles, contribute to the phenomenon. However, regional historical studies indicate that increased sulfate particle concentration can be roughly correlated, if averaged over several years, with decreased visibility in both western and eastern USA.

CONCLUSIONS

The causes and effects of acid rain are now reasonably well understood. Its net effects are positive for crops and trees, and its impact on human health appears negligible. On materials and monuments, it is slightly negative. Its effects on surface waters and visibility are slight to modest.

Emissions of SO_2 will be much less in the future. Consequently, the effects, which are already small, will be reduced to negligible levels in a few decades. Thus the potential threat from acid rain to the environment will become a mere historical footnote.

The acid rain story supports three propositions that should undergird rational environmental policy: (1) Once an environmental problem is scientifically defined, a technological solution can be developed. (2) Rushing to adopt control policy and regulation, before adequately understanding the causes and effects of the suggested environmental threat and the cost/benefit ratio of any proposed solution, can be enormously wasteful to societal assets. (3) Increased use of energy need not be accompanied by increased pollution.

The net benefits from reducing the emissions of the precursors of acid rain by 50 percent (table 47.2) as stipulated in the US Clean Air Act of 1990 appear to be negligible compared to the costs (NAPAP, 1991).

REFERENCES

Blank, L. W., T. M. Roberts, and R. A. Skeffington (1988): "New Perspectives on Forest Decline." *Nature*, 336 (3 November).

EPRI (1991): *Analysis of Alternative SO$_2$ Reduction Strategies*. Report EN/GS-7132. Palo Alto, CA: Electric Power Research Institute.

Irving, Patricia M. (1987): "Effects on Agricultural Crops." In NAPAP Interim Assessment (1987) (below).

Kulp, J. L. (1987): "Effects on Forests." Chapter 7 in NAPAP Interim Assessment (1987) (below).

—— (1990): "Acid Rain: Causes, Effects, and Control." *Regulation*, Winter.

Lefohn, A. S., and S. U. Krupa, eds (1988): *Acid Precipitation, A Technical Amplification of NAPAP's Findings, Proceedings of An Air Pollution and Control Association International Conference*. Pittsburgh: APCA.

Likens, G. E., R. F. Wright, J. F. Galloway, and T. J. Butler (1979): "Acid Rain." *Scientific American* 241 (4), 43–51.

National Acid Precipitation Assessment Program [NAPAP] (1987): *The Causes and Effects of Acidic Deposition*, Interim Assessment, eds J. L. Kulp and C. N. Herrick. Washington: US Government Printing Office.

National Acid Precipitation Assessment Program [NAPAP] (1991): *Acidic Deposition: State of Science and Technology*, Final Report, ed. Patricia M. Irving. Washington: US Government Printing Office.

Roberts, T. M., R. A. Skeffington, and L. W. Blank (1989): "Causes of Type 1 Spruce Decline in Europe." *Forestry*, 62 (3).

48

Stratospheric Ozone:
Science and Policy

S. Fred Singer

The announcement, in 1985, of the existence of a "hole" in the atmospheric ozone layer near the South Pole has focused worldwide interest on what is happening to ozone, a minor yet vitally important constituent of the earth's atmosphere. It has also raised concern about possible depletion of ozone on a global scale and resulting health effects, particularly an increase in the skin cancer rate.

An issue dating back to the 1970 controversy about the effects of supersonic transport aircraft has resurfaced: To what extent are human activities producing ozone changes? In particular, the emission into the atmosphere of chlorofluorocarbons (CFCs) has raised fears of "destroying" the ozone layer and has led to demands that the production of these chemicals be curtailed or even abolished (see Rowland, 1989).

The case against CFCs is based on a plausible but still incomplete theory, whose predictions are in a state of flux; on observations of the Antarctic ozone hole (AOH), whose future is uncertain; and on an assertion, as yet unverified, that ozone has been declining on a global scale.

THEORY

As understanding of the complicated ozone photochemistry has improved – it involves over 150 simultaneous reactions – estimates of the effects of the CFCs have fluctuated. In recent years, the calculated effects have diminished; for example, the National Academy of Sciences in 1979 calculated an 18 percent ozone depletion due to CFCs; in 1982 it calculated a 9 percent effect; and in 1984 only a 3 percent effect. But more recent calculations suggest between 5 and 7 percent ozone depletion for a CFC scenario of continued production. We have also learned that other polluting gases released by human activities, such as NO_x, methane, and carbon dioxide, all tend to *diminish* the CFC effects – an interesting and entirely fortuitous circumstance.

A major criticism of the calculations has been that they do not consider the input of chlorine and bromine – from various natural sources – which would dilute the effects of CFCs: volcanoes (Symonds, Rose, and Reed, 1988), ocean biota (Manley and Dastoor, 1987; Singh, Salas, and Stiles, 1983) and salt particles (Finlayson-Pitts, Ezell, and Pitts, 1989).

Data have now become available (Rinsland, et al., 1991) that show both hydrochloric acid and hydrofluoric acid increasing in the stratosphere, but at different rates – 5 percent and 11 percent per year, respectively. This indicates that chlorofluorocarbons are an important source for chlorine but that there are natural sources as well.

ANTARCTIC OZONE HOLE

The Antarctic ozone hole was not predicted by current theory. It was discovered by British scientists operating an observing station on the Antarctic continent. After they reported their findings in 1985, NASA scientists searching their records of satellite data confirmed the effect. Indeed, the "hole" has been around since the mid-1970s, and getting larger every year, reaching a depletion of about 50 percent (Rowland, 1989). Concern has centered on the rapid increase and the fear that it may grow to engulf the whole globe.

(Of course, the "hole" is not really a hole at all, but a temporary thinning in the stratospheric ozone layer. This phenomenon takes place for a few weeks, around October, in the region of the Antarctic. As far as we can tell, there have been no long-term changes in the ozone elsewhere, although the evidence is not conclusive.)

In spite of recent discoveries related to the mechanism of the AOH, we do not yet have a sufficient scientific base to answer the important policy questions: Is the AOH a completely new phenomenon? What is its likely future behavior? Will it persist, grow, or weaken? And what can and should be done about it?

There is little doubt that chlorine chemistry is the *immediate* cause of the seasonal (October) ozone decrease at around 18 km in the southern polar regions – rather than purely meteorological effects based on dynamics or direct solar influences related to the solar cycle (Rowland, 1989). It is also probable that the major source of the chlorine is man-made chlorofluorocarbons [CFCs] – although no precise estimate exists of the chlorine contributed by various natural sources (see above).

Yet how does one explain the sudden onset and rapid growth of the AOH phenomenon? Starting from essentially zero in the mid-1970s, the thinning reached, within a few years, about 50 percent of the vertical ozone column – and essentially saturation in the lower stratosphere. This rapid change presents an important clue. The CFC content of the atmosphere has not risen quite so rapidly, nor should one expect any trigger effect related to the chlorine concentration. The research results suggest that, in addition to the chlorine, ozone destruction requires the presence of ice particles, "polar stratospheric clouds," that can form in the coldest part of the earth's atmosphere, the lower Antarctic stratosphere (see Rowland, 1989). But it is highly unlikely that the water vapor content could have increased so suddenly within a few years' time – although increased emissions of methane should lead to increased injection of water vapor into the stratosphere (Singer, 1971). Recent measurements have confirmed and extended this hypothesis (Blake and Rowland, 1988).[1]

This line of reasoning led me to propose that the *trigger* for the AOH has been a gradual cooling of the stratosphere that took the temperature below the freezing point; this cooling could have taken place as a part of a general climate fluctuation of the earth (Singer, 1988). And indeed, there has been an unusual surface temperature increase since about 1975; under some theoretical models of climate change, such a surface warming should be accompanied by a cooling of the upper atmosphere (Ramanathan, 1988).

If this hypothesis is borne out by appropriate measurements, then the AOH should disappear – or at least become less pronounced – if the stratosphere warms again, perhaps in conjunction with a cooling of the earth's surface. Conversely, a further cooling of the stratosphere could induce an Arctic ozone hole and a larger Antarctic hole.

The policy implication is that the AOH would not be much affected by further slow increases of atmospheric CFC, nor could it be removed if the CFC concentration were to decrease. In other words, the AOH phenomenon should be reasonably insensitive to stratospheric chlorine concentration, and extremely dependent on the exact value of the temperature minimum in the lower Antarctic stratosphere (Singer, 1988).

GLOBAL OZONE DECREASES?

In March 1988, the Ozone Trends Panel of NASA, after a massive re-analysis of data from ground stations and satellites, announced the existence of a declining trend in northern hemisphere ozone over the period of 1970 to 1986. A press release was issued, but the underlying analysis was not released for independent examination until 1991. A news story in *Science* (Kerr, 1988) quotes an average decline of −0.2 percent per year, which is greater than predicted from the current CFC-ozone theory. Since the decline was judged "worse than expected," the not very logical conclusion was reached that CFCs must now be phased out completely and rapidly. (A more logical conclusion might have been that the analysis, or the theory, or possibly both, are incorrect.)

While the NASA panel's report was not available, a parallel report from the Center for Applied Mathematics of Allied-Signal, Inc. was

distributed at a United Nations Environmental Programme Ozone
Science Meeting at the Hague in October 1988. The Allied study
(Bishop, Hill, and Marcucci, 1988) carries out a sophisticated regression analysis of the same data as the NASA study. After correcting for
many natural variations, including the 11-year solar cycle, they derive a
decline of −1.9 percent over the period 1970 to 1986 – which is only
1½ solar cycles. But their sensitivity analyses show that the result
depends on the time interval under consideration, suggesting therefore
that the solar cycle correction was not adequate and that the decline is
at least partly an artifact of the analysis (Singer, 1990).

Independently, it was suggested that the ozone data themselves, used
to support the ozone depletion analysis, were flawed, being contaminated by a trend in the concentration of atmospheric sulfur dioxide (De
Muer and De Backer, 1992). Satellite data should be superior to the
ground-based data, free of such contamination and global in nature. But
the satellite data series goes back only to 1979 – too short to establish a
trend, in view of the large solar-cycle variation. (See figure 48.1.) In
addition, the stability of long-term calibration of the satellite instrument
has always been a problem (Mims, 1993).

There is an additional problem: As can be seen from figure 48.2, the
mean sunspot number shows long-term variations that should reflect in

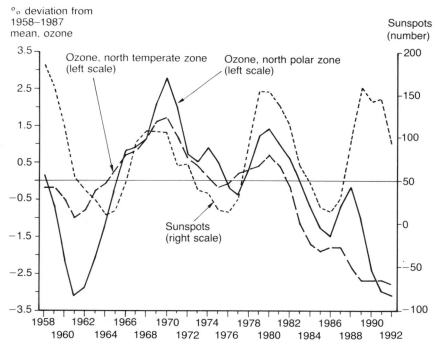

Figure 48.1 Deviations from mean concentration of ozone, and sunspot activity, 1958–92
Source: National Oceanic and Atmospheric Administration.

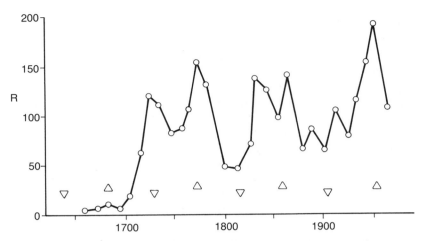

Figure 48.2 Annual mean sunspot number R at maxima of the 11-year cycle, 1645 to present, to demonstrate long-term trends in solar activity.
Note: Evident is the 80-year "Gleissberg cycle" (extrema shown as triangles) imposed on a persistent rise since the Maunder Minimum.
Source: Professor John Eddy.

long-term, natural fluctuating trends of ozone. It is interesting to reflect that during the Maunder sunspot minimum, in the seventeenth century, atmospheric ozone might have had a similar minimum – but without any obvious effects on humans, animals, or plants.

THE SKIN CANCER ISSUE

The possible connection of skin cancer with stratospheric ozone first gained public attention during the supersonic transport controversy in 1970. Certain forms of common skin tumors, basal cell and squamous cell carcinomas, have a much greater incidence at lower latitudes. They are more than twice as common in south Texas than in Minnesota – presumably because of the greater UV exposure in Texas due to the steeper sun angle there. With ozone weakened, more UV would reach the earth's surface everywhere, causing more tumors. It was this emotional skin cancer issue, more than anything else, that persuaded Congress to cancel the SST program. It is ironic that on the basis of current theory, SSTs flying in the lower stratosphere are believed to enhance ozone there rather than destroy it (Kinnison and Wuebbles, 1988).

Further relevant information adds important perspectives to the skin cancer discussion.

1.   For example, the increase of UV-B radiation that is feared to result from the thinning of ozone does in fact occur simply as a result of moving closer to the

equator. (The reason is the steeper sun angle, leading to less absorption of UV by the horizontal ozone layer.) Thus ozone decreases of the order of 5 percent would cause the same UV increase as a move of less that 100 miles.

2. Also omitted is the observation that UV-B radiation has decreased at all eight locations where measured during the last 11 years (Scotto, et al., 1988). The widely publicized claim (Kerr and McElroy, 1993) that there is an upward trend of UV-B radiation (280–320 nanometers) has been shown to be spurious and based on a faulty statistical analysis (Michaels, et al., 1994).

3. Melanoma is said to have increased at an "alarming" rate from 1974 to 1983. Actually, its incidence has increased some 700 percent since 1935, when proper records were first taken – long before CFCs came into use. A pioneering laboratory experiment (Setlow, et al., 1993) has demonstrated that melanoma incidence is related to exposure of UV-A radiation (320–400 nanometers), a region where ozone does not absorb. If confirmed, this startling result will undercut the main health concern about a possible depletion of ozone.

4. Non-melanoma skin tumors (nearly 100 percent curable) do show an increase towards lower latitudes – the only one of the effects to show a latitude dependence. But it is not sure that all of the increase is due to UV-B intensity, as generally assumed when predicting future increases due to stratospheric ozone depletion. Much of the increase probably comes from greater exposure and other lifestyle factors at the warmer locations.

SUMMARY

The science of stratospheric ozone is at an interesting crossroads. The CFC-theory is not yet good enough to explain the observations, and the observations are not yet good enough to confirm the theory. The policy question is whether drastic worldwide controls should be instituted immediately or whether one should wait for a better scientific understanding.

NOTE

1 By an analogous mechanism anthropogenic activities may contribute atmospheric sulfur compounds and thus stratospheric aerosols: D. J. Hoffman, "Increase in the stratospheric background sulfuric acid aerosol mass in the past 10 years," *Science* 248, 996–1000 (1990).

REFERENCES

Blake, D. R., and F. S. Rowland (1988): "Worldwide increase in tropospheric methane," *Science* 239, 1129.

Bishop, L., W. J. Hill, and M. A. Marcucci (1988): "An Analysis of the NASA Ozone Trends Panel Dobson Total Ozone Data over Northern Hemisphere," Centre for Applied Mathematics, Allied-Signal, Inc., Aug. 3.

De Muer, D. and H. De Backer (1991): "Revision of 20 Years Dobson Total Ozone Data at UCCHE (Belgium): Fictitious Dobson Total Ozone Trends Induced by Sulfur Dioxide Trends." *Journal of Geophysical Research*, April 15.

Finlayson-Pitts, B. J., M. J. Ezell, and J. N. Pitts, Jr (1989): "Formation of chemically active chlorine compounds by reactions of atmospheric NaCl particles with gaseous NO and ClONO." *Nature* 337: 241.

Kerr, J. B. and McElroy, C. T. (1993): "Evidence for Large Upward Trends of Ultraviolet-B Radiation to Ozone Depletion," *Science* 262, 1032–4.

Kerr, R. A. (1988) "Research News." *Science*, March 25.

Kinnison, D. E., and D. J. Wuebbles (1988): "A Study of the Sensitivity of Stratospheric Ozone to Hypersonic Aircraft Emissions." Livermore National Laboratory, UCRL-98314 (preprint), September.

Manley, S. L., and M. N. Dastoor (1987): "Methyl Halide (CH₃X) Production from the Giant Kelp, *Macrocystis*, and Estimates of Global CH₃X Production by Kelp." *Limnol Ocean*, 32(3), 709–15.

Michaels, P. J., Singer, S. F., and Knappenberger, P. C. (1994): "Analyzing Ultraviolet-B Radiation: Is There a Trend?" *Science* 264, 1341–2.

Mims, F. M. (1993): "Satellite Ozone Monitoring Error." *Nature* 361 (February 11), 505.

Ramanathan, V. (1988): "The greenhouse theory of climate change: A test by an inadvertent global experiment." *Science* 240, 293.

Rinsland, C. P. et al. (1991): "Infrared Measurements of HF and HCl Total Column Abundances Above Kitt Peak 1977–1990." *Journal of Geophysical Research*, 96, no. D8 (August 20), 15523–40.

Rowland, F. S. (1989): "Chlorofluorocarbons and Stratospheric Ozone." In *Global Climate Change*, ed. S. F. Singer. New York: Paragon House.

Scotto, J., et al. (1988): "Biologically effective UV radiation: Surface Measurements in the US, 1974 to 1985." *Science* 239, 762.

Setlow, R. B., et al. (1993): "Wavelengths Effective in Induction of Malignant Melanoma, " *Proceedings of the National Academy of Science USA* 90, 6666–70.

Singer, S. F. (1971): "Stratospheric water vapor increase due to human activities." *Nature* 223, 543.

——— (1988): "Does the Antarctic Ozone Hole have a future?" *Eos* 69, 47 November 22.

——— (1990): "What Could Be Causing Global Ozone Depletion?" In *Climate Impact of Solar Variability*, eds K. H. Schatten and A. Arking. NASA Conference Publ. 3086.

Singh, H. B., L. J. Salas, and R. E. Stiles (1983): "Methyl Halides in and over the Eastern Pacific" (40N–32S). *Journal of Geophysical Research* 88, 3684–90.

Symonds, R. B., W. I. Rose, and M. H. Reed (1988): "Contributions of Cl- and F-bearing gases in the atmosphere by volcanoes." *Nature* 334, 415.

EDITOR'S NOTE (E. CALVIN BEISNER)

When people hear of depletion of the ozone layer, often they assume that very large variations, persistently downward, are involved. Figure 48.3 depicts, to scale, the percent variation in stratospheric ozone concentrations in the north polar and north temperate regions for 1958–92. The greatest variations from the 1958–87 mean were both in the polar region, 3.1 percent below the mean in 1961 and 1992. The highest upward variation was also in the north polar region, 2.8 percent above the mean in 1970.

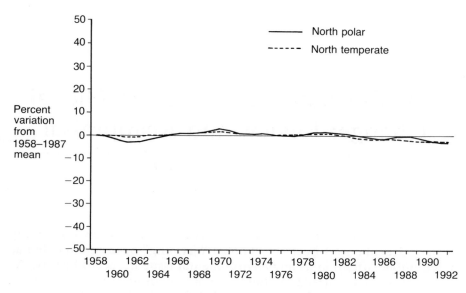

Figure 48.3 Trends in stratospheric ozone, variations from mean, 1958–92
Source: National Oceanic and Atmospheric Administration.

49

The Greenhouse Effect and Global Change: Review and Reappraisal

Patrick J. Michaels

It is doubtless that large-scale environmental changes known popularly as the "Greenhouse Effect" or "Global Warming" have commanded as much public attention as any environmental issue in recent history. As an example, *Time* painted the following picture in its 1988 "Man of the Year" issue, featuring Planet Earth:[1] If left unchecked, anthropogenic alterations of the atmosphere will bring, by the years 2030–2050, a global temperature rise of 4°C, ecological chaos including famine, related civil strife, and tidal waves crashing through a Manhattan landscaped with palm trees. As a rhetorical vehicle, I will refer to this scenario as the "Popular Vision" throughout this article.

Such scenarios also abound in the writings of politically active environmental scientists, environmental lobbyists, and newswriters.[2-5] United States Elected Officials have compared the current situation to that of fascist Germany, where several events, such as Kristallnacht, presaged the holocaust. Those that do not recognize this have been labeled modern Neville Chamberlains.[6] Draconian interventionist legislation has been proposed,[7] and the implementation of policies to counter "global warming" is now a touchstone of US foreign policy.[8]

Clearly, deep emotional commitments now guide this issue. In such a highly charged light, it would behoove us to examine the current state of the issue of anthropogenic global warming.[9]

The balance of this essay will concentrate on some of the scientific uncertainties and inconsistencies that should be factored into policy decisions. Specifically, I will address the problems of (1) global and hemispheric temperature histories *vis-à-vis* trace gas concentrations, (2) Popular Vision climate models and their successors, (3) artificial warming from urban heat islands, (4) high latitude and diurnal temperatures, and (5) the holistic nature of the global change problem and negative feedbacks in the pollution system.

This compact review is a reappraisal of the Popular Vision as opposed to a revisionist view. The latter, in the sense of the social sciences, occurs primarily in the context of societal change, rather than in the light of new information. This reappraisal is tendered in the light of more refined scientific findings.

TRACE GAS CONCENTRATIONS AND TEMPERATURE HISTORIES

While there are several thermally active trace gases that have increased as a result of human activity, almost all of the radiative forcing is associated with (in descending order) carbon dioxide, methane, nitrogen oxides, and the chlorofluorocarbons. The total radiative forcing of the last three is approximately equal to 80 percent of the effect of a change in CO_2 from 279 to 350 ppm.[10, 11]

Intensive intrumental records of CO_2 concentration date from the late 1950s at Mauna Loa Observatory, where the 1958 annual average was 315 ppm. The concentration is now very close to 352 ppm.

"Pre-industrial" (circa 1800) concentrations were initially assumed to be in the range of 295 ppm,[12] giving a net increase of 19 percent over the last 190 years. Initial ice-core studies gave a background value of 270–90, with a most likely value of 279 ppm.[13] Another analysis obtained a lower bound of 260 ppm.[14] The highly publicized Soviet/French work on the long Vostok Station ice core appears to corroborate the lower values.[15, 16] A background of 260 ppm implies an anthropogenic rise of 35 percent. For a base year of 1800, Idso's comprehensive review uses a value of 281 ppm.[17]

Background CH_4 concentration, again calculated from ice cores, appears to be around 800 ppb,[18] compared to a current value of nearly 1,700 ppb.[19] Indirect measurements give a concentration of 1,140 in 1951.[20] The primary sources for this increase appear to be rice paddy agriculture and bovine flatulence. Neither of these seem likely to be reduced in the forseeable future. An increasing fraction now comes from biomass burning.[21] Recent evidence suggests that both high intensity agriculture common over much of the industrialized world and "green revolution" culture in developing countries reduce the rate at which methane is microbially resequestered from the atmosphere.[22]

It is clear that both CO_2 and CH_4 concentrations have risen exponentially in the last 40 years. As of 1982, the climate forcing effect of the methane concentrations was 38 percent of that of increased carbon dioxide.[23] A figure of 40 percent certainly seems reasonable for 1990.

Precise knowledge of the sources and sinks of the nitrogen oxides is unavailable, and background concentration estimates are much less reliable than those for the other trace gases. The historical estimate of 285 ppb should be taken with some caution, as should future projections. The current value is 298–308 ppb.[19]

Virtually all chlorofluorocarbons (CFC's) are anthropogenerated. Concentrations in 1950 are estimated at 0.001 ppb for CFC-11, and 0.005 for CFC-12. Current values are 0.219 ppb for CFC-11 and 0.378 for CFC-12.[24] While the contribution of CFC's to the overall anthropogenic radiative forcing is relatively small, they are highly efficient absorbers in the infrared; Lovelock[25] estimates that, on a molecule-for-molecule basis, the net warming effect of CFC's is 14 times greater than that of CO_2.

The 1987 estimate that the combined current radiative effect of the non-CO_2 trace gases is 80 percent of that caused by a change in CO_2[10] concentration from 279 to 350 ppm implies the current atmosphere can be viewed as having an *effective* CO_2 concentration of 411 ppm (assuming no other increases in thermally active gases), or 147–58 percent of a background range of 260–79 ppm. In other words, because of the combined effects of the various trace gases, *we have already gone half way to an effective doubling of the preindustrial CO_2 concentration. This will not be reversed in our lifetimes.*

GLOBAL VERSUS HEMISPHERIC TEMPERATURE HISTORIES

It is customary to present the time history of global mean temperatures as "at least not contradictory to"[26] climate model projections. Calculations based upon the assumed radiative forcing in the penultimate suite of General Circulation Climate Models[27–31] (those that have given credence to the Popular Vision), produce an expected warming, after thermal equilibrium is established between the land, air, and water, of 1.7°C to date.[9] The observed mean warming has been 0.4–0.6°C since dense records began in 1860; these values are themselves subject to some reduction because of the urban warming detailed below. If a 260 ppm background is used instead, an analogous calculation gives an expected mean global warming of 2.0°C to date. Inspection of the hemispheric temperature histories shown in figure 49.1 reveals that much of the warming in the southern hemisphere, and virtually *all* of the warming in the northern hemisphere, was *prior* to the major increases in the trace gases.[32]

These discrepancies from the penultimate ("Popular Vision") suite of General Circulation Models (GCMs) are well known, although their magnitude is seldom emphasized in public discussion or in the scientific literature; notable exceptions to this include testimony to the US Congress and the writings of Idso.[17, 33]

The discrepancies are only partially explained as a function of oceanic thermal lag, whose time to equilibrium is estimated at from 10 to 70 years. Nonetheless, even liberal estimates of this lag still imply an expected warming to date of 1.0–1.2°C,[10] suggesting that the globe has

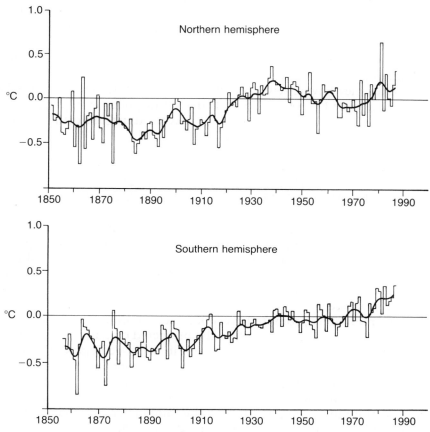

Figure 49.1 Hemispheric temperature trends, 1850–1990
Note: Southern hemisphere temperature behavior for the last half century resembles what one would expect from a greenhouse alteration, except the magnitude of the rise appears to be low (top). Virtually all of the warming in the northern hemisphere record took place prior to the major trace gas emissions; there has been no net warming for the last half century (bottom).
Source: Data of P. D. Jones and T. M. L. Wigley, University of East Anglia.

warmed up between a factor of two and three times (0.4–0.6°C) less than combinations of the penultimate generation of climate models coupled to rough approximations of oceanic lag. Why this much warming is "missing" is simply unknown,[33] although we will speculate on causes here.

Intrahemispheric comparisons support the contention that oceanic thermal lag itself may be overestimated (figure 49.1). The southern hemisphere, with the disproportionate share of ocean surface, displays a warming whose functional form (but not magnitude) is what might be expected from a straightforward interpretation of model output. The northern hemisphere shows most of its warming prior to the major input of trace gases, and shows no net change over the last 55 years[32] – when CO_2 concentrations went from approximately 300 to 352 ppm and the other thermally active gases were in their steepest growth phases, giving rise to the current effective concentration of 411 ppm.

Stratospheric temperatures, which should fall in a trace-gas enriched atmosphere, have dropped considerably more in the southern hemisphere than they have elsewhere, with the greatest declines in the polar zone. These may be associated with the south polar springtime ozone minimum.[17] There is no statistically significant trend in northern hemisphere stratospheric temperatures.[34]

"POPULAR VISION" AND NEW GENERATION CLIMATE MODELS

The penultimate generation of general circulation models (GCMs)[27–31] calculated a mean equilibrium warming for a doubling of atmospheric CO_2 of 4.2°C, with maximum warmings (on the order of 15°C) during north polar winter, and less significant warming (approximately 2°C) over tropical oceans. Figures 49.2 and 49.3 detail "typical" output from this suite of models.

These simulations prompted considerable attention from politicians, environmentalists, and the press, as the projected climatic changes were doubtless of sufficient magnitude to result in major ecological disruption and significant rises in sea level. Further, even though many agriculturally important midlatitude areas were projected to see increases in precipitation, concomitant temperature changes resulted in evaporation rates that more than compensated for the rainfall, leading to projections of increased drought frequency. One modeler made a public "forecast" of increased drought frequency in the early 1990s.[35, 36]

Figure 49.2 is included to demonstrate that, after some initial start-up, the temperature response of GCMs can be quite linear with respect to CO_2 or combined trace gas forcing, which has also been noted by Schneider.[37] The implication is that, if the atmosphere and ocean were approaching some equilibrium, enhanced warming should now be observed in polar regions, and midlatitude regions should be nearly 2°C warmer than background temperatures; obviously, neither of these events has occurred.

Major shortcomings in this generation of models included unrealistic ocean dynamics and ocean–atmosphere coupling, inadequate cloud parameterization (which included debate on the *sign* and the magnitude of surface temperature forcing by clouds), and with one exception[28] the use of stepwise (instantaneous) doubling of CO_2 rather than the low-order exponential increase that exists in the real world.

Ramanathan, et al.[38] demonstrated, using satellite observations, that the net effect of terrestrial cloudiness is a cooling of the surface. Since

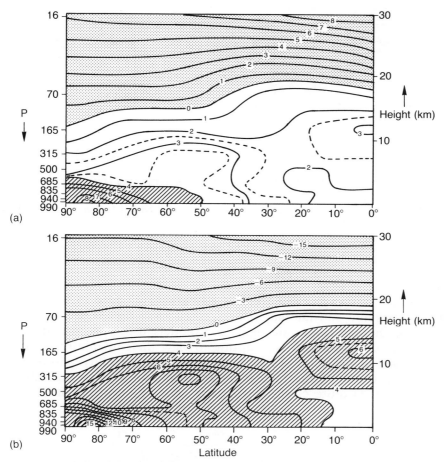

Figure 49.2 Typical output of General Circulation Models
Note: Latitude/altitude plot of temperature change projected for a doubling (top) and a quadrupling (bottom) of CO_2 in one of the penultimate, or "Popular Vision," suites of climate models.[27] Because of the combined effect of other trace gases, we are now over half way to an equivalent doubling of the background concentration.
Source: Manabe and Wetherald, 1980.

then, a more realistic cloud parameterization was introduced into the United Kingdom Meteorological Office (UKMO) model, and projected mean global warming dropped in that model from 5.5°C to 2.7°C.[39]

Two of the other four Popular Vision GCMs have recently included a more realistic coupling between the atmosphere and the ocean. The Geophysical Fluid Dynamics Laboratory model now projects a mean global warming (for a doubling of CO_2) of 1.8°C, compared to 4.3°C previously.[35] A similar recalculation in the National Center for Atmospheric Research (NCAR) model now predicts a warming of 1.6°C,

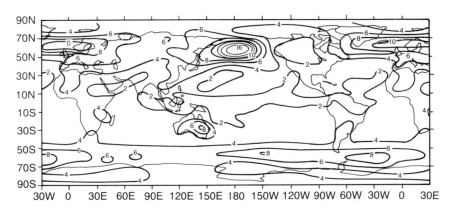

Figure 49.3 Sample General Circulation Model output, National Center for Atmospheric Research, $2 \times CO_2 - 1 \times CO_2$
Note: Winter (Dec.–Feb.) equilibrium temperature changes projected by another of the penultimate generation of models[31] (The National Center for Atmospheric Research Community Climate Model). Compare to figure 49.4.
Source: Washington and Meehl, 1983.

instead of 3.7°C in its earlier run.[40] Comparison of figure 49.4 and figure 49.3 shows how substantially the projected warming has been reduced.

All of these revisions occurred for models using an instantaneous doubling of CO_2, and it is tempting to view their unanimity of major reductions in projected warming as a sign of reliability. However, a transient version of the NCAR model, using an increase in radiative forcing that is quite close to what has actually occurred since 1950,

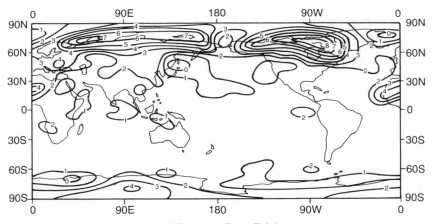

Figure 49.4 Surface air temperature differences (Dec.–Feb.)
Note: The new version of the NCAR-CCM using more realistic ocean and ocean–atmosphere dynamics. Compare figure 49.3.
Source: Washington and Meehl, 1989.

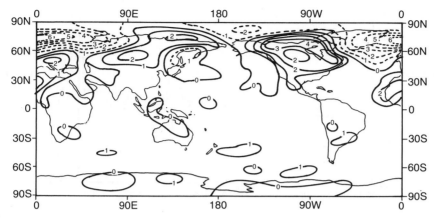

Figure 49.5 ΔT_{991} (Dec.–Feb.) transient minus control (26–30)
Note: Calculated regional temperature changes from background in the years 26–30 average of a simulation that realistically simulates the trace gas forcing increase since 1950. Magnitude of both the projected warming (over northern North America) and the cooling (over northern Eurasia) have not been observed.
Source: Washington and Meehl, 1989.

suggests that current temperatures should be very nearly 1°C above their 1950 value; global temperatures have in fact risen only 0.22°C in that time, with the majority of the warming in the southern hemisphere. Thus, even conservative, transient models appear to be grossly overestimating prospective warming.

An additional problem is that this version, which is certainly as sophisticated as any, clearly predicts unrealistic regional temperatures for the current increase in trace gases. Figure 49.5 details the geographic distribution of December–February temperature changes expected in the five-year average of years 26–30, in a thirty-year transient run in which trace gas forcing was increased in a realistic manner. Roughly speaking, the plot is analogous to any five-year aggregate from 1975 through 1985. Neither the 2–4°C warming of the northern half of North America, nor the 3–6°C *cooling* of much of northern Eurasia, which together comprise the major signal in the model, occurred in reality.

URBAN HEAT ISLANDS AND LONG-TERM SITE BIAS

Much has been written on the subject of urban contamination of long-term climate records, but its true magnitude remains elusive.

The most comprehensive recent studies are by Karl, et al.,[41] which compared climate records of NASA and the East Anglia research group[32, 42] – the two most cited records – to an unurbanized subsample over the United States. The NASA record makes no attempt to directly

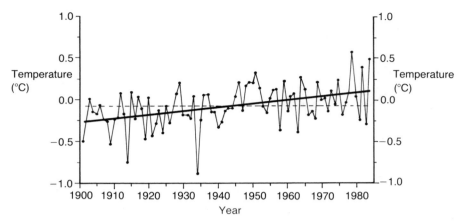

Figure 49.6 Annual average temperature anomaly differences over USA (Hansen minus historical climate network)

Note: NASA's record over the coterminous US warms up approximately 0.4°C compared to urbanization-adjusted Historical Climate Record. The discrepancy may result from a programming error.

Source: Karl and Jones, 1989.

remove urban bias, while the East Anglia record does so subjectively. Karl, et al. found an artificial warming in the NASA record of 0.4° in the twentieth century, and 0.10–0.15°C in the East Anglia record; see Figure 49.6. There is no a priori reason to suspect that the urban bias is appreciably different in other parts of the industrialized world. Therefore, a simple extension to the globe – which is not possible to rigorously defend at this time because the exhaustive research has not been performed – would yield a residual warming in the NASA record warming of 0.2–0.3°C. The residual warming in the East Anglia record would be 0.25–0.40°C during the twentieth century.

While the considerable disparity between the NASA record and the analyses of Karl, et al. remains unresolved, unpublished accounts now say it is related in part to a programming error when NASA supplied the data to NOAA for analysis.[43] Unpublished reanalyses place the bias in the 0.1–0.2°C range. This is significant because it brings *both* the NASA and East Anglia curves into agreement that the net warming of this century is slightly less than a half of one degree. All records agree that much of this warming was prior to 1940, or before the major emissions of the thermally active trace gases. Yet if the pre-1940 warming is ascribed to human activity, then the post-1940 global temperatures should have risen dramatically. They have not.

It should be noted parenthetically that statements about "99% confidence" of "cause and effect" between observed temperatures and anthropogenic greenhouse alterations[44] did not take into consideration the very common and understandable human factors that may induce errors in the analysis. Figure 49.6 details the disparity between the NASA and Karl, et al. records.

An additional, and more insidious, problem may be introduced into long-term climate records by a *general* site bias. Over what was the "developing world" of the late nineteenth century, the longest-standing records have tended to originate at points of commerce. Because of the predominance of water power during that era, it is likely that the longest records are from near-riverine environments. These sites are therefore located preferentially in zones where cold-air drainage predominates. Such locations may be buffered from anthropogenic climate changes, particularly in their nighttime temperatures.

In an attempt to compensate for riverine site bias, which doubtlessly occurs over peripolar regions such as Alaska, Michaels, et al. calculated mean annual temperatures from the ideal gas law, based upon surface pressure and 500 mb. heights, and purposefully neglecting thermometric measurements.[45] No net warming was detected since records began in 1948.

An analogous study calculated temperatures from 1,000–500 mb thickness values over the USA, southern Canada, and surrounding ocean regions.[46] While observations began in 1948, the record was extended back to 1885 by using an expanded version of Hayden's comprehensive cyclone track record.[47] The resulting temperature curve resembles the unurbanized mean annual US temperature history of Karl, et al.,[48] inasmuch as there is a rise from ca. 1900 values to mid-century, followed by a fall through 1975, and a consequent rise through the 1980s. While the net temperature change is approximately +0.3°C, the magnitude of the century-scale swing is twice that measured thermometrically (figure 49.7). It is unknown whether this is related to the noise in the statistical transfer functions between the cyclone track record and the 1,000–500 mb thickness, or whether it truly reflects free-atmosphere temperature variability. If the latter is true, then large, agriculturally important regions have experienced climatic changes in the twentieth century that are not appreciably different from those projected by the improved GCMs.[35, 39–40] GCM temperature calculations are based more on free atmosphere conditions than they are on local site considerations, which they explicitly ignore. Thus this temperature analysis is more directly comparable to model output than locally measured temperatures.

HIGH LATITUDE AND DIURNAL TEMPERATURES

Both the Popular Vision models and the new generation continue to predict amplified warming during winter in high latitudes, although the transient model of Washington and Meehl[40] has areas of both strong warming and cooling for a period roughly analogous to 1975–85 (figure 49.5).

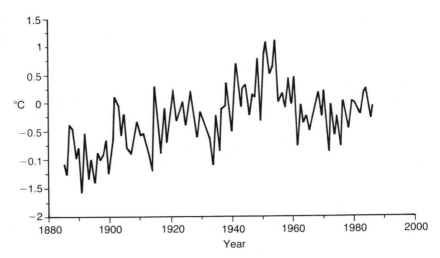

Figure 49.7 USA, Canada, surrounding oceans temperature pattern

Note: The temperature history of the US, southern Canada, and surrounding oceans calculated from the 1000–500 mb proxy record[46] shows climatic variation not appreciably different from that projected by the new family of GCMs.[35, 39–40]

Sources: Michaels et al., 1989.

As noted above, because of the combined effect of the thermally active trace gases, we have effectively gone beyond half way to a doubling of CO_2. It is quite apparent from figure 49.8 that Arctic temperatures have simply not responded in the fashion predicted by the Popular Vision. In fact, the data indicate a rapid rise in temperature *prior* to the major emissions, followed by an equivalent *decline*.[49] The NCAR transient model run[40] is much more consistent with the observation of no net change in high latitude temperatures, although the strong cellular patterns indicated in that model have clearly not developed.

Observed antarctic surface temperatures show no trend whatsoever.[50] This is particularly noteworthy because some remote locations, such as Amundsen-Scott at the South Pole, have minimum problems that might compound the records, such as marine advection, cloudiness, urban warming, or topographic effects that might induce local cold air drainage. In particular, the record during polar night is interesting: even though the effective change in trace gases has been considerable since the record began in 1957, there is no change whatsoever in the winter temperatures (figure 49.9). How this could occur is surely mysterious, although it may indicate that the depth of the winter anticyclone is sufficient to buffer the station from any anthropogenic alterations of atmospheric composition. If this is the cause, then the problem of general site bias in long-term records, noted above, is quite severe, as shallow cold air pooling may have the same effect worldwide.

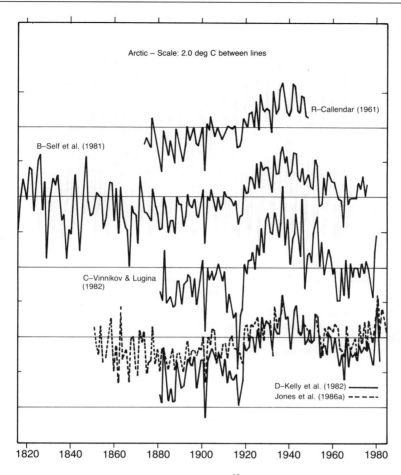

Figure 49.8 Arctic temperature patterns, 1820–1980[49]
Source: Elsaesser et al. 1986.

Recent studies on diurnal temperatures do not bode well for the Popular Vision. Karl, et al.[48] examined day and night temperatures using the US Historical Climate Network (HCN), a record that contains a minimum of urban heating. Careful analyses indicate that daily temperature ranges have declined precipitously, concurrent with the major trace gas emissions. Interestingly, daytime high temperatures have actually declined, while night temperatures are rising relative to day values; see figure 49.10. This behavior is consistent with both an increase in the trace gases and increases in cloudiness that have been documented across the country.[51]

If anthropogenic warming takes place primarily at night, the Popular Vision is simply wrong. Evaporation rate increases, which are the primary cause of projected increases in drought frequency, are minimized.

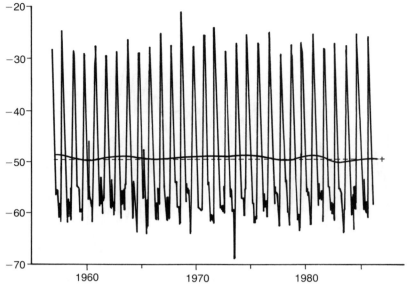

Figure 49.9 Amundsen-Scott temperature patterns, 1957–87

Note: The record at Amundsen-Scott (South Pole), where compounding influences are likely to be absolutely minimized, shows no change in polar night temperatures in spite of the massive increase in infrared-absorbing trace gases.[50]

Source: Sansom, 1989.

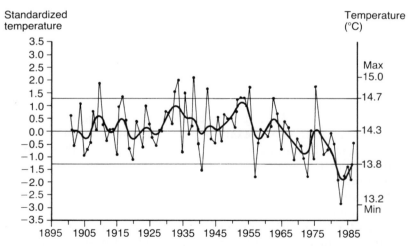

Figure 49.10 Areally weighted temperature range national (urbanization adjusted) annual (Jan.–Dec.)

Note: The difference between US daily high and low temperatures. The narrowing of the range is consistent with a benign expression of the trace gas changes.

Source: Karl et al., 1988.

The growing season is longer, because that period is primarily determined by night low temperatures. If cloudiness of any type continued to increase, the incidence of skin cancer would decline (after adjusting for age and behavior). Many plants, including several agriculturally important species, will show enhanced growth uninhibited by drought, because of the well-known "fertilizer" effect of CO_2.

Also, if warming is primarily at night, the component of sea level rise resulting from large-scale melting is drastically diminished, as virtually all polar melting takes place during polar day. At night, temperatures are so far below freezing that anthropogenic warming is insufficient to induce melting.

Such a "positive" vision of the future climate has been promulgated by Russian academician Mikhail Budyko,[52] whose position in Russia's national science hierarchy is analogous to those of, say, Steven Schneider or James Hansen in the United States. Clearly there is a very broad range of future climate expectations, some of which are diametrically opposed to the Popular Vision.

THE HOLISTIC NATURE OF GLOBAL CHANGE AND NEGATIVE FEEDBACKS IN THE POLLUTION SYSTEM

One of the arguments often tendered concerning the certainty that disastrous warming either has begun or looms in the immediate future is based upon the coincidence of climate changes and CO_2 in antarctic ice cores dating back through the last glacial cycle. In fact, the resolution of those data is insufficient to determine whether changes in CO_2 presage or follow the climate change. However, even if they "cause" the change, the usefulness of the pleistocene analogy is severely limited by the cause of the current trace gas excursion: man.

It is clear that human activity, besides altering the concentration of thermally-active trace gases, also produces substances that can serve to counter that effect. These include particulates, which serve to scatter radiation, and sulfur dioxide molecules, which in their oxidized state can serve as cloud condensation nuclei. Anthropogenic pollutants can therefore serve to "brighten" clouds, reflecting away increasing amounts of solar radiation and possibly compensating for greenhouse warming. A recent calculation demonstrates that the magnitude of this effect could indeed explain the the fact that the northern hemisphere, where most sulfate emissions occur, shows no net warming during the last half-century.[53]

Analyses of a limited set of satellite data confirm that ocean-surface stratocumulus – one of the most common clouds on earth, and the variety most likely to be affected by increasing numbers of condensation nuclei – indeed are considerably brighter in the lee of regions of major anthropogenic sulfur dioxide emissions.[54] The brightening persists for thousands of miles downstream from the continental source regions. A control study, over the clean South Pacific, yields no strong trend in cloud brightness. Some of these results are shown in figure 49.11.

Thus we are faced with the possibility that the same emission that causes acid rain may in fact be protecting the northern hemisphere from the disastrous greenhouse warming:

> The effects of SO_2 associated with acidic precipitation and urban pollution are clearly detrimental, and measures to reduce emissions are being implemented widely. However, if we were successful in halting or reversing the increase in SO_2 emissions we could, as a by-product, accelerate the rate of greenhouse-gas-induced warming, so reducing one problem at the expense of another.[53]

While it is plausible that we have in fact compensated for trace-gas warming with sulfur oxides, it is unclear if or when increased trace gas loading will overcome this effect.

If human activity is indeed brightening clouds for thousands of miles in the lee of industrialized continents, energy balance considerations would suggest some localized net cooling in these midlatitude oceanic regions. It seems likely there would be some compensation for altered surface warming in the position of the jet stream. In particular, one would hypothesize that there would be a tendency for more frequent jet stream troughs in the regions of relative cooling, such as downstream from Eurasia and North America, with a compensating tendency for ridge development in the eastern North Pacific ocean.

One consequence of this would be an increase in the magnitude of northwesterly flow over eastern North America, which has in fact been documented.[55, 56] This could explain the cooling of the United States daytime temperatures from the mid 1930s to the mid 1980s,[48] or the dramatic drop (3°C) in southeastern mean winter temperatures over the last 60 years,[57] in the face of the trace gas increase.

THE CRISIS ON THE HORIZON

The intense politicization of the global change problem almost guarantees that some type of action will in fact be taken, as proposed by former President George Bush, to drastically limit sulfur dioxide emissions. Nonetheless, the interrelatedness of the components of global change dictates that any single remedial activity – such as a reduction in sulfate emissions in an attempt to reduce acid rain – can have unforseen or negative consequences that could in fact serve to exacerbate the much more serious problem of potential anthropogenic warming.

A crisis will soon occur in environmental politics when the public realizes that, regardless of the remedial attempts that are invoked, con-

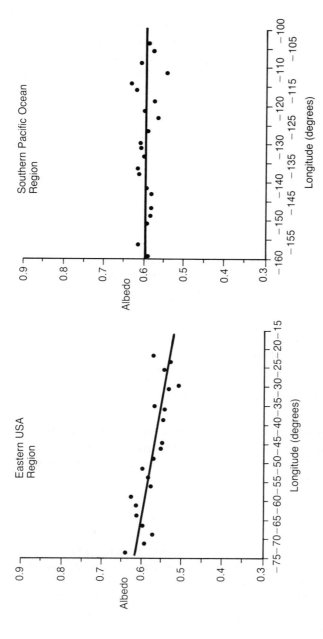

Figure 49.11 Effect of sulfate emissions on downwind albedo effect

Note: Albedo of ocean-surface stratocumulus clouds (left) over a long transect downstream from sulfate emission regions of the US and (right) over the more pristine southern Pacific Ocean. A brightening similar to that observed over the Atlantic has also been observed in the lee of Eurasia.[54]

Source: Cess, 1989.

centrations of most greenhouse gases will continue to increase. In particular, reduction of methane or CO_2 to the point that atmospheric concentrations stabilize seems unlikely for several decades. Responses to large-scale environmental disturbances that do not take into consideration the interrelatedness between environmental problems can make the specter of climatic change much more imminent.

If global warming continues to be expressed primarily in night temperatures (if indeed that is what is occurring now), will a public that has endured an intense campaign suitable enough to ensconce the Popular Vision suddenly accept climate change as possibly benign? What will become of the credibility of the scientists and politicians who have in all earnestness and out of genuine concern promoted that Popular Vision?

Alternatively, if the Popular Vision indeed begins to verify – which may require a *reversal* of trends that have developed since the major increases in trace gases – the public will have to choose a mix of two actions: direct and purposeful intervention in the climate system (as proposed by Budyko) or attempts at adaptation. According to academician Budyko:

> Another approach to limiting global warming . . . (is) accessible even to modern technology [and] was proposed long before the onset of wide international studies on anthropogenic climate changes. The main idea of this method is to increase the stratospheric aerosol concentration by burning sulfur delivered by aircraft into the lower stratosphere.

> It can be noted that this method will require *incomparably* [emphasis ours] less expense than those due to damage caused by drastic reductions in carbon fuel consumption aimed at retarding global warming. One other most important advantage of this method is the possibility to considerably change climate to cooling for a short period of time . . .[58]

It seems highly improbable that a public that is abnormally risk-averse (see chapter 53 in this volume by Whelan and chapter 56 by Zeckhauser and Viscusi), as evinced by the recent ALAR controversy in the United States, will agree to this remedy.

CONCLUSION

The Popular Vision of the world influenced by increasing concentrations of trace gases is one of ecological disaster brought on by rising temperatures, sea level, and evaporation rates. This vision, which developed from the penultimate suite of climatic models, persists, even though the newer, more physically realistic models have all lowered their estimates of warming by more than a factor of two. The vision continues to be espoused despite several other lines of observational evidence that indicate such a "carbon dioxide in/disastrous warming out" concept is immature at this time, given the complexity of the global change problem. Nonetheless, politicians have compared the severity of the expected paroxsysms to those of the Nazi holocaust. Global warming is now one touchstone of US foreign policy.

This evolved in the light of remarkable inconsistencies. The northern hemisphere, which should be the first to warm, is no warmer than it was 55 years ago. One very careful study shows relative warming at night, which may in fact be beneficial. The amount of measured global warming is at least a factor of three less than predicted by our most sophisticated models. High latitude temperatures, which some calculations indicated should be most sensitive, show no "greenhouse" signal. If findings about urban contamination of climate records over the USA persist worldwide, the amount of warming over the last 120 years is even less than the generally accepted 0.5°C. Finally, there is evidence that other anthropogenic compounds may in fact be mitigating the expected warming. Nonetheless, we have now embarked on a road to eliminate some compensating emissions, even while there is universal agreement that the concentrations of greenhouse gases will continue to increase, despite our best efforts.

A crisis in environmental politics lurks on the horizon. If the greenhouse expression is most likely to be primarily benign, it will be very difficult to convince the public, now so highly sensitized, to accept it. If in fact the Popular Vision appears likely – something that is becoming more remote as climate simulations become more realistic – it will be equally difficult to convince the public that some type of direct intervention and adaptation may be the only viable remedy.

NOTES

1   *Time* Magazine's 1988 "Man of the Year" issue was changed to "Planet of the Year" to underscore their perception of the importance of this current crisis.

2   As an example of a politically active environmental scientist espousing this view, see Titus, J. G., "Rising Sea Levels: The Impact they Pose," *EPA Journal*, 1986.

3   Lobbyist example: Speth, G., "Can the Human Race be Saved?" *EPA Journal*, 1989, 47–50.

4   See a sidebar that appeared in the *Sacramento* (California) *Bee* on July 3, 1988, titled "Now for a worst-case scenario," written by Philip Shabecoff of the *New York Times*.

5   An excellent compilation of these reports can be found in "The Atmospheric Crisis: The Greenhouse Effect and Ozone Depletion," Social Issues Resource Series, POB 2348, Boca Raton, Florida 33427.

6   Gore, A. V., Jr (1989). "An Ecological Kristallnacht. Listen." *New York Times*, March 18.

7   See Senate Bill 603, sponsored by R. M. Boschwitz and A. V. Gore, Jr.

8   Goshko, S. M., *The Washington Post*, January 31, 1989.

9   It is important to distinguish anthropogenic global warming from ahomic or "natural" global warming that appears to accompany the exit from the "little ice age" that ended in the nineteenth century.

10 Wigley, T. M. L. (1987): "Relative Contributions of Different Trace Gases to the Greenhouse Effect," *Climate Monitor* 16, 14–28.

11 Warrick, R. A., and P. D. Jones (1988): "The Greenhouse Effect: Impacts and Policies." *Forum for Applied Research and Public Policy* 3, 48–62.

12 Schlesinger, M. E. (1982): "The Climate Change Induced by Increasing Atmospheric Carbon Dioxide," in *Carbon Dioxide Proliferation: Will the Icecaps Melt?* special publication #21 of the IEEE Power Engineering Society, 9–18.

13 Neftel, A., E. Moor, H. Oeschger, and B. Stauffer (1985): "Evidence from Polar Ice Cores for the Increase in Atmospheric CO_2 in the Past Two Centuries." *Nature* 315, 45–7.

14 Raynaud, D., and J. M. Barnola (1985): "An Antarctic Ice Core Reveals Atmospheric CO_2 Variations over the Past Few Centuries." *Nature* 315, 309–11.

15 Lorius, C., J. Jouzel, C. Ritz, L. Merlivat, N. I. Markov, Y. S. Korotkevich, and V. M. Kotlyakov (1987): "A 160,000 Year Climatic Record from Antarctic Ice." *Nature* 239. 591–6.

16 Jouzel, J., C. Lorius, J. R. Petit, C. Genthon, N. I. Barkov, V. M. Kotlyakov, and V. M. Petrov (1987): "Vostok Ice Core: A Continuous Isotope Temperature Record over the Last Climatic Cycle (160,000 years)." *Nature* 239, 403–8.

17 Idso, S. B. (1989) *Carbon Dioxide and Global Change: Earth in Transition.* Tempe, Arizona: IBR Press.

18 Pearman, G. I., D. Etheridge, F. deSilva, and P. D. Fraser (1986): "Evidence of Changing Concentrations of Atmospheric CO_2, N_2O and CH_4 from Air Bubbles in Antarctic Ice." *Nature* 320, 248–50.

19 Rasmussen, R. A., and M. A. K. Kahlil (1986): "Atmospheric Trace Gases: Trends and Distributions over the Last Decade." *Science* 232, 1623–4.

20 Rinsland, C. P., J. S. Levine, and T. Miles (1985): "Concentration of Methane in the Troposphere Deduced from 1951 Infrared Solar Spectra." *Nature* 318, 245–9.

21 Craig, H., C. C. Chou, J. A. Whelan, C. M. Stevens, and A. Engelmeier, (1988): "The Isotopic Composition of Methane in Polar Ice Cores." *Science* 242, 1535–9.

22 Stueudler, P. A., R. D. Bowden, J. M. Melillo, and J. Aber (1989): "Influence of Nitrogen Fertilization on Methane Uptake in Temperate Forest Soils." *Nature* 341, 314–16.

23 Craig, H., and C. C. Chou (1982): "Methane: The Record in Polar Ice Cores." *Geophysical Research Letter* 9, 1221–4.

24 Hansen, J. E., A. Lacis, D. Rind, G. Russell, P. Stone, I. Fung, R. Ruedy, and J. Lerner (1984): "Climate Sensitivity: Analysis of Feedback Mechanisms." *Geophys. Mono. Ser.* 29, 130–63.

25 Lovelock, J. (1988): *The Ages of Gaia: A Biography of Life on Earth.* New York: Norton.

26 MacCracken, M. C., and G. J. Kukla (1985): "Detecting the Climatic Effects of Carbon Dioxide: Volume Summary," in *Detecting the Climatic Effects of Increasing Carbon Dioxide.* Washington: United States Department of Energy, Publication DOE/ER-0235, 163–76.

27 Manabe, S., and R. T. Wetherald (1980): "On the Distribution of Climatic Change Resulting from an Increase in the CO_2 Content of the Atmosphere." *Journal of Atmospheric Science* 377, 99–118.

28 Hansen, J., I. Fung, A. Lacis, D. Rind, S. Lebedeff, F. Ruedy, and G. Russell (1988): "Global Climate Changes as Forecast by the Goddard Institute for

Space Studies 3-Dimensional Climate Model." *Journal of Geophysical Research*, 93, 9341–64.

29 Schlesinger, M. E. (1984): "Climate Model Simulation of CO_2-Induced Climatic Change." *Advances in Geophysics* 26, 141–235.

30 Mitchell, J. F. B. (1983): "The Seasonal Response of a General Circulation Model to Changes in CO_2 and Sea Temperature." *Quarterly Journal of the Royal Meteorological Society* 109, 113–52.

31 Washington, W. M., and G. A. Meehl (1983): "General Circulation Model Experiments on the Climatic Effects Due to a Doubling and a Quadrupling of Carbon Dioxide Concentration." *Journal of Geophysical Research* 88, 6600–10.

32 Graphics supplied by H. F. Diaz, University of Massachusetts, from data of P. D. Jones and T. M. L. Wigley, University of East Anglia.

33 Testimony of Patrick J. Michaels to the Committee on Foreign Relations, United States Senate, April 20, 1989.

34 Angell, J. K. (1986): "Annual and Seasonal Global Temperature Changes in the Troposphere and Low Stratosphere." *Monthly Weather Review* 107, 1922–30.

35 This was reported at the Second meeting of the Climate Trends Panel, held in Amherst, Massachusetts in May, 1989. An article about this meeting appears in Kerr, R. A. (1989). "Hansen vs. the World on the Greenhouse Threat." *Science* 244, 1041–3.

36 This forecast forms a testable hypothesis. If one takes the definition of the "early 1990's" strictly, this comprises the period January 1, 1990 through December 31, 1993. Where available, Palmer Drought values should be calculated during this period and compared to a long-term mean. The comparison should not be made over mountainous areas, as the Palmer Index is insensitive to snowpack and snowpack irrigation.

37 Schneider, S. H. (1989): "The Greenhouse Effect: Science and Policy." *Science* 243, 771–81.

38 Ramanathan, V., R. D. Cess, E. F. Harrison, P. Minnis, B. R. Barkstrom, E. Ahmad, and D. Hartman (1989): "Cloud-Radiative Forcing and Climate: Results from the Earth Radiation Budget Experiment." *Science* 243, 57–63.

39 Mitchell, J. F. B., C. A. Senior, and W. J. Ingrahm (1989): "CO_2 and Climate: A Missing Feedback?" *Nature* 341, 132–4.

40 Washington, W. M., and G. A. Meehl (1989): "Climate Sensitivity due to increased CO_2: Experiments with a Coupled Atmosphere and Ocean General Circulation Model." *Climate Dynamics* 4, 1–38.

41 Karl, T. R., and P. D. Jones (1989): "Urban Bias in Area-Averaged Surface Air Temperature Trends." *Bulletin of the American Meteorological Society* 70, 265–70.

42 Hansen, J., and S. Lebedeff (1988): "Global Surface Air Temperatures: Update through 1987." *Geophysics Research Letter* 15, 323–6.

43 T. Karl, personal communication to P. J. Michaels, May 1989.

44 Testimony of J. E. Hansen to US House of Representatives, May 1988.

45 Michaels, P. J., D. E. Sappington, and D. E. Stooksbury (1988): "Anthropogenic Warming in North Alaska?" *Journal of Climate* 1, 942–5.

46 Michaels, P. J., D. E. Sappington, and D. E. Stooksbury (1989): "Regional 500mb Heights Prior to the Radiosonde Era." Proceedings, 6th Conference on Applied Climatology, American Meteorological Society, Charleston, SC, 184–7.

47 Hayden, B. P. (1981): "Secular Variation in Atlantic Coast Extratropical Cyclones." *Monthly Weather Review* 109, 159–67.

48 Karl, T. R., R. G. Baldwin, and M. G. Burgin (1988): "Time Series of Regional Season Averages of Maximum, Minimum, and Average Temperature, and Diurnal Temperature Range across the United States, 1901–1984." US Dept. of Commerce, Historical Climatology Series 4–5, National Climatic Data Center, Asheville, North Carolina.

49 Elsaesser, H. W., M. C. MacCracken, J. J. Walton, and S. L. Grotch (1986): "Global Climatic Trends as Revealed by the Recorded Data." *Review of Geophysics* 24, 745–92.

50 Sansom, J. (1989): "Antarctic Surface Temperature Time Series." *Journal of Climate* 2, 1164–72.

51 Seaver, W. L., and J. E. Lee (1987): "A Statistical Examination of Sky Cover Changes in the Contiguous United States." *Journal of Climatology and Applied Meteorology* 26, 88–95.

52 Budyko, M. I (1988). Paper read to the first meeting of the Climate Trends Panel, held at National Research Council, September.

53 Wigley, T. M. L. (1989): "Possible Climate Change Due to SO_2-Derived Cloud Condensation Nuclei." *Nature* 338, 365–7.

54 Cess, R. E. (1989): Presentation to US Department of Energy meeting on research planning for Global Change, Germantown, MD, April.

55 Michaels, P. J., P. J. Stenger, and D. E. Sappington (1989): "Atmospheric Pollutant Transport: Take it or Leave it." Proceedings, Sixth Conference on Applied Climatology, American Meteorological Society, Charleston, South Carolina, 184–7.

56 Wallace, J. M., and D. S. Gutzler (1981): "Teleconnections in the Geopotential Height Field during Northern Hemisphere Winter." *Monthly Weather Review* 109, 784–812.

57 Balling R. C., and S. B. Idso (1989): "Historical Temperature Trends in the United States and the Effect of Urban Population Growth." *Journal of Geophysical Research* 84, 3359–63.

58 Budyko, M. I., and Yu. S. Sudunov (1988): "Anthropogenic Climatic Changes." World Congress, Climate and Development, Hamburg, FRG.

50

Greenhouse Scenarios to Inform Decision Makers

Lester Lave

The United States and other nations must decide how much resources to devote to reducing greenhouse gas emissions as well as how much to devote to adapting to a changed climate. Good policy decisions require reliable answers to the following questions: (1) How large will be the changes in temperature and precipitation in each region? (2) What will be the effects on people and the environment we care about, after people have had a chance to adjust to the new climate? (3) What will be the cost of reducing greenhouse gas emissions to prevent climate change? Uncertainty dominates analysis of greenhouse issues, beginning with the extent to which greenhouse gas emissions will change climate. Individual modelers may have unique judgments as to the warming that would be associated with a doubling of atmospheric concentrations of carbon dioxide (CO_2) (National Academy of Sciences, 1991).

A great many resources are being invested in atmospheric science research to answer the first question. Some models use historical analogies to investigate periods when there was more carbon dioxide in the atmosphere and the Earth was warmer (see Intergovernmental Panel on Climate Change, 1990 [IPCC], Environmental Protection Agency, National Academy of Sciences, Schneider). Other atmospheric scientists depend on relatively simple models to investigate the issues. Finally, there are more than half a dozen general circulation models (GCMs), computer simulations that work on only the largest computers. The three methods don't produce identical results; indeed, there isn't agreement on predictions such as temperature increase within a method, such as among the GCMs (Marshall Institute (1989); Schneider (1989)). Because of the complexity of the atmospheric system, definitive models relating climate to greenhouse gases probably will not be available for many years.

A large source of uncertainty is the emissions rate for greenhouse gases. In general, the various projections have assumed that emissions will continue to increase at past rates. If so, doubling could occur as early as 2025.

Until the uncertainty is resolved, scientists have been trying to generate estimates that receive general agreement, or at least are not subject to major objections. A 1991 report from the National Academy of Sciences (NAS, 1991) estimates that a doubling of carbon dioxide (or its equivalent increase in other greenhouse gases) might raise temperature 1°–5°C. Other scientific groups have come to slightly different conclusions; e.g., the IPCC used a narrower range (2°–4°C). Some noted scientists think these estimates are too large (Marshall Institute (1989); Lindzen). There is even more difference among scientists with regard to the implications of a changed climate, e.g., effects on ocean currents, clouds, and rainfall.

THE COSTS OF ABATEMENT AND OF ADJUSTING TO A WARMER CLIMATE

Some energy models crafted to explore the effects of curtailed supplies of energy have been adapted to examine greenhouse effects (Manne and Richels (1990); Nordhaus (1990); Jorgenson and Wilcoxin, (1990)). Some of these estimate the cost of curtailing the use of fossil fuels (Nordhaus, Manne, and Richels), while others attempt to model the adjustments associated with less fossil fuel use (Jorgenson and Wilcoxin). For example, Jorgenson and Wilcoxin examine the extent to which economic and productivity growth would be reduced by constraining fossil fuel use.

Relatively little scientific attention has been devoted to the effects of a warmer climate. Nordhaus has estimated that approximately a 2.5°C warming would curtail US GNP by perhaps 0.4 percent, which he rounds up to 1 percent. Lave and Vickland agree that the effect in developed countries is likely to be small, but they conclude that the results in developing countries could be much larger. Unlike the USA, in which only 4 percent of GNP comes from the production of food and fiber, the developing countries can have more than two-thirds of GNP originating in food and fiber production. Thus, the effects on developing countries could be much larger than those of the developed nations. While the effect of a 2.5°C warming on world GNP is estimated to be small, some countries could suffer a large loss.

The scientific discussions have focused on the adverse outcomes of climate change: sea level rise, droughts, possible increase in large storms, etc. But climate change need not be disruptive; warming could be beneficial, at least in human perceptions. If the warming occurs mostly in winter and at night, as the models suggest, the growing season in the high latitudes would be extended, allowing for more food production (if the soils aren't already a constraint). Migration patterns in the USA and Canada suggest that people desire warmer climates.

A SIMPLE MODEL OF THE THREE UNCERTAINTIES

The overwhelming uncertainty associated with all aspects of global climate change has led to confusion in public and private debates. The following discussion focuses on a doubling of greenhouse gases in the atmosphere assumed to occur as early as 2040. The full effect on temperatures would be delayed some years. I focus on the three essential areas of uncertainty to achieve clarity.

(1) For the range of climate change, I choose three values: less than 1°C, 1°–5°C, and more than 5°C (corresponding to the National Academy of Sciences range). Since there is general agreement among scientists that 1°–5°C is the likely range, I assume that a temperature increase in this range has a 60 percent likelihood. The remaining 40 percent likelihood is evenly divided between higher and lower temperatures.

(2) For the effects on people and the environment, an even greater simplification is necessary. I assume that a temperature increase of less than 1°C would have a tiny effect on the economy and environment. I also assume that a temperature increase greater than 5°C would have a massive, disruptive effect on the economy and environment. Within the National Academy of Sciences range, the effects could either be benign or harmful, depending on when and where the warming takes place, what happens to precipitation, and how people adjust to the change. Thus, there are four cases: warming of less than 1°C, 1°–5°C (either predominantly good or bad), and greater than 5°C.

(3) The cost of abatement depends on improvements in the technologies for nuclear energy, carbon sequestration, and renewable energy sources, as well as technologies that conserve energy. The cost of reducing greenhouse gas emissions is nonlinear, since energy conservation could handle small emissions reductions. Massive disruptions and cost increases are likely if emissions of CO_2 per dollar of GNP must be reduced to one-half to one-fourth of current levels. I have chosen some estimates of the costs of greenhouse gas abatement that are consistent with the various studies (see table 50.1).

The CO_2 abatement strategies are characterized as "no abatement," "moderate abatement" (reducing greenhouse gas emissions per dollar of GNP to 80 percent of 1988 levels, but allowing emissions to rise as GNP rises), and "stringent abatement" (reducing greenhouse gas emissions to 80 percent of 1988 levels). The "moderate abatement" level requires only a 20 percent abatement of greenhouse gases per dollar of GNP, no matter how much GNP rises. The "stringent abatement" level is much more difficult to achieve since aggregate greenhouse gas emissions are held to 80 percent of their 1988 level. If GNP grows 2.5 percent per year, in 2040 it will be 3.6 times larger than it was in 1988. If so, reducing emissions to 80 percent of 1988 levels would require emissions per dollar of GNP to fall to 22 percent of the 1988 level, a 78 percent

Table 50.1 Costs of climate change and effects and costs of greenhouse gas emissions abatement in the United States by 2040

Effect of abatement on costs caused by climate change	
No abatement:	0 percent reduction
Moderate abatement:	25 percent reduction
Stringent abatement:	100 percent reduction

Cost of abating CO_2 emissions in United States by 2040 (percent of GNP)			
Degree of abatement	*Optimistic*	*Moderate*	*Pessimistic*
No abatement:	0	0	0
Moderate (CO_2/GNP is 80 percent of 1988 level):	− 1	0.4	1
Stringent (CO_2 is 80 percent of 1988 level):	5	10	60

Note: GNP is assumed to grow 2.5 percent per year; thus the 2040 GNP is 3.61 times the 1988 level and so CO_2 emissions per dollar of GNP fall from 1 in 1988 to 0.19 in 2040.

Effect of climate change on United States economy in 2040 (percent of GNP)			
Temperature change	*Optimistic*	*Moderate*	*Pessimistic*
Small (<1°C):	0	0	0
Moderate (1°–5°C):			
Good effect	1	− 0.5	− 3
Bad effect	− 1	− 3	− 10
Large (>5°C):	− 10	− 30	− 60

reduction in emissions per dollar. Stringent abatement by all nations is assumed to be sufficient to prevent further global climate change; moderate abatement would be assumed to lower damage by 25 percent.

Costs of abatement are uncertain. Nordhaus estimates that 20 percent (moderate) abatement would cost about 0.4 percent of US GNP. An optimistic view is that small or even moderate abatement would actually lower costs by 1 percent. A pessimistic view is that the costs of 20 percent abatement might be perhaps 1 percent of GNP.

There are no estimates for the cost of a 78 percent reduction in greenhouse gases by 2040. Using currently available technologies, the maximum attainable level of abatement is perhaps 25–30 percent, given the current output mix and level (Office of Technology Assessment). By 2040 there will be new technologies for conserving energy and new technologies that generate energy without emitting greenhouse gases. Nonetheless, replacing less efficient technologies will be difficult and expensive. In order to provide a wide range of estimates, I assume that the cost of this degree of abatement would range from 5 percent of GNP to 10 percent or even up to 60 percent.

THE SOCIAL BENEFITS AND COSTS OF
ACTION AND INACTION

The next step in the analysis is to examine the social costs and benefits of each climate outcome and abatement policy. As noted above, I have assumed that a doubling of greenhouse gases will lead to probabilities of the four outcomes of 0.20, 0.30, 0.30, and 0.20, respectively.

The effects of climate change are even less certain than the costs of abatement. A climate change of less than 1°C is assumed to have negligible social and environmental costs. For moderate and large climate changes, I have defined three views, expressing the social and environmental costs in terms of reduction in GNP. An optimistic view of a good outcome from a moderate temperature increase is that people would be benefited equivalent to an increase in GNP of 1 percent. In the median view, moderate-but-good climate change would be good in the long run, but might involve some short-term costs, say 0.5 percent of GNP each year. A pessimistic view is that this moderate temperate change that was predominantly good might still cost the equivalent of 3 percent of GNP.

Moderate temperature changes that were predominantly good are assumed to affect the public in a way that translates into a GNP change of 1 percent, –0.5 percent, and –3 percent; moderate but predominantly bad GNP changes would be –1 percent, –3 percent, or –10 percent of GNP, depending on whether one is optimistic, moderate, or pessimistic. No one has analyzed the effect of a temperature change larger than 5°C. In order to get some numbers for the analysis, I assume such a temperature change would cost 10 percent, 30 percent, or 60 percent of GNP. These are very large estimates that assume wholesale disruption of the economy and large decreases in the standard of living and quality of the environment.

These effects of climate change are measured in terms of GNP, although they are meant to include all effects, not just changes in GNP. For example, they might be thought of in terms of the amount of income people would be willing to give up in order not to have an environmental effect occur.

THREE SCENARIOS OF CLIMATE CHANGE

The three views of the cost of abatement and three views of the effects of climate change are combined to describe three cases or scenarios, shown in table 50.2. The first scenario is defined by optimism concerning the cost of climate change and pessimism concerning abatement cost. Some industrialists appear to have this view. This scenario should lead to the least expenditure on abatement and the greatest climate change. The second scenario is a middle view of climate change cost and middle view of abatement cost. At this time, there are few people who

Table 50.2   Climate change and abatement policy scenarios

Scenario 1: optimistic cost of climate change and pessimistic abatement cost (percent of GNP)

		Temperature change			
	Small	Moderate-good	Moderate-bad	Large	Expected cost
Abatement policy	(<1°C)	(1°–5°C)	(1°–5°C)	(>5°C)	
No abatement:	0	– 1	1	10	2.0
Moderate abatement:	1	0.25	1.75	8.5	2.5
Stringent abatement:	60	60	60	60	60.0

Scenario 2: moderate cost of climate change and moderate abatement cost (percent of GNP)

		Temperature change			
	Small	Moderate-good	Moderate-bad	Large	Expected cost
Abatement policy	(<1°C)	(1°–5°C)	(1°–5°C)	(>5°C)	
No abatement:	0	0.5	3	30	7.05
Moderate abatement:	0.4	0.78	2.65	22.9	6.41
Stringent abatement:	10	10	10	10	10.00

Scenario 3: pessimistic cost of climate change and optimistic abatement cost (percent of GNP)

		Temperature change			
	Small	Moderate-good	Moderate-bad	Large	Expected cost
No abatement	(<1°C)	(1°–5°C)	(1°–5°C)	(>5°C)	
No abatement	0	3	10	60	15.90
Moderate abatement	– 1	1.25	6.5	44	10.93
Stringent abatement	5	5	5	5	5.00

have this moderate view. This lack of a large center means that the greenhouse debate swings between polar extremes when relatively tiny bits of new scientific data become available. The third scenario combines a pessimistic view concerning climate change cost with an optimistic view concerning abatement cost. This view is held by many environmentalists, who think that energy conservation is essentially free and that any anthropogenic climate change is bad. This third scenario reflects the greatest risk aversion concerning the future environment and should lead to the greatest abatement.

The three scenarios are shown as matrices with abatement strategy as the rows, climate change as the columns, and percent reduction in GNP as the entries in the matrices. They can be thought of in terms of game theory as "games" in normal form. Humans get to choose a row (an abatement strategy) and nature gets to choose a column (the amount of climate change). The entries in the matrix refer to the human loss should a particular combination occur. The costs are expressed as a proportion of GNP, although GNP is used as a surrogate for all costs.

This formulation helps to clarify a basic debate about policy. Should the USA act now, with little knowledge of the extent of global climate change, or should we wait for more certainty, knowing that adjustment might be more difficult and expensive after continuing to increase greenhouse gas emissions? If we must act now, we have a large chance of committing to a strategy that could be inappropriate and needlessly costly. In the matrix, if nature chooses first, revealing the extent of climate change, the uncertainty is greatly diminished. For example, if nature chooses a small climate change in response to a doubling of CO_2, humans should choose the no abatement policy. If nature chooses a moderate warming, humans would either choose no abatement or stringent abatement, depending on whether the effect of this climate change was benign or harmful.

This observation illustrates the second controversy. Even if people wait for resolution to the issue of how great will be climate change, the resolution will still leave uncertainty concerning whether the effect of the climate change will be good or bad. To resolve that uncertainty, people would have to wait until the adjustment to the new climate had largely been completed.

If people waited until adjustment to the new climate had been accomplished, there would be nearly complete knowledge concerning what abatement strategy was best. However, the Earth would have to bear with the consequences of the climate change for many decades or centuries to come. If, for example, the climate change were large and harmful, the USA could suffer the equivalent of a 60 percent reduction in GNP. Furthermore, emissions of greenhouse gases would have continued to rise, leading to even greater warming.

This discussion clarifies the effects of uncertainty regarding climate change and its effects. Although uncertainty concerning the sensitivity of the climate to greenhouse gases could be determined by some future date, we cannot know whether that effect will be good or bad or whether the cost of abatement will be high or low unless there is extensive experience. Thus, a decision concerning abatement must necessarily be made in the midst of overwhelming uncertainty. Therefore, I return to the matrices setting out the three scenarios assuming that humans must make an abatement decision without knowing the extent of climate change, whether it will be predominantly good or bad, or even what the costs of abatement will be.

EXPLORING THE SCENARIOS

Industrialists, optimistic about the effects of climate change and pessimistic about abatement cost (scenario 1), would welcome small to moderate (good) climate change. If this climate change occurred, taking no action to decrease greenhouse gas emissions would lead to a 1 percent gain (or possibly no change) in GNP. For someone with a moderate view about the effects of climate change and of abatement costs (scenario 2), climate change couldn't help, and so a small change would be best. If nature obliged with a small climate change, a no abatement policy would cost nothing. An environmentalist, optimistic about the costs of abatement and pessimistic about the effects of climate change (scenario 3), would see that climate change couldn't help; like the moderate, an environmentalist would hope that the climate change would be small.

For now, science doesn't provide answers as to whether climate change will be large or small, whether moderate climate change will be good or bad, or whether the costs of abatement will be large or small. But each decision maker should know whether he or she is optimistic, moderate, or pessimistic about these issues. Thus, each decision maker should identify with one of the three scenarios. The matrix would help decide what actions to take.

Knowing what column nature was going to choose (which climate change to expect), a decision maker could choose the best abatement strategy. If nature chose a *small climate change*, moderate and pessimistic decision makers would choose a "no abatement" policy; an optimistic decision maker would choose moderate abatement because it would save energy at no cost. For a *moderate climate change*, either good or bad in effect, a decision maker whose views are summarized by the first scenario should choose a no abatement strategy. A decision maker whose views were summarized by the second or third scenarios would choose more stringent abatement strategy for "moderate-bad" than for "moderate-good" (no and moderate abatement in scenario 2 and moderate and stringent abatement for scenario 3). For a *large climate change*, a decision maker represented by the first scenario would choose moderate abatement; a decision maker represented by scenario 2 or 3 would choose stringent abatement. This is a case where decision makers with an optimistic view of the costs of climate change and a pessimistic view of abatement cost would never agree to stringent abatement. Decision makers with less optimistic views of the costs of climate change and less pessimistic views of abatement cost would refuse any policy except stringent abatement. This is a situation of maximum conflict.

CHOOSING AN ABATEMENT POLICY UNDER UNCERTAINTY

The Min-Max Solution

But we don't know how large climate change will be. One criterion used in game theory is min-max: Choose the row that will give the smallest loss, no matter how much the climate changes (that is, assuming large temperature change). The min-max criterion would lead to choosing stringent abatement in scenarios 2 and 3 and moderate abatement in scenario 1. In scenario 1, the social cost is lower for living with climate change, even if it is large, than for reducing emissions stringently. In the other two scenarios, stringent abatement is the least costly strategy for dealing with a large climate change (10 and 5, respectively).

Much of the discussion of greenhouse policy adopts the philosophy behind such a min-max choice: Humans should act so that they minimize the chance that the Earth will be spoiled for future generations. In this one-sided discussion, the possibility that climate change may turn out to be beneficial is irrelevant. The min-max view is that even if spoiling the Earth is viewed as extremely unlikely, the possibility that it might occur is sufficient to warrant protection.

The Minimize-Expected-Loss Solution

A second solution approach is to compute the expected loss for each abatement strategy, assuming that the probabilities of climate change are 0.2, 0.3, 0.3, and 0.2, respectively. The expected values are shown in the last column of each matrix.

For the first scenario, the strategy of "no abatement" has an expected loss of 2, compared to an expected loss of 2.5 for "moderate abatement" and 60 for "stringent abatement." The expected loss solution would lead to choosing "no abatement." However, we might also want to consider that the likelihood of large climate change, with its large losses, could be greater than assumed. This is particularly relevant for the moderate abatement, since its expected cost is only 0.5 higher than no abatement. Choosing "moderate abatement" would mean giving up 0.5 percent of GNP (2 vs. 2.5) in expected cost in order to gain a reduction of 1.5 percent of GNP (10 vs. 8.5) in the maximum loss. Some risk-averse people might decide this tradeoff was worthwhile.

For the second scenario, the moderate abatement strategy has the lowest expected loss: 6.41 percent of GNP. This abatement strategy also has the advantage of a much smaller loss (22.9 vs. 30) should the climate change be large. However, even the 22.9 loss for a large climate change is daunting. If society were willing to give up 3.59 in expected loss (6.41 vs. 10), it could reduce the maximum loss from 22.9 to 10. Risk averse people might prefer the "stringent abatement" strategy.

For the third scenario, "stringent abatement" has both the lowest expected loss and the smallest maximum loss: 5 percent of GNP.

CONCLUSION

Several insights can be inferred from these three scenarios. First, these simple matrices are a helpful way of looking at the rather complicated issues. They provide insight into the implications of having different views of the cost of abatement, of the magnitude of climate change, and of the effects of climate change. The most important questions that science cannot currently answer are translated into three scenarios that represent these different views. The scenarios push aside a great deal of unessential material and focus rather narrowly.

Second, a small number of scenarios is sufficient to explore the most important scientific uncertainties. They also manage to explore the relevant abatement policies.

Third, different solution approaches lead to vastly different solutions, as might be expected. The solution concepts, min-max or minimize expected cost, have their analogies in the ways in which various interest groups approach the issues.

Fourth, the best abatement policy depends on the scenario (assumptions about the cost of abatement, extent of climate change, and effect of climate change). No wonder people of different views conclude that quite different abatement policies ought to be pursued!

The amount of climate change that would take place if atmospheric concentrations of carbon dioxide were doubled is only one of three crucial issues. It is less important than whether the effects of climate change will be predominantly good or bad and the cost of abatement. Research has focused on one part of the problem, and there is more than a suggestion that the focus has not been on the most important aspect.

REFERENCES

Environmental Protection Agency (1989): *The Potential Effects of Global Climate Change on the United States: A Report to Congress.* Washington: EPA.

Intergovernmental Panel on Climate Change (1990): *Policymakers Summary.* New York: United Nations Environmental Program.

Jorgenson, Dale W., and Peter J. Wilcoxin (1990): "Environmental Regulation and U.S. Economic Growth." *The Rand Journal of Economics.*

Lave, Lester B., and Katherine H. Vickland (1989): "Adjusting to Greenhouse Effects: The Demise of Traditional Cultures and the Cost to the USA." *Risk Analysis.*

Lindzen, Richard (1990): "Some Coolness Concerning Global Warming." *Bulletin of the American Meteorological Society.*

Manne, Alan S., and Richard G. Richels (1990): "CO_2 Emission Limits: An Economic Cost Analysis for the USA." *Energy Journal.*

Marshall Institute (F. Seitz, K. Bendelsen, R. Jastrow, and W. Nierenberg) (1989): *Scientific Perspectives on the Greenhouse Problem.* Washington: George C. Marshall Institute.

National Academy of Sciences (1991): *Policy Implications of Global Warming.* Washington: National Academy Press.

Nordhaus, William D. (1990): "A Survey of the Costs of Reduction of Greenhouse Gases." *Energy Journal.*

Office of Technology Assessment (1991): *Changing by Degrees: Steps to Reduce Greenhouse Gases.* Washington, DC: Government Printing Office.

Schneider, Stephen H. (1989): "The Greenhouse Effect: Science and Policy." *Science.*

51

The Hazards of Nuclear Power[1]

Bernard L. Cohen

About 80 percent of the American public believes it is more dangerous to generate electricity from nuclear power than from coal.[2] Scientific studies show, however, that coal is many times more dangerous.[3] Even Henry Kendall,[4] director of the anti-nuclear lobbying group Union of Concerned Scientists, and anti-nuclear activist Ralph Nader, in private,[5] concede this.

The enormous public misunderstanding about nuclear power may be largely attributed to (1) a widespread and exaggerated fear of radiation, (2) a highly distorted picture of reactor accidents, (3) grossly unjustified fears about disposal of radioactive waste, and (4) failure to understand and quantify risk.

RADIATION

How Dangerous Is Radiation?

Is being struck by a particle of radiation a terrible tragedy? No. Every person is struck by about a million particles of radiation every minute from natural sources. (The rate varies with geography and other factors.) This rate is hundreds of times greater than our exposure to radiation from the nuclear power industry. So is our average exposure to radiation from medical X-rays.

Although a single particle of radiation can cause cancer, the chance it will do so is only about one in 30 quadrillion. Hence, the million particles that strike us each minute have only one chance in 30 billion of causing a cancer. A human lifespan is about 40 million minutes; thus, all of the natural radiation to which we are exposed has about one chance in 700 of causing a cancer. Since our overall chance of dying from cancer is one in five, only one in 140 of all cancers may be due to natural radiation. The average exposure from a nuclear power plant to those who live closest to it is about 1 percent of their exposure to natural radiation; hence, if they live there for a lifetime, there is perhaps one chance in 70,000 (1/100th the chance from natural radiation and 1/14,000 the chance from all causes) that they will die of cancer as a result of exposure to its radiation.

Scientific Basis for Risk Estimates

How do we know these risks so quantitatively? Several prestigious scientific groups provide frequent summaries and evaluations of available data. In the past few years, the US National Academy of Sciences Committee on Biological Effects of Ionizing Radiation (BEIR)[6] and the United Nations Scientific Committee on Effects of Atomic Radiation (UNSCEAR)[7] have issued reports assessing the cancer risk from low-level radiation. There should be one extra cancer for every 2.6 million mrem (1 millirem = approximately 5 billion particles of radiation) of exposure to humans; i.e. every mrem of exposure produces a 1-in-2.6-million risk of fatal cancer.

The health effects of *high-level radiation* are well known and serious.[6,7] But there is little direct evidence on effects of low-level radiation. The simplest way to make estimates is to derive them from data on high-level radiation by assuming a linear dose-effect relationship. For example, if high-level dose D causes a cancer risk R, we assume that a dose 0.01 D will cause a risk 0.01 R. This assumption is nearly always used, with relatively minor variations. It was used to derive the estimate of one cancer death per 2.6 million mrem given above. However, this is more likely to overestimate than to underestimate the effects of low-level radiation, and the minor variations from linear behavior often used[6] are such as to reduce, moderately, the estimated effects at low levels. This may be described graphically by stating that the curve of cancer risk vs radiation dose is concave upward, i.e., it curves upward at high dose from a simple straight line. There is abundant evidence supporting this viewpoint.[8]

Routine Emissions from the Nuclear Industry

In operation, nuclear power plants routinely release small quantities of radioactive gases and contaminants in water into the environment. More importantly, when reactor fuel is chemically reprocessed, more radioactive gases are released at the reprocessing plant. Extensive studies[7] predict that, with current technology, routine releases of radiation due to operation of one large power plant for one year, including reprocessing, will cause 0.25 cancer deaths over the next 500 years. Since we are not reprocessing fuel now, effects of current operations are only about 20 percent of this. Available technologies can drastically reduce releases from reprocessing plants.

REACTOR ACCIDENTS[9]

Power plants include many levels of protection against radioactivity releases, based on a defense in depth design philosophy. For example, an accident could be initiated by a sudden rupture in the system, allowing the cooling water to escape. Levels of protection against this are:

1 the highest quality standards on materials and equipment in which such a rupture might occur;
2 elaborate inspection programs to detect flaws in the system using X-ray, ultrasonic, and visual techniques;
3 leak-detection systems (Normally a rupture starts out as a small crack, allowing water to leak out slowly. Such leaks would be detected by these systems and repaired before a rupture could occur.);
4 an emergency cooling system, which would rapidly replace the water lost in such a rupture accident, restoring cooling to the reactor fuel (In this type of accident there are several different pumping systems, any one of which would provide sufficient water to avert a meltdown if all the others somehow failed.);
5 the containment, a strong building in which the reactor is housed, which would normally hold the released radioactivity inside even if there were a meltdown.

Occasionally there is a failure at some power plant in one of these lines of defense, e.g., a valve that should be open is found closed. The media report this as a near miss or a disaster, apparently not understanding defense in depth. While it is *possible* for each line of defense to fail, one after another, the probability of such a sequence is extremely low. Moreover, the same reasoning applies to almost any other technology. In pumping gasoline, for example, a highly improbable sequence of highly improbable events could burn down a city, killing most of its inhabitants, yet the media do not trumpet warnings of these impending catastrophes.

How large is the risk to the public from a reactor accident? The scientific approach to this problem is through Probabilistic Risk Analysis (PRA), which, in principle, considers every combination of events that can lead to health impacts, along with their probability of occurrence. The individual events are failures of valves, pumps, welds, and such, for which there are abundant data on frequency. It is therefore not difficult to determine the probability for any given sequence of events. PRA have been completed on about 30 plants and will soon be available for all US plants. We give typical results here.[10] We also give results from a study by the anti-nuclear activist organization, Union of Concerned Scientists.[11]

For the frequency, the PRA give one meltdown per 10,000 plant-years, whereas the UCS estimate is one per 2,000 plant-years. After about 4,000 plant-years of commercial operation of this type reactor around the world and about 3,000 equivalent plant-years of naval reactor operation, there has been only one fuel damage accident, Three Mile Island.* Thus, the UCS estimate implies that we have been extremely lucky, although these results are not surprising based on PRA.

There is widespread misunderstanding of the consequences of a fuel meltdown. We often hear that it would kill tens of thousands of people and contaminate a whole state, but such statements are grossly misleading. The containment building would ordinarily contain the radioactive dust inside long enough (about a day) to clean it out of the air. For example, investigators of the Three Mile Island accident[12] all agree that even if there had been a complete meltdown, there would have been little harm to the public, because there is no reason to believe that the integrity of the containment would have been compromised. In most meltdowns, no fatalities are to be expected.

Some events could break open the containment building, releasing radioactive dust into the environment. If this happens, the consequences depend on timing and weather. In the most unfavorable condition, with a large containment break early in the accident, PRA estimates 50,000 fatalities in some cases, but this unusual combination is expected only once in 100,000 meltdowns, i.e., about once in a billion plant years.

According to PRA, the average number of fatalities to be expected in a reactor meltdown is 400; according to UCS, it is 5,000. Estimates of the fatality rate due to air pollution from coal burning are at least 10,000 each year.[13] For reactor meltdowns to be as harmful as coal-burning, we would therefore need a meltdown every two weeks according to the PRA, or every six months according to UCS. No one has suggested that meltdowns will occur anywhere near that frequently.

When the frequency and consequence estimates are combined, the Nuclear Regulatory Commission concludes that we may expect an average of 0.02 fatalities per plant-year; UCS predicts 2.4. Even the latter figure is less than one-tenth the 25 fatalities per plant-year due to air pollution from coal-burning electricity generation.[13] Of course these fatalities from air pollution are not detectable in the US population, in which two million people die every year. But the same would be true of 98 percent of the predicted fatalities from reactor meltdown accidents. For example, the worst such accident normally considered would cause 45,000 extra cancer deaths in a population of ten million over 50 years. For each of these ten million, the risk of dying from cancer would rise by only 2.5 percent, which would hardly be detectable. The present risk in different states varies between 16 percent and 24 percent, so the added cancer risk in moving from one state to another is often many times larger than the added risk from such a nuclear accident.

Detectable fatalities – occurring shortly after the hypothetical accident and clearly attributable to it – would be rather rare. According to PRA,

---

* The Chernobyl reactor was of a very different type from those used in the West. Its safety problems are very different and much more severe. A reactor of that type could never be licensed in the United States.

98 percent of all meltdowns would cause no detectable fatalities, the average number for all meltdowns is ten, and the worst meltdown in the analyses (a 1-in-100,000 occurrence) would cause 3,500. The largest coal-related incident to date was an air pollution episode in London in 1952 that caused 3,500 fatalities within a few days.[13] Thus, as far as detectable fatalities are concerned, the worst nuclear accident in 100,000 meltdowns has already been equaled by coal burning.

The extent of land contamination in a reactor meltdown accident depends on one's definition of contamination. The whole earth can be said to be contaminated because there is natural radioactivity everywhere. But if we use the internationally accepted definition of the level that calls for remedial action, the worst meltdown normally considered (one in 10,000) would contaminate an area equal to a circle 60 miles in diameter. About 90 percent of this could be easily decontaminated by use of fire hoses and plowing open fields, so the area where relocation of people is necessary would be equal to that of a circle 20 miles across. Forced relocation of people is not catastrophic. It occurs when new dams permanently flood large areas, in highway construction, in urban redevelopment, etc. In such situations the major consideration is the cost of relocating people. Therefore, it seems reasonable to consider land contamination by a nuclear accident on the basis of its monetary cost.

In the worst 0.01 percent of accidents, the cost could exceed $30 billion, but the *average* cost for all meltdowns would be $200 million. Air pollution from coal burning also does property damage, estimated in the range of $1 billion per year.[14] At an average of $200 million per meltdown, it would take a reactor meltdown every two months to be as costly as the property damage from coal burning.

RADIOACTIVE WASTE

There are several types of radioactive waste generated by the nuclear industry, but we will concentrate largely on the two most important and potentially dangerous, high-level waste and radon.

High-Level Waste[15]

In a rationally planned and developed nuclear power program, spent reactor fuel would be shipped to a reprocessing plant to remove valuable components. The residue, containing nearly all of the radioactivity produced in the reactor, is called high-level waste. (In the US program, where spent fuel is simply potted and encased, reprocessing is not currently done or even planned, but it is done in other countries.) Following reprocessing, the waste can be converted into a rock-like form and buried deep underground in a carefully selected geological formation.

One important aspect of high-level waste disposal is the small quantities involved. The waste generated by one large nuclear power plant in one year and prepared for burial is about six cubic yards, roughly one truckload. This is two million times smaller by weight, and billions of times smaller by volume, than wastes from a coal plant. The electricity generated in a year sells for more than $400 million, so if only 1 percent of the sales price were diverted to waste disposal, $4 million might be spent to bury this one truckload of waste, enough to pay for very elaborate protective measures.

To understand the very long-term (millions of years) hazard, natural radioactivity in the ground is a good comparison. The ground is full of naturally radioactive materials. By adding nuclear waste to it, the total radioactivity in the top 2,000 feet of US soil would increase by only one part in ten million per plant-year. Moreover, the radioactivity in the ground (except perhaps very near the surface) does virtually no harm.

The principal concern about buried waste is that it might dissolve in groundwater and contaminate food and drinking water supplies. How dangerous is this material to eat or drink? When first buried, it is highly toxic, and a fatal dose (yielding a 50 percent chance of death) is only 0.01 ounce. However, after 100 years the fatal dose is 0.1 ounce, and after 600 years it is one ounce, making it no more toxic than many common household chemicals and medicines. After 10,000 years a lethal dose is ten ounces.

Some people are frightened on hearing that nuclear waste must be isolated for a few hundred years. But 2,000 feet below earth's surface, things remain essentially unchanged for millions of years. Waste burial plans include measures to delay the release of the waste to the environment for a very long time, yielding near-perfect protection from the short-term (several hundred or thousand years) problem. First, wastes will be buried in rock formations isolated from groundwater and expected to remain so for at least 1,000 years. Second, if water did enter, it would be no more likely to dissolve the waste than the surrounding rock; for the latter the time required is ordinarily millions of years. Third, clay backfill materials surrounding the waste package swell up to seal very tightly when wet, keeping out any appreciable amount of water. Fourth, if groundwater did reach and dissolve some of the waste, these clays would effectively filter radioactive material out of solution before it could escape with the water. Also, waste will be sealed in corrosion-resistant casings that would not dissolve even if soaked in groundwater for many thousands of years. Finally, if radioactivity did reach surface waters, it would be detected easily – a millionth of the amounts that can be very harmful are readily detected – and measures could be taken to prevent it from getting into drinking water or food.

With all these safeguards it seems almost impossible for much harm to result during the first few hundred years, while the waste is highly toxic, and there is substantial protection over the long term.

It can be helpful to compare radioactive waste with ordinary rock. An average atom of rock has about one chance in a billion per year of escaping into surface waters and one in a thousand of getting into a human body afterward. If these probabilities are combined and applied to buried radioactive waste, the result indicates that the waste would eventually cause 0.02 fatalities per plant-year – a thousand times less than the health effects of air pollution from coal burning.

Rational risk assessment considers alternatives. Nuclear wastes remain radioactive for thousands or even millions of years, but some cancer-causing solid wastes released in coal burning – like arsenic, beryllium, cadmium, chromium, and nickel – last forever. They can be expected to cause about 70 eventual fatalities per plant-year, thousands of times more than nuclear waste. Also, solar electricity technologies require vast amounts of materials, and deriving these requires burning large quantities of coal. The wastes from solar technologies are many times more harmful than nuclear wastes. In addition, some solar technologies use large quantities of cadmium, which increases the health consequences considerably.[15]

Radon Problems[16]

Another aspect of nuclear waste may involve important health impacts: the release of radon, a radioactive gas that naturally evolves from uranium. There has been some concern over increased releases of radon due to uranium mining and milling operations. These problems have now been substantially reduced by cleaning up those operations and covering the residues with several feet of soil. The health effects of this radon are several times larger than those from other nuclear wastes, such as the high-level waste discussed above, but they are still much smaller than the effects of coal burning.

However, a far more important impact of the nuclear industry on radon is that by mining uranium out of the ground, we avert future radon emissions and thus avoid future health impacts. Most of the uranium is mined from deep underground, so one might think the radon could not escape. However, the ground surface is constantly eroding away, so eventually the uranium mined would have been near the surface, where its radon emissions could cause lung cancers. When these effects are calculated, the result is an eventual *saving* of 450 lives per plant-year of operation. This saving is thousands of times larger than the lives calculated to be lost from radioactive waste. Also, coal burning releases small amounts of uranium into the environment, eventually causing 30 fatalities per plant-year through radon released.

Summary on Waste

The number of deaths per plant-year estimated in the preceding discussion (plus a few others) is summarized in table 51.1. Since many people

Table 51.1 Eventual deaths per 1,000 MW plant per year of operation due to wastes

Source	Next 500 years	Millions of years
Nuclear:		
high-level waste	0.0001	0.02
radon emissions	– 0.0650	– 450
low-level waste	0.0001	0.0004
total	– 0.0648	– 450
Coal:		
air pollution	25	25
radon emissions	0.11	30
cancer-causing chemicals	0.5	70
total	25.61	125
Solar photovoltaics[a]:		
coal for materials	0.8	3.7
cadmium sulfide (if used)	0.8	80
total	1.6	83.7

[a] Results are those from producing the same amount of electricity as is generated by a large nuclear or coal plant in one year.

(including myself) feel that it is meaningless to consider effects over many millions of years, a column has also been included summarizing effects realized over the next 500 years. It should be understood that the minus signs on the numbers for radon from nuclear power indicate lives *saved* rather than lost. Note that there are three types of waste from coal burning, *each* of which is thousands of times more harmful than the nuclear waste.

RISKS IN PERSPECTIVE[17]

Risks are commonly stated in terms of probabilities of death at various ages, but to make them more understandable we express them as loss of life expectancy (LLE) (see figure 51.1). (An LLE of one day does *not* mean that each person will die one day sooner but that the *average* shortening of each life is one day – true if one person in a thousand dies 1,000 days earlier while 999 are unaffected.) The LLE for nuclear power is about one hour (0.04 day) according to most scientific estimates, or 1.5 days according to the Union of Concerned Scientists. The LLE for coal-burning air pollution is about 13 days, from oil burning about 4.5 days, and from natural gas about 2.5 days. This makes the LLE from nuclear generation from 8 to 300 times less than from coal, 3 to 100 times less than from oil, and 1.7 to 60 times less than from natural gas. To put these numbers in perspective, we show in figure 51.1 the LLE from some other common risks we face.[17] Notice that even energy

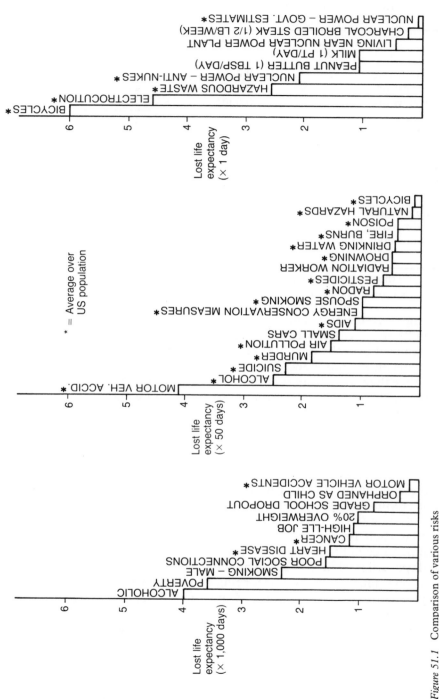

Figure 51.1 Comparison of various risks

Note: Height of bar is LLE (Lost Life Expectancy). Asterisk designates average risk spread over the total US population; other bars refer to risks of those exposed or participating in particular risky activities.

conservation bears risks: radon trapped by tighter housing construction; smaller, riskier cars; more crimes and accidents in reduced lighting; etc. Any one of these items makes energy conservation much more risky than nuclear power.

Far more important is the danger that over-zealous conservation may reduce our wealth. In the United States, well-to-do people live ten years longer than poor people. Producing wealth requires a lot of energy. Therefore, if conserving energy reduces wealth, the ultimate health risks from conservation would dwarf the few hours LLE from nuclear power.

Another way to put the risks of nuclear energy into perspective is to show what other risks they are equivalent to. To make this non-controversial, we use both the PRA and UCS estimates for risks of nuclear power, the latter in parentheses. The risks from having all electricity in the United States generated by nuclear power are equivalent to the following risks:

1   a regular smoker smoking one extra cigarette every 15 years (3 months);
2   an overweight person increasing his weight by 0.012 ounce (0.8 ounce);
3   raising the highway speed limit from 55 to 55.006 (55.6) miles per hour.

WHY THE PUBLIC MISUNDERSTANDING?

Why are the risks of nuclear power so grossly exaggerated in the public mind? The public gets most of its information from the news media, so the media are responsible. Stories about dangers of radiation are exciting and therefore get wide coverage. Yet there has not been a single fatal accident in the United States involving radiation for 25 years, during which there have been three million fatalities from other accidents. Members of the media generally do not read the scientific literature. Their contact with science is often through a few publicity-seeking scientists who tell them what they want to hear. Any scientist who reports the slightest evidence that makes radiation seem dangerous gets tremendous coverage, while contrary evidence is generally ignored.

As a consequence, all new power plants ordered for the past several years have been coal burners. Yet every time a coal-burning plant is built instead of a nuclear plant, many hundreds of people are condemned to premature death.

NOTES

1   This chapter is based on B. L. Cohen, *The Nuclear Energy Option*. New York: Plenum, 1990.
2   Opinion Research Corp. "Public Attitudes Towards Nuclear Power vs Other Energy Sources," *ORC Public Opinion Index*, 38, 17 (September 1980).

3 National Academy of Sciences Committee on Nuclear and Alternative Energy Systems, *Energy in Transition*, 1985–2010. San Francisco: W. H. Freeman, 1980; American Medical Association Council on Scientific Affairs, "Health evaluation of energy generating sources." *Journal of the American Medical Association*, 240 (1978), 2193; Nuclear Energy Policy Study Group, *Nuclear Power – Issues and Choices*. Cambridge, MA: Ballinger, 1977; Union of Concerned Scientists, *The Risks of Nuclear Power Reactors*; United Kingdom Health and Safety Executive, *Comparative Risks of Electricity Production Systems* (1980). Norwegian Ministry of Oil and Energy, *Nuclear Power and Safety* (1978); Science Advisory Office, State of Maryland, *Coal and Nuclear Power* (1980); Legislative Office of Science Advisor, State of Michigan, *Coal and Nuclear Power* (1980); H. Fisher, et al., "Comparative Effects of Different Energy Technologies," Brookhaven National Laboratory Report BNL 51491 (1981); American Medical Association Council on Scientific Affairs, "Medical Perspective on Nuclear Power" (1989).

4 H. Kendall, Physics Colloquium, at Carnegie-Mellon University (1980).

5 In answer to my question, Nader replied, "Maybe we can clean up coal, or maybe we shouldn't burn coal either."

6 National Academy of Sciences Committee on Biological Effects of Ionizing Radiation (BEIR-V), *Health effects of Exposure to Low Levels of Ionizing Radiation*. Washington: 1990. A large number of references is given.

7 United Nations Scientific Committee on Effects of Atomic Radiation (UNSCEAR), *Sources, Effects and Risks of Ionizing Radiation*. New York: United Nations, 1988. A large number of references is given.

8 B. L. Cohen. "The Cancer Risk from Low Level Radiation." *Health Physics*, 39 (1980), 659. Many references are given. (A few papers have purported to give evidence that the cancer risk vs dose curve is concave downward, which would indicate that the linear assumption underestimates the risk of low-level radiation. Each of these, however, has been severely criticized in the scientific literature and has been rejected by all groups charged with responsibilities in radiation protection.)

9 Many people think that a nuclear reactor can explode like an atomic bomb, but technical arguments easily prove that this is absolutely impossible.

10 Reactor Safety Study, Nuclear Regulatory Commission Document WASH-1400, NUREG 75/014 (1975); "Severe Accident Risks: An Assessment for Five U.S. Nuclear Power Plants," Nuclear Regulatory Commission Document NUREG-1150 (1989).

11 Union of Concerned Scientists. *The Risks of Nuclear Power Reactors* (Cambridge, MA: 1977). They give 2.4 deaths/GWe-year, which, multiplied by 100 GWe, the total amount generated in the USA, gives 240 deaths/year. While this study is rather old, no new information has developed in recent years that would increase risk estimates.

12 *Report of the President's Commission on The Accident at Three Mile Island*, J. B. Kemeny, chairman. Washington: October 1979; *Three Mile Island, A Report to the Commissioners and to the Public*, M. Rogovin, director. Washington: January 1980.

13 R. Wilson, S. D. Colome, J. D. Spengler, and D. G. Wilson, *Health Effect of Fossil Fuel Burning*. Cambridge, MA: Ballinger, 1980; H. Ozkaynak and J. C. Spengler, "Analyses of Health Effects Resulting from Population Exposure to Acid Precipitation," *Environmental Health Perspectives*, 63 (1985), 45.

14 W. Ramsay, *The Unpaid Costs of Electrical Energy*. Baltimore: Johns Hopkins University Press, 1979.

15 This discussion is based on a group of papers reviewed in B. L. Cohen, "Risk Analysis of Buried Waste from Electricity Generation." *American Journal of Physics*, 54 (1986), 38.

16 B. L. Cohen, "The Role of Radon in Comparisons of Environmental Effects of Nuclear Energy, Coal Burning, and Phosphate Mining." *Health Physics*, 40 (1981), 19; "Health Effects of Radon from Coal Burning." *Health Physics*, 42 (1982), 725.

17 B. L. Cohen, "Catalog of Risks Extended and Updated." *Health Physics*, 61 (1991), 317.

52

Pesticides, Cancer, and Misconceptions

Bruce N. Ames

> There was once a town in the heart of America where all life seemed in harmony with its surroundings.
>
> The harmony of all life in this idyllic town followed a biological balance of Nature, a balance which man had not yet learned to disturb by drastic intervention on his own behalf.
>
> As the Sun went down, the buzzing of mosquitos could be heard in the town; the malaria parasites in their salivary glands were about to continue their life cycle in the red blood cells of human victims.
>
> The last slanting rays of the Sun lingered on the small headstones in the town graveyards. Here slept the children who had perished from diphtheria, scarlet fever, and whooping cough. Beside them lay the bodies of those who had lived and died in harmony with proliferant typhoid germs.
>
> From "A Town in Harmony," Thomas H. Jukes, 1962

In the last several decades there has been a persistent widespread belief among many groups in this country that nature is benign and that man-made things – that is, modern technology – have destroyed our benevolent relationship with nature. This yearning for a time when humans were happily in harmony with nature is a yearning for a time that never existed: in reality, life before the modern industrial era was, for most people, even in Thomas Hobbes's time, "nasty, brutish, and short." Diseases – such as plague, typhus, and malaria – and malnutrition caused by shortages of affordable crops ensured a very short life expectancy, an early end to the misery of life in a natural world.

The history of agriculture is one of a nonending contest with pests such as insects and fungi. Fields of crops, which often have been bred to have low levels of natural plant defensive chemicals in order to be more edible for human consumption, are easy sources of food for thousands of species of insects and fungi. Infestation of crops by pests can have dramatic impacts on human life: last century, the potato fungus *Phytophthora infestans* wiped out the potato crop of Ireland, which led to the deaths of over a million people due to malnutrition (which made people susceptible to disease) and starvation. The relationship between pesticides and disease is significant. DDT, the first synthetic pesticide, eradicated malaria from many parts of the world, including the USA. It was so effective against many diseases because (1) it was lethal to many vectors of disease, for example, mosquitos, tsetse flies, lice, ticks, and fleas; and (2) it was lethal to many crop pests, and so significantly increased the supply of food and lowered the cost of food, making fresh nutritious foods accessible even to relatively poor people. It was also remarkably non-toxic to people. In Sri Lanka, in less than 20 years of DDT use, the number of cases of malaria decreased from 2.8 million per year to 17. (After Sri Lanka stopped using DDT, the number of malaria cases increased again.)

Some people have twisted the story of pesticides: instead of pesticides freeing us from disease, they assert, pesticides are bringing us disease. Various misconceptions about the relationship between environmental pollution and human disease, particularly cancer, both stem from and drive the maligning of pesticides. In the following sections, I will highlight seven of these misconceptions and briefly present the scientific evidence that undermines each.

MYTHS AND FACTS ABOUT CARCINOGENS IN FOODS

The first misconception is that cancer rates are soaring. Cancer death rates overall in the United States (after adjusting the rates for age and smoking) have decreased 14 percent since 1950. The types of cancer deaths that have been decreasing since 1950 are primarily stomach, cervical, uterine, and rectal. The types that have been increasing are primarily lung cancer (due to smoking, as are 30 percent of all cancer deaths in the United States), melanoma (possibly due to sunburns), and non-Hodgkin's lymphoma. (Cancer incidence rates are also of interest, although, as cancer epidemiologists Richard Doll and Richard Peto point out, they should not be taken in isolation, because trends in the recorded incidence rates are biased by improvements in the level of registration and diagnosis.) Cancer is fundamentally a degenerative disease of old age, although external factors can increase cancer rates (e.g., cigarette smoking in humans) or decrease them (e.g., caloric restriction in rodents). As life expectancy continues to rise, cancer rates (unadjusted for age) also rise.

The second misconception is that cancer risks to humans can be assessed by high-dose animal cancer tests. Approximately half of all chemicals – whether natural or synthetic – that have been tested in standard animal cancer tests have turned out to be carcinogenic. Standard animal cancer tests of chemicals are conducted at chronic, near-toxic doses – the maximum tolerated dose – and evidence is accumulating that it may be cell division caused by the high dose itself, rather than the chemical per se, that is the major risk factor for cancer. (This is because high doses cause chronic wounding of tissues, cell death, and consequent chronic cell division of neighboring cells, which is a risk factor for cancer.) At the low exposures to which humans are normally exposed, such increased cell division does not occur. Therefore, the very low levels of chemicals to which humans are exposed through water pollution or synthetic pesticide residues are likely to pose no or minimal cancer risks.

The third misconception is that human exposures to carcinogens and other toxins are nearly all to synthetic chemicals. On the contrary, the amounts of synthetic pesticide residues in plant foods are insignificant compared to the amount of natural pesticides produced by plants themselves. Of all dietary pesticides humans ingest, 99.99 percent are natural: they are toxins produced by plants to defend themselves against fungi, insects, and other animal predators. Because each plant produces a different array of toxins, I estimate that on average Americans ingest roughly 5,000 to 10,000 different natural pesticides and their breakdown products. I also estimate that Americans eat about 1,500 mg of natural pesticides per person per day, which is about 10,000 times more than they consume of synthetic pesticide residues.

Even though only a small proportion of natural chemicals has been tested for carcinogenicity, half of those tested are carcinogens, and naturally occurring pesticides that are rodent "carcinogens" have been detected in numerous fruits, vegetables, herbs, and spices, including: apple, banana, basil, broccoli, cabbage, carrot, cauliflower, celery, cherry, cinnamon, cloves, cocoa, coffee, comfrey tea, dill, eggplant, grapefruit juice, grape, honeydew melon, lettuce, mango, mushroom, mustard, nutmeg, orange juice, parsley, peach, pear, pepper, pineapple, plum, potato, radish, raspberry, strawberry, tarragon, and thyme. In addition, cooking foods produces about 2,000 mg per person per day of burnt material that contains many rodent carcinogens and many mutagens.

By contrast, the FDA found that residues of 200 synthetic chemicals, including the synthetic pesticides thought to be of greatest importance, average only about 0.09 mg per person per day. The known natural rodent carcinogens in a single cup of coffee are about equal in weight to an entire year's worth of carcinogenic synthetic pesticide residues (assuming about half the untested synthetic residues turn out to be carcinogenic), even though only 2 percent of the natural chemicals in roasted coffee have been tested for carcinogenicity.

The fourth misconception is that synthetic toxins pose greater carcinogenic hazards than natural toxins. If either is a hazard at low doses, the possible carcinogenic hazards from synthetic pesticides (at normal exposures) are minimal compared to the background of nature's pesticides. The overwhelming number and weight of chemicals that humans eat are natural, and the proportion of natural chemicals that is carcinogenic when tested in both rats and mice is the same as for synthetic chemicals – roughly half. The assumption that synthetic chemicals are more hazardous has led to a bias in testing, such that synthetic chemicals account for 82 percent of the 427 chemicals tested chronically at high doses in both rats and mice. The natural world of chemicals has never been tested systematically. But if one ranks the possible carcinogenic hazards of various natural and synthetic toxins according to the results of animal cancer tests, one finds, for example, that eating the natural carcinogens of a daily mushroom poses a much greater possible hazard than drinking a daily glass of juice made from Alar-tainted apples. I doubt, however, that either is worth concern. To cite another example, cabbage and broccoli contain a chemical whose breakdown products appear to act on the body like dioxin – one of the most feared industrial contaminants that is a rodent carcinogen. At ordinary (low) doses, it is not clear that either dioxin or the natural dioxin simulators in cabbage and broccoli are any hazard; cabbage and broccoli have many anticarcinogenic nutrients as well.

The fifth misconception is that the toxicology of man-made chemicals is different from that of natural chemicals. It is often assumed that, because plants are part of human evolutionary history whereas synthetic chemicals are recent, the mechanisms that have evolved in animals to cope with the toxicity of natural chemicals will fail to protect us against synthetic chemicals. I find this assumption flawed for several reasons.

1 Humans have many natural defenses that make us well buffered against normal exposures to toxins. These defenses are usually general, rather than tailored for each specific chemical, and thus they work against both natural and synthetic chemicals. Examples of general defenses include the continuous shedding of cells exposed to toxins – the surface layers of the mouth, esophagus, stomach, intestine, colon, skin, and lungs are discarded every few days; DNA repairs enzymes, which repair DNA that was damaged from many different exogenous and endogenous sources; and detoxification enzymes of the liver and other organs, which generally target classes of toxins rather than individual toxins. That defenses are usually general, rather than specific for each chemical, makes good evolutionary sense. The reason that predators of plants evolved general defenses against toxins is presumably to be prepared to counter a diverse and ever-changing array of plant toxins in an evolving world; if an herbivore had defenses against only a set of specific toxins, it would be at a great disadvantage in obtaining new foods when favored foods became scarce or evolved new toxins.

2 Various natural toxins, some of which have been present throughout vertebrate evolutionary history, nevertheless cause cancer in vertebrates. Mold aflatoxins, for example, have been shown to cause cancer in trout, rats, mice, monkeys, and possibly humans. Eleven of 16 mold toxins tested have been reported to be carcinogenic, and 19 mold toxins have been shown to break chromosomes. Many of the common elements are carcinogenic (e.g., salts of lead, cadmium, beryllium,

nickel, chromium, selenium, and arsenic) at high doses, despite their presence throughout evolution. Furthermore, epidemiological studies from various parts of the world show that certain natural chemicals in food may be carcinogenic risks to humans; for example, the chewing of betel nuts with tobacco has been correlated with oral cancer world-wide.

Plants have been evolving and refining their chemical weapons for at least 500 million years and incur large fitness costs in producing these chemicals. If these chemicals were not effective in deterring predators, plants would not have been naturally selected to produce them.

3   Humans have not had time to evolve into a "toxic harmony" with all of the plants in their diet. Indeed, very few of the plants that humans eat would have been present in an African hunter-gatherer's diet. The human diet has changed dramatically in the last few thousand years, and most humans are eating many recently introduced plants that their ancestors did not – such as coffee, cocoa, tea, potatoes, tomatoes, corn, avocados, mangoes, olives, and kiwi fruit. Natural selection works far too slowly for humans to have evolved specific resistance to the food toxins in these newly introduced plants.

4   DDT is often viewed as the typically dangerous synthetic pesticide because it concentrates in the tissues and persists for years, being slowly released into the bloodstream. There is no convincing epidemiological evidence, nor is there much toxicological plausibility, that the levels normally found in the environment are likely to be a significant contriutor to cancer. Although DDT was unusual with respect to bioconcentration, natural pesticides also bioconcentrate if they are fat soluble. Potatoes, for example, naturally contain the fat soluble neurotoxins solanine and chaconine, which can be detected in the bloodstream of all potato eaters. High levels of these potato neurotoxins have been shown to cause birth defects in rodents.

The sixth misconception is that correlation by itself implies causation. The number of storks in Germany has been decreasing for decades. At the same time, the German birth rate also has been decreasing. We would be foolish to accept this high correlation as evidence that storks bring babies. The field of epidemiology tries to sort out the meaningful correlations from the numerous chance correlations. Cancer clusters in small areas are expected to occur by chance alone, and epidemiology lacks the power to establish causality in these cases. It is important to show that a pollution exposure that purportedly causes a cancer cluster is significantly greater than the background of exposures to naturally occurring rodent carcinogens. There is no persuasive evidence from either epidemiology or toxicology that pollution is a significant cause of cancer.

The seventh misconception is that there are no trade-offs involved in elimi-nating pesticides. On the contrary, fruits and vegetables are very import-ant for reducing cancer; making them more expensive by reducing use of synthetic pesticides is likely to increase cancer. People with low incomes eat fewer fruits and vegetables and spend a higher percentage of their income on food.

Since no plot of land is immune to attack by insects, plants need chemical defenses – either natural or synthetic – in order to survive pest attack. Thus, there is a trade-off between naturally occurring pesticides and synthetic pesticides. One consequence of disproportionate concern about synthetic pesticide residues is that some plant breeders are currently developing plants to be more insect-resistant by making them higher in natural toxins. A recent case illustrates the potential hazards of this approach to pest control: When a major grower introduced a new variety of highly insect-resistant celery into commerce, complaints flooded the Centers for Disease Control from all over the country because people who handled the celery developed rashes when they were subsequently exposed to sunlight. Some detective work found that the pest-resistant celery contained 6,200 parts per billion (ppb) of carcinogenic (and mutagenic) psoralens instead of the 800 ppb present in normal celery. The celery is still on the market. Certain cultivated crops have become popular in developing countries because they thrive without costly synthetic pesticides. However, the trade-offs of cultivating some of these naturally pest-resistant crops are that they are highly toxic and require extensive processing to detoxify them. For example, cassava root, a major food crop in Africa and South America, is quite resistant to pests and disease; however, it contains cyanide at such high levels that only a laborious process of washing, grinding, fermenting, and heating can make it edible; ataxia due to chronic cyanide poisoning is endemic in many of the cassava-eating areas of Africa.

SUMMARY

Synthetic pesticides were a great advance for human health as they markedly lowered the cost of plant foods, thus making them more available to customers. Eating more fruits and vegetables is thought to be the best way to lower risks from cancer and heart disease other than giving up smoking; our vitamins, antioxidants, and fiber come from plants and are important anticarcinogens. Inadequate intake of fruit and vegetables compared to adequate intake (about five fruits and vegetables daily) doubles the rate for most types of cancer (lung, larynx, oral cavity, esophagus, stomach, colon and rectum, bladder, pancreas, cervix, and ovary). The data on those cancer types known to be associated with hormone levels are not as consistent and show less of a protective effect: for breast cancer the protective effect was 30 percent. There is also a large literature on the protective effect of fruit and vegetable consumption on heart disease. Only 9 percent of Americans eat the recommended two fruits and three vegetables per day.

The effort to eliminate synthetic pesticides because of unsubstantiated fears about residues in food will make fruits and vegetables more expensive, decrease consumption, and thus increase cancer rates. The levels of synthetic pesticide residues are trivial in comparison to natural chemicals, and thus their potential for cancer causation is extremely low.

NOTE

This paper was adapted from B. N. Ames (1992) *J. Assoc. Off. Anal. Chem.* 75. For references see Ames, B. N. and Gold, L. S. (1990) *Proc. Natl. Acad. Sci. USA* 87, 7772–6; Ames, B. N. et al. (1990) *Proc. Natl. Acad. Sci. USA* 87, 7777–81; ibid., pp. 7782–6; Ames, B. N., Shigenaga, M. K., and Hagen, T. M. (1993), *Proc. Natl. Acad. Sci. USA*, 90, 7915–22. Gold, L. S., Slone, T. H. Stern, B. R., Manley, N. B., and Ames, B. N. (1992), *Science*, 258, 261–5.

53

The Carcinogen or Toxin of the Week Phenomenon: The Facts Behind the Scares

Elizabeth M. Whelan

"Life causes cancer." Or so says the bumper sticker I saw recently.

This gloom and doom stands in stark contrast to the scientific reality that we in the United States are living longer than ever before. Life expectancy in the United States is 29 years longer in 1991 than it was in 1900.[1,2] We have controlled nearly the full spectrum of infectious diseases so prevalent at the turn of the century, reducing deaths from influenza and pneumonia, for example, by 93 percent.[3] Equally important, we have identified major lifestyle factors – such as smoking, alcohol and drug abuse, and failure to use life-saving technology like seat belts and smoke detectors – which are largely responsible for premature deaths in this country.[4-9]

Most health scares can be juxtaposed with scientific realities. To illustrate, I here present fourteen well publicized health scares (many covered in my book *Toxic Terror*) and the scientific realities that lie behind them.

ALAR

In February 1989, Americans tuned in to *60 Minutes* heard Ed Bradley declare, "The most potent cancer-causing agent in our food supply is a substance sprayed on apples to keep them on the trees longer and make them look better." Fifty million viewers were told that Alar posed an "intolerable" human cancer risk, particularly to children. Panic ensued.

60 Minutes failed to report two crucial, related points. The scare was a shrewdly orchestrated media extravaganza based on a Natural Resources Defense Council study on mice using excessively high doses of a byproduct of Alar. The study was unusual for two reasons: (1) the Environmental Protection Agency studied this byproduct only after several studies found Alar itself incapable of causing cancer; and (2) the doses were so high as to kill many of the mice by poisoning alone. The study was neither peer reviewed nor printed in a reputable scientific journal.

How, then, did it reach the press with such force? The NRDC hired a public relations consultant, who struck a deal with CBS whereby the network would present the results of the NRDC study so long as they were promised exclusivity. A media barrage ensued. The public relations consultant admitted, "Our goal was to create so many repetitions of NRDC's message that the average American consumer could not avoid hearing it – from many different media outlets within a short period of time. . . ."[10]

What the average consumer probably didn't know was that:

1   The EPA's Scientific Advisory Panel on Alar had repeatedly refused to ban Alar because it thought the rodent data inadequate. The EPA's experts did not think Alar posed a threat to human health.

2   The EPA sets tolerance levels – the minuscule residues allowed – on all pesticides, and the Food and Drug Administration makes sure these tolerances are not exceeded. Alar residues were well within the tolerance.

3   Other than the one high-dose mouse study on the Alar byproduct, no evidence indicates that Alar posed any health hazard to humans – children or adults. (The rodent-to-human extrapolation is at the basis of most claims of carcinogenesis.) It is erroneous to imply that a substance is a probable human carcinogen based on a carcinogenic response in a single animal study using high doses. Further, the major problem with animal cancer tests lies not with the numerous limitations in animal cancer bioassay but with the legal and regulatory use the tests serve.[11]

Near the third anniversary of the Alar-apple scare, Dr Richard Adamson, director of the Division of Cancer Etiology at the National Cancer Institute, told me, ". . . the risk posed by Alar was no greater than eating a peanut butter sandwich." Dr C. Everett Koop, former US Surgeon General and chairman of the National Safe Kids Campaign, also commented in 1992 about Alar: "As a pediatric surgeon, as well as the nation's former Surgeon General, I care deeply about the health of children, and if Alar ever posed a health hazard I would have said so then and would say so now. But the truth is that Alar never did pose a health hazard. The American food supply is not only the most abundant in the world, but it is also the safest. Paradoxically – it has achieved that position in the world market just because of chemicals like Alar that have made it possible."[12]

BST

Claims have been made that hormones given to cows to increase milk production contaminate milk and make people sick.

BST, or Bovine Somatotropin, is a naturally occurring growth hormone in cows. Synthetically produced BST, almost identical to naturally occurring BST, is fed to cows to increase milk yields by as much as 15 to 25 percent. BST is a protein that, consumed orally, is broken down into inactive fragments in the gastrointestinal tract during digestion. Studies in the 1950s proved BST's inactivity in humans when attempts were made to treat human dwarfism in children by injecting them with BST – to no effect. Finally, tests show that cows receiving BST supplement have no higher BST levels in their milk than those not receiving the supplement.[13]

The real opposition to BST is not health-related, but economic, with opponents claiming that BST is an economic threat to the family farm. Animal rights activists claim that BST use can over stress dairy cows, a charge unsupported by scientific evidence.

FOOD IRRADIATION

Some people think that irradiating food threatens public health because it depletes the nutritional value of food and renders it radioactive.

Food products have been preserved through irradiation since 1958 in over twenty countries worldwide. The process has been studied more extensively than any other food preservation process, such as canning, freezing, and dehydration. When food is irradiated under FDA-approved conditions, it does not become radioactive; neither are radiolytic products formed; the process does not generate radioactive waste.

Food irradiation offers many significant advantages. Foremost among them are reducing the incidence of food-borne, disease-causing organisms and effectively extending the shelf life of food without adversely affecting its taste and appearance. On May 2, 1990, the FDA issued a rule saying that irradiation is a safe and effective means to control Salmonella, a bacterium that contaminates up to 60 percent of poultry sold in the United States, and Campylobacter, another bacterium that may contaminate all chicken.[14] The US Department of Agriculture estimates that American consumers will receive approximately $2 in benefits (reduced spoilage, less illness) for each $1 spent on irradiating food.[15]

ANTIBIOTICS IN COW'S MILK

Recently, a Ralph Nader-affiliated consumer group – in conjunction with the *Wall Street Journal* – reported that trace levels of antibiotics were found in milk, suggesting a human hazard, particularly a risk of human resistance to antibiotics. The alarming story drew on screening assays that were never confirmed (screening assays merely indicate that some residues of a drug class may be present but must be confirmed by further tests). The FDA properly criticized these two organizations for their irresponsible forays into science and then conducted a nation-wide sampling and testing program using the same screening tests followed by confirmatory assays. It found no residues of antibiotics.[16]

Antibiotics are used in two different ways in livestock and poultry. Therapeutic levels are used to treat diseases and subtherapeutic levels to increase the rate of weight gain in growing animals or birds, to increase the amount of meat that can be produced from a given amount of feed (feed efficiency), and to help prevent bacterial diseases.

Theoretically, feeding antibiotics to farm animals could increase the prevalence of antibiotic-resistant bacteria, which might cause human diseases, which in turn could not be treated successfully with the same drugs. Logically, we would see this occur in the most exposed populations: the animals themselves, livestock producers, and slaughterhouse workers. Additionally, the effectiveness of the drugs as animal growth promotants would also be expected to decrease. These events have not occurred in 30 years of treating livestock and poultry with antibiotics, and there is no evidence to suggest that this practice poses a threat to human health. Moreover, there is no known human benefit to be gained from discontinuing the feed additive use of antibiotics.[17]

DIOXIN

Some scientists have stated that dioxin is the most powerful toxin/carcinogen known to man, causing disease and death.

Like most chemicals, dioxin can be hazardous at sufficiently high doses. But there is no evidence that the levels of exposure in our environment pose a hazard to human health.

There has been sufficient experience with dioxin to study possible health effects. Dioxin has been found as a contaminant of waste oil sprayed on unpaved roads to retard dust; as a contaminant of herbicides used here and in the Vietnam War; and as a byproduct of the paper and pulp industry. In all cases, however, the only two observed health effects – both reversible – were chloracne and a general "not feeling well" syndrome that includes sleeplessness, headache, nausea, and irritability. An accident at a chemical manufacturing plant in Seveso, Italy, that exposed local residents and workers to the largest known single doses of dioxin resulted only in chloracne and minor reversible nerve damage.[18]

Although public and political pressure forced the federal government to buy out the town of Times Beach, Missouri, when roadside dioxin contamination was discovered in 1983, there was no documented case of death or serious illness caused by dioxin among residents.

In 1991, scientists made several discoveries about dioxin that significantly downgraded its estimated cancer-causing potential.[19] In an unprecedented action, EPA Administrator William K. Reilly ordered a reevaluation of dioxin's risks, commenting, "I don't want to prejudge the issue, but we are seeing new information on dioxin that suggests a lower risk assessment for dioxin should be applied. . . . There isn't much precedent in the Federal establishment for pulling back from a judgment of toxicity. But we need to be prepared to adjust, to raise or lower standards, as new science becomes available.[20] Vernon Houk, Assistant Surgeon General and the US public health official who ordered the Times Beach evacuation, said, "Given what we know about this chemical's toxicity and its effects on human health, it looks as though the evacuation was unnecessary. Times Beach was an over-reaction. . . ."[21]

The dioxin reappraisal has been fraught with political problems. The Banbury Conference on Dioxin, co-sponsored by the Chlorine Institute and the EPA, was accused of being a "public relations exercise" for the Chlorine Institute. George L. Carlo, chairman of the Health and Environmental Sciences Group, addressed the poor reporting and false accusations:

> It is neither justifiable nor realistic to assume that the U.S. [EPA], which co-sponsored the conference, is in the habit of paying for the outreach activities of trade associations. Had you seriously considered the publish-ed Banbury proceedings, you would have noticed that the chairmen of the conference considered it to be an important and necessary gathering, giving in the preface to the proceedings the rationale that "much of the new biological work and its implications for human health risk [has] yet to make its way into the regulatory arena and, in particular, into quantitative risk assessment." To suggest that the Banbury conference was a manipu-lation in view of the powerful science considered and the high caliber of the participants is both naive and an insult to all those involved.

Dr Carlo added,

> Irresponsible reporting such as yours is hindering the ability of our gov-ernment to make important regulatory decisions based on science rather than hysteria. The reevaluation of the EPA dioxin risk assessment is a step in the right direction. . . .[22]

PCBs

Some studies have indicated that environmental exposure to PCBs causes miscarriages, birth defects, impaired infant development, and other health problems.

Discovered over 100 years ago, PCBs solved the urgent problem of fire hazards in electrical equipment. Because they enjoyed such widespread use and are so chemically stable, PCBs became a significant part of the environment. They have attracted the greatest attention as a contaminant in rivers, streams, and the fish in them.

PCBs caught the public eye in the late 1960s when over 1,000 people on the island of Kyushu, Japan, became ill from consuming rice oil contaminated with a PCB heat transfer agent. In the years following the incident, it became increasingly likely that it was not PCB but an industrial contaminant that was responsible. Most researchers who have studied this incident believe that it has little relevance to potential health effects from minute, intermittent PCB exposures in this country.[23]

Although PCBs have been detected in human fat tissue and in mothers' milk, no evidence exists that the mere presence of the trace levels causes adverse effects. Indeed, certain industrial workers were exposed to relatively high levels of PCB for as long as 35 years, and none showed ill health effects.[24]

There is no evidence that PCBs cause human death or disease in the United States. They have, in fact, saved lives by minimizing the risks of fires by replacing combustible insulating fluids. Not only have PCBs been banned in this country as a result of public outcry – a real loss in a technological sense – but also, billions of dollars have been spent to replace and clean up PCB residues. The best action, however, is no action. Time alone will allow PCBs to biodegrade naturally.[25-30] We don't need to capture them, destroy them, or fear them.

ASPARTAME

The non-nutritive sweetener Aspartame has been declared dangerous because it causes behavioral changes.

Introduced in the USA in 1981, Aspartame is a low calorie sweetener best known to the public by the trade names Nutrasweet and Equal. An amino acid mixture, Aspartame underwent years of testing and premarket scrutiny. This extensive testing concluded that Aspartame is safe, even when used in large amounts.[31] (Phenylketonurics and other individuals who must limit their intake of phenylalanine must restrict their use of Aspartame.) Aspartame's greatest virtue is its taste, considered by most people to be very close to sugar, without an unpleasant aftertaste.

Some very preliminary scientific evidence suggests that certain amino acids, carbohydrates, and other dietary components could influence mood and behavior. But there is little evidence, even in laboratory animals, to indicate that consumption of Aspartame has such an effect. The safety of Aspartame has been tested in studies with human volunteers consuming the sweetener at abusive levels. These tests have shown that Aspartame is safe for human consumption even at these extremely high levels.[32]

SACCHARIN

In March 1977, one Canadian study concluded that saccharin in exceedingly high doses caused bladder cancer in male rats. The FDA moved to ban it on this evidence. Congress, however, intervened, and saccharin is still with us but now carries a warning that it causes cancer in rodents.

Discovered in 1879, saccharin is a white crystalline powder several hundred times sweeter than sugar. The sweetener has been used extensively in America, particularly during World War II as a substitute for sugar and in the 1950s when diet drinks became increasingly popular. Saccharin has a reassuring track record with humans who have used large quantities of it, such as diabetics. Studies of diabetics who have used saccharin for most of their lifetimes reveal no unusual cancer patterns that can be linked with its use.[33]

FLUORIDE

Fluoride – like all substances, including water – is toxic at excessive levels. Strict drinking water regulations, however, ensure that fluoride levels in water are safe.[34-36] But for years, misguided antifluoridation activists have persuaded some local governments and voters to reject fluoridation. They claim that fluoride causes cancer, sickle cell anemia, Down's syndrome, and even AIDS. To date, however, there is no scientific evidence that properly fluoridated water has ever caused an adverse health effect.[37,38]

Fluoride reduces tooth decay by hardening tooth enamel and protecting it from deterioration caused by acid-producing bacteria in the mouth. Children exposed to optimally fluoridated water from birth experience 50 to 70 percent fewer cavities than those not exposed; such benefits last a lifetime.[39]

Recent laboratory findings indicating that male rats fed high doses of sodium fluoride have a higher than expected incidence of bone cancers have fueled the anti-fluoride movement and put considerable pressure on the EPA to classify fluoride as a "probable human carcinogen."[40] These results, however, were incomplete and released prematurely. The Centers for Disease Control have described them as "inappropriate to speculate on."[41] Nonetheless, Phoenix, Arizona, recently suspended a water fluoridation project on the basis of those data.

This is a travesty from the public health view, returning us to the era of dental caries with no public health benefits. The fluoride story demonstrates that our current regulatory apparatus not only fails to consider the dose (e.g., our low exposure to fluoride) but also excludes consideration of its benefits and the risks of not taking those hypothetical risks.

MICROWAVE OVENS

Any radiation, in excessive amounts, can be dangerous. Excessive sound waves can damage hearing; too bright a light can cause blindness; too much ultraviolet light will produce sunburn and may, after prolonged exposure, cause skin cancer; enough heat from any source will cause burns; X-rays can cause radiation burns and increase the risk of cancer after repeated exposure. And excessive exposure to microwaves may cause some health damage at extremely high doses.

But the FDA has set very conservative limits on radiation exposure from microwave ovens – the so-called leakage limits. These levels are well below accepted danger levels. Contrary to some people's fears, it is very unlikely that any harm will come from this minuscule amount of leaked microwave radiation. Indeed, 20 years of microwave oven use, together with research studies, has failed to substantiate any claims of harm, including claims of cataracts allegedly due to microwave radiation.[42]

DDT

The DDT controversy began when the pesticide was found in lakes and streams. Environmentalists charged that it was responsible for the near extinction of select bird populations. On these assertions, activists quickly claimed that DDT might cause human disease, including cancer. They clamored for and quickly won the banning of DDT in 1972.

DDT was an effective and low-cost pesticide and was solely responsible for drastically reducing the number of deaths caused by malaria. (The number of deaths in Ceylon [now Sri Lanka] exceeded two million per year before DDT; with DDT spraying, it was reduced to a mere 17. Five years after the spraying was stopped, however, the malarial deaths again exceeded two million.)[43]

DDT is found in the environment, but only in trace amounts. It is decomposed by ultraviolet light in the air, and over 93 percent of DDT and its metabolites are broken down in sea water in 38 days. Extensive research and analysis of bird counts before, during, and after the years of DDT use fail to show that DDT is responsible for the extinction of any birds. A prime example is the robin, the bird the environmentalists (including Rachel Carson) frequently said was doomed because of DDT, which became the most abundant bird in North America during the DDT years.[44]

Finally, extensive research fails to find any evidence that DDT is a cause of human cancer or any other human disease.[45] DDT poisoning was rare, in fact, and the few incidents were caused by massive accidental ingestion of the insecticide. There were never any toxic effects in humans in areas properly sprayed with DDT. (One would assume that if the pesticide was inherently dangerous, workers in constant contact with it would be the first affected, but not a single related case of chronic illness or fatality was reported among occupationally exposed individuals.)

Banning DDT was the first victory for environmental alarmists. It eliminated a lifesaving chemical by means of politics, emotion, and pseudoscience. The DDT conflict set a pattern for environmental decisions based less on science than on histrionics, mudslinging, hyperbole, and dishonest science. In the 1990s we are still paying for the 1972 ban on DDT.

EDB

Concern arose about the safety of EDB – ethylene dibromide – when the pesticide was found as a contaminant in groundwater near areas where it had been used as a soil fumigant. Soon Florida issued an emergency stop-sale order on all food products that contained any detectable amount of EDB.

The pesticide was effectively used for 50 years to protect stored wheat, corn, and other grains from contamination by molds and fungi and from destruction by insects. But following Florida's order, hysteria reigned. Suddenly, EDB was alleged to pose a substantial carcinogenic risk. Panic led to the February, 1984, EPA banning of the further use of EDB for treating grain and grain machinery.

Again, there was no evidence to support this action. EDB does cause cancer in laboratory animals under specific conditions: the equivalent of absurdly high levels of human food intake. More significantly, over 50 years of experience with the chemical gives us rather extensive information about the health effects of prolonged exposures. Some workers engaged in the manufacture of the pesticide were exposed to doses some five to ten thousand times higher than consumers, for periods of up to 16 years, yet showed no evidence of elevated cancer rates.[46]

Based solely on rodent-to-human extrapolation, EDB was banned. Ironically, it was replaced with chemicals that did not offer a greater safety or health advantage to the public.

TOXIC WASTE: LOVE CANAL

Residents of an area known as Love Canal in Niagara Falls, New York, were exposed to toxic chemicals. Some claim that this caused serious human illness and a higher than normal incidence of cancer and death.

The Love Canal was an ideal waste site for the Hooker Electrochemical Company. The clay canal was filled with waste chemicals and appropriately capped. In 1950 the site was donated to the Board of Education with the stipulation that underground excavation would most certainly disrupt the chemical repository. This warning was not heeded, and ensuing excavation led to extensive chemical leakage into the Love Canal community. Nonetheless, there is no evidence to date that the resultant human exposure is responsible for human disease or death.[47]

Nonetheless, unconfirmed studies leaked to the press prior to the normal process of scientific peer review were used to support charges that led to an expensive evacuation of the area. In 1991, Love Canal remains a ghost town, despite continuing studies that fail to confirm any ill health effects.

The panic over Love Canal exemplifies the questionable – indeed immoral and dishonest – tactics of some environmentalists. It is a classic story of half truths, distorted historical facts, unprecedented media exaggeration, and misguided government intervention, all of which caused substantially more human upset and misery than did even the most toxic of Hooker's chemicals.

CYCLAMATES

Cyclamates are a non-caloric substance about 30 times sweeter than sugar. When very limited and equivocal Canadian evidence suggested that they caused cancer in laboratory animals, cyclamates were banned in the United States. This 1970 ban has not been reversed, despite follow-up studies that failed to confirm the carcinogenic response.[48]

A petition is currently under review by the FDA to reconsider the evidence for the safety of cyclamates. The petition cites scientific evidence that the ban may have relied on unsound scientific and statistical judgments.

SO WHAT'S A CONSUMER TO DO?

Unfortunately, scientists in the United States are strangely unwilling to step forward and correct claims by environmental hyperbolists echoed by the media. Consumers left to their own resources to sort out claims about the current "carcinogen of the week" should keep the following points foremost in mind:

1 Only the dose makes the poison. While extremely high exposure levels can make almost any substance a health hazard, low levels may be quite safe. There are safe ways to use potentially dangerous materials and processes.

2 The word *carcinogen* is one of the most overused words in our vocabulary. To assess its use, determine whether it refers to human or animal cancer-causing agents. Most often it is the latter, and there is little or no relevance to human cancer risk.

3 Beware the risks of not taking risks. For example, when the banning of a pesticide is proposed, question the alternatives. What are the risks in abandon-

ing the techniques of modern agriculture? Is it worth poorer food quality and a diminished food supply?

4   When in doubt, note the obvious: we are living longer than ever before. Cancer death rates (except for lung cancer, caused largely by cigarette smoking, and melanoma, mostly from excessive sun exposure) have stabilized or declined in the past 40 years. We must be doing something right!

NOTES

Note: This chapter summarizes in brief many issues addressed in E. Whelan, *Toxic Terror*. Buffalo: Prometheus Press, 1993; and in publications of the American Council on Science and Health. (ACSH). ACSH publications can be obtained by calling 212 362 7044, or writing to ACSH, 1995 Broadway, New York, NY 10023.

1   James F. Fries, "Aging, Natural Death and the Compression of Morbidity." *New England Journal of Medicine* 306 (1982), 131.

2   National Center for Health Statistics, Prevention Profile, *Health United States, 1987*. Hyattsville, MD: Public Health Service, 1990, 1.

3   Ibid. 123.

4   Ibid.

5   National Cancer Institute and National Heart, Lung and Blood Institute, *Smoking and Health: A Program to Reduce the Risk of Disease in Smokers, Status Report*, Dec. 1978.

6   US Department of Health and Human Services, Public Health Service, Office on Smoking and Health, *Health Consequences of Smoking for Women: A Report of the Surgeon General*. Washington: HHS, 1980, 10–11.

7   *Healthy People: The Surgeon General's Report on Health Promotion and Disease Prevention*. Washington: Government Printing Office, 1979, 74.

8   *Promoting Health, Preventing Disease: Objectives for the Nation: The Surgeon General's Report on Health Promotion and Disease Prevention*. Washington: Government Printing Office, 1979, 45.

9   US Department of Transportation, National Highway Traffic Safety Administration, "Drunk Driving Facts," September 1990.

10   Kenneth Smith, "Alar: One Year Later. A Media Analysis of a Hypothetical Health Risk." New York: American Council on Science and Health, 1990.

11   H. F. Kraybill and L. T. Flynn, *From Mice to Men: The Benefits and Limitations of Animal Testing in Predicting Human Cancer Risk*. New York: American Council on Science and Health, 1990.

12   American Council on Science and Health press conference, Washington National Press Club, February 26, 1992.

13   Evidence cited or reviewed in: R. W. Rhein, *BST: A Safe, More Plentiful Milk Supply*. New York: American Council on Science and Health, 1990.

14   D. Blumenthal, "Food Irradiation: Toxic to Bacteria, Safe for Humans." *FDA Consumer* reprint, Washington: Food and Drug Administration, 1990.

15   Evidence cited or reviewed in American Council on Science and Health, *Irradiated Foods*. New York: American Council on Science and Health, 1988.

16   Evidence cited or reviewed in American Council on Science and Health, *Antibiotics in Animal Feed: A Threat to Human Health*. New York: American Council on Science and Health, 1985.

17 Ibid.

18 Evidence cited or reviewed in R. W. M. Letts, *Dioxin in the Environment.* New York: American Council on Science and Health, 1991.

19 V. N. Houk, *Dioxin Risk Assessment for Human Health: Scientifically Defensible or Fantasy?* Columbia, MO: 25th Annual Conference on Trace Substances in Environmental Health, University of Missouri, May 21, 1991. Dawn Goldman, presentation to the Florida Department of Environmental Regulation, May 1, 1991; R. A. Squire and J. H. Levitt, "Report of a workshop on classification of specific hepatocellular lesions in rats." *Cancer Research* 35 (1975): 3214–24; Robert A. Squire, letter to EPA assistant administrator LaJuana S. Wilcher, May 24, 1990, regarding need for reassessment of potency factor in light of new evidence, submitted in support of Georgia water quality standard for dioxin.

20 K. Schneider, "U.S. Backing Away from Saying Dioxin is a Deadly Peril." *New York Times*, August 15, 1991.

21 *New York Times*, August 15, 1991.

22 George L. Carlo, letter to the editor, *The Wall Street Journal*, March 27, 1992.

23 L. T. Flynn, *PCBs: Is the Cure Worth the Cost?* New York: American Council on Science and Health, 1989.

24 Ibid.

25 R. V. Thomann, J. A. Mueller, R. P. Winfield, and C. F. Huang, "Mathematical Model of the Long-term Behavior of PCBs in the Hudson River Estuary." Prepared for the Hudson River Foundation, June 1989.

26 Personal communication, Karl Berger, Citizen Participation Specialist at NYDEC, November 11, 1990.

27 J. F. Brown, Jr, et al., "Polychlorinated Biphenyl Dechlorination in Aquatic Sediments." *Science* 236, 709–12.

28 J. F. Quensen III, J. M. Tiedje, and S. A. Boynd, "Reductive Dechlorination of Polychlorinated Biphenyls by Anaerobic Microorganisms from Sediments." *Science* 242, 752–4.

29 "PCBs are Biodegraded in Nature." *Research and Development*, July 1985.

30 "Industrious Bacteria." *Scientific American* 253 (1985), 66–8.

31 Evidence cited or reviewed in *Low Calorie Sweeteners*. New York: American Council on Science and Health, 1986.

32 Ibid.

33 Ibid.

34 40 Code of Federal Regulations (CFR) 141.11.

35 40 CFR 141.51.

36 40 CFR 143.3.

37 Nature Resource Council, Drinking Water and Health, Washington, DC, vol. 1, 1977; vol. 2, 1980; vol. 3, 1980; vol. 4, 1982; vol. 5, 1983.

38 WHO, *Guidelines for Drinking Water, Volume 2: Health Criteria and Other Supporting Information.* Geneva: World Health Organization, 1984.

39 Evidence cited or reviewed in F. J. Stare, J. H. Shaw, S. J. Moss, *Fluoridation.* New York: American Council on Science and Health, 1990.

40 Evidence cited or reviewed in W. T. Jarvis, "Fluoridation and the EPA." *Priorities for Your Good Health*, Summer 1990.

41 Ibid.

42 Evidence cited or reviewed in *Microwave Ovens*. New York: American Council on Science and Health, 1985.

43 Whelan, *Toxic Terror* (1993) (Prometheus Books).

44 Ibid.

45 Ibid.
46 W. R. Havender, *Ethylene Dibromide (EDB)*. New York: American Council on Science and Health, 1984.
47 Whelan, *Toxic Terror* (1993).
48 Evidence cited or reviewed in *Low Calorie Sweeteners*.

PART VI

Thinking about the Issues

American Public Opinion: Environment and Energy

William M. Lunch and Stanley Rothman

This chapter explores the background to, and changes in, public opinion concerning the environment and energy over time and questions the conventional interpretation of opinion concerning these topics. Assessments of public opinion play an important role in the political system; elected and other officials use polls as a central guide to public attitudes. For example, in April 1990 the *New York Times* published a front page article entitled "Oratory of Environmentalism Becomes the Sound of Politics." The lead sentence of the article read, "The environment, once dismissed as a fringe cause by many politicians, has reached the forefront of American politics. . . ."[1] That judgment, based on polling, by the leading national newspaper has influence in policy debate, even if the questions asked and the issues to be resolved are quite different.

There are at least three levels at which public opinion can be measured. First, there is a symbolic level; broad questions asking if citizens favor clean air or accessible parks are the contemporary equivalent of "motherhood and apple pie" questions in a bygone era. Symbolic questions can appear to be specific. For example, do Americans believe acid rain should be controlled? (Yes, they do.) Second, many questions ask respondents where in the panoply of public issues environmental concerns are – that is, how salient are environmental issues compared to others? Third, a few questions ask specifically about the costs associated with environmental or energy policy. It is important to distinguish these three levels, because policy-makers do, even if they do not know methodological or social scientific jargon.

On the first two measures, the public appears to favor protection of the environment, broadly defined. This problem of definition is one to which we will return. When citizens tell pollsters that "protecting the environment is so important that requirements and standards cannot be too high and continuing environmental protection must be made regardless of cost" (a *New York Times/CBS* poll question), what do they have in mind, specifically? Of course, they may have nothing specific in mind;

broad, largely symbolic questions are often asked in polls. But symbols count in politics, and the symbolic success of the environmental movement has been substantial. When members of Congress are considering legislative proposals, reference to broadly worded questions expressing largely symbolic preferences – about national defense, housing, the environment or other policies – can have a significant impact. Moreover, many members of Congress (and even some state legislators) regularly poll their own constituents, attempting to assess public opinion. Thus, though some questions may lack specificity, they are not unimportant.

Salient questions measure the degree to which an issue is visible and important among issues before the public. One notable measure of salience is the fraction of citizens who declare that the environment or aspects of environmental policy are among the "most important problems" facing the nation. Salience is heavily influenced by events and, at times, by political and opinion leaders.

A question related to salience probes the preferences of the public when two or more policies are compared. For example, it is no surprise that in the aftermath of the oil shocks of the seventies, comparisons of energy production versus environmental protection showed a tilt in favor of production instead of protection. But by the late eighties, when concerns about energy availability had faded, the environmental side of the equation was preferred again.

ENVIRONMENTAL OPINIONS OVER TIME

We begin our review with a retrospective on changes over past decades. We consulted the Roper organization (hereafter Roper), the *New York Times/CBS* poll (hereafter *NYT/CBS*), and other sources. Broadly, we found strong support for symbolic environmental questions in the late sixties and early seventies, some decline during the late seventies and early eighties, and finally a rebound during the mid to later eighties.

Over the past two decades, for example, Roper has repeatedly asked, "Do you think environmental protection laws and regulations have gone too far, not far enough, or have struck about the right balance?" This question was not asked every year, but it was asked regularly.

* In 1973, 34 percent said environmental protection had not gone far enough; 32 percent felt it was about right; only 13 percent felt it had gone too far. To put this another way, between two to three times as many respondents felt that protection had not gone far enough as felt it had gone too far.

* By 1979, this relationship had changed. In that year, 24 percent felt that environmental protection had gone too far, while 29 percent felt it had not gone far enough. To be sure, those who felt protection was insufficient still outnumbered those who felt it had gone too far, but the numbers were very close compared to 1973.

* By 1989, however, those who felt that environmental protection needed to go farther had reasserted themselves – 55 percent felt that way, compared to only 11 percent who felt it had gone too far (figure 54.1).

The symbolic resonance of environmental concerns can also be measured through a simple closed-end question (figure 54.2). The *NYT/CBS* poll over the past decade has asked respondents whether they agree or disagree with the following: "Protecting the environment is so important that requirements and standards cannot be too high and continuing environmental protection must be made regardless of cost."

One important reason for at least the initial surge in support for environmental protection may have been press attention paid to it. For example, a review of stories on air pollution in the *New York Times* shows a substantial increase between 1968 and 1971, followed by a decline – but even after the decline, more stories on air pollution were published in later years than prior to the start of increased attention.[2]

Whatever the cause for the attention, in a period often dominated by discussions of the budget and public spending levels, the cost of protecting the environment is often debated. The Roper organization asked a question probing levels of spending for the environment through the period under discussion (figure 54.3).

The political appeal of the environment seems to have been growing, particularly as the decade of the eighties ended. But asking somewhat different questions, pollsters got rather different results. Perhaps most

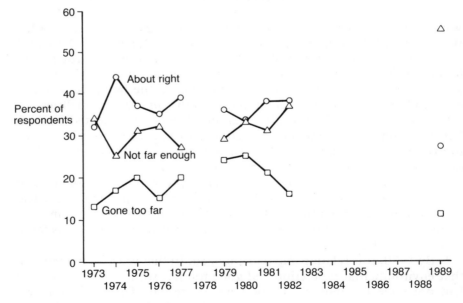

Figure 54.1 How far should environmental protection go?
Source: *Roper Reports*, Sept. 1989, 111–13.

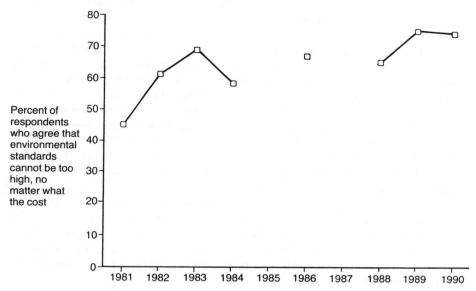

Figure 54.2 Importance of environmental protection and cost
Sources: New York Times, July 2, 1989, p. A1; April 17, 1990, p. A1; and *Congressional Quarterly Weekly Reports*, Jan. 20, 1990, 142.

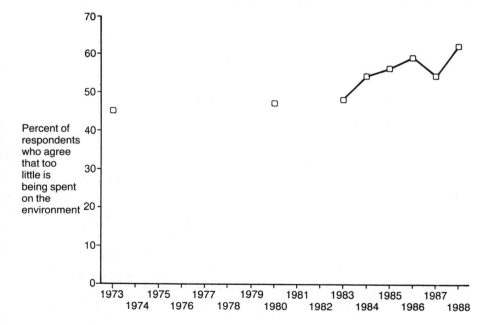

Figure 54.3 Spending levels for the environment
Source: Roper Reports, Jan. 1989, 63.

notably, given the strong symbolic support for the environment, the level of salience for environmental concerns seems oddly low. The public quite carefully distinguishes between question wording; at the same time that the environment ranked highly as a *public* issue, relatively small proportions of the public selected "air and water pollution," as the issues that most concerned them *personally* (these were the environmental issues offered in a long list of alternatives). To be sure, the fraction concerned about pollution of air and water rose during periods when generalized environmental protection sentiment was high, but the numbers were still rather low. Fifteen percent expressed this concern in 1987, 14 percent in 1988, rising to 19 percent in 1989 and 21 percent in 1990. By contrast, during the seventies, personal concern about "the fuel and energy shortage" reached relatively high levels; in 1974, 46 percent of respondents put this on their short list of public issues of personal concern; in 1979, though the numbers were declining, 25 percent mentioned energy in this context, compared to 10 percent for air and water pollution. Then in the eighties, personal concern about energy declined notably while concern about pollution increased modestly (figure 54.4).

These findings are consistent with other soundings that indicate sharp but brief surges in concern about energy. As figure 54.5 indicates, energy policy was listed by large numbers of Americans as the "most important problem" facing the nation only during the periods when

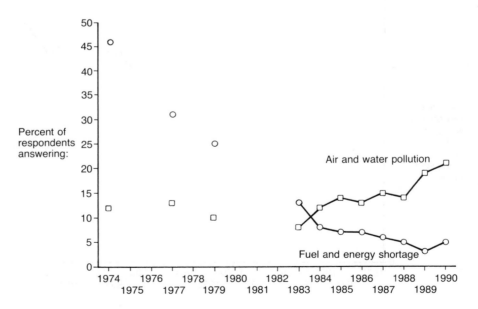

Figure 54.4 Salience of environmental issues, air and water pollution versus energy
Source: *Roper Reports*, Jan. 1990, 102–3.

energy shortages were manifest. There is one other surge of concern about energy in 1977 and 1978, when President Carter was trying to persuade Congress and the American people of the need for a comprehensive energy policy. Evidently, the bully pulpit can focus attention, at least for a while.

Mainly, however, concern about energy – and a willingness among some citizens to reduce environmental protection to secure energy supplies – coincides with shortages during the seventies, when the price shot up due to the Arab oil embargoes. The public evidently uses something of a "balancing test" when the issue is framed this way. That is, environmental protection is weighed against higher prices for gasoline, home heating oil, and so forth. There is, however, a fraction of the public that consistently rejects the notion that environmental protection and sufficient energy supplies are mutually exclusive; such respondents say that there is no conflict between adequate energy and environmental protection.

A significant fraction refuses to accept the alternatives as framed in energy *versus* environment questions. Indeed, much of the public rejects

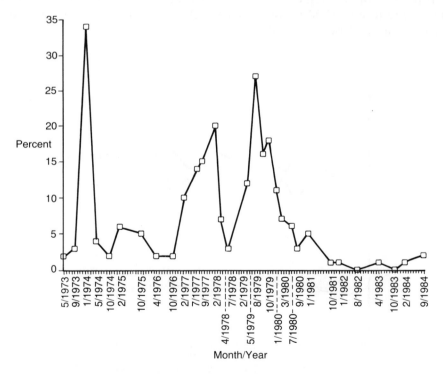

Figure 54.5 Energy as the "most important problem"
Source: Gallup polls, reported by R. G. Niemi, J. Mueller, and T. W. Smith, *Trends in Public Opinion* (Westport, CT: Greenwood Press, 1989), 44–5.

the dichotomous approach frequently taken in public opinion polls on environmental and energy issues. For example, Everett C. Ladd reports that when respondents were offered a middle way between protecting the environment and economic growth by Opinion Dynamics in December, 1989, almost two-thirds of the sample chose "have both."[3]

Without a stated alternative favoring both energy and the environment, a segment of the public nonetheless volunteers it. For example, when Roper has presented respondents with the statement below, between 10 and 14 percent have rejected the premise implicit in it by saying there is no conflict between the goals:

> Some people say that the progress of this nation depends on an adequate supply of energy and we have to have it even though it means taking some risks with the environment. Others say the important thing is the environment and that it is better to risk not having enough energy than to risk spoiling our environment. Are you more on the side of adequate energy or more on the side of protecting the environment?

(See figure 54.6.)

But when the Gallup organization asked in 1989 if respondents would be willing to pay $200 more in taxes each year to increase spending on air pollution, 71 percent said no.

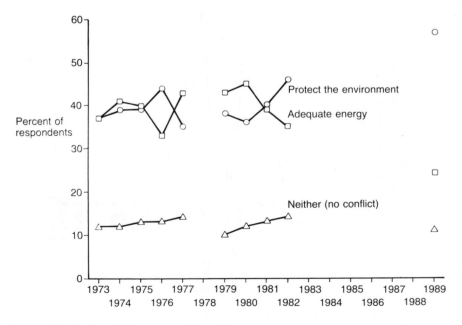

Figure 54.6 Environment versus energy
Source: Roper Reports, Sept. 1989, 111.

Moreover, in a poll conducted for the *Christian Science Monitor* (CSM) in 1990, Roper found that attitudes about the environment and life-styles interact in more complex patterns than the responses to either symbolic or salient questions would suggest. As the *New York Times* poll found in the same year, there is close to unanimous support for symbolic statements favoring environmental protection. But the CSM poll went further to ask how respondents reflected their environmental beliefs in day-to-day activities. There is a gap between attitudes and action. Eleven percent, the "true-blue greens," both believe in environmental protection and act on those beliefs by, for example, using biodegradable products and recycling cans. The next group, the "green-back greens," also espouse environmental views, but are busier and have less time to act on their beliefs. Two additional groups – the "sprouts" and "grou-sers" – say they are concerned about the environment but are less likely to act on such views than the first two groups. These two groups account for 50 percent of the respondents. Finally, the "basic browns" (28 percent) are indifferent to the environment in both attitudes and beha-vior. The article reporting these findings was entitled, "On The Envi-ronment, Americans' Words Are Louder Than Deeds."[4]

As the analysis by Ladd and the CSM poll suggest, while environmen-tal protection has achieved wide symbolic appeal, there is a difference between announced attitudes and action. Most Americans are reluctant to give up conveniences associated with life in an advanced industrial society. Their politicians, much maligned but ever sensitive to the nuances of public opinion – and its relation to voting behavior – have shown reluctance to accept the most advanced positions offered by environmental groups. In this, they are reflecting the ambivalent feelings of their constituents, who genuinely favor environmental protection – as James Watt learned, to his chagrin – but who are not willing to make radical changes in their lives to accomplish them – as environmental group leaders have been frustrated to discover.

Neither the symbolic nor the salient questions fully describe the range of public views on environmental issues. Clearly, there are deep feelings favoring environmental protection, and these extend beyond symbolism. The strength of environmental issues in elections in both 1988 and 1990 should serve as a warning to those inclined to dismiss the issue. At the same time, careful review of public opinion data reveals that Americans are not willing to sacrifice all other values in order to realize an ambi-tious environmental agenda.

NOTES

1   Richard L. Berke, "Oratory of Environmentalism Becomes the Sound of Poli-tics." *New York Times*, April 17, 1990, 1+.
2   See Marc K. Landy, et al. *The Environmental Protection Agency: Asking the Wrong Questions.* New York: Oxford University Press, 1989, 23–4, and *The New York Times Index.*

3 See Everett C. Ladd, "What Do Americans Really Think About the Environment?" *The Public Perspective*, May/June 1990, 11–13.
4 Marshall Ingwerson, "On the Environment, Americans' Words Are Louder than Deeds." *Christian Science Monitor*, August 2, 1990, 1–2.

55

Public Opinion About, and Media Coverage of, Population Growth

Rita J. Simon[1]

National opinion polls about population growth going back 45 years show that the majority of the American public has been and continues to be fearful and worried about population growth. They believe that population growth in the world, and in the USA particularly, is a serious problem and that the US population should be smaller than it is, rather than larger. Concern about population size and growth are associated in the public's mind with increased levels of air and water pollution, greater energy scarcities, and too many immigrants coming to the United States.

In 1947, a Gallup poll asked whether the country would be "better off or worse off if there were more people living here." Fifty-five percent said "worse off," 16 percent said "better off," 14 percent said "the same," and 15 percent had no opinion.[2]

Between 1960 and 1967, the following question appeared on four national polls:[3]

Have you read or heard about the great increase in population which is predicted for the world during the next few years? (A1PO)

Followed by:

Are you worried or not worried about this population increase? (A1PO)

Responses to the second part of the question showed that from 1960 to 1967, between 21 and 30 percent of the respondents were worried about the predicted increase in worldwide population.

In 1965, a Gallup poll solicited US opinion about the rate of world population growth and the rate of US population growth with the following results.[4]

The rate of world population growth is a serious problem	62%
The rate of US population growth is a serious problem	54%

In 1971, two national polls included items about population. In January, Planned Parenthood commissioned a poll that asked whether population growth is:[5]

A major problem now	41%
Somewhat of a problem now	27%
Not a problem now, but will be	19%
Not a problem now and will not become one	8%
No opinion	5%

The public was also asked whether population growth would affect "the quality of the respondent's life," to which the public responded as follows:

Will seriously affect quality of life	25%
Could affect quality of life and concerned	29%
Could affect quality of life but not concerned	27%
Will not affect quality of life	14%
No opinion	5%

In April 1971, "the US Commission on Population Growth and the American Future" commissioned a national poll that contained the following items:[6]

Do you feel that the growth of the United States population is a serious problem, a problem but not so serious, or no problem at all?

The responses were distributed as follows:

A serious problem	65%
Not so serious a problem	26%
No problem at all	7%
No opinion	2%

Do you think the present size of the United States population is about right or do you think it should be smaller or larger?

About right	57%
Should be smaller	22%
Should be larger	8%
No opinion	13%

The percentages in table 55.1 describe the extent to which respondents agreed or disagreed with the impact that population growth was having on other issues.

In 1974, a Gallup poll asked:[7] "Do you wish there were more people in this country?" Eighty-three percent answered no.

The data show that between 1947 and 1974, public opinion became more concerned about population growth and its negative implications.

Table 55.1 Percent who agree/disagree about impact of population growth (1971)

Impact of population growth	Agree	Disagree	No opinion
Population growth is causing the country to use up its natural resources too fast.	57	35	8
Population growth is the main reason for air and water pollution.	48	47	5
Population growth helps keep the economy prosperous.	36	52	12
Population growth is producing a lot of social unrest and dissatisfaction.	64	28	8
Population growth is important in keeping up the nation's military strength.	36	56	8
People should limit the size of their families even though they can afford a large number of children.	57	32	10

In the polls conducted in the 1960s, no more than 30 percent said they were worried about population growth, although 54 percent saw it as a serious problem for the USA. In 1971, 41 percent saw it as a "major" and "immediate" problem; and later in the same year, 65 percent said they thought population growth was a "serious" problem. In 1947, 55 percent said the United States would be worse off if it had more people; in 1974, 83 percent answered no when asked whether they wished there were more people in the country. Some explanation about why the American people perceive population growth negatively is shown in their responses to the relationship between population growth and social and economic issues. Population growth was perceived as increasing air and water pollution and social unrest and depleting natural resources. It was not perceived as having a positive impact on the economy or on military strength.

The following question was asked between 1974 and 1988 on seven national polls:[8]

Here is a list of some different kinds of problems people might or might not be facing 25 to 30 years from now. Would you go down that list, and for each one tell me whether you think it will or will not be a serious problem your children or grandchildren will be facing 25 to 50 years from now?

"Overpopulation" appeared on that list in all eight polls and the percent who answered "it will be a serious problem" ranged from a low of 44 percent in 1978 to a high of 65 percent in 1991. A comparison of the percent who perceived "overpopulation" as a serious problem as opposed to air and water pollution and shortage of water and energy supplies is shown in table 55.2.

Concerns about air and water pollution and shortages of water increased between 1974 and 1991, while concerns about food shortages and energy supplies remained fairly stable or declined over the 17-year time span. Concern about overpopulation declined between 1976 and 1984, but increased in 1988 and again in 1991 to its highest level.

Roper also asked the following question on three national polls in 1977, 1980, and 1985:[9]

> At its present rate of growth, the population of the United States will double in about 50 years to about 475 million people. Some people are concerned about this and others are not. Here are some different points of view as to what, if anything, should be done with respect to population growth. Would you go down this list and for each item on the list tell me whether you agree or disagree with it.

The solutions to issues with which at least 75 percent of the respondents agreed are shown in table 55.3. In contrast, no more than 26 percent said:

	1977	*1980*	*1985*
Don't do anything to slow population growth. Let nature take its course.	21	20	26

The most recent poll, commissioned in April 1990 by the Federation for American Immigration Reform, included this item:[10]

> I am going to read you three statements. Please tell me which statement most clearly reflects your opinion.
> a   The U.S. is a large country and it will be a long time before we have to worry about population growth problems.
> b   The U.S. is a large country, but we are already having some population problems. We should probably start thinking about ways to control population growth now.
> c   The U.S. is a large country, but we already have major problems with population growth that will only get worse in the future and we *must* try to solve the problems now.
> d   Don't know.

Fifty-five percent selected alternative c; 32 percent selected b; and 10 percent selected alternative a. Three percent said they did not know.

The 55 percent in 1990 may be compared against the 54 percent in 1965, the 41 percent in 1971, and the 65 percent in 1971 who answered

Table 55.2 Percent who perceive the issues as serious problems

Issues	1974	1976	1978	1980	1982	1984	1988	1991
Some air pollution	68	73	68	68	72	70	82	79
Severe water pollution	69	72	66	69	71	71	82	75
Shortage of water supplies	53	57	50	57	59	53	66	66
Overpopulation	60	50	44	52	52	56	61	65
Shortage of energy supplies	62	68	63	68	59	50	54	57
Shortage of food	65	50	46	54	49	46	48	54

Table 55.3 Percent who agree actions should be taken to reduce population growth

Actions recommended	1977	1980	1985
Make an all-out effort to stop the illegal entry into the United States of many foreigners who don't have entry visas.	91	91	89
Make birth control information and devices widely available at low cost.	87	86	85
Publicize the fact that population will reach 475 million in about 50 years so that people will be aware of the growth.	86	85	85
Reduce the quotas of the number of legal immigrants who can enter the United States each year.	75	80	77

that the rate of population growth was a serious problem. From all of these responses, it is clear that from the mid-1960s, population growth has been considered a serious, immediate, and major problem by a majority of the American people.

National polling began in the mid-1930s, and from that time until the present, one item concerning population has appeared on at least 19 national surveys. It asked: "What do you think is the ideal number of children for a family to have?" The results are shown in figure 55.1.

Note that between 1968 and 1971 there was a sharp increase in the percentage of families who chose two children and a sharp decrease in the percentage who chose four children. Remember that the poll responses in 1971 showed that 64 percent thought the growth of the US population was a serious problem and 41 percent said it was a major problem. Before that, from 1960 through 1967, between 21 and 30 percent of the American public said they were worried about population growth. Figure 55.1 shows that in the 1960s, about 40 percent of the

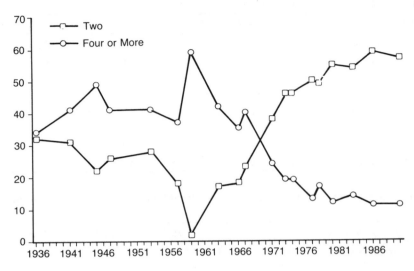

Figure 55.1 Ideal number of children
Source: Gallup.

respondents thought four was the ideal number of children, and only 20 percent thought two was the ideal number for a family to have.

POPULATION AND THE MEDIA

One strategy for gauging the saliency of "population" as a public issue is noting the frequency of references to population growth and decline in the print media. In the index of the *New York Times*, we tallied all of the news stories, editorials, and letters to the editor on population growth and decline printed between 1930 and 1990. Stories reporting census data, vital statistics, population movements within or between cities or regions, differences in birth and death rates among different populations were not included. But stories and comments about population growth, decline, or stability at a national, regional, or global level and discussion of the social and economic implications of the population levels were counted. Other newspapers included for shorter time periods were the *Wall Street Journal* from 1955 to 1990, the *Washington Star* from 1930 until it ceased publishing in 1972, and the *Washington Post* from 1970 to 1990.[11] In addition to counting the number of stories and other comments that appeared, we made an assessment about the tone of the report: a "+" indicates that the report believed population was increasing and that was bad, that is, food supply could not keep up, social and economic problems would increase, and so on; a "–" indicates that population was on the decline or growing at a slower rate than earlier years and that was bad or dangerous because of the social and economic ramifications; a "0" indicates that no conclusions or warning appeared in the report. Table 55.4 describes the number of stories per year from the 1930s through the 1980s and their tone.

The number of references to population trends increased steadily from the 1930s through the 1970s. For the *New York Times*, there was more than a tenfold increase from 4.1 references per year in the 1930s to 42.0 in the 1970s. In the 1980s the *New York Times* showed a marked decline from the previous decade and the *Washington Post* showed a slight decline.

Except for the wartime decade of the 1940s, the tone of a large percentage of the *New York Times* stories was negative *vis-à-vis* population growth. More population meant more social and economic problems, less food, and more pollution. In the 1950s and 1960s, the relatively few stories that appeared in the *Wall Street Journal* were strongly antipopulation growth. The 1970s and 1980s witnessed an increase in stories about population issues and a sharp drop in their negative implications. Both the *Washington Star* and in the 1970s, the *Washington Post*, published fewer negatively slanted stories *vis-à-vis* population growth than did the *Times* and the *Journal*.

There was a large enough number of editorials and letters in the *New York Times* and the *Washington Post* from the 1960s through the 1980s to make a distinction between the slant that appeared in the news stories and those that characterized the editorials and letters.

As shown in table 55.5, the direction of the editorials and news stories in the *New York Times* was quite consistent and less negative about population growth than was the direction of the letters. The letter pattern was more erratic. In the 1970s, a smaller proportion of letters, as opposed to articles and editorials, expressed negative sentiments. In the 1980s, with many fewer letters and editorials, all of the former voiced negative sentiments. For the *Washington Post*, the letters and editorials were more likely to express similar sentiments than were the

Table 55.4 Number of reports per year and percentage that characterizes population growth as dangerous and troublesome

Year	Newspapers							
	NY Times		*Washington Star*		*Washington Post*		*Wall Street Journal*	
	Reports per year	*%*	*Reports per year*	*%*	*Reports per year*	*%*	*Reports per year*	*%*
1930s	4.1	68	0.7	14				
1940s	4.9	47	1.4	7				
1950s	11.7	81	2.6	46			1.4	71
1960s	40.4	93	5.4	47			1.2	100
1970s	42.0	76			22.7	34	2.9	59
1980s	8.2	60			15.2	61	3.3	36

Table 55.5 Percentage of news stories, editorials, and letters that characterize population growth as dangerous

| Year | New York Times | | | Washington Post | | |
	News stories	Editorials	Letters	News stories	Editorials	Letters
1960s	93	100	86			
1970s	77	82	61	32	50	56
1980s	56	57	100	64	48	40

news stories. In the 1970s, the news stories were less negative *vis-à-vis* population growth than were the letters and editorials. In the 1980s, the pattern was reversed; the news stories were more negative, the editorials showed little change, and the letters were less negative.

Overall, the general pattern that emerges is that from the mid-1960s to the end of the 1970s, "population" made news. The 1980s witnessed a sharp decline in coverage of the topic in the *New York Times*, less so in the *Washington Post*, and a steady pattern in the *Wall Street Journal*. In the 1970s, the *Washington Post* differed from both the *Times* and the *Journal* in the way they perceived population growth. In the 1980s, both the *Times* and the *Journal* had a smaller percentage of coverage describing population growth in negative terms. The *Washington Post*'s coverage shifted in the opposite direction. The media data suggest that they may have played an important role in sensitizing the American public to the perceived dangers of population growth. The *New York Times* especially increased its coverage of population issues in the 1960s and 1970s, and at least three quarters of the entries characterized population growth in negative and dangerous terms.

NOTES

1 Angela Musolino and Jeffrey Bagato helped collect and organize the data for this piece.
2 Reported in *ZPG Reporter*, Oct./Nov./Dec. 1979.
3 Hazel Gaudet Erskine, 1966. "The Polls: The Population Explosion, Birth Control and Sex Education." *Public Opinion Quarterly* 30 (3): 490–1; 1967, 31 (2): 303–4.
4 The Population Council, "American Attitudes on Population Policy," *Studies in Family Planning*. 9 (January 1966): 5–8.
5 Richard Pomeroy and Morton Silver, "Public Opinion, Population and Family Planning." Research Department of Planned Parenthood – World Population, (mimeo) 1972.
6 Robert Parke, Jr, and Charles F. Westoff (eds) *Commission on Population Growth and the American Future* (Washington, DC: 1971).
7 Reported in *ZPG Reporter*, op. cit.
8 Roper Report: 89–1.
9 Roper Report: 85–6.

10 "American Attitudes Toward Immigration," *The Roper Organization*, June 1990.

11 Indices for the *Wall Street Journal* and the *Washington Post* were available only for those years.

56

Risk Within Reason

Richard J. Zeckhauser and W. Kip Viscusi

Society's system for managing risks to life and limb is deeply flawed. We overreact to some risks and virtually ignore others. Often too much weight is placed on risks of low probability but high salience (such as those posed by trace carcinogens or terrorist action); risks of commission rather than omission; and risks, such as those associated with frontier technologies, whose magnitude is difficult to estimate. Too little effort is spent ameliorating voluntary risks, such as those involving automobiles and diet. When the bearers of risk do not share in the costs of reduction, moreover, extravagance is likely.

Part of the problem is that we rely on a mix of individual, corporate, and government decisions to respond to risk. Our traditional coordinating mechanisms – markets and government action – are crippled by inadequate information, costly decision-making processes, and the need to accommodate citizens' misperceptions, sometimes arising from imbalances in media attention.

Risk can never be entirely eliminated from life, and risk reductions come at a price (in dollars, forgone pleasures, or both). Our current muddled approach makes it difficult to reach wise, well-informed decisions as to the preferred balance of risk and cost. Some risks we ignore; some small ones we regulate stringently. Worse, our overreaction to very small risks impedes the kind of technological progress that has historically brought dramatic improvements in both health and material well-being. In addition, we are likely to misdirect our efforts, for example, by focusing on risks that command attention in the political process, such as newly identified carcinogens, rather than those that would permit the greatest gains in well-being, such as individual lifestyle choices.[1]

Our regulatory efforts focus too much on equipment and physical processes, too little on human error and venality. We may set stringent emission standards, which impose high costs per unit of environmental quality gained, yet ignore the haphazard operation of nuclear weapons plants.

HUMAN FALLIBILITY IN RESPONSE TO RISKS

Decisions involving risks reveal the limits of human rationality, as a substantial recent literature documents.[2] One fundamental problem is that individuals have great difficulty comprehending extremely low probability events. When comparing a risk of 10^{-7} with a risk 100 times as large, 10^{-5}, for example, even scientists may not appreciate how much greater a payoff might be gained by addressing the larger probability.

Few decision-making problems that arise with respect to small probabilities are by themselves of consequence. Expected welfare loss from any single error may well be negligible. Aggregated, however, low-probability events make up a large part of an individual's risk level. Even truly substantial risks, such as the chance of death from a stroke – roughly one in 2,000 annually averaged over the population – are usually influenced by a myriad of decisions, each of which has only a small probabilistic impact on our longevity. Systematic errors in many of these decisions might have an enormous cumulative effect.

Mistakes in Estimation

Whereas people generally overestimate the likelihood of low-probability events (death by tornado), they underestimate higher risk levels (heart disease or stroke).[3] We are particularly likely to overestimate previously unrecognized risks in the aftermath of an unfavorable outcome.[4] Such perceptional biases account for the emotional public response to such events as the Three Mile Island nuclear accident, or to occasional incidents of deliberate poisoning of foodstuffs or medicines.

Risk perceptions are also affected by the visibility of a risk, the fear associated with it, and the extent to which individuals believe they can exercise control over it.[5] Consider the greenhouse effect. Although global warming is a prime concern of the Environmental Protection Agency (EPA), it ranks only twenty-third among the US public's environmental concerns.[6] The high risk of automobile fatality – car accidents kill one in 5,000 Americans each year – might be reduced significantly if drivers, informed with a more realistic sense of what they can and cannot do to control the risk, drank less alcohol and wore seat belts more often.

Studies of consumers show that many individuals would be willing to pay a premium for the assured elimination of a risk.[8] The Russian roulette problem illustrates. Consider two alternative scenarios for a forced round of play. In the first, you have the option to purchase and remove one bullet from a gun that has three bullets in its six chambers. How much would you pay for this reduction in risk? (Assume you are unmarried, with no children.) In the second situation, the gun has only a single bullet. How much would you pay to buy back this bullet? From an economic standpoint, you should always be willing to pay at least as much and typically more in the first situation since there is some chance you will be killed by one of the remaining bullets, in which case money is worthless (or worth less).[9] However, experiments find respondents are typically willing to pay more when a single bullet is in the gun, because its removal will ensure survival.[10]

The Chilean grape scare of 1989 provides a recent example of a risk that did not lend itself to statistical estimation or scientific assessment. Neither the government nor consumers could estimate how much consumers' risk was increased by the discovery of traces of cyanide in two Chilean grapes in Philadelphia.[11] When precise scientific judgments concerning probabilities are elusive, concerns about regret[12] are likely to influence decision making significantly. If societal norms were flouted, regret would be greater still. (Few of us would leave a baby sleeping alone in a house while we drove off on a ten-minute errand, even though car-crash risks are much greater than home risks.) Hindsight is frequently able to identify why an individual or society should have known certain risk estimates were far too low, as we "learned" with the highway collapse during the San Francisco earthquake of 1989.

Distortion in Valuation

The valuation of a risk is likely to depend on how the risk is generated. We tolerate voluntarily assumed risks more than those, such as environmental hazards, over which we have no control. We regard acts of commission as much more serious than acts of omission. In pharmaceutical screening, for example, the Food and Drug Administration (FDA) worries more about introducing harmful new drugs than about missing opportunities for risk reduction offered by new pharmaceutical products.[13]

Society's pattern of lopsided trade-offs between errors of omission and commission persists for at least two reasons. First, even apart from possible effects on people's chosen levels of risk, their consumption of information is relevant. When a federal agency demonstrates that it will not take chances with individual health, that reassurance alone enhances individual welfare. Conversely, a perception that the government tolerates risks to the public might be more damaging than the risks themselves. Second, it is easier to observe the costs of bad drugs that are approved than to assess the forgone benefits of good drugs that were not introduced. (Note, however, that potential beneficiaries, such as saccharin users and AIDS victims, sometimes put substantial pressure on the FDA to compromise normally stringent procedures for approving food additives and drugs.[14])

How should we proceed once we admit that individuals do not react to many risks "correctly"? We might ask the government to make more decisions. It is not clear, however, that the government itself is well equipped to compute certain risks accurately or to make sensible decisions once that information is obtained. Alternatively, we could shift decision-making authority to those best qualified to make particular kinds of choices; however, the preferences of those making a decision might not be the same as those affected by it. A third possibility would be to develop processes enabling both agents and principals to participate in risk-related decisions, but there is little evidence that such processes would produce convergence. Finally, we might try to improve the quality of individuals' decisions by providing them, for example, with expert-certified information, much as accounting firms verify the accuracy of reported financial data.

The Informational Approach

Society's objective should be to foster informed consumer choice. With respect to cigarette smoking, for example, this may not be the same as seeking a tobacco-free society. (Note that research linking aflatoxin and cancer risks[15] has not moved the Surgeon General to call for a peanut-butter-free society.) Politically, of course, the passive smoking concern may be a trump card, rendering it irrelevant whether the risk imposed on others is substantial or negligible.

Hazard warnings are often used to convey risk information. Congress has mandated labels for cigarettes, artificial sweeteners, and alcoholic beverages. Federal agencies impose labeling requirements for consumer products, workplace risks, and pesticides. Informational efforts work in conjunction with market forces rather than attempting to supersede them.

Individuals may have difficulty processing risk information, however.[16] Overambitious information efforts may outstrip decision-making capabilities (e.g., California Proposition 65,[17] which requires warnings for products that expose consumers to annual risks of cancer of one in seven million). The dangers are underreaction, overreaction, and nonreaction – a complete dismissal of the risk information effort. Sound decisions are unlikely to result. Indeed, the supposition of informed consent is called into question.

More general human cognitive limitations also work against detailed informational efforts. If a warning label contains more than a handful of items, or if warnings proliferate, problems of information overload arise.[18]

In a democratic society one should hesitate to override the legitimate preferences of segments of the population, taking care not to dismiss diversity of taste as mere nonrational choice.[19] Where there is broad consensus on a rational course of action, however, and either the cost of providing information is high or individuals cannot process the information adequately, then mandatory requirements may be preferable to risk information efforts. Laws requiring the use of seat belts are one possible example.

Individuals often fail to interpret risks or value their consequences accurately. Government efforts may escape some of these biases, but are often thrown off course by political pressures and agency losses. The consequence is that our risk portfolio enjoys no legitimacy and satisfies no one. The first step toward a remedy is to develop a broad-based understanding of the nature of risk.

TOWARD REASONABLE RISK POLICIES

Regulatory Efforts and Misplaced Conservatism

Governmental efforts at developing risk information are not guided by the formal statistical properties of the risk but rather by administrative procedures incorporating various types of "conservatism." Although risk assessment biases may operate in both directions,[20] most approved procedures tend to overstate the actual risk.[21] In regulating toxic substances, for example, results from the most sensitive animal species are often used, and government agencies such as the EPA routinely focus on the upper end of the 95 percent confidence interval as the risk level, rather than use the mean of the distribution. A series of such conservative assumptions – for example, on exposure or focusing on the most sensitive human beings – can overstate the mean probability of an unfavorable outcome by several orders of magnitude.

If lives are at stake, should we not be conservative when risk estimates are known to be uncertain? In fact, conservatism of this nature is undesirable for three reasons. First, because these conservative biases often are not uniform across risks, they may distort comparative risk judgments. If we focus on reducing risks for which standard errors are large with respect to their level, then we will save fewer expected lives than if we were guided by the mean of our probability distribution on the risk level. In effect, society will be curtailing the wrong risks, ones that offer less expected health improvement than other available options, for the resources and benefits forgone. The bias that results will cut against new technologies and innovative products. Second, stringent regulation of uncertain risks destroys opportunities for learning. Third and most fundamental, tilting risk assessments in a conservative direction confuses the informational and decision aspects of research about risks.[22] A conceptually sound form of conservatism would have the decision maker (not the risk estimator) adjust the weights on the consequences. Adjusting the probabilities amounts to lying to ourselves about what we expect.

Striking the Balance

Restrictions on a risky activity, such as exposure limits, should be based on the relative gains and losses of the activity as compared with its alternatives. In thinking about these trade-offs, one should remember that improvements in mortality and morbidity have come primarily from technological progress and a higher standard of living, not from government regulation or private forbearance.[23] A dramatic case in point is that of postwar Japan, where mortality rates have fallen for all age groups. Over the period 1955–75, with a rapid rise in the standard of living, mortality rates for men aged 65–9 fell 32 percent; men aged 25–9 had a 64 percent drop.[24] Sustained economic development also seems to be the principal factor in explaining increased life expectancy in the United States. In contrast, risk regulation policies often provide few major dividends.[25]

It is useful to think about risk-averting policy in terms of the rates of trade-off involved, such as the cost per expected life saved. Using this lives-saved standard of value highlights the most effective means of promoting our risk reduction objective.[26] [For example, see chapter 51, "The Hazards of Nuclear Power" in this volume.] The cost-effectiveness of existing regulations ranges widely, from $200,000 per life saved for airplane cabin fire protection to as much as $132 million per life saved for a 1979 regulation of DES in cattle feed.[27] These wide discrepancies reflect differences among agencies in their risk-cost balancing as well as differences in the character of risk-reducing opportunities. The Federal Aviation Administration has traditionally undervalued lives, looking only at lost earnings, whereas food additive regulations and EPA ambient air quality standards are set without consideration of cost. Elimination of such interagency imbalances would foster better control of risks at less cost.

The fundamental policy question is how far to proceed with life-saving expenditures. Economists are accused, sometimes with justification, of concluding too quickly that choosing policies to save lives is merely a question of setting an appropriate price. In contrast, society often is insensitive to the trade-offs that must be made. Indeed, 80 percent of respondents polled two months after the Exxon *Valdez* oil spill indicated a willingness to pursue greater environmental protection "regardless of cost."[28] Ultimately, however, society must decide how much of a resource commitment it will make.

Learning from Market Outcomes

Market outcomes provide a natural starting point for obtaining information on how risk reduction policies are valued by their beneficiaries. Health risks are important components of goods and services sold on markets, providing an approach to valuation. Wage differentials for high-risk occupations imply a value of several million dollars for each expected death in the workplace.[29]

Market data for many risk outcomes are not available, in part because government policies are largely directed at situations in which the market is believed not to function effectively, if at all. Thus, we have little price information to guide us when deciding, for example, whether society's resources would be better used to reduce rates of birth defects, to promote better nutrition, or to reduce oil spills from tankers.

The policies for which no market reference is possible are the very ones in which current practice may be furthest from the optimum. How much, for example, is it worth to prevent a low-level risk of genetic damage? Such valuation questions have received little careful consideration. When risks are received collectively, as when a sewage treatment plant or prison is placed in a community, little is learned about valuation since compensation is rarely paid.[30] The result has been severe inequity for the unfortunate few, and a democratic society that cannot find places to site essential though noxious facilities.

Finding Appropriate Roles

The government's responsibility in generating and using risk information involves structuring a decision process in which individuals and societal institutions work together. Policy choice in a democratic society is, however, complicated by discrepancies between lay and expert opinion. In some situations, the government must decide whether to intervene to overcome apparent limitations on individual choices. But it can be difficult to distinguish irrationality from legitimate citizen preferences. Are people who do not wear seat belts irrational? What about those who consume animal fats? Analogous questions arise with respect to policy emphasis. To what extent should the government focus on risks that are of particular concern to its citizens, who may be misinformed and subject to severe errors in perceptions and valuation of risk? Government agencies, subject to political pressures, may find it difficult to set their course in the direction indicated by dispassionate analysis of risks and overall benefits to society.

As science advances and our ability to detect risks improves, our opportunities for influencing risks have proliferated. To date we have proceeded haphazardly, responding to each risk in turn, whether it arises from a new technology, is revealed by scientific investigation, or is catapulted to prominence by media attention. This is not a sensible strategy for making balanced decisions across the entire spectrum of risks.

We need to acknowledge that risks to life and limb are inherent in modern society – indeed in life itself – and that systematic strategies for assessing and responding to risks are overdue. Such strategies will involve significant reassignment of decision-making responsibilities. Individuals should do more for themselves, paying greater attention, for example, to their diets and driving habits. Governments should focus less on microscopic contingencies and more on human mistakes and misdeeds – the source of far greater risks.

NOTES

1 The role of lifestyle is discussed by Fuchs, *Who Shall Live? Health, Economics, and Social Choice* (New York: Basic Books, 1974), particularly 52–4. He assesses the stark differences in mortality between high-living Nevada and sober Utah.

2 See H. Kunreuther, et al., *Disaster Insurance Protection.* (New York: Wiley, 1978); A. Tversky and D. Kahneman, "The Framing of Decisions and the Psychology of Choice." *Science,* 211 (1981); P. Slovic, "Perception of Risk." *Science,* 236 (1987); M. J. Machina, "Decision-Making in the Presence of Risk." *Science,* 236 (1987): 537–43; B. Fischhoff, et al., *Acceptable Risk.* Cambridge: Cambridge University Press, 1981; and W. K. Viscusi and W. A. Magat, *Learning About Risk: Consumer and Worker Responses to Hazard Information.* Cambridge: Harvard University Press, 1987.

3 See B. Fischhoff, et al. (1981), op. cit.

4 See M. J. Machina (1987), op. cit.

5 See H. Kunreuther, et al. (1978), op. cit.; A. Tversky and D. Kahneman (1981), op. cit.; P. Slovic (1987), op. cit.; and B. Fischhoff, et al. (1981), op. cit.

6 See R. H. Baxter and F. W. Allen, "Assessing Environmental Risks: The Public's Views Compared with Those of the Environmental Protection Agency." Annual Conference of the American Association for Public Opinion Research, May 20, 1989, cited in the *New York Times,* May 22, 1989, p. B7.

7 National Safety Council, *Accident Facts* (Chicago: National Safety Council, 1985), 15.

8 W. K. Viscusi, W. A. Magat, and J. Huber, "An Investigation of the Rationality of Consumer Valuations of Multiple Health Risks." *Rand Journal of Economics* 18, 4 (1987): 465–79, report an experiment involving household chemicals in which there is a reference to risk effect and a premium for certain elimination of the risk.

9 Let p be the initial probability of survival, q be the increased probability of survival, purchased at cost Z, $U(Y)$ be the utility of income if alive, where $U'(Y) > 0$, and $U(\text{Death}) = 0$, independent of income. Assuming expected utility maximization, by definition Z satisfies $pU(Y) + (1 - p)U(\text{Death}) = (p + q)U(Y - Z) + (1 - p - q)U(\text{Death})$ or settling $U(\text{Death}) = 0$, $pU(Y) = (p + q)U(Y - Z)$. Totally differentiating, one has

$$\frac{dZ}{dp} = \frac{U(Y - Z) - U(Y)}{(p + q)U'(Y - Z)} < 0$$

D. Kahneman and A. Tversky use prospect theory to explain a range of anomalies including this problem, which they attribute to Zeckhauser. They find that where normative theory employs probabilities, human beings employ decision weights that are not linearly related to them. D. Kahneman and A. Tversky, "Prospect Theory: An Analysis of Decision under Risk." *Econometrica* 47 (1979).

10 See D. Kahneman and A. Tversky (1979), op. cit.

11 *Wall Street Journal,* March 17, 1989, p. B12.

12 D. Bell, "Regret in Decision Making under Uncertainty." *Operations Research* 30 (1982): 961–81.

13 H. Grabowski and J. Vernon, *The Regulation of Pharmaceuticals*. Washington, DC: American Enterprise Institute, 1983.

14 W. Booth, "FDA Looks to Speed up Drug Approval Process." *Science* 241 (1988).

15 B. N. Ames, R. Magaw, and L. S. Gold, "Ranking Possible Carcinogenic Hazards." *Science*, 236 (1987).

16 W. K. Viscusi and W. A. Magat (1987), op. cit.

17 L. Roberts, "A Corrosive Fight Over California's Toxics Law." *Science* 243 (1989), 306–9.

18 W. K. Viscusi and W. A. Magat (1987), op. cit.

19 The requirement that motorcyclists wear helmets, despite their known preferences, is justified because it copes with a significant financial externality (i.e., high medical treatment costs and liability costs imposed on others). It also saves their lives at a reasonable cost.

20 L. Roberts, "Is Risk Assessment Conservative?" *Science* 243 (1989).

21 A. Nichols and R. Zeckhauser, "OSHA After a Decade: A Time for Reason." *Regulation* 10 (1986), 11–24.

22 W. K. Viscusi, *Fatal Tradeoffs: Public and Private Responsibilities for Risk*. New York: Oxford University Press, 1992.

23 A. Wildavsky, *Searching for Safety*. New Brunswick: Transaction Publishers, 1988.

24 Data provided by Ministry of Health and Welfare, Japan.

25 The President's Council of Economic Advisors recently concluded that "In many cases, government control of risk is neither efficient nor effective." *Economic Report of the President*. Washington, DC: US Government Printing Office, 1987, 207.

26 If very good information is available, one can employ the more refined measure of cost per quality-adjusted life year (QALY) saved, thus taking into account both the number of person-years gained and their quality. See R. J. Zeckhauser and D. S. Shepard, "Where Now for Saving Lives?" *Law and Contemporary Problems* 39 (1976), 5–45.

27 J. F. Morrall, "A Review of the Record." *Regulation* 30 (1986).

28 A *New York Times*/CBS News Poll asked people if they agreed with the statement "Protecting the environment is so important that requirements and standards cannot be too high, and continuing environmental improvements must be made regardless of cost." Seventy-four percent of the public supported the statement in April 1989, shortly after the Exxon *Valdez* spill, while 80 percent agreed with it two months later. The *Times* concluded that "Public support for greater environmental efforts regardless of cost has soared since the Exxon *Valdez* oil spill in Alaska." *New York Times*, July 2, 1989, 18.

29 W. K. Viscusi (1992), op. cit.

30 Little heed has been paid to innovative suggestions, such as the proposal of M. O'Hare, L. Bacow, and D. Sanderson, *Facility Siting and Public Opposition*. New York: Van Nostrand, 1983, that communities submit negative bids for accepting noxious facilities.

57

Natural Ecology Today and in the Future: A Personal View

Kenneth Mellanby

The term *ecology* was first used by Ernst Haeckel in 1873 for the branch of biology that deals with the interrelationships between organisms and their environment. During recent years the concept has spread rapidly, and today the genuine, scientific ecologist strives to obtain an accurate picture of the world as it really is. The ecologist's competence is to describe the natural world, and to evaluate the effects of man's actions on that world. Thus, the ecologist may study problems arising from pollution and from changes in human and other populations. He may legitimately use his observations as a base for speculations as to future developments, but he must always make it clear that these speculations, particularly when they relate to changes in the distant future, are only speculations and must not be mistaken for facts. Provided that the ecologist does not claim too much for his studies and his extrapolations from them, he can play a very important part in helping mankind to understand the global changes for which mankind is largely responsible.

Unfortunately the term *ecology*, as used by the media and by many of the general population, including most politicians, has been greatly debased. Most publicity is given to those described by J. Valroff in his report to the French Assembly (1985) as "the charlatans of ecology." These include journalists and publicists who always exaggerate the harmful effects of any economic development, and who often give quite erroneous descriptions of the way the environment is affected. Examples of this are given below. The term is also trivialized, so we hear of university students involved in "the ecology of litter." This apparently means the highly desirable process of cleaning up the campus. The term may be misused even by the most worthy environmentalists. Thus, Max Nicholson, the distinguished Director General of the British Nature Conservancy (the government agency responsible for wildlife conservation), says that "ecology is the study of plants and animals in relation to their environment and to one another. But it is also more than that: it is the main intellectual discipline and tool which enables us to hope that human evolution can be mutated, can be shifted on to a new course, so that man will cease to knock hell out of the environment on which his own future depends" (Nicholson, 1970). Because of these debased uses it has been suggested that scientific ecologists should find a new term for their activities. However, the term is still with us, and we must try to see that it is used properly.

No ecologist would claim that every part of the earth is in perfect condition, and that serious damage may not occur from industry and the pollution industry may generate. But he should give us an accurate picture of the real situation, and one that will enable us to identify the causes of damage and so plan to rectify the situation. A false and exaggerated picture makes control of real damage much more difficult.

Thus, there is no doubt that acid rain, resulting mainly from the transformation of sulfur dioxide (emitted by industry and particularly by coal-burning electricity generating stations) to sulfuric acid, may acidify fresh water in areas many hundreds of miles from the source of the pollution. Nearer to these sources, the sulfur dioxide itself may, when the concentration is high enough, directly damage trees and forests. But many reports, particularly from so-called environmentalist organizations, are most misleading. Thus Greenpeace, in publicity material widely circulated in Britain in October 1990, says: "Our chimneys continue to belch out sulphur dioxide just when we have realised that acid rain is killing Europe's forests. Hundreds of millions of trees in Britain are now affected; nearly 70 percent of our oaks are unhealthy, as are 96 percent of yews. A recent UN report says that forest condition in Britain is worse than in any other Western European country."

I was particularly interested in this statement, which has been widely publicized in the popular press because I have, for the last nine years, acted as chairman of the working group on acid rain of the Watt Committee on Energy. The Watt Committee is an umbrella organization of 61 professional institutions (chemists, physicists, biologists, engineers, etc.) in Britain. My working group included 60 other scientists, all actively involved in the study of pollution related to acid rain. In 1989 we published our second report, which I edited, and the following summarizes our findings regarding British trees:

> Surveys in Britain show that today, as in the past, trees may be affected by climatic factors including drought and cold winters, by pests and diseases, and by pollution. In our woods it is always possible to find some unhealthy trees, and to observe premature leaf-fall, particularly after a dry spell. But we can find no evidence that there is serious damage to our woods and forest today that can be clearly attributed to air pollution or "acid rain." All the evidence available suggests that our woods are "normal".[2]

Friends of the Earth has issued reports as misleading as those of Greenpeace. Thus in 1986 they told us that British beech trees were so seriously damaged by air pollution that most of them were doomed. This finding occurred at the end of an unusually hot and very dry summer. The beech trees lost their leaves in August and September, when normally they would remain firmly attached much later in the autumn. Fortunately the next year, 1987, was much wetter, and most of the beech trees appeared normal and healthy. In fact they were very healthy, and the leaves remained green and firmly attached even later than normal, so that when we had an abnormal gale on October 17, the wind resistance provided by the leaves was such that a high proportion of these shallow-rooted trees were blown down. After a dry summer most would have survived.

There is an apocryphal saying attributed to Chairman Mao, to the effect that "prophesy is very difficult, particularly when it concerns the future." This is very true when we consider some of the more alarmist environmental forecasts, some made by otherwise reputable scientists who should have known better. They might at least have been wise enough to forecast situations for a period sufficiently far ahead for them not to be there when they proved to be false. Thus in 1969, Paul Ehrlich published a paper that said that before the year 1980 the world's oceans would be dead, and that the Chinese and Japanese would suffer from starvation from the disappearance of fish and other seafood from their diet (Ehrlich, 1969). It is true that in some parts of the world fish stocks have become scarce, generally due to over fishing, and when this stops they recover. In a recent visit to the Far East, I found no shortage of seafood in the restaurants.

In 1972, "A Blueprint for Survival" appeared in Britain. This was the work of a group of so-called ecologists, and it received wide publicity. It suggested that, because of the damage modern farming techniques were doing to the soil, the productivity of British farms (which had increased greatly during the last 30 years) had now reached its peak, and that serious falls in productivity, with the risk of serious food shortages, would occur in the near future. In fact, productivity has increased as never before, and the yields from farms in Britain, in Western Europe, and in many countries have increased so much that Europe is faced with the problem of overproduction and mountains of unwanted wheat, butter, and meat, and government policy is now aimed at taking as much land as possible out of agriculture to try to get rid of these surpluses.

The "charlatans of ecology" and many environmental publicists are trying to scare the world's population with a false picture of environmental crisis, which they allege is rapidly getting worse and worse. I have quoted certain instances where they have gotten things seriously wrong. The other writers of chapters in this book describe more accurately the situations in their fields of interest, and mostly give a much more hopeful picture of the future.

I believe that the genuine ecologists have the important task of giving an accurate picture of the natural world as it is at the present time. They must draw attention to environmental damage when and if it occurs;

they may then be able to suggest ways in which the situation may be improved. Equally important, they must show where improvements have occurred, and where the human ecosystem has been saved from damage that might otherwise have proved to be disastrous.

REFERENCES

Anon (1972): "A Blueprint for Survival". *The Ecologist*, 2, 1–43.

Ehrlich, P. (1969): "Eco-catastrophe". *Ramparts*, 24–8.

Mellanby, K., ed. (1989): Air Pollution, Acid Rain and the Environment. Elsevier Applied Science Publishers for the Watt Committee on Energy.

Nicholson, E. M. (1970): *The Environmental Revolution*. London: Cambridge University Press.

Valroff, J. (1985): *Pollution atmosphérique et pluies acides*. Paris: La Documentation Français.

PART VII

Conclusion:
From the Past to the Future

58

What Does the Future Hold?
The Forecast in a Nutshell

Julian L. Simon

> No food, one problem. Much food, many problems.
> (Anonymous)

This is my long-run forecast in brief: The material conditions of life will continue to get better for most people, in most countries, most of the time, indefinitely. Within a century or two, all nations and most of humanity will be at or above today's Western living standards. The basis for this forecast is the set of trends contained in this volume, together with the simple economic theory stated in the Introduction.

I also speculate, however, that many people will continue to *think and say* that the material conditions of life are getting *worse*. This assessment will only become more cheerful when (or if) humanity invents or evolves or stumbles into an invigorating set of new challenges that will (a) capture peoples' imaginations and hearts and wills, and (b) replace the intergroup political struggles that now increasingly supplant the struggle against nature for a better material life.

This pessimistic outlook for our *world* does not mean, however, that people will be less "happy" about their own lives; about that I have no prediction. I do not predict how the changed material conditions will affect struggles between good and evil, or how increased affluence will change life in the future emotionally, sexually, socially, or spiritually.

Why should you believe this forecast rather than the forecasts made by the doomsayers? Three reasons:

1 My colleagues and I have been right across the board in the forecasts we have made in the past few decades, whereas the doomsayers have been wrong across the board.

2 Throughout the long sweep of history, forecasts of resource scarcity have always been heard, and – just as now – the doomsayers have always claimed that the past was no guide to the future because they stood at a turning point in history. But the turning–point forecasts have been wrong; there have been ups and downs, but no permanent

reversals. In every period those who would have bet on improvement rather than deterioration in fundamental aspects of material life – such as the availability of natural resources – would usually have been right.

3   I'll bet my money and my reputation on these forecasts, whereas the doomsayers back off from putting their money where their mouths are; they refuse to put either their cash or their names on the line to back what they say. (Indeed, the credibility of the most famous of the doomsayers suffered greatly when in 1980 his group actually did wager on some of his forecasts, which turned out to be wrong at the expiration of the wager period in 1990.) Their unwillingness to wager should call into question whether (a) they really believe the dire forecasts they make, or (b) they just make statements they don't believe in order to scare the public and mobilize the government to do their will.

THE PURPOSE AND THE METHOD OF THE FORECASTS

The recorded past of the human enterprise presented in this volume can help us forecast the future. And sound forecasts can help us evolve wise policies so that the future will be to our liking.

The rationale for statistical evidence far back into the past was given by (or attributed to) Winston Churchill: The further backward you look, the further forward you can see.

A crucial premise for using the past to forecast the future is the constancy of human nature. Along with David Hume and Adam Smith and their Scottish colleagues, and with William James and Friedrich Hayek, I assume that human propensities and appetites, as well as human relationships, will continue to be much the same as they have been; people will change their behavior when changes in conditions give them strong incentive to do so, but only then. This is a fundamental difference from many social commentators of the 1960s who forecast the disappearance of the nuclear family, the decline in the importance of physical beauty and other major changes in the relationship of the sexes, rejection of formal traditions such as the senior prom and evening dress, and the like. This is also unlike the assumption made by Karl Marx – an assumption which, as part of the general theory of communism, might rank as the worst intellectual blunder of all time – that human motivation can be altered so that collective incentives would substitute effectively for individual incentives and private property in inducing hard work and cooperation in impersonal relationships.

It is largely because of these differences in views of human nature that the forecasts to follow differ sharply from the forecasts of most of those who call themselves "futurists." The "futurists" base their predictions mainly on theories drawn from physics, biology, and social science while paying little or no attention to the long time series of history.

It is commonly believed that human activities are less predictable than are phenomena in the physical and biological sciences – that is, that unlike natural-scientific events, human events cannot be forecast accurately. But this common observation is entirely unfounded – and indeed, it is baffling that this is said by anyone who has ever confidently expected a friend to show up for a scheduled date, or gone to a movie whose time and place were advertised in the newspaper. It is true that *some* human events cannot be predicted well – the winner of the next World Series, and tomorrow's interest rate. But some *physical* events cannot be predicted well, either – which side a well-flipped coin will fall on, or what the weather will be a year from today.

It also is commonly believed that long-run prediction is more difficult than short-run prediction. Martin Gardner, famous for his writings about mathematical puzzles and scientific fallacies, says that prediction "is like a chess game. You can predict a couple of moves ahead, but it's almost impossible to predict 30 moves ahead." "If it's so hard to be right about a decade, imagine the howlers in store a century hence," says the *Wall Street Journal*. The *Wall Street Journal*'s columnist Lindley H. Clark put the matter thusly: "Economists have a great deal of trouble predicting the future, and it's unlikely that this unhappy situation ever will change."

It is true that economists cannot predict *short-run trends* of interest rates, exchange rates, and security prices. The incapacity to forecast short-run economic events is well established scientifically, and there is sound reason *in principle* for the incapacity.

It is, however, possible to forecast *long-run trends* with great reliability. Indeed, the most important long-run economic predictions – those I make below – are almost a sure thing, subject only to the qualification that there be no global war or political upheaval.

The method which underlies most of my forecasts is as follows: (1) Array the longest available time series of the phenomenon, and decide whether there is a convincing reason not to consider those data to be a representative sample of the "universe" of experience from which the future experience also is likely to be derived. (2) If there is no compelling reason to reject this past experience as a basis for forecasting the future, consider whether there is a convincing theory to explain the trends it shows. (3) If the long-run data seem relevant, and there is a sound theory to explain them – or even if there is no theory but the data are very many and very consistent – extrapolate the long-run trend as the prediction.

FORECASTS ABOUT HUMAN WELFARE

These are my most important long-run predictions, contingent on there being no global war or political upheaval: (1) People will live longer lives than now; fewer will die young. (2) Families all over the world will have higher incomes and better standards of living than now. (3) The costs of natural resources will be lower than at present. (4) Agricultural land will continue to become less and less important as an economic asset, relative to the total value of all other economic assets. These four predictions are quite certain because the very same predictions, made at all earlier times in history, would have turned out to be right. And sound theory explains these benign trends, as discussed in the Introduction to the book.

Almost as certain is that (5) the environment will be healthier than now – that is, the air and water people consume will be cleaner – because as nations continue to get richer, they will increasingly buy more cleanliness as one of the good things that wealth can purchase. People will probably continue to be worried about pollution nevertheless, both because new sorts of pollutions will occur as new kinds of economic activities develop, and because ability to detect pollutions increases. But the danger of the pollutions that catch our attention will diminish, because we address the worst pollutions first and leave the lesser ones for later, and because our capacity to foresee newly created pollutions in advance will increase. And (6) not only will accidents such as fires continue to diminish in number, but losses to natural disasters such as hurricanes and earthquakes will get smaller, as our buildings become stronger and our methods of mitigating disasters improve.

Perhaps the easiest and surest prediction is that (7) nuclear power from fission will account for a growing proportion of our electricity supply, and probably our total energy supply as well, until it is displaced by some other cheaper source of energy (perhaps fusion). (8) Nuclear power will never be displaced by solar energy using the kinds of technology that are currently available, or by any ordinary development of those kinds of technology.

Can one really make almost surefire long-run predictions? Check for yourself: Though the stock market gyrates from day to day and week to week, its course from decade to decade has almost always been upward. The story is the same in reverse with natural resource prices. Copper, iron, wheat, rice, sugar, and every other natural resource have fallen in price, and therefore risen in availability, throughout the two hundred years of US history, and over the thousands of years of human history wherever records exist. Indeed, the history of civilization is a history of increased knowledge to produce goods more efficiently and cheaply. This goes hand in hand with liberty becoming more widespread, and with increased mobility and communication. All this progress is reflected in the long-run trends of human welfare.

The reader may remark on the absence of forecasts about the ozone layer, the warmth of the surface of the earth, and related atmospheric issues, despite the high current interest in them. There are several reasons: (1) My interest is about human welfare and not about physical conditions – that is, skin cancers but not the ozone layer, agriculture but not the warmth of the earth. And I predict that each of the related human welfare measures will show improvement. If the ozone layer or the warmth of the earth truly does threaten aspects of human life, society will use its large modern capacities to alter those conditions, either in the atmosphere or by protecting individuals directly. (2) The record of the doomsayers in forecasting such matters is atrocious. Remember that only a decade or so before the global warming scare got going – in the middle 1970s – the very persons and institutions that now scold us about taking action to reduce global warming were raising the alarm about global cooling. That is, it took only about a decade for the switch from one scenario of doom to the opposite. The worriers about cooling included *Science*, the most influential scientific journal in the world, quoting an official of the World Meteorological Organization; the National Academy of Sciences worrying about the onset of a 10,000 year ice age; *Newsweek*, warning that food production could be adversely affected within a decade; the *New York Times* quoting an official of the National Center for Atmospheric Research; and *Science Digest*, the science periodical with the largest circulation, writing that

> [T]he world's climatologists are agreed on only two things: That we do not have the comfortable distance of tens of thousands of years to prepare for the next ice age, and that how carefully we monitor our atmospheric pollution will have direct bearing on the arrival and nature of this weather crisis. The sooner man confronts these facts, these scientists say, the safer he'll be. Once the freeze starts, it will be too late. (Cited in Bray, 1991.)

Now ask yourself: How reliable could the evidence for the cooling alarm have been? And in connection with that answer, how reliable could be the evidence for the warming alarm of the 1990s, given that it is mostly composed of exactly the same records over many centuries that made up the evidence for the earlier cooling alarm? (Of course the doomsters err by ignoring most of this history and focusing only on the past few years or decades.) And most important, what would have been the results if we had acted on the recommendations made in connection with the cooling scare?

WILL PROGRESS CONTINUE?

Some wonder: How can we be sure that scientific and technical progress will continue indefinitely? Related to this question is another: What will be the rate of economic growth in the future? Elsewhere (Simon, 1992, chapter 19) I conclude that no sensible answer about future rates of advance is possible in principle, in considerable part because of the inherent impossibility of comparative measurement of the value of progress in science from one period to the next.

But uncertainty about the rate of advance in technology does not imply uncertainty about the direction of the future of humankind. Our future material welfare is already assured by our knowledge of how to obtain energy from nuclear fission in unlimited quantities at constant or declining cost, even if no other source of energy is discovered until fissionable material runs out at some almost infinitely distant time. And energy is the only strong constraint on the supplies of all other raw materials.

Assurance about our raw-material future does not imply absence of need for more technology in the short run or the long run, however. There are, and always will be, endless ways to improve human life, and plenty of pressing problems to challenge us. But we can confidently face the future without worrying about threats to the end of civilization from "over-consumption" and raw-materials shortages that technology is unable to deal with.

The rate of technical advance with respect to raw materials is not crucial for the long run, because the world's raw-material problems have been resolved for all time with technology already developed. (In brief, if energy is sufficiently cheap, all other raw materials can be obtained at low prices, because energy allows extraordinary transformations of many kinds (see Goeller and Weinberg, 1978). And nuclear fission with the breeder – and even more so, nuclear fusion if it becomes practical – provides an unlimited amount of energy at constant or declining cost forever (or at least for billions of years beyond the horizon of any conceivable contemporary social decision).

Space for living and working is the only other resource requiring attention here. Construction technology now provides us space in huge quantity relative to the amount used until now, by building multistory buildings and by heating and cooling areas of the earth heretofore considered unusable for human habitation because of their extreme climates. If we wish to imagine a bit further, the sea and outer space can provide vast additional living space, and even now they are not impracticably costly. An evaluation of future technical advance might tell us how fast the costs of space and energy will fall, but those rates are not crucial to any decision about population growth. For more details, see Barnett and Morse (1963), or Simon (1981, and second edition forthcoming).

Of course it was not always so. In past eras, natural resources constrained human progress. In the present and immediate future, too, additional technology can improve the standard of living more rapidly than otherwise. Nothing said here implies that our future *economic* problems have been solved, or can be solved for all time. But the problems are less and less those of physical resources.

This leads us to ask what kinds of needs will make technical advance important in the very long run. Health and life come first to mind, of course. But if we accept the contention that our bodies inevitably wear out around age 90 no matter how effectively individual diseases are prevented or controlled, then we are already almost as far as we can go, without much possibility of further advance (see Fries and Crapo, 1981). Of course, biogenetics might engineer different bodies for us, making us a different sort of species. It is not obvious that we would consider this an advance, however, and it is too complex and controversial a matter to discuss here.

We certainly would value advances that would help us live our lives with more serenity, more excitement, and more enjoyment, in greater harmony with our fellows. We also would greatly value advances that would improve teaching and learning in such fashion that individuals could more fully take advantage of the talents with which they are born, in order to make a greater contribution to others and to live more satisfying lives than otherwise. Science may be able to help. But such knowledge is likely to come from fields other than physical science. Once we enlarge the concept of technology to include social and psychological knowledge, we move to a different sphere of discourse – one in which, for example, the concept of "breakthrough" must have a very different meaning than it has in the physical sciences.

The argument, then, boils down to this: The crucial contributions to living that advances in productive technique might make in the long future differ fundamentally from those that it has made in the past and in the immediate future. We now possess knowledge about resource locations and materials processing that allows us to satisfy our physical needs and desires for food, drink, heat, light, clothing, longevity, transportation, and the recording and transmission of information and entertainment. We can perform these tasks sufficiently well that additional knowledge on these subjects will not revolutionize life on earth. It still remains to us to organize our institutions, economies, and societies in such fashion that the benefits of this knowledge are available to the vast majority rather than a minority of all people. And our desires for (among other things) leisure, wisdom, love, spirituality, sexuality, adventure, and personal beauty are quite unsatiated, and perhaps must always be so. But the sort of advances in productive knowledge that in the past brought us the possibility of satisfaction for our physical needs cannot sensibly be measured in a fashion comparable to future advances in beneficial knowledge, given our present skills in measurement. Therefore, we should not concern ourselves about the rate of future advances in physical knowledge compared with the rates in the past.

Though I predict that the future of physical discovery will not be like its past, I do not believe that we are at a turning point now. The shift I describe has been going on for at least a century and perhaps much longer, depending on how you view it, and should continue indefinitely. There is no discontinuity to be seen here.

This is not an argument for neglect of scientific and engineering research, of course. I hope that we vigorously continue to increase our technology, and thereby reduce the cost and increase the distribution of the means of satisfaction of our physical needs as, for example, in agricultural research. Furthermore, science is a great human adventure, worthwhile for the observers as well as the participants; space exploration may serve as an example. And even if we do not "need" the technical advances that may occur in the future, we may well find that they are worth far more to us than we would individually be willing to pay for their fruits, in which case there is justification for government support of such activities; space exploration, with the economic benefits it already has begun to provide, may again serve as an example.

One more qualification: Some writers such as Robert Higgs (1987) believe that governments will increase the sizes of their roles in modern economies and societies, which in turn will choke economic progress and even reverse it. In contrast, such writers as George Gilder (1984) and Richard McKenzie and Dwight Lee (1991), believe that technical progress and competition will force governments to play a smaller role. In choosing between these assessments about the future of government there is little historical experience to rely upon. It seems to me relevant, as Stanley Engerman's essay in this volume documents, that slavery has diminished over the centuries. Less reliably, the evidence seems to point toward more freedom and democracy throughout the world since World War II, but the record is scrappy. Hence in this matter we are forced to rely heavily on analysis – that is, the combination of various theoretical arguments and selected supporting factual evidence. And in my judgment, the analysis of Gilder and McKenzie–Lee is more convincing, given the evidence since the mid-1980s.

The two main reasons that I think government will become less powerful are these: (1) The rapid movement across borders of capital and information facilitates sharp competition for resources among firms in different countries, which reduces the power of governments to tax captive businesses and individuals. (2) These new technologies of movement strengthen small enterprises relative to large enterprises, continuing the process that began when electricity became available as a substitute for steam or water power, and the truck became available as a substitute for the railroad. Smaller enterprises are harder for governments to control – both their locations and their activities.

Future fertility in modern countries is another important element about which we lack historical evidence. Will affluent couples continue to have children at a rate that will not increase the population? Or will procreation in rich countries increase sufficiently, while the proportion of the population in now poor countries falls, to continue the long-run expansion of population on Earth and perhaps elsewhere? More about this below.

WHERE ARE WE IN THE LONG SWEEP
OF HUMAN EXISTENCE?

People alive now are living in the midst of what may well be the most extraordinary three or four centuries in human history. The "industrial revolution" and its technical aftermath – even including the spectacular rise in living standards for most human beings from near subsistence to the level of today's modern nations – is only part of the upheaval. The process has already been completed for perhaps a third of humanity, as described in various chapters in this book, and within a century or two (unless there is a holocaust) the rest of humanity is almost sure to attain the amenities of modern living standards; the worst holocaust imaginable could only delay the process by a century or so.

The most spectacular development, and by far the most meaningful in both human and economic terms, is the revolution in health that we are witnessing in the second part of the twentieth century. Barring catastrophic surprises in the first half of the twenty-first century, most of humanity will come to share the long healthy life that is now enjoyed by the middle-class contemporary residents of the advanced countries.

The technical developments of the past two centuries certainly depended on earlier discoveries. But the knowledge that emerged before the last two centuries was only infrastructure. Until the most recent generations, most people could not observe the effects or gain the benefits of this progress in their own lives. Now we are reaping the full fruits of that earlier investment.

The spreading of a high level of living will be speeded by another phenomenon that can be predicted for sure: increased migration from poor to rich countries. The lure of a higher standard of living pushes migrants from their native countries and pulls them to richer places. And the felt need in the richer countries for youthful persons in the labor force to balance the ever greater concentrations in the older age cohorts fuels the deman for them. One can see the drama of this process in the youthful medical and custodial staffs in big-city hospitals who came from abroad to tend to the aged natives and the veterans of earlier migrations.

Might there not be even more and faster and more radical change in future? In human terms, I doubt it. Life expectancy and child mortality cannot fall much faster unless we change genetically. Concorde-like supersonic speed of travel demonstrably does not matter very much (and the trip to and from airport will take a long time to speed up). Communications cannot become much faster, many being at the speed of light now. The distances one can travel – to the planets and beyond – will eventually increase greatly, and this may alter life significantly, though I cannot imagine how.

The only impending shortage is a shortage of economic shortages. (For investors this implies that profitable investments lie in the economic sectors that profit from affluence. Stay away from commodities; their prices will fall, as they have been falling for hundreds and thousands of years. Sell marginal farm land; it will become less valuable as productivity per acre increases. Buy acreage that was wasteland in the past because it is mountainous or inaccessible; it will now become more profitable for those very characteristics, its recreational value.)

In *The Next Two Hundred Years*, Herman Kahn, my coauthor of the predecessor volume, decades ago foresaw this four-century emergence from isolated subsistence farming in which most of humanity has lived throughout history (Kahn, Brown, and Martel, 1976). Herman was frequently "accused" of being an optimist. He would reply, "I'm not an optimist, I'm a realist." Indeed, it is realistic to forecast improving long-run material trends for humanity, forever.

THE FUTURE GROWTH IN EDUCATION AND OPPORTUNITY

Chapter 21 shows the astounding increase in the amount of education that young people in the poorer countries of the world have been acquiring. This education implies an increase in opportunities to use their talents for their own and their families' benefits; the realization of these talents benefits others in society as well as those persons. This trend is to me one of the most important, and one of the most happy, of all trends in the human enterprise. One can see the results in the nameplates on professors' doors in departments of computer science and chemistry (for example) in universities all over the United States – Asian and African names that would not have been there a decade or two ago. Less and less often will people of genius and strong character live out their entire lives in isolated villages where they cannot contribute to civilization.

Quantitative evidence for the spread of education and of access to knowledge can be seen in the statistics of world education in chapter 21. But the most compelling evidence is found in the stories of individuals such as William Owens (see his astounding autobiography), who grew up just after the turn of the century in Pin Hook, Texas, so poor that he could not get more than a few months of schooling in each of the few years when he got any at all, and could obtain literally nothing to read–some old newspapers pasted onto the walls of a shack in which he lived, to keep out drafts, were the most he could find for a while. By the time Owens had miraculously become a professor of literature and folklore at Columbia University, access to reading material had become universal in the United States, and there were good schools and even wonderful junior colleges within the reach of just about every American.

Yes, Gutenberg's invention of printing was crucial. But it was the rise in economic productivity that has brought the benefits of that invention within reach of humanity at large. Aside from our victory against premature death in the past century, this spread of education and knowledge may be the most important alteration of all time in human welfare.

As with other aspects of the globalization of a modern standard of living, the process of providing enough education to liberate all young people and to empower them to exercise their talents to the fullest is far from complete. One can still see children sharing rickety desks and scarce books in a near-subsistence Colombian fishing village, just two miles from a busy international airport. Yet the situation there is better than it was just a few years ago when there was no school at all. We can be confident that a century from now scenes like that poor school will be few and far between. The children will have become too valuable to others to allow them to grow up that way.

Increased education does not imply that the public will be more enlightened on crucial issues than now. For example, I do not expect that most people and their political leaders will be much more in favor of truly free trade than now; the idea is simply too counter-intuitive for wide belief; for 200 years the public has shown that it does not rapidly learn this idea. I do expect that trade will become freer anyway, however, simply because of the pressure of international competition among countries. This is one of the many issues where we can expect that the inherent advantages of certain sorts of regimes – free trade, nuclear power, personal freedom – will gradually win against ideas and desires that run in the opposite direction. Ideas do have consequences, even if they are bad ideas, but the counterproductive consequences are limited and are eventually overcome by economic reality.

THE LIKELIHOOD OF CATASTROPHIC DISEASE

What about the possibility of a catastrophic disease that could devastate humanity? Many thoughtful people worry that increased human mobility might raise the chance of global disease transmission.

Before considering this possibility, we should note that even a disease of greater magnitude than has ever occurred would only reverse contemporary human progress for a relatively short time. The demographic and economic losses from the worst disaster in history – the Black Death, which killed perhaps a quarter of the population of major European countries – had been recovered after only a century or so; the population size and the standard of living soon were back almost to where they would have been otherwise.

Furthermore, even a disaster of unprecedented scale – say, a devastation of 90 percent or even 99 percent of humanity – would not have permanent effects. The only essential element for a modern economy and society is the knowledge that resides in libraries. With the books housed there, a small number of people could create what they would need in a matter of decades.

The Necessary Characteristics of a Catastrophic Disease

These are the characteristics that a disease would require to be catastrophic: (1) There must be a "vector" – a mechanism that spreads the disease – of great rapidity and efficiency. (2) The disease must kill or debilitate a large proportion of those who become infected. (3) Most important, the disease must show symptoms and be diagnosed only many years after people become infected.

The importance of a time lag between the infection and the appearance of symptoms is that humanity now has enormous capacity to protect itself against almost any conceivable vector once the disease is known and the vector is sought. The causes of new diseases of the sorts that have occurred in the past – bacterial, viral, environmental, even genetic – nowadays can be determined quickly; it would seem that most imaginable (and even unimaginable) diseases also would reveal their causes to scientists within a very few years. (The main mode of transmission of AIDS was discovered within weeks or months of the diagnosis of the first cases.) And once the vector has been identified – whether it be by air or water or insect bite or whatever – nowadays it is within our capabilities to block that transmission effectively. (Sexual transmission is the most difficult to prevent, but because a large proportion of the population is either monogamous or not sexually active, a sexually-transmitted disease could not kill a majority of the human population, let alone almost everyone.) Therefore, only if there is a long lag could a disease transmitted through the air or water manage to infect a large proportion of humanity. (If the elapsed time between being infected and showing symptoms were not so long in the case of AIDS, it would have done less harm.)

Our Growing Ability to Deal with Catastrophic Disease

It is also relevant that our capacity to learn quickly and deal rapidly with new diseases has expanded enormously over the centuries and the last few decades. The first AIDS case was diagnosed only a decade before the time of writing this in 1992. And we should note the beneficial spin-offs of new knowledge due to the onset of new diseases that will help check diseases in the future.

ABSOLUTE AND RELATIVE PROGRESS

The predictions in this chapter are for the *absolute* progress of humanity as a whole, in keeping with the spirit and title of this volume. The public and the press are often – much too often, in my view – interested in the *relative* achievements of particular countries and groups, as discussed in the Introduction to the volume. The historical trends that are the foundations for the predictions in this chapter do not support solid predictions for particular countries and groups except in one respect: In light of data in the appendix to chapter 15, communities may be expected to converge in their standards of living in the long run.

Nevertheless, I hazard a few weak predictions about relative progress: With respect to Eastern Europe, the former Soviet Union has the benefit of a large educated class. But even so, it and the other Eastern European nations will need decades to create the legal and institutional infrastructure to support solid economic growth; this is the key element, in my view. Indeed, I expect the former Soviet Union republics to lag far behind several of the other Eastern European countries.

A hundred years from now China or India could be the leading nation in the world. If freedom wins out fairly soon in China, and if people in those two countries are reasonably wise about reconstructing their economic institutions to enable enterprise to flourish, a century could be more than enough time for them to nearly catch up economically with the leading countries in the world. After all, Hong Kong caught up more than halfway in less than half a century, and events are likely to move even faster in the future. If so, the sheer demographic weight of China and India is likely to dominate all other countries. They are then likely to exert leadership in various international forums.

Bets are off if India and China break up into smaller nations, of course. To my mind, there is little bad about such breakups, and it may benefit the individuals in the smaller units. But as Europe has shown, such splitting does not make for world dominance. Perhaps with enough splitting, no nation will be a super power. That might be the best world of all.

(Is international leadership a good thing? Maybe not. But if one does want the leadership role for the United States or another country, one should wish that its population increases rapidly rather than slowly, because that may be the only way to delay India's and China's surge to leadership.)

A shocking last prediction: The economic reasons for war diminish as land becomes less important relative to other assets and therefore less worth spending money and lives to annex. I predict that this trend will reduce the incidence of war in the long run, though I agree with writers who argue that the economic element is not the only crucial determinant of war.

CONDITIONS GETTING BETTER, PERCEPTIONS GETTING WORSE

Even though the material conditions of life have been getting better, many people believe that conditions have been getting worse. The Introduction began by illustrating this process. The title of a news article conveys the flavor of the matter: "Down in the Dumps – The Glooming of America: If the numbers aren't so bad, why is the country feeling so lousy?" (*Newsweek*, January 13, 1992).

A curiosity: My mother's life spanned almost half of the past two-century revolutionary period. She was born in 1900 and saw her only child saved from death at age seven by the first new wonder drug. In her eighties she knew that her friends had mostly lived extraordinarily long lives. She recognized the convenience and comfort provided by such modern inventions as the telephone, air conditioning, and airplanes. And yet she disagreed when I said the conditions of life had markedly improved. When I asked Mother why she still thought things have gotten worse, she replied: "The headlines in the newspaper are all bad."

Journalism will get worse for at least a while into the future, as it covers a smaller proportion of its traditional stories – fires, politics, and local events – and covers more events for which its traditional techniques are not fitted – scientific developments, especially concerning the environment, social scientific trends, and other matters that require more than first-hand observation and individual interviews. Where this will end is not clear to me. Training of journalists in nontraditional techniques may be in the cards, but probably will never cure this problem because it is inherent in covering stories quickly as "news" without historical digging.

Scientific research will become more and more involved in advanced technique, which will mean that fewer exciting problems will be worked on, and science will become a less attractive field for creative persons.

We will always find grounds for worry. Apparently it is a built-in property of our mental systems that no matter how good things become, our aspiration levels ratchet up so that our anxiety levels decline hardly at all, and we focus on ever smaller actual dangers. Parents manage to worry about their kids' health and safety even though the mortality of children is spectacularly lower than in prior decades and centuries. And orthodox Jews and Muslims in the United States continue to worry about whether their food is ritually pure even though the protections against ritual contamination are remarkably better than in the past. (Once upon a time orthodox Jews said that "A Jew eats a small pig every year without knowing it." Nowadays, with plastic wrapping at the factory, and the microscopic examination techniques of modern science, the level of purity is much higher than in the past. But the level of concern does not abate.)

Remember please that there are fewer life-threatening disasters from decade by decade. Some evidence for this is found in statistics of accidents (see chapter 9). Other evidence may be seen in the headlines of newspapers, which less and less frequently concern earthquakes, fires, and floods, and more and more are about political and social issues. Surely this trend will continue for the foreseeable future. Fewer and fewer of our struggles will be against nature, and more and more will be battles of one group versus another. This suggests that until we find new challenges – such as terraforming other planets – we will be caught up in zero-sum issues that are less likely to satisfy the spirit than are the battles against nature that society tends to win.

THE BELIEF IN PROGRESS

A hundred years ago belief in continued progress and a bright future was commonplace. Why not now?

One reason for the loss of belief in progress is the focus of the media on bad things happening, a phenomenon discussed in the Introduction and the section above. But another reason might well be the shattering blow to people's confidence that World War II inflicted. As the novelist Stefan Zweig wrote in his autobiography, *The World of Yesterday*:

> Against my will, I have witnessed the most terrible defeat of reason and the wildest triumph of brutality in the chronicle of the ages. Never – and I say this without pride, but rather with shame – has any generation experienced such a moral retrogression from such a spiritual height as our generation has. . . .
>
> In the short interval between the time when my beard began to sprout and now, when it is beginning to turn gray, in this half-century more radical changes and transformations have taken place than in ten generations of mankind; and each of us feels it is almost too much. (p. xviii, University of Nebraska Press/Viking Press, 1943/1964)

Since Zweig wrote those lines during World War II, just about every change important to humankind has been for the better–and the trends have been dramatic in their extents. But this apparently has not been enough to convince people that World War II was not an anomaly and is not likely to happen again in the foreseeable future.

But this may be wrong, too. A generally thoughtful columnist recently wrote: "We have lost our conviction that things will always get better" (Richard Cohen, "Progress Ain't What It Used to Be." *Washington Post Magazine*, October 13, 1991, p. 7). Maybe people *never* had the conviction that things would always get better.

THE CRUCIAL NEED FOR IMPORTANT CHALLENGES

Frank Knight once wrote that what people most seek is not to have their wants fulfilled, but bigger and better wants. This may not be true of those who are cold and hungry at the moment, or of adolescents driven wild by their hormones, or of harried parents trying to be in three places at once. But it certainly is true of many middle-class people who have acquired the physical appurtenances they consider they need, and who have succeeded in their own eyes in their professions. And it is true of many young people who do not see on their horizons exciting challenges to their talents and ambitions – no crusades or jihads to defeat an infidel, no new continents to discover, no frontier from which to hack out a homestead. As James Buchanan put it, "I do not envy the youngsters in modern suburbia, who lack a sense of scarcity along any [material] dimension" (1992, 23).

Anecdotal history shows the need for challenging problems. I heard Dutch people say about themselves in the 1990s that they are gloomy now but that in the 1950s, when Holland was still rebuilding after the war, the national mood was very much more cheery. Today every corner of the Netherlands seems already to have been improved and neatened. And there is discussion there of removing from agricultural use some of the fertile land that earlier was taken at such great effort from the sea, because the agricultural produce from that acreage is no longer needed. The Netherlands' worst problem, they say, is the disposal of millions of tons of pig excrement that they have not yet found uses for. This is not the sort of problem that fires the imaginations of the young.

More and more, as affluence spreads throughout humanity, our species' biggest problem will be a lack of satisfying challenges – opportunities to sacrifice, to make a large contribution to a larger cause, to be part of a team, to achieve nobility – truly William James's "moral equivalent of war." That's what the environmental activities now seem to offer. But they are flawed because they are mainly retentive, rather than creative (a subject I discuss in more detail in Simon, 1981; 1995).

A vast expansion of the human population could present such a challenge and might lead to buccaneering expeditions to conquer outer space. But the future course of fertility (upon which population growth will depend almost solely, because further mortality reduction will not be great) is not predictable as of now.

Only such new challenges, I believe, can prevent us from descending into C. S. Lewis's vision of the netherworld: "We must picture Hell as a state where everyone is perpetually concerned about his own dignity and advancement [and I would add, the dignity and advancement of the racial, religious, ethnic, national, and other groups one is a member of], where everyone has a grievance, and where everyone lives the deadly serious passion of envy, self-importance and resentment." (From *The Screwtape Letters*, quoted in *Washington Post*, January 17, 1993, C4.)

In the past, family resources constrained family size. But the evidence suggests that as average income continues to grow, it is likely – though by no means certain – that fertility and money income are becoming increasingly separated. That is, family size at incomes much higher than now in the USA might still be of the same order as in average American or European families today. Indeed, the family size of those on the highest income today is not particularly large. Families in the super-wealthy future may be bigger, or smaller, than now. Income is not likely to be a good predictor of what would happen if incomes climb to the point at which additional income does not matter. And if income and fertility do become increasingly separated, the biggest changes in fertility may then be associated with one or the other sort of fear: either the fear of depopulation, as in the 1930s, or the fear of overpopulation, as in the late 1960s and into the 1970s, in the United States and China. The overall result may be a series of rises and falls in fertility, triggered by, and in turn triggering, these fears. This sequence might bear little or no relationship to basic economic conditions, just as in fact there was no basic economic difference in the two periods that would explain the depopulation and overpopulation fears in the 1930s and 1960s.

It may be that *relative* income is the key factor, as Richard Easterlin (1968) has argued. If so, fertility may continue to be affected by fluctuations over time and by the shape of income distribution. Only the far future will give the answers to these speculations, however.

WHAT WILL DETERMINE THE FUTURE?

> The history of mankind is the history of ideas. For it is ideas, theories, and doctrines that guide human action, determine the ultimate ends men aim at, and the choice of the means employed for the attainment of these ends. (Ludwig von Mises, quoted in *The Freeman*, Feb 1993, p. 42.)

Though I greatly admire much of von Mises' economics, I consider the above remark as exemplifying the megalomania of the intellectual class. Curiously, von Mises agrees on this point with J. M. Keynes. And Hayek in 1944 said, "I agree with Lord Keynes that 'the ideas of economists and political philosophers, both when they are right and when they are wrong, are more powerful than is commonly understood. Indeed, the world is ruled by little else' " (Hayek, 1991, p. 36).

All of us like to think we are important – intellectuals certainly as much as anyone. And yes, ideas can be powerful in the intermediate run – witness the disaster caused by Marxist thinking for 70 years in Eastern Europe and China. But in the longer run, the elemental forces of peoples' desires to carve out a good living for themselves and for their families, to have children and raise them happy and well-educated, to employ one's talents and energies and to enjoy their fruits – these forces will eventuate in government policies that allow people these fundamental freedoms. There probably always will be temporary reversals and reversions to totalitarianism for a while. But we can hope that in an ever-ramified world where people can move ever more freely, these reversions will be of shorter duration and of lesser magnitude.

If ideas were all-important, the horde of more-or-less sensible economists descending (literally) on Eastern Europe should be able to put things straight in a hurry. But we will see that decades of evolutionary institutional rebuilding will be necessary; there is no single set of grand ideas that can replace that necessary work.

All that is required of philosophers and economists is that they manage to prise from rulers a small chink of living and operating room for individuals, and some degree of rule of law. All the rest will be done by individuals themselves, without recourse to any grand ideas. Indeed, this is the very doctrine of von Mises and Hayek – and Friedman, Smith, and Hume. In this rare respect, Hayek and von Mises contradict themselves. They all know the power of black markets even in the teeth of the most fearsome sanctions. Such markets are not driven by grand ideas, but by human desires and economic incentives.

The more that humanity progresses, the less these ideas will matter, because the variety of régimes offered competitively by the various countries, and the easy mobility among them, will provide the necessary opportunities for entrepreneurs and other talented persons. The need grows less for philosophers of liberty to eke out a bit of freedom for society from the clutches of politicians intoxicated with the chance to put into operation their delusions of improvement by central control (theirs). People will achieve it anyway.

On the other hand, with greater progress comes greater freedom from pressing survival needs, which in turn enables people to indulge themselves in foolish, irrational, and counter-productive thinking, and can lead to mass movements that impede progress. (We might note that farmers and small retailers, who are of necessity in exceedingly close touch with economic reality, are – at least I so think – relatively free of foolish economic thinking, if only as the result of a Darwinian process.)

THE BRIEF FORECAST AGAIN

Whatever nature has produced that we make use of – food, oil, diamonds – humankind now can also produce, and faster than nature. An expectancy of health and a standard of living higher than that which any prince enjoyed 200 years ago is the birthright of every middle-class and working-class person in developed countries, and of most people in poverty as well. What is still to come is to bring these material gains to all groups of humanity. That may take half a century or a century. Yet that benign outcome may be predicted with a high degree of likelihood – a happy vision, indeed.

The future for the correct perception of these trends looks bleak, due to their portrayal in the press, however. The techniques that journalists use so well to cover fires and local politics do not work well for matters that go beyond first-hand observation. This includes scientific matters, as illustrated nowadays by environmental questions. And the bad-news bias in journalism transforms every story into a negative one, even if the underlying facts are positive. This in turn leads the public to think that conditions in general are getting worse. The press then reports this as pessimism. All this could have increasingly dire effects upon the public mood.

As to the non-material aspects of human existence – good luck to us.

REFERENCES

Barnett, Harold J., and Chandler Morse (1963): *Scarcity and Growth: The Economics of Natural Resource Availability*. Baltimore: Johns Hopkins.

Bray, Anna J. (1991): "The Ice Age Cometh." *Policy Review*, Fall, 82–4.

Buchanan, James (1992): *Better Than Plowing, And Other Essays*. Chicago: University of Chicago Press.

Easterlin, Richard A. (1968): *Population, Labor Force, and Long Swings in Economic Growth*. New York: National Bureau of Economic Research.

Fries, James F., and Lawrence M. Crapo (1981): *Vitality and Aging*. San Francisco: W. H. Freeman and Company.

Gilder, George (1984): *The Spirit of Enterprise*. New York: Simon and Schuster.

Goeller, H. E., and A. M. Weinberg (1978): "The Age of Substitutability," *Science* 191, 683–9.

Hayek, Friedrich (1991): "On Becoming an Economist." In Hayek's *The Trend of Economic Thinking*. Chicago: University of Chicago Press.

Higgs, Robert (1987): *Crisis and Leviathan*. New York, Oxford: Oxford University Press.

Kahn, Herman, with William Brown and Leon Martel (1976): *The Next 200 Years*. New York: Morrow.

McKenzie, Richard B., and Dwight R. Lee (1991): *Quicksilver Capital*. New York: Free Press.

Owens, William (1986): *This Stubborn Soil*. New York: Vintage.

Simon, Julian L. (1981): *The Ultimate Resource*. Princeton: Princeton University Press. Second edn. forthcoming 1995.

—— (1992): *Population Growth and Economic Development in Poor Countries*. Princeton: Princeton University Press.

Weisbrod, Burton A. (1962): "An Expected-Income Measure of Economic Welfare." *The Journal of Political Economy*, August, 355–67.

List of Figures

1.1   Infant mortality rate, total and by race, USA, 1915–89

1.2   Ratings of own life, country, and economy, 1959–89

1.3a   Trends in life expectancy over the millennia

1.3b   Life expectancy, England, Sweden, France, and China, 1541–1985

1.4   Child mortality, 1751 to present

1.5   Percent of population employed in agriculture, Great Britain, 1600 to present

1.6a   Wheat prices indexed by consumer price index, USA, 1801–1900

1.6b   Wheat prices indexed by wages, USA, 1801–1990

1.7a   Emissions of major air pollutants in the USA, 1940–90

1.7b   Air quality trends in major urban areas, USA (number of PSI days greater than 100)

1.7c   Pollutants in the air, USA, 1960–90

1.7d   Smoke level and mean hours of winter sunshine in London, 1923–84

2.1   Life expectancy, England and Sweden, 1541–1985

2.2   Life expectancy around the world, 1950–5, 1985–90, and gains

2.3   Estimation of world population, 1600 BC to 2000 AD

3.1   Infant mortality rate, Sweden, 1751–1988, with smoothed median trend

3.2   The relations between probabilities of dying in infancy and between ages one and five in the nineteenth and twentieth centuries by sex: the Netherlands, France and Sweden

3.3a and 3.3b   Indices of mortality rates in infancy and childhood by major cause of death group; males, England and Wales

3.4   Infant mortality rate, Chile, 1901–87, with smoothed median trend

3.5   Trends in the probability of dying by age five in the developing world; by region, 1950–85

4.1   Causes of death, percent distribution, model populations

4.2   Respiratory tuberculosis, standardized death rates, 1861–1964

4.3   Other infectious disease, unstandardized death rates, 1861–1964

5.1   Crude death rates in France and England, 1740–1870

5.2 The impact of crisis mortality on the average crude death rate in England

5.3a and 5.3b A comparison of the relationship between body height and relative risk in two populations. 5.3a: Relative mortality rates among Norwegian men aged 40–59, between 1963 and 1979. 5.3b: Relative rejection rates for chronic conditions in a sample of 4,245 men aged 23–49 examined for the Union Army

5.4 The relationship between BMI and prospective risk among Norwegian adults aged 50–64 at risk between 1963 and 1979

5.5 Estimated average final heights, European males

5.6 A comparison between the trend in the mean final height of native-born white males and the trend in their life expectation at age 10 a_{10}^0 (height by birth cohort; a_{10}^0 by period)

6.1 Persons reporting limited activity due to health, by age; USA, 1957–89

6.2 Persons reporting limited activity due to health, by degree of limitation

6.3 Annual days of restricted activity, per person, by age

6.4 Annual bed disability days per person, by age; USA, 1961–89

7.1 Causes of death, USSR and the West, able-bodied ages, by sex, 1987

7.2 Crude death rates, by sex, USSR, 1958–90

7.3 Life expectancy at birth, by sex, USSR, 1897–1989

7.4 Mortality by cause of death (rate), USSR, 1985–9

7.5 Infant and maternal mortality rates, USSR, 1913–89

8.1 Homicide rates since 1850

8.2 Suicide rates since 1750

8.3 Age-standardized suicide rates, selected countries, late 1980s

8.4a Suicide trends in the USA

8.4b Suicide trends in the USA

8.5 Deaths from suicide, homicide, and execution in England since the thirteenth century

9.1 US accidental death rate

9.2 Accidental death rates by age, 1903–12 and 1989, with percent change

9.3 Accidental death rates by cause; motor vehicle, work, and home

9.4 Death rates from catastrophic accidents

9.5a Accidental death rates, other industrialized countries

9.5b Accidental death rates, other industrialized countries

10.1 Number of cigarettes consumed annually per adult (age 18 and older): USA, 1900–88

10.2 Percent of adults smoking regularly, by sex, USA, 1945–85

10.3 Cigarette smoking quit ratio, by sex and race; USA, 1965–88

11.1 Apparent absolute alcohol consumption, USA, US gallons per capita aged 15 and over

11.2 Apparent absolute alcohol consumption, Germany, US gallons per capita aged 15 and over

11.3 Apparent absolute alcohol consumption, France, US gallons per capita aged 15 and over

11.4 Apparent absolute alcohol consumption, United Kingdom, US gallons per capita aged 15 and over

12.1 Agricultural labor force, England, the Netherlands, and France

12.2 Agricultural labor force, USA

12.3 Agricultural labor force, Belgium and Finland

12.4 Agricultural labor force, Japan, Italy, and Spain

12.5 Agricultural labor force, Brazil and Mexico

12.6 Agricultural labor force, Turkey, Egypt, and Colombia

12.7 Agricultural labor force, Philippines, Indonesia, and Thailand

12.8 Agricultural labor force, India

13.1 Two estimates of economic growth in Britain 1675–1900

13.2 Longevity, mortality, and height in Britain, 1700–1988

13.3 Trends in output of beer in four European countries

13.4 Growth in cotton consumption in four European countries

13.5 Growth in sugar consumption around the world

13.6 Trends in height in selected developed countries

13.7 Indices of real GDP per employed person, purchasing power parity exchange rates

14.1a Personal consumption expenditures, per capita, constant dollars, 1900–87

14.1b Personal consumption expenditures, per capita, constant dollars, 1900–87

14.2 Comparative consumption per capita, selected countries, 1985

14.3 Percent of families with consumption changes, 1900–79, USA

15.1 Total labour and factor productivity, USA, 1840–1980

15.2 Increasing speed of digital computers

15.3 Productivity growth in spinning cotton thread

15.4 Improvement in output per man-hour from mechanization

15.5 Light output per watt for different devices

15.6 Illumination per watt by different technologies

15.7 Top speed of ground transport of humans, 1784–1967

15.8 Top speed of air transport of humans, 1905–1965

15.9 Comparative levels of productivity, 1870–1987 (US GDP per man-hour = 100)

15.10 Comparative levels of productivity, 1870–1987 (US GDP per man-hour = 100)

15.11 Productivity levels as percent of USA, France, Germany, and Japan, 1950–90

15.12 Overall labor productivity, 1989

15.13 Number of human genes mapped, 1973–92

17.1 Life expectancy at birth, USA, by race, 1900–89

17.2 Infant mortality rates, USA, 1915–89, by race

17.3 Black income per capita, 1867–1990

17.4   Unemployment rates in the USA, by race, 1890–1990

17.5   Percent owner-occupied housing units, USA, by race, 1890–1990

17.6   Ratio of black-to-white family income, USA, 1967–90

17.7   Lynchings and prevented lynchings, by race, USA, 1889–1932

18.1   Shares of wealth held by top wealth holders in America, 1647–1969

18.2   The Kuznets curve: international 60-country cross-section from the 1960s and 1970s

18.3   The Kuznets curve: historical time series from five European countries and America

19.1   Unemployment rates in the USA, 5-year averages, 1870–1989

20.1   Index of real housing prices (1967 = 1.0)

20.2   Index of real structure prices (1967 = 1.0)

20.3   Per capita real housing consumption

20.4   New privately owned one-family houses completed 1970–91

21.1   Enrollment in public elementary and secondary school, USA, 1850–1990, proportion of children ages 5–17, by race

21.2   Amount of schooling per child, USA, 1870–1990, public elementary and secondary school

21.3   Pupil/teacher ratios, USA, 1870–1991, public and private, elementary and secondary school

21.4   High school enrollment, graduates and dropouts, USA, 1890–1990

21.5   Bachelor's and first professional degrees conferred, USA, 1900–1990

21.6   Higher education, selected enrollment rates, 1870–1990

21.7   School years completed, by race, 25 to 29 years old, 1920–1990

21.8   Median school years completed, 25 to 29 years old, by race, 1920–90

21.9   Proportion of world's children, ages 6–11, enrolled in school, 1960–90

21.10   Proportion of world's children, ages 12–17, enrolled in school, 1960–90

21.11   Proportion of world's children, ages 18–23, enrolled in school, 1960–90

21.12   Adult illiteracy, developing countries, by birth year, pre-1925 to 1973

21.13   Literacy rates per thousand population, India, 1901–81

21.14a Enrollment in primary and secondary school, selected nations, 1850–1970

21.14b Enrollment in primary and secondary school, selected nations, 1900–70

21.15   Number of persons ages 18–23 enrolled in school worldwide, 1960–90

21.16   Expenditures per student for primary education from public and private sources

21.17   Expenditures per student for secondary education from public and private sources

21.18   Expenditures per student for tertiary education from public and private sources

21.19a Percent of 25- to 64-year-old population with higher education, university and non-university

21.19b Percent of 25- to 64-year-old population with higher education, university

21.19c Percent of 25- to 64-year-old population with primary education only

22.1   Length of average workweek, USA, 1850–1988

22.2   Annual hours worked per person, selected countries, 1870–1989

23.1   Historical trends in USA poverty rates

23.2   Official USA poverty rate, and adjusted for noncash benefits

23.3 Poverty rates for elderly, children, and single female-headed households

23.4 Official USA poverty rates, 1959–90, white, black, and Hispanic races

23.5 Percent of families in poverty, USA and other developed countries

24.1 Housing space, square feet per capita (1987)

24.2a Declining crowding in American housing, persons per room, 1900–87

24.2b Declining crowding in American housing, persons per room, 1900–87

24.3a Amenities in American housing, 1900–87

24.3b Amenities in American housing, 1900–87

24.4a Consumer durables in US households, 1900–87

24.4b Consumer durables in US households, 1900–87

25.1 Number of homeless families in New York City

26.1 Employment, productivity, and output, USA, 1954–92

26.2 Annual increase in manufacturing output, average over 10-year period

26.3 Average real family income, USA, 1970–90

26.4 The gap between investment and savings

26.5 Real federal receipts and spending, 1967–92

26.6 Budget deficits (–) and surpluses

26.7 Real private net worth, households and businesses

27.1 Long-term energy prices: wood and coal in Britain and the USA

27.2 US energy prices, 1800–1988

27.3 US energy consumption by type of fuel, 1850–1989

28.1 Real gasoline prices, 1920–89; index: 1982 = 100

28.2 Crude oil prices, 1869–1990

28.3 Crude reserves and 10-year production, USA, 1903–89

28.4 Reserves and 10-year cumulative production, world excluding the USA, 1943–89

28.5 Crude oil reserve value, 1946–86, and post-tax development cost, 1955–86

28.6 Ratio, reserve value/development cost, crude oil reserves, USA, 1955–86 (development cost is post-tax)

28.7 Non-communist world supply curve (excluding N. America and W. Europe)

28.8 Crude oil, world known reserves/annual world production

28.9 Estimated crude oil reserves

29.1 Estimated capital cost per kW for US nuclear plants, 1968–90

30.1 Copper price indexes, 1801–1990

30.2 Change in real wages over time

30.3 Lead price indexes, 1801–1990

30.4 Nickel price indexes, 1840–1990

30.5 Mercury price indexes, 1850–1990

30.6 Silver price indexes, 1850–1990

30.7 Zinc price indexes, 1853–1990

30.8   Platinum price indexes, 1880–1990

30.9   Tin price indexes, 1880–1990

30.10   Aluminum price indexes, 1896–1990

30.11   Antimony price indexes, 1900–89

30.12   Tungsten price indexes, 1900–90

30.13   Magnesium price indexes, 1915–89

31.1   World zinc resources, production use, recycle and losses since 1880

32.1   Public recreation lands, 1900–88

32.2   Recreational use of public land systems, 1910–88

33.1   Forest products, deflated prices

33.2   Growing commercial timber in the USA, net volume, all species

33.3a   Forested area in temperate regions, selected areas

33.3b   World forested area

34.1   Myers–Lovejoy estimates of species extinction and their extrapolations to the year 2000

35.1   Cereal yields in the developed world since 1800

35.2   Yields in the county of Norfolk, 1250–1850

35.3   Conjectural wheat yields, prehistory to early twentieth century

35.4   Estimated North American corn yields, 1490–1990

35.5   Trends in rice yields in selected Asian countries

35.6   Index of US farm labor productivity in corn, wheat, and cotton, 1800–1967

36.1a   World cereal production, 1950–90

36.1b   World cereal production, 1950–90

36.2a   Per capita grain production, 1950–90

36.2b   Per capita grain production, 1950–90

36.3a   World cereal yields, 1950–90

36.3b   World cereal yields, 1950–90

36.4   China grain production, 1960–90

36.5   China grain yield, 1960–90

36.6a   Index of world agricultural production

36.6b   Index of world agricultural production

36.7   Index of crop productivity, USA, 1960–4 to 1982–6

36.8   Index of livestock productivity, USA, 1960–4 to 1982–6

37.1   Indices of total and per capita food production, 1951–89 (1967–71 = 100)

37.2   Northeast Brazil: apparent per capita daily consumption of major starchy staples among low-income classes, 1974–5 (calories)

37.3   Relationship between starchy staple ratio and per capita GNP, 1964–6 and 1984–6

37.4   Relationship between infant mortality rate and per capita GNP, 1965 and 1988

38.1   World grain stocks, 1963–90

38.2 World total grain stocks vs. price

38.3 World wheat stocks, 1970–90, actual and foregone by acreage reduction

38.4 World coarse grain stocks, 1970–90, actual and foregone by acreage reduction

38.5 World grain stocks by principal holders, major importers and exporters, and others

39.1 World fisheries landings, 1938–90

39.2 Selected world fisheries landings, 1938–90

39.3 World fish landings per capita, 1950–90

40.1 US agricultural productivity

41.1 US water withdrawal, 1960–85, by public supply systems

41.2 Reservoir storage in the USA, 1880–1985

41.3 Total water consumption per dwelling, Boulder, Colorado, 1955–68

42.1 Trends in major uses of land, 48 contiguous states, 1880–1987

42.2 Cropland used for crops, 1949–90, index of land use and productivity

42.3 Urban land use, 48 contiguous states, 1960–90

42.4 Trends in special land uses, 48 contiguous states, 1950–87

43.1 Oxygen content of the Baltic Sea (in milliters per liter), Station F 74 (depth of approximately 150 meters)

43.2 Oxygen content (in milliters per liter) at 100 meters depth at Station F 12 in the Bothnian Bay of the Baltic Sea between 1900 and 1968

43.3 Changes in the concentrations of total dissolved solids in Lakes Erie and Ontario

43.4 Changes in the concentrations of sulfate in Lakes Erie and Ontario

43.5 Changes in the concentrations of calcium in Lakes Erie and Ontario

43.6 Changes in the concentrations of sodium-plus-potassium, chloride, calcium, and sulfate in each of the Great Lakes

43.7 Concentrations of total dissolved solids in the Great Lakes (circled points are averages of 12 or more determinations)

43.8 Five-year moving averages of the annual dissolved oxygen (as percentage of saturation) for the main branches of New York Harbor

43.9 Settleable particulate matter for five New York City boroughs (milligrams per square centimeter)

43.10 SO_2 air quality data for six US cities (micrograms of SO_2 per cubic meter)

43.11 National air pollutant emissions

43.12 Analysis of water of river Thames: percentage saturation with dissolved oxygen at high water (average July–Sept.)

43.13 Carbon monoxide concentrations in New York City, 1958–75 (parts per million)

43.14 Total suspended particulates for six cities, 1967–72 (micrograms per cubic meter)

43.15 Industrial lead pollution at Camp Century, Greenland, since 800 BC, micrograms per kilogram

43.16 Total tonnage of solid waste received at city disposal sites, Cincinnati, Ohio, 1931–76

43.17 Five-year moving averages of the minimum height of the Nile river, 810–50

43.18 Five-year moving averages of the maximum height of the Nile river, 810–50

43.19 Five-year moving averages of the minimum height of the Nile river, 808–78

43.20 Five-year moving averages of the maximum height of the Nile river, 810–90

43.21 Variations in height of Nile Flood. The sloping line indicates the secular raising of the bed of the Nile by the deposition of silt

43.22 Estimated phosphorus loadings to the Great Lakes

43.23 DDT levels in the Great Lakes

43.24 Dieldrin levels in the Great Lakes

43.25 PCB levels in the Great Lakes

43.26 Trends in the quality of drinking water in the USA

43.27 Ambient air pollutant concentrations, USA, 1976–90

43.28 Municipal waste recovery and incineration, US, 1960–90

44.1 Smoke and sulfur dioxide levels, London, 1585–1940

44.2 Mean winter SO_2 concentrations, selected sites in the United Kingdom

44.3 Smoke level and mean hours of winter sunshine in London

44.4 SO_2 and smoke emission and concentration, United Kingdom, 1962–88

44.5 Air pollution and bronchitis deaths, Manchester, 1959–60 to 1983–4

44.6 Acidity of Round Loch, Scotland, 1300–1980

44.7 Sulfate and nitrate concentrations in atmospheric aerosols, Harwell, United Kingdom

44.8 Ozone concentrations, central London

44.9 Nitrogen oxides and lead levels, central London, 1978–89

45.1 Trends in settleable dust in six cities

45.2 Number of days per year with highest concentration equal to or greater than levels shown

45.3 Air pollutant trends, 1960–88

45.4 Ambient air pollutant concentrations, USA, 1976–90

45.5 Ambient air pollutant concentrations, index of deviations from standards

45.6 Emissions of major air pollutants in the USA

45.7 Air quality trends in major urban areas (number of days greater than the PSI index level)

45.8 Pollutants in the air, USA, 1960–90

46.1 Emissions of air pollutants, nine market economies, 1970–85

46.2 Emissions of air pollutants, by sources, USA, 1940–88

46.3 Indexes of the real gross national product, energy use, and emissions of air pollutants, USA, 1940–88

46.4 Emissions of air pollutants and energy use per capita and per 1982 $1,000, USA, 1940–88

46.5 Trends in water quality in developed countries, pollution indicators in rivers, USA and five European countries, 1979–86

46.6 Pesticide residue levels in human adipose tissue and human milk, USA 1970–83, United Kingdom 1963–83, Japan 1976–85

46.7 Concentrations of three air pollutants, 16 countries, 1985–7

46.8 The Japanese indexes of concentrations of air pollutants, eight countries, 1980s, latest available year

46.9 Emissions of air pollutants from fuel combustion and industrial processes, by types, USA, USSR, and China, 1988

46.10 Emissions of air pollutants, USA and USSR, 1970–88 (selected years), different types of measurement

46.11 Emissions of air pollutants per capita, market and socialist economies, 12 countries, 1985

46.12 Emissions of air pollutants per $1,000 of GDP, market and socialist economies, 12 countries, 1985

46.13 Inland water pollution, USSR, 1980–9, and ten market economies, ca 1985

46.14 The contraction and pollution of the Aral Sea, 1960–89

47.1 pH scale, acidity of common substances and current rain/clouds/fog

47.2 Distribution of acid rain over the USA in 1990 (NAPAP data)

47.3 Historical trends of emissions of SO_2 and NO in the USA, 1900–90

47.4 Effect of simulated acid rain over the growing season on the yield of several important agricultural crops

47.5 Diameter growth of 2- to 5-year-old seedlings exposed to different concentrations of simulated acid rain over 30 months

48.1 Deviations from mean concentration of ozone, and sunspot activity, 1958–92

48.2 Annual mean sunspot number R at maxima of the 11-year cycle, 1645 to present, to demonstrate long-term trends in solar activity

48.3 Trends in stratospheric ozone, variations from mean, 1958–92

49.1 Hemispheric temperature trends, 1850–1990

49.2 Typical output of General Circulation Models

49.3 Sample General Circulation Model output, National Center for Atmospheric Research, $2 \times CO_2 - 1 \times CO_2$

49.4 Surface air temperature differences (Dec.–Feb.)

49.5 ΔT_{991} (Dec.–Feb.) transient minus control (26–30)

49.6 Annual average temperature anomaly differences over USA (Hansen minus historical climate network)

49.7 USA, Canada, surrounding oceans temperature pattern

49.8 Arctic temperature patterns, 1820–1980[49]

49.9 Amundsen-Scott temperature patterns, 1957–87

49.10 Areally weighted temperature range national (urbanization adjusted) annual (Jan.–Dec.)

49.11 Effect of sulfate emissions on downwind albedo effect

51.1 Comparison of various risks

54.1 How far should environmental protection go?

54.2 Importance of environmental protection and cost

54.3 Spending levels for the environment

54.4 Salience of environmental issues, air and water pollution versus energy

54.5 Energy as the "most important problem"

54.6 Environment versus energy

55.1 Ideal number of children

List of Tables

3.1 Child mortality in Europe, 1550 to 1849

4.1 Causes of death, model populations

4.2 Death rates from respiratory tuberculosis

4.3 Death rates from infectious diseases, excluding tuberculosis

5.1 Provisional French and English daily consumption of kcals per consuming unit toward the end of the eighteenth century

9.1 Largest US disasters by category

19.1 Adult per capita consumption of manufactured cigarettes

10.2 Worldwide antitobacco activities

13.1 Annual growth rates of per capita income, selected countries

13.2 Daily per capita calories, selected countries

14.1 Percentage of households with appliances

14.2 Nutrients available

14.3 Food consumption

15.1 Growth of factor inputs and output and decomposition of growth rate by epoch, 1840–1980

19.1 Annual incidence of unemployment and unemployment durations, 1885–1989

19.2 Unemployment by race and by place of birth

19.3 Unemployment among different age groups

22.1 Differences in total weekly hours of free time, by year and demographic factors

22.2 How free time is distributed across activities, by year and gender

24.1 Households without modern amenities

24.2 Consumer durables owned by "poor" American households: 1987

24.3 Food consumption: low-income persons compared to upper-middle income persons

24.4 Average nutrients consumed as percentage of 1980 recommended dietary standards

24.5 Meat consumption by average citizens in various nations

25.1 Estimated total number of homeless in the USA

25.2 Estimated number of homeless in selected cities

25.3 Characteristics of homeless persons estimated in various local studies

29.1 Breakdown of labor costs for nuclear power plants and coal-burning plants from the 1987 EEDB

29.2 Cost per kilowatt-hour for various types of power plants

31.1 Comparison of current consumption and current and ultimate resources of elements expected to eventually have very large to infinite resources

31.2a Elements expected to always have limited resources

31.2b Extended to current resource ratios for elements with currently limited resources

31.3 Byproduct elements and their sources

33.1 Estimates of tropical forest area and rate of deforestation

33.2 World forested area: past and current estimates

33.3 Deforestation by country, according to three different sources

36.1 World crop yields, 1980–91

39.1 World marine landings, 1990

40.1 Soil loss per cropland acre and erodibility, 1987

41.1 Value of water in agriculture compared to price of water

41.2 Comparison of metered versus flat-rate use rates in the United States

41.3 Variance in industrial unit water withdrawals around the world, 1960

47.1 Summary of US National Lake Survey

47.2 Potential benefits (costs) of reducing acid rain by 50 percent

50.1 Costs of climate change and effects and costs of greenhouse gas emissions abatement in the US by 2040

50.2 Climate change and abatement policy scenarios

51.1 Eventual deaths per 1,000 MW plant per year of operation due to wastes

55.1 Percent who agree/disagree about impact of population growth (1971)

55.2 Percent who perceive the issues as serious problems

55.3 Percent who agree actions should be taken to reduce population growth

55.4 Number of reports per year and percentage that characterizes population growth as dangerous and troublesome

55.5 Percentage of news stories, editorials, and letters that characterize population growth as dangerous

List of Contributors

Morris A. Adelman, Professor Emeritus, Department of Economics, MIT, Cambridge, Massachusetts.

Bruce Ames, Professor, Department of Molecular and Cell Biology, University of California, Berkeley.

Terry Anderson, Professor, Political Economy Research Center, Bozeman, Montana.

Jeremy Atack, Professor, Department of Economics, Vanderbilt University, Nashville, Tennessee, and Research Associate, National Bureau of Economic Research.

Dennis Avery, The Hudson Institute, Swoope, Virginia.

William Baumol, Professor, C. V. Starr Center for Applied Economics, New York, New York.

Calvin Beisner, Professor, Covenant College, Lookout Mountain, Georgia.

Michael S. Bernstam, Professor, Hoover Institution, Stanford, California.

Rebecca Blank, Professor, Department of Economics, Northwestern University, Evanston, Illinois.

Allan M. Brandt, Professor, Department of Social Medicine, Harvard Medical School, Boston, Massachusetts.

Joyce Burnette, Professor, Department of Economics, Loyola University of Chicago, Chicago, Illinois.

Jean-Claude Chesnais, Professor, Institut National d'Etudes Démographiques, Paris.

Marion Clawson, Resources for the Future, Washington, DC.

Bernard L. Cohen, Professor, Department of Physics, University of Pittsburgh, Pittsburgh, Pennsylvania.

Eileen M. Crimmins, Professor, Ethel Percy Andrus Gerontology Center, University of Southern California, Los Angeles, California.

Hugh W. Ellsaesser, Participating Guest Scientist, Atmospheric and Geophysical Sciences Division, Lawrence Livermore National Laboratory, Livermore, California.

Derek Elsom, Professor, Geography Unit, Oxford Brookes University, Oxford, UK.

Stanley L. Engerman, Professor, Department of Economics, University of Rochester, Rochester, New York.

Murray Feshbach, Professor, Department of Demography, Georgetown University, Washington, D.C.

Randall K. Filer, Department of Economics, Hunter College, City University of New York, New York.

Robert Fogel, Professor, Graduate School of Business, Center for Population Economics, University of Chicago, Chicago, Illinois.

H. Thomas Frey, geographer, retired from the Economic Research Service, US Department of Agriculture, Washington, D.C.

Bruce F. Gardner, Professor, Department of Agriculture and Resource Economics, University of Maryland, College Park, Maryland.

H. E. Goeller, Knoxville, Tennessee.

George Grantham, Professor, Department of Economics, McGill University, Montreal, Quebec.

Michael Haines, Professor, Department of Economics, Colgate University, Hamilton, New York.

William J. Hausman, Professor, Department of Economics, The College of William and Mary, Williamsburg, Virginia.

Robert Higgs, Research Director for the Independent Institute of Oakland, California, and Visiting Scholar in the Department of Economics and Finance, Seattle University, Washington.

Kenneth Hill, Professor, School of Hygiene and Public Health, Johns Hopkins University, Baltimore, Maryland.

Arlene Holen, Commissioner, Federal Mine Safety and Health Review Commission, Washington, D.C.

William Hudson, independent consultant in world agriculture and economics, based in Maumee, Ohio.

Dominique Ingegneri, Professor, Gerontology Institute, University of Massachusetts, Boston, Massachusetts.

Alexander Keyssar, Professor, Department of History, Duke University, Durham, North Carolina.

J. Laurence Kulp, Professor of Civil Engineering, University of Washington, Seattle, Washington.

Lester Lave, Professor, Department of Economics, Carnegie-Mellon University, Pittsburgh, Pennsylvania.

Stanley Lebergott, Professor, Department of Economics, Wesleyan University, Middletown, Connecticut.

Peter Lindert, Professor, Department of Economics, University of California, Davis, California.

William Lunch, Associate Professor, Political Science Department, and Coordinator, Environmental Policy Program, Oregon State University, Carvallis, Oregon. Political Analyst, Oregon Public Broadcast.

Robert A. Margo, Professor of Economics, Vanderbilt University, Nashville, Tennessee, and Research Associate, National Bureau of Economic Research.

Patrick J. Michaels, Professor, Department of Environmental Studies, University of Virginia, Charlottesville, Virginia.

Joel Mokyr, Professor, Department of Economics, College of Arts and Sciences, Northwestern University, Evanston, Illinois.

Stephen Moore, The Cato Institute, Washington, DC.

Richard Muth, Professor, Department of Economics, Emory University, Atlanta, Georgia.

John Myers, Professor (retired), Southern Illinois University, Carbondale, Illinois.

Robert H. Nelson, Professor, College of Public Affairs, University of Maryland, College Park, Maryland.

Wallace Oates, Professor, Department of Economics, University of Maryland, College Park, Maryland.

Thomas T. Poleman, Professor, Department of Agricultural, Resource, and Managerial Economics, Cornell University, Ithaca, New York.

Samuel H. Preston, Professor, Population Studies Center, University of Pennsylvania, Philadelphia, Pennsylvania.

Robert Rector, Senior Policy Analyst, The Heritage Foundation, Washington, D.C.

Alan Reynolds, The Hudson Institute, Indianapolis, Indiana.

James S. Roberts, Professor, Office of the Provost, Duke University, Durham, North Carolina.

John Robinson, Professor, Department of Sociology, University of Maryland, College Park, Maryland.

Stanley Rothman, Director, Center for the Study of Social and Political Change, Mary Huggins Gamble Professor Emeritus of Government, Department of Government, Smith College, Northampton, Massachusetts.

Roger A. Sedjo, Resources for the Future, Washington, DC.

Julian L. Simon, Professor, College of Business and Management, University of Maryland, College Park, Maryland.

Rita J. Simon, Professor, Department of Criminal Justice, American University, Washington, DC.

S. Fred Singer, Director, The Science and Environment Policy Project, Fairfax, Virginia.

Richard J. Sullivan, Economist, Policy and Special Studies, Division of Bank Supervision and Structure, Federal Reserve Bank of Kansas City, Missouri.

W. Kip Viscusi, Professor, Department of Economics, Duke University, Durham, North Carolina.

Elizabeth M. Whelan, President, American Council on Science and Health, New York, New York.

Aaron Wildavsky, Professor, Survey Research Center, University of California, Berkeley, California.

Jeffrey G. Williamson, Professor, Department of Economics, Harvard University, Cambridge, Massachusetts.

John P. Wise, Consultant in Marine Affairs, 4545 Connecticut Avenue N.W., Washington, DC.

Richard J. Zeckhauser, Professor, Kennedy School of Government, Harvard University, Cambridge, Massachusetts.

Index

Italic numbers indicate information contained in figures or tables only; page locators with superscript numbers refer to information contained in the numbered chapter notes.

accidents
  nuclear reactors 578–80, 585
  predictions about 655
  US rates 98–105
  see also risk management
acid rain 17–18, 483–6, 523–34
  causes 523–5
  cost benefits of reduction
    533, 534
  distribution and trends
    483–6, 525–7
  effects 484, 527–34, 638–9
  pH scale 483, 524
Acreage Reduction Programs
  (ARPs), USA 406–8
Acsadi, George, life expectancy
  30, 38
Adams, O. B., health 77
Adamson, Richard, Alar-apple
  scare 596
Adelman, M. A., oil 287–93
advertising, cigarettes 110,
  111–12
Africa
  agriculture 13, 376: alley
    cropping 385; cassava
    380, 382, 385, 387, 593;
    cereal yields 377, 381;
    corn 378, 383, 387;
    farming technology 386–7;
    grain production 379, 387;
    nutritional well-being 395,
    396, 401, 402, 403; ox-
    drawn plows 383;
    pesticides 385; population
    growth and 403; rice 383;
    sorghum hybrids 382;
    starchy staple ratios 399;
    total and per capita
    productivity 397
  economic growth: daily per
    capita calories 142, 143;
    starchy staple ratio and 400
  famine 386–7, 396, 403

forests 329, 335, 336, 338, 340
  life expectancy 33–4: of
    infants and children 45,
    46, 47, 48, 401, 402
  slavery 172, 174, 175
  smoking 108, 109, 111
  see also named countries
African-Americans 178–87,
  236–7, 250
  consumer spending 180
  education 182, 209, 213–14
  employment 181–2, 185, 199
  housing 180, 181, 182, 185
  infant and child mortality
    rates 5–6, 179, 181
  life expectancy 179, 180, 181
  slavery 171–5
  voting 181
  wealth and income
    distribution 179–83, 186,
    236–7, 250
agricultural productivity
  364–409, 639
  acid rain's effects 528
  capital investment 386
  cereal cultivation origins
    365–6
  estimates 369–71, 376–7
  farming technology 381,
    386–7
  fertilizers 364–5, 371, 385
  global warming 566
  grain stock trends 405–9
  markets 367–9, 386, 406–9
  medieval revolution 366–7
  nutritional well-being
    394–403
  population growth and
    367–8, 379, 388–9, 403
  potential yield gains 377–80
  scientific contributions to
    381–6
  soil erosion 418, 419, 420
  transport 368, 386

agriculture
  human disease and 51–2
  irrigation systems 386, 427,
    428, 432, 519
  labor 10, 11, 124–34, 225
  land use trends 434, 435–7,
    440, 650
  methane production 488, 545
  nonrenewable resources 319
  pesticides 385–6: cancer and
    588–94; Great Lakes 470,
    471, 472; residue levels
    508, 510; soil runoff 419,
    519
  soil erosion 381, 416–23
  soil runoff 418–19, 422, 432
  water management 427, 428,
    429, 432, 519
  see also agricultural
    productivity; livestock
    farming
AIDS 653
air pollution 14, 15–16, 17–18
  acid rain 17–18, 483–6,
    523–34, 638–9
  economic systems compared
    503–8, 511–17
  greenhouse effect 17–18,
    342, 488, 544–61, 565–74,
    645–6
  history of environmental
    damage 446–8
  long-term trends 446–7, 448,
    454–6, 457–60
  nuclear power compared with
    coal 579–80, 582, 583, 585
  public opinion about 612,
    614, 616
  UK 18, 455–6, 476–89: acid
    deposition 481–6;
    greenhouse gases 488;
    photochemical pollution
    and vehicle exhaust
    emissions 486–8; pre-20th

air pollution (*Cont.*)
  century 447, 448, 464[3], 466[3], 476–8; smoke and sulfur dioxide decline in twentieth century 478–83, 485–6
  USA 491–502: data issues 492–4; general trends 494–7, 498–502; long-term trends 446–7, 454–5, 457, 458, 460
  see also specified pollutants
air transport
  productivity growth 167
  protective regulations 633
  US fatal accidents 101
Alar, health scare 595–6
albedo effects, sulfate emissions 557–8, 559, 560
alcohol
  beer output 141
  consumption 114–20, 158: homelessness and 263–5, 266
algae (diatoms), acid deposition and 483, 484
Algeria
  education 218
  nutrition 400, 402
  smoking 109, 110
Allen, J. C., forests 336, 337, 338, 340–2
alley cropping 385
aluminum
  consumption and availability 313, 316, 320, 321
  prices 309, 310
  recycling 315
Ames, Bruce N., pesticides 588–94
Anderson, J. R., land use 434, 435
Anderson, Terry L., water 425–32
animals, cancer-causing agents 589–90, 591–2, 595–6, 601, 603, 604
  see also livestock farming; species loss
Antarctic ozone hole (AOH) 537–8
antibiotics, in cows' milk 597–8
antimony 309, 310, 311, 312, 318, 321
Aral Sea, pollution of 428–9, 519
Argentina
  agricultural productivity 382, 388–9
  deforestation 340
  education 218
  smoking 109, 110
ARPs (Acreage Reduction Programs), USA 406–8
arsenic
  availability 318, 321
  carcinogenic effects 582, 592
Ashby, E., air pollution 478
Asia
  forests 329, 333, 336, 338, 340
  rice production 374

smoking 108, 109, 110, 111, 112
  see also named countries
aspartame, health effects 600
Atack, Jeremy, productivity trends 161–70
Atkins, D. H. F., air pollution 485
atmospheric pollution *see* air pollution
Auliciems, A., air pollution 480
Australia
  accident fatalities 103
  education 219, 220, 222
  forests 335
  health 77
  labor productivity 168
  poverty 238, 239
  smoking 109, 112
  work time 230
Austria
  consumption per capita 159
  education 219, 220, 222
  income per capita 139
  labor productivity 168
  nutrition 400
  serfdom 173
  smoking 109, 111
  suicide rates 93
  water pollution 518
  work time 230
automobiles *see* motor vehicles
Avery, Dennis T., food productivity 376–89

Bailey, M. N., US health 80
Bairoch, Paul
  cereal yields 369
  economic growth 139, 141
  urbanization 133[6]
Baltic Sea, pollution 445–6, 466[9]
Bangladesh
  agricultural productivity 374, 386, 398, 399, 403
  deforestation 340
  infant mortality 401
  nutrition 142, 398, 399, 402, 403
  smoking 109
Barber, F. R., air pollution 479
Barclay, George, life expectancy 33, 45
Barnes, Douglas F., forests 336, 337, 338, 340–2
Barsky, Arthur J. III, "Worried Sick" syndrome 80
basal metabolic rate (BMR) 64
Baumol, William J.
  environmental quality 444–75
  productivity 169
Beauchamp, Dan E., alcohol 114
Bebbington, A. C., health 77
beer
  consumption 114–20
  output trends 141
Beeton, A. M., water pollution 449, 451, 452
Beisner, E. Calvin
  living standards 147–8
  ozone layer 542, 543

Bekaert, Geert, nutrition 136, 143
Belgium
  agricultural labor force 126, 127
  agricultural productivity 365, 366–7, 368
  beer output 141
  cotton consumption 142
  daily per capita calories 142
  education 219, 220, 222
  housing 244
  income per capita 139
  labor productivity 168
  nuclear power 301
  pollution 509, 512
  smoking 109, 111
  sugar consumption 143
  work time 230
beliefs, importance of 645–6
Benedictine monks, mortality rates 31
Bennett, Merrill K., nutrition 398
Bennett, William, tobacco industry 107
Bergson, Abram, USSR consumption 158, 159
Bernstam, Mikhail S., resource use and pollution 503–20
Bernstein, H. T., air pollution 478
beryllium
  availability 318
  carcinogenic effects 582, 592
Bhat, Mari, India, life expectancy 33
BHC (HCH), in human tissue and milk 510
Bills, Nelson L., soil erosion 417
biodiversity 15, 346–60
  biologists' goals 356–7
  commercialization of 343
  DDT's effects 602
  defining species 347–8
  loss estimates 348–55
  tropical deforestation and 339, 351, 352, 354–5, 356, 359–60
  value of species-saving 357–8
biogeography, species loss 354, 360
Biraben, J. N., mortality 53
birth rates, life expectancy and 30
Bishop, L., ozone layer 539
Biswanger, Hans P., deforestation 329
black Americans *see* African-Americans
Blake, D. R., Antarctic ozone hole 538
Blanchet, M., health 77
Blank, Rebecca M., US poverty trends 231–9
Blum, Jerome, serf emancipation 173
BMR (basal metabolic rate) 64
body mass index (BMI), mortality risks and 63, 65–9, 70–1

Boeckh index, construction costs 204–5
Boggs, Rebecca, education 208–23
Bolino, A. C., unemployment *198*
bonds, US economic performance and 276–7
boredom 20–1
Bostrom, Ann, work time 227
Bound, J., health 77, 80
Bourgeois-Pichat, J., population estimates *35, 36*
bovine somatotropin (BST), health scare 596–7
Boyd-Orr, Lord John, malnutrition 394
Brake, G. T., alcohol *117*
Brandt, Allan M., smoking 106–13
Braudel, Fernand, living standards 137
Brauman, John I., substitutability 321
Bray, Anna J., global cooling 646
Brazil
 agricultural labor force 128, 129, 133–4[8]
 agricultural productivity 377, 386, 398, 399
 education *218*
 forests 335, 336, *340*
 nutrition: daily per capita calories *142*; dietary adequacy 398, 399, *402*; meat consumption *250*; starchy staple ratio *400*
 slavery 174
 smoking *109, 110*
Breakey, William R., homelessness *263*
Brimblecombe, P., air pollution *18*, 476, 477, 478, *480*
Britain *see* UK
Brobst, D. A., mineral resources 319
Brooks, C. P., Nile floods *465*
Brothwell, D., life expectancy 38
Brown, Carl, homelessness 264–5
Brown, G. G., species loss 353
Brown, John, economic growth 139
Brown, K. S., species loss 353
Brown, Lester, grain production 406
Bryce-Smith, D., aerosol pollutants 459
BST *see* bovine somatotropin
Buchanan, James, challenges 657
Budyko, Mikhail, global warming 557, 560
buildings
 environmental damage 446–7, 448, 531, 580
 predictions for 647
Burnette, Joyce, living standards 135–45
Burt, Martha R., homelessness *264*

Burton, I., air pollution 480

cadmium
 availability *318, 321*
 carcinogenic effects 582, *583*, 591
calcium
 consumption and availability *316*
 Great Lakes 449, *450*
Campbell, Bruce M. S., crop yields *370*
camping, USA 325
Canada
 accident fatalities 103
 acid rain 529, 533
 agricultural productivity 382, *397*
 air pollution *504, 512, 516–17*, 529, 533
 economic growth *147*
 education *222*
 forests 331, 333, *334*
 health 77
 labor productivity *168, 169*
 poverty 238, 239
 smoking 108, *109, 110*
 water pollution 449–51, *452*, 467[18], *470–2, 518*
 work time *230*
cancer
 carcinogen scares 588–94, 595–6, 597, 598–9, 601–5
 nuclear power 576–7, 582
 ozone layer 540–1
 USSR mortality rates 89
capital
 movement of, government role and 649
 productivity and 165, *166, 170*, 386
 US investment and saving 274, 276–7
capital punishment, England 96–7
carbon
 consumption and availability *316*
 petrochemical substitutability 320
 water used in production of *431*
carbon dioxide
 CFCs and 536
 as greenhouse gas: abatement costs 567–73; climate models 548, *549*, 550; concentrations 545, 546, 560; deforestation 342; temperature changes 548, 566; trace gas equivalence 546, 554; UK emissions 488
carbon monoxide
 concentrations: Europe 457; USA *17, 455, 457, 474, 494, 496, 498–500, 502, 514*
 danger of 491
carcinogens *see* cancer
Carlo, George L., dioxin risks 599
cars *see* motor vehicles

Carter, Lawrence, life expectancy 32
Case, R. A. M., life expectancy 31
Caselli, Graziella, disease 56–7
cassava *380*, 382, 385, 387, 398, 593
catastrophes
 disease 32, 51–2, 53, 57, 59, 652–3
 nuclear reactor accidents 578–80
 predictions 655
 USA 100–2
cattle *see* livestock farming
Census Bureau (USA) 23–4
 income *183, 186*
 life expectancy 32, *180, 181*
 poverty 234, 241, 251–2
 suicide trends 95
 work time 225
Central America
 forests 329, 335, 339, *340*, 343
 homicide rates 92
 nutrition *142, 400, 402*
 smoking *109, 110*
cereal cultivation *see* crop yields
cesium *318*
Cess, R. E., albedo effect of sulfate 559
CFCs (chlorofluorocarbons) 488, 536–8, 546
challenges, need for 656–8
Chandler, L. V., unemployment *198, 200*
Chandler, W. U., smoking 110
Chesnais, Jean-Claude, murder and suicide 91–7
children
 education 217–23
 free time and 228
 infanticide 92
 mortality *see* infant and child mortality
 nutrition 250, 401
 poverty *235, 236, 250, 401*
 slavery 175
 suicide rates 94
 unemployed teenagers 199–200
Chile
 agricultural labor force 129
 education *218*
 forests 335
 life expectancy 33, 44–5, 48
 smoking *109, 110*
China
 agricultural productivity 377, 388: corn 382; grain stocks 408–9; nutritional well-being 401; rice *374, 383, 384*, 388; total and per capita *397*; wheat *383, 384*, 388
 deforestation *340*
 economic growth 136: daily per capita calories *142*; nutritional well-being 401, *402*; pollution and 511, *512–13, 514–15*; starchy staple ratio *400*
 forests 333, 334–5

China (*Cont.*)
life expectancy 8, 9, 33:
infants 45, 46, 47, 401, *402*
predictions for 654
smoking *109*, 110
Chirikos, T. N., health 77
chlorine
chloride pollution 449, 451,
518
chlorofluorocarbons and 537
consumption and availability
316
Chlorine Institute, dioxin risks
599
chlorofluorocarbons *see* CFCs
cholera 53, 58
Christian Science Monitor poll,
environment 617
Christy, Francis T., wood
resources 330
chromium *317*, 319, 582, 592
cigarette smoking 106–13, *158*,
631
Cipolla, Carlo M., disease and
mortality 51
Civil Rights movement, USA
181, 182
Clark, Colin, poverty 240
Clark, William C., pollution
342
class, social, unemployment
and 200
Clawson, Marion, forests
328–43
Clement, J., forest plantations
335, 336
Cleopatra's obelisk, New York
446–7
climatic changes, global
warming *see* greenhouse
effect
clouds, greenhouse effect and
548–9, 557–8, *559*
coal
development of 25–6
fossil fuel pollution 342: acid
rain 523, 525, 527; long-
term trends in UK 447,
455–6, 476–8; methane
emissions 488; nuclear
power compared 579–80,
582, 583, 585; UK smoke
controls 476–81
price and availability 25,
280–2, *283*, 284, 476
substitutability 319–20
cobalt *317*, *318*, 319,
320
Cohen, Barbara E.,
homelessness *264*
Cohen, Bernard L., nuclear
power 294–301, 576–85
Cole, W. A., economic growth
138
Colinvaux, Paul A., species loss
354
Colombia
agricultural labor force 128,
130, 133–4[8]
homicide rates 92
nutrition *400*, *402*
smoking *109*, *110*
Colvez, A., health 77

Commerce index, construction
costs 204
computers, productivity and
162, 164–5
conservation
farming technology and 381
soil erosion: conservation
tillage 381; US
Conservation Reserve
Program 420–2
species loss 15, 346–60:
biologists' goals 356–7;
commercialization of
biodiversity 343; DDT
and 602; defining species
347–8; loss estimates
348–55; tropical
deforestation and 339,
351, 352, 354–5, 356,
359–60; value of
species-saving 357–8
tropical forests 342–3
see also risk management
consumers
protection: fatal accidents
104–5; risk estimation
629–30
US expenditure 156, *158–60*:
African-Americans 180;
housing 205–6; poverty
levels and 254–7, 251–2
want fulfilment and future
challenges 656–7
warning labels 631
contract labor 175
copper 11, 303, *318*, 319, 320,
321
corn
price 406, *407*, 408
production 378, *405*, 406;
acid rain and *528*; Africa
378, 383, 387; genetic
engineering and 382, 383;
population growth and
389; USA *373*, *375*, 378,
392, 406–8
see also wheat
cotton
consumption *142*, 145
production *163*, 164, *375*,
392, 436–7: Aral Sea
pollution and 428–9, 519
cows *see* livestock farming
Crafts, N. F. R., income
growth 138–9
crime
capital punishment 96–7
homicide rates 91–2, 96
lynchings *186*
Crimmins, Eileen M., US
health 72–82
crop yields 364–409
acid rain 528
capital investment and 386
cereal cultivation origins
365–6
estimates 369–71, 376–7
estimates of 369–71
farming technology 381,
386–7
grain stock trends 405–9
markets and 367–9, 386,
406–9

medieval revolution 366–7
nutritional well-being and
394–403
organic fertilizing 364–5
population growth and
367–8, 379, 388–9
potential yield gains 377–80
scientific contributions to
381–6
Crosby, Alfred W., Jr, disease
and mortality 51
Crystal, Stephen, homelessness
264
Cuba
deforestation *340*
slavery 174
smoking *109*
Cummins, Gaylord, nutrition
136
Cuomo Commission,
homelessness *264*
cyclamates, health effects 604
Czechoslovakia
forest damage 484
life expectancy 32
pollution 511, *512–13*,
516–17
smoking *109*, *111*

Danziger, S., poverty *232*, 233
Dastoor, M. N., CFC estimates
537
Daugherty, A. B., land use *435*,
437, 439
Davies, T. D., air pollution 482
Davis, Lana, productivity *161*
DDT
Great Lakes *470*
health effects 589, 592, 602–3
in human tissue and milk *510*
De Backer, H., ozone layer 539
De Muer, D., ozone layer 539
Deane, agricultural labor force
11
Deane, Phyllis, economic
growth *138*
death rates *see* mortality
debt bondage 175
debt-for-nature swaps, forests
342–3
Deer, Noel, sugar consumption
143
defense
incidence of war 654
US land use 439
US spending 275
deforestation 328–30, 331,
336–43
species loss 339, 351, 352,
354–5, 356, 359–60
Denmark
agricultural productivity 365
education *219*, *220*, 222
height trends 67, *144*
income *139*, *194*
labor productivity *168*
life expectancy and height 67
slavery 174
smoking *109*, *111*
sugar consumption *143*
suicide rates 92–3
work time *230*
Dension, E. F., productivity *161*

Desmond, Anabelle, population 36
Detwiler, R. P., atmospheric pollution 342
Deutsch, Albert, homelessness 260
developing countries
  agricultural labor force 127–30, 130–1, 133[6,7]
  agricultural productivity 378–9, *384*, 397
  debt-for-nature swaps 342–3
  economic growth 141, 142–4, 566
  education *215–17*
  life expectancy 33–4: infants and children 44–7, 401, *402*
  smoking 108, *109*, 110, *111–12*
  see also named countries
Dewhurst, J. F., work time 225
Diamond, Jared, species loss 346, 355, 356, 357
diary data, leisure time 226–9
diatoms, acid deposition and 483, *484*
dieldrin *471*, *510*
diet, definition 63
  see also nutrition
dioxin 591, 598–9
diphtheria 57, 85, 89
disasters *see* catastrophes
disease 51–9
  AIDS 653
  cancer: carcinogen scares 588–94, 595–6, 597, 598–9, 601–5; nuclear power 576–7, 582; ozone layer 540–1; USSR *89*
  cholera 53, 58
  control of 57–8, 59
  diphtheria 57, 85, 89
  forecasts 647–8, 652–3
  life expectancy and 32, 53–7: children *43*, 44, 46–7, *53*; epidemics 32, 51–2, 53
  malaria 589, 602
  malnutrition and 63
  measles 52, 53, 57, 90
  polio 90
  risk predictors, body mass index (BMI) 63, 65–9, 70–1
  scarlet fever 57
  smallpox 52, 53, 57
  smoking and 107–8, 110, *111*, 112–13
  tuberculosis 54–7, 85, 89
  typhoid 90, 150
  US trends: explanations 78–81; indicators 73; medical care 153, *159*
  USSR 85, *86*, 87, 89–90
Doll, R.
  cancer incidence rates 589
  smoking 107
Dovring, Folke., agricultural labor 126, 127, 130
Drake, Edwin L., oilwells 25
drought
  African food production 386–7

climate predictions 548
soil erosion 418
drug abuse, homelessness and *263–5*, 266
Drury, T., US health 80

earthquakes, fatalities from *101*
East Anglia research group, temperature trends *547*, 551–2
Easterlin, Richard, population 658
Eastern Europe
  agricultural productivity *397*
  education *218*
  forecasts for 654, 658
  life expectancy trends 32
  meat consumption *250*
  serfdom 173
  smoking *109*, *111*, *112*
  starchy staple ratio *400*
  see also Czechoslovakia; Hungary; Poland
Eberstadt, Nicholas, life expectancy trends 32
EC *see* European Community
ecology, concept of 637–40
  see also environment; wildlife
economic performance
  agriculture and 124, 126, 368–9, 371
  alcohol consumption 116, 117, 118–19
  child mortality 47
  forecasts 654, 657–9
  health and disease 58–9
  living standards 135–45, 147–8: absolute and relative progress 6–7, 653–4
  pollution and: economic systems compared 503–20; global warming 566, 567–73
  population growth and 16–17, 24–6, 657–8
  productivity 161–70
  USA 271–7: income 191–2, 193–4; poverty 235
economists, role of 658–9
EDB (ethyl dibromide), health effects 603
Edirisinghe, Neville, nutrition 398, *399*
education 19–20
  predictions about 651–2
  USA 208–23:
    African-Americans 182, 209, *213–14*; dropout rates 208–9, 211, *212*
  of women, child mortality and 47
EEC *see* European Community
Egypt
  agricultural labor force 128, 129, *130*, 133–4[7, 8]
  agricultural productivity 366
  education *218*
  Nile river 461–4, *465*
  nutrition *400*, *402*
  smoking *109*, *111*
Ehrlich, Anne, starvation 394

Ehrlich, L. R., carbon monoxide 491, 496
Ehrlich, Paul
  species loss 347
  starvation 394, 639
  water 425
EI (erodibility index), soil erosion 416–17, 420
Eisenbud, M., carbon monoxide pollution 491, 496
electricity consumption, USA 151, 152–3, *159*, 160
electricity generation
  nuclear 294–301, 576–85, 645, 647
  solar 582, *583*, 645
  water use in *431*
  see also fossil fuel pollution
Ellsaesser, H. W.
  Arctic temperature patterns 555
  US air pollution 491–7
Elsom, Derek M., UK atmospheric pollution 476–89
Emigh, G. D., phosphate resources 319
employment *see* labor force
energy (nutrition) 63–5
  daily per capita calories 142–4
  economic growth and 136
energy resources
  invention and discovery 26
  nuclear power 294–301, 576–85, 645, 647
  past crises 25, 476
  price and availability 280–6
  public opinion about 610, 614–17
  solar power 582, *583*, 645
  substitutability 319–20
  US use of 150–1, 152–3, *159*, *506*, 507–8, 520
  waste incineration *475*
  water use in production of *431*
  see also coal; oil
Engel, Ernst, alcohol 116–17
Engerman, Stanley L., slavery 171–5
England *see* UK
environment
  costs of protection: public opinion on 610, 612, *613*, 616, 633; *see also* risk management
  deforestation 328–30, 331, 336–43: species loss and 339, 351, 352, 354–5, 356, 359–60
  farming technology and 381
  greenhouse effect 17–18, 342, 488, 544–61, 565–74
  ozone layer 17–18, 536–41, 542, *543*
  predictions about 645
  public opinion 558–60, 561, 610–17, 629, 633
  soil erosion 381, 416–23
  species loss 15, 339, 343, 346–60, 602
  water management 425–32

environment (*Cont.*)
  see also natural resources;
    pollution
EPA (Environmental Protection
    Agency)
  air pollution 494–7, 498, *499*,
    505–7
  Alar 595–6
  dioxin risks 599
  fluoride 601
  risk management 632
epidemics
  control of 57, 59
  life expectancy trends and 32,
    51–2, 53
erodibility index *see* EI
Ethiopia
  daily per capita calories *142*
  deforestation *340*
  famine 396
  ox-drawn plows 383
  smoking *109*
ethyl dibromide *see* EDB
Europe *see named countries*
European Community
  air quality 481, 485–6, 487,
    488
  nuclear power 301
  see also named member countries
Eveleth, Phyllis, height trends
  144
Evelyn, John, air quality 476–7
executions
  England 96–7
  US lynchings *186*
explosions, fatalities *102*
Ezell, M. J., CFC estimates 537

factories, boring tasks 20–1
famine
  Africa 386–7, 396, 403
  crisis mortality 61–3
  doomsayers' forecasts 22
  economic growth and 136
  Irish potato famine
    588–9
  USA 154
FAO (Food and Agriculture
    Organization)
  cereal production *377, 378,
    381, 383, 384*
  deforestation 328, 333, *334,
    335*, 338, *340–2*
  fish harvests *411*
  hunger quantification 394–5,
    399, *400*
  per capita grain production
    379, 380
farming *see* agriculture; livestock
    farming
Farr, Rodger, homelessness *263*
fatal accidents
  forecasts 655
  nuclear power hazards
    579–80, 585
  USA 98–105
FDA (Food and Drug
    Administration)
  Alar-apple scare 596
  antibiotics in cows' milk 598
  cyclamates 604
  food irradiation 597
  microwave ovens 602

risk management 630
saccharin 601
Feldman, J. J., health 77
fertility (human), income
    growth and 657–8
fertilizers 319, 364–5, 371,
    385
Feshbach, Murray, USSR
    health and mortality 85–90
Filer, Randall K., homelessness
    257–68
Finland
  agricultural labor force 126,
    127
  consumption per capita *159*
  education *219, 220, 222*
  labor productivity *168*
  smoking *109, 111*
  tuberculosis 56
  water pollution *518*: Baltic
    Sea 445–6, 466[9]
  work time *230*
Finlayson-Pitts, B. J., CFC
    estimates 537
Finley, Moses I., slave societies
    172
fires, fatalities *102*
Firket, J., air pollution 491
fish harvests 411–15
  water pollution and 428, 532,
    533, 639
Fisher, Irving, life expectancy
    34
Flieger, Wilhelm, life
    expectancy *31*, 32, 33, *40*
Flinn, Michael W., mortality
    140
floods, fatalities *101*
Floud, Roderick, body height
    69, *140*, *144*
fluorine
  availability *318*, 319
  fluoridation 601
Fogel, Robert W. 31, 61–9,
    136, *140*, *143*, *144*
Fonselius, Stig H., water
    pollution 446
food
  availability trends 11, *12*, 13,
    364–415, 639: cereal
    cultivation origins 365–6;
    doomsayers' forecasts 22;
    estimates 369–71; global
    warming and 566; grain
    stocks 405–9; markets and
    367–9, 386, 406–9;
    medieval revolution
    366–7; nutritional well-
    being 394–403; sea food
    411–15, 428, 639
  effect on economic growth 136
  energy requirements 63–5,
    142–4
  health scares 589–98, 600–2,
    604–5, 630
  irradiation 597
  mortality and 61–9, 70–1
  US living standards 150,
    154–6: daily per capita
    calories *142*; low-income
    Americans 247–51; slaves
    178–9; sugar consumption
    143, 155

Food and Agriculture
    Organization (UN) *see* FAO
Food and Drug Administration
    (US) *see* FDA
forests 328–43
  acidification effects 484,
    528–31, 638–9
  debt-for-nature swaps 342–3
  extent and characteristics of
    332–3
  forest products 330–1, *332,
    343*
  outdoor recreation 324, 325,
    326
  ownership problems 342–3
  plantation forestry 328, 330,
    334–6
  in temperate regions 328,
    333–4: plantations 334–5;
    USA 331–2, 352, 434,
    435, 436
  tropical deforestation
    328–30, 335–43: species
    loss 339, 351, 352, 354–5,
    356, 359–60
fossil fuel pollution 342
  acid rain 523, 525, 527
  economic systems compared
    504, 505, 511, 514–17, 520
  nuclear power compared
    579–80, 582, 583, 585
  UK: long-term trends 447,
    455–6, 476–8; methane
    emissions 488; smoke
    controls 476–81
France
  agricultural labor force 124,
    125, 126, 127
  agricultural productivity 366,
    368, 377
  air pollution *504, 512, 516–17*
  alcohol consumption 115,
    116, 117–20
  beer output *141*
  capital punishment 97
  consumption per capita *159*
  cotton consumption *142*
  economic growth *139, 147*
  education *219, 220, 222*
  height trends 67, 68, *144*
  housing *244*
  infanticide 92
  labor productivity *168, 169*
  life expectancy 8, *9*, 31
  mortality: accident fatalities
    103; calorie consumption
    and 64–5; causes of death
    86; children 42; crisis
    mortality 61–3; height and
    67, 68; suicide rates 93
  nuclear power 300, 301
  nutrition 64–5, *142, 143, 250*
  serfdom and slavery 173, 174
  smoking *109, 111*
  water pollution *518*
  work time *230*
Frasché, D. F., mineral
    resources 315
free time 20, 21, 224–30
  outdoor recreation 323–7,
    650
Freeman, Richard B.,
    homelessness 257, *264*

Freudenberger, Herman,
  nutrition 136
Frey, H. Thomas, US land use
  434–40
Friends of the Earth, acid rain
  638–9
fuel resources *see* energy
  resources; gas; oil

Gallaway, L., unemployment
  185
gallium, consumption and
  availability *317, 321*
Gallup polls
  adult smokers *107*
  environmental issues *615*, 616
  population growth 619, 620,
    624
game hunting, USA 325–6
garbage *see* waste
Gardner, Bruce L., soil erosion
  416–23
gas
  fossil fuel pollution 342
  petrochemical substitutability
    319–20
  price and consumption *283*
gasoline *see* motor vehicles,
  exhaust emissions; oil
GDP and GNP
  nutrition and *400, 402*
  per capita growth rates
    136–9, 141–2, 147–8
  pollution and *506, 507–8*,
    520: global warming 566,
    567–73
  US deficits and surpluses and
    275–6
General Circulation Models
  (GCMs), radiative forcing
    546, 548–51, 553, 565
genetic engineering
  agricultural productivity and
    381–3
  commercialization of
    biodiversity 343
  potential hazards 593
genetic resources
  gene mapping *170*
  germplasm banks 339
  see also species loss
germanium *318, 321*
Germany
  air pollution *504, 512, 516–17*
  alcohol consumption 115,
    116–17, 118–20
  beer output *141*
  consumption per capita *159*
  cotton consumption *142*
  education *218*, 219–20, *222*
  housing *244*
  income *139, 194*
  international rates of
    economic growth *147*
  mortality: accident fatalities
    103; causes of death *86*;
    suicide rates 93
  nuclear power 301
  nutrition: meat *250*; sugar
    143
  pollution and economic
    growth 511, *512–13*
  poverty *238*, 239

pre-industrial living
  standards 137
productivity: labor *168, 169*;
  manufacturing industry
    272, *273*
  serfdom 173
  smoking *109, 111*
  water pollution 517, *518*
  work time *230*
Gershuny, Jonathan, work time
  227
Ghana
  agricultural productivity 387
  nutritional well-being
    402
  smoking *109*
Gilder, George, government
  649
Gillis, Malcolm, deforestation
  329
Gleissberg cycle *540*
Global 2000 Report (GTR) 1
  fish harvests 411
  forests 328
  species loss 346–7, 348, 350,
    351
  water 425
global cooling 646
global firms, government role
  and 649
global warming *see* greenhouse
  effect
GNP *see* GDP and GNP
Goeller, H. E., mineral
  resources 313–22
gold, availability *318*
Goldsmith, J. R., photochemical
  pollution 491
Golze, Alfred R., US water
  supply 427
Gottschalk, Peter, poverty *232*,
  233
government
  future role 649, 658–9
  regulatory efforts for risk
    management 629, 630–4
  US federal spending 275–6
  welfare programs 234–5,
    238, 239, 248, *249*, 251,
    252
grain production *see* crop yields
Grantham, George W.,
  agricultural productivity
    364–75
Gray, Thomas, *Elegy Written in
  a Country Churchyard* 208
Great Britain *see* UK
Great Lakes, pollution 449–51,
  452, 467[18], *470–2*
Greece
  consumption per capita *159*
  housing *244*
  nutrition *400*
  slavery 172
  smoking 108, *109, 111*
greenhouse effect 17–18,
  544–61, 565–74
  abatement: costs and benefits
    567–72; policy selection
    572–3; public opinion and
    558–60, 561
  adjustment costs in warmer
    climates 566

clouds and 548–9, 557–8, *559*
deforestation 342
future and 645–6
temperature measurements
  and projections:
  compensating emissions
    557–8, *559*, 561; day or
    night warming 555, 557,
    560, 561; gas
    concentrations and 545–6,
    548, 554, 555, 558, 560,
    565–6; General Circulation
    Models (GCMs) 546,
    548–51, 553, 565; global
    versus hemispheric 546–8;
    latitude effects 553–4, *555,
    556*, 561; site bias 553,
    554; urban heat islands
    551–2, 555, 561
  UK contribution to 488
  uncertainties modelled 567–8
Greenpeace, acid rain 638
Griggs, Tim, biodiversity 339
Gruenberg, E. M., health 78–9
Gulland, J. A., fish harvests
  413, 414
Gupta, P. C., tobacco-related
  disease 110

Haagen-Smit, A. J.,
  photochemical pollution
    491, 492
hafnium, availability *318, 321*
Haines, Michael R., disease and
  health 32, 51–9
Haiti
  deforestation *340*
  slavery 174
  smoking *109*
Hall, Brian, homelessness *257,
  264*
Hall, C. A. S., pollution 342
Hammond, E. C., smoking 107
Hanke, S. H., water demand
  430
Hanley, Susan, Japanese living
  standards 136
Harley, C. Knick, income
  growth 138
Harrington, Walt, racism 184,
  187
Harrington, Winston,
  deforestation 339
Hastings, Alan, species loss
  356
Hausman, William J., energy
  prices 280–4
Haveman, R., health 77, 80
Hayek, Friedrich, ideas 658,
  659
hazard warnings 631
HCH (BHC) levels, in human
  tissue and milk *510*
HCN (Historical Climate
  Network) 555
Heady, soil erosion 418, 419
health 51–9
  AIDS 653
  antibiotics in cows' milk
    597–8
  aspartame 600
  bovine somatotropin (BST)
    and 596–7

health (*Cont.*)
  cancer: carcinogen scares
    588–94, 595–6, 597,
    598–9, 601–5; nuclear
    power 576–7, 582; ozone
    layer 540–1 USSR *89*
  Chilean grape scare 630
  DDT 589, 592, 602–3
  fertility forecasts 649, 657–8
  fluoridation 601
  forecasts 647–8, 650, 652–3
  indicators of 73, 82
  life expectancy: children *43*,
    44, 46–8; practices in 34;
    risk comparisons 583–5;
    urbanization 32
  malnutrition and 63: *see also*
    nutrition
  mental illness and
    homelessness *263–5*, 266
  PCBs 599–60
  pollution and: acid rain 531;
    atmospheric lead 459,
    460; coal-burning or
    nuclear power 579–80,
    582, 583, 585; ozone layer
    536, 540–1; pesticides
    588–94, 602–3; smoke and
    sulfur oxides 468[27], 476,
    477–9, *482*; toxic waste
    603–4
  smoking and 107–8, 110,
    111, 112–13
  USA 72–82: medical care
    153, *159*, 234
  USSR 85–90
  see also disease
health aids, living standards
  18–19
heating, US living standards
  150–1, 152–3, 160
Heckman, James,
  African-Americans 182
height
  economic growth and *144*,
    145
  mortality risk prediction and
    65, *66–7*, 68–9, 70–1
  slaves 178–9
Heimlich, Ralph E., soil
  erosion 417, 419
Heise, Lori, forests 336
helium, availability *318*
Henry, L., child mortality 39,
  40
hepatitis, USSR 89
Heywood, V. H., species loss
  352, 353
Higgs, Robert
  African-Americans' living
    standards 178–87
  government 649
higher education
  USA 212–13, 220–1, *222*
  worldwide 214, *217*, *219*, *222*
Hill, A. B., smoking 107
Hill, K., child mortality *10*,
  37–48
Hill, W. J., ozone layer 539
Hirshleifer, Jack, industrial
  water use *431*
Historical Climate Network
  (HCN) 555

Hobcraft, J. N., infant
  mortality 401
Hoch, Charles, homelessness
  260–1
Hoffmann, W. G., alcohol *115*
Holden, Constance, species
  loss 350
Holen, Arlene, US accident
  rates 98–105
Hollingsworth, T. H., child
  mortality 39, *40*
home accidents, USA 98, 100
home ownership *see* housing
homelessness, USA 237,
  257–68
  among families 261–2, *263–5*,
    266
  causes of 259–60, 266–7
  characteristics of homeless
    persons 262, *263–5*, 266
  estimates of 257–8
  history 260, 261
  policy on 260–1
  public shelters 260–1
  variation across cities
    258–60, 262, *263–5*, 266
homicide rates 91–2, 96
  lynchings *186*
Honig, Marjorie, homelessness
  259–60
Hooker Electrochemical
  Company, toxic waste
  603–4
Hoover, Herbert, poverty 231
Hopkins, M. K., child mortality
  39
Hopper, W. D., rice yields *374*
Horn, D., smoking 107
Houk, Vernon, dioxin risks 599
household appliances, US living
  standards 152–3, *159*, 160,
  245–7
housing
  home ownership 23
  predictions for 647
  USA 151, 160:
    African-Americans 180,
    181, 182, *185*; cost and
    quality trends 203–7;
    economic performance and
    276–7; homelessness 237,
    257–68; poverty levels and
    241–5; subsidies 234
Howell, N., life expectancy 38
Huck, Paul, economic growth,
  in Britain 139
HUD estimates, homelessness
  257, *259*
Hudson, William J., grain
  stocks 405–9
Hume, David, attitudes to the
  past 3
Hungary
  body height *67*, *144*
  pollution and economic
    growth *509*, 511, *512–13*
  serfdom 173
  smoking *109*, 111
  starchy staple ratio *400*
  suicide rates 93
Hunter, Robert, poverty 231–2
hunting, USA 325–6
hurricanes, US fatalities *101*

hydrogen
  consumption and availability
    316
  iron substitutability and 319
hydrogen peroxide, acid rain
  524
Hyman, Merton M., alcohol
  114

Iceland
  accident fatalities 103
  smoking 108, *109*, 111
ideas, importance of 654–6,
  658–9
Idso, S. B., global warming
  545, 546
illiteracy rates 214, *216*
illness *see* disease; health
incineration, of waste 461, *475*
income
  changes over time *305*
  forecasts 644, 657–8
  USA 188–94: 1970s–1990
    272–4; African-Americans
    179–83, *186*, 236–7; one-
    and two-earner families
    271–2; poverty 231–7,
    241–52
  see also economic performance
income tax
  census data and poverty
    estimates 251–2
  home ownership 205
  US economic performance
    275–6
India
  agricultural labor force
    129–30, *132*
  agricultural productivity 377:
    grain stocks 408, 409;
    irrigation 386; nutritional
    well-being 403; ox-drawn
    plows 383; population
    growth and 389, 403; rice
    374
  forecasts 654
  forests 335, *340*
  housing space 243
  infant mortality 401
  life expectancy 33
  literacy rates *217*
  nutrition *142*, *143*, 403
  smoking *109*, 110, *111*
indium, availability *318*, *321*
Indonesia
  agricultural labor force
    129–30, *131*
  agricultural productivity *374*,
    377, *398*, 399
  forests 335, *340*
  nutrition *398*, 399, *400*, *402*
  smoking *109*
industrialization
  economic growth with 135–45
  health risks 56, 59
  pollution: acid rain 482–6,
    523–5; air quality trends
    454–7, 480–6; albedo
    effects 557–8, *559*;
    dispersion of pollutants
    482–5, 487; economic
    systems compared 503–20;
    Great Lakes 449–51,

industrialization (*Cont.*)
467[18], *470–2*; historical
phenomenon 446–7, 448,
464[3], 466[3]; lead 457–9;
New York City Harbor
451–4
infant and child mortality rates
38–48
African-Americans 5–6, 179,
181
causes of death *43*, 44, 46–7,
53, 57
data issues 38–9
developing countries 44–7,
401, *402*
Europe 39–44, *140*
explanations for US trends
78–9
fatal accidents in USA 98–9
news slant 5–6
nutrition and 401, *402*
overall life expectancy trends
and 9, *10*
in pre-and early history 38–9
USSR 87–8, *90*
infants
infanticide 92
nutrition, low-income
families 250, 401
infectious diseases 51–2
AIDS 653
cholera 53, 58
control of 57–8, 59
diphtheria 57, 85, 89
influenza 32, 52, 90
life expectancy trends and
53–7: children *43*, 44,
46–7, *53*; influenza
epidemics 32
measles 52, 53, 57, 90
polio 90
scarlet fever 57
smallpox 52, 53, 57
tuberculosis 54–7, 85, 89
typhoid 90, 150
US medical care 153
USSR 87, 89–90
influenza 32, 52, 90
Ingegneri, Dominique G., US
health 72–82
international firms, government
role and 649
invention and discovery 25–6
continued progress 646–9
journalistic bias and 655
productivity and 161–70
iodine, consumption and
availability *317*, 319
IPPC, greenhouse gases 566
Iran, smoking *109*
Iraq
slavery 174
smoking *109*
Ireland
agricultural productivity 365
consumption per capita *159*
education *219*, *220*, 222
famine 588–9
housing *244*
smoking *109*, *111*
television ownership 253[20]
iron *316*, *318*, 319
irradiation *see* radiation

irrigation systems 386, 427, 432
Africa 387
Aral Sea 428–9, 519
Irving, Patricia M., acid rain
528
island biogeography, species
loss 354, 360
Israel
nutrition *400*
smoking *109*, *111*
Italy
agricultural labor force 127,
128
air pollution *504*, *516–17*
consumption per capita *159*
education *219*, *220*, 222
homicide rates 91
housing *244*
income *139*
international growth rates
compared *147*
labor productivity *168*, *169*
life expectancy 32, *55*, 56–8
meat consumption *250*
nuclear power 301
Seveso disaster 598
smoking *109*, *111*
sugar consumption *143*
suicide rates 93
see also Roman empire
IUCN (International Union for
the Conservation of Nature
and Natural Resources),
species loss 339, 346, 351,
352
Ivory Coast
deforestation *340*
nutrition *402*
smoking *109*, *111*

Jamaica
deforestation *340*
nutrition *402*
smoking *109*
James, Franklin, J.,
homelessness *263*
Japan
agricultural labor force 127,
128
consumption per capita *159*
economic growth 136: life
expectancy and 136, 633;
pollution and 511, *512–13*;
starchy staple ratio and *400*
education *218*, 219–20, 222
health 77
housing 207, *242*, *243*, *244*
income 139
infectious diseases *55*, 56, 58
international comparisons of
living standards *147*, *160*
labor productivity *168*, *169*
manufacturing productivity
272, *273*
mortality *55*, 56, 58, *86*, 102,
136, 633
nutrition *142*, *143*, 250
PCB health scare 600
pesticides *510*
pollution 505
rice production *374*
smoking 108, *109*, 110, *112*
suicide rates 93

Jevons, Stanley, coal crisis 25
joblessness
trends in 196–202
see also labor force
Johnston, F. E., life expectancy
38
Jones, Eric, economic growth
136, 137
Jones, P. D.
temperature trends *547*
urban heat islands *552*
Jorgenson, Dale W., greenhouse
effect 566
journalism
news bias 4–5, 6, 208–9,
218–22, 654–5, 659
newspaper readership 22
public opinion and 612,
624–6
use of term 'ecology' 637
Jukes, Thomas H., nature 588

Kahn, Herman, realism of 651
Karl, T. R., global warming
551–2, 555, *556*
Kendall, Henry, nuclear power
576
Kenya
agricultural productivity 377
deforestation *340*
nutrition *402*
smoking *109*, *111*
Kerr, R. A., ozone layer 538
Keyfitz, Nathan, life
expectancy *31*, 32, 33, *40*
Keyssar, Alexander, US
unemployment 196–201
Kinnison, D. E., supersonic
transport and ozone 540
Knight, Frank, wants 656
Koop, C. Everett
Alar-apple scare 596
smoking 107
Korea
deforestation *340*
international growth
comparisons *147*
nutrition *400*
rice production *374*
smoking *109*
Kovar, M. G., health surveys 73
Kramer, J. R., water pollution
449
Krutzfeldt, Janette, grain
production 408–9
kudzu 381
Kulp, J. Laurence, acid rain
523–34
Kuznets curve, income
distribution 193–4
Kuznets, Simon, economic and
population growth 16

La Gory, Mark, homelessness
263
labor force
accidents at work 98, 100,
105
African-Americans 171–5,
181–2, *185*, 199
in agriculture 10, *11*, 124–34,
225
boring tasks 20–1

labor force (*Cont.*)
  energy requirements of 63–5:
    slave diet 178–9
  poverty and 235–6
  productivity 161, *163*, 164,
    165, 166–7, *168–70*, 271,
    272
  slavery and serfdom 171–5,
    178–9, 190
  trends in make up of 224,
    271–2
  unemployment trends
    196–202
  work space predictions 647
  work time 20–1, 224–30
  working years 225–6
  workplace smoking *110–11*,
    112
Ladd, Everett C., environment
  protection 616, 617
lakes
  acid rain's effects on 532–3
  Great Lakes pollution
    449–51, *452*, 467[18], *470–2*
Lamb, W. H., child mortality 47
Lancaster, H. O., infectious
  diseases 57
land contamination
  nuclear power and 580–3
  toxic waste 603–4
land use
  forecasts 650
  US trends 434–40
Landes, David, economic
  growth 136
Lanly, J. P., forests 335, 336,
  338, *340–2*
lanthanum, availability *318*
Latin America *see* Central
  America; South America
Lave, Lester, greenhouse
  scenarios 565–74
Laxen, D. P. H., air pollution
  480, 484
lead
  availability *318*, 321
  prices 303–4, *305*
lead pollution 457–60
  USA *17*, 455, 460, *474*, *494*,
    496–7, *498–500*, *502*
  vehicle exhaust emissions 488
Lebergott, Stanley
  unemployment 201[3]
  US living standards 149–56,
    242, *243*, *244*, *245*
Ledermann, Sully, alcohol *116*,
  117
Lee, D. O., air pollution 485
Lee, Dwight, government 649
Lee, Ronald, life expectancy 9,
  32, 63
Lefohn, A. S., acid rain 523
leguminous crops 367, 381
leisure time 20, 21, 224–30
  outdoor recreation 323–7, 650
Lewis, C. S., Hell 657
life expectancy 8–10, 14, 30–4
  African-Americans 179, *180*,
    181
  children *see* infant and child
    mortality rates
  economic growth and 139,
    140

forecasts 644, 647
health and disease and 51–7
risk comparisons 583–5
USSR 85–8
light sources, innovations in
  164, *165*
"Limitation of Activity", health
  indicators 73, 82
Linaweaver, F. P., water use
  430
Lindert, Peter
  economic growth 139
  poverty rates *232*
  US wealth distribution
    188–95
Lindzen, Richard, greenhouse
  effect 566
literacy rates 214, *216*
lithium *317*
livestock farming
  antibiotics in cows' milk
    597–8
  BST health scare 596–7
  dietary adequacy 398
  fish meal 414
  methane production 488, 545
  population growth and 388–9
  USA 391, *393*
living standards 13–19
  absolute and relative progress
    6–7, 653–4
  accident rates and 105
  agricultural labor force
    124–34
  disease and 58–9
  economic growth and
    135–45, 147–8
  environment 14–15, *16–18*,
    17–19
  forecasts summarized 642–59
  health aids 18–19
  income and wealth
    distribution 179–83, *186*,
    188–94, 231–7, 241–52
  population growth 16–17
  purchasing power 13, 147–8
  USA 149–60:
    African-Americans
      178–83, 184–7, 236–7,
      250; consumer spending
      156, *158–60*; food 150,
      154–6, 247–51; free time
      224–30; heat 150–1,
      152–3, 160; housework
      and household appliances
      152–3, 155, *159*, 160,
      245–7; housing *see* housing;
      medical care 153, *159*,
      234; outdoor recreation
      323–7; poverty 151,
      231–40, 241–52; water
      149–50, *160*; wealth and
      income distribution
      179–83, *186*, 188–94,
      236–7
Logan, W. P. D., infant
  mortality 44
Lovejoy, species loss 348,
  350–1, 356–7
Lovelock, J., CFC warming
  effects 546
Low Countries *see* Belgium;
  Netherlands

Lowdermilk, W. C., soil
  erosion 418
Ludwig, J. H., air pollution
  492, 493, 495
Lugo, Ariel E., species loss
  351, 353–4
lumber
  net volumes *332*
  price trends 330–1
Lunch, William M.,
  environment and energy
  610–17
lynchings, USA *186*

McCarroll, Thomas,
  substitutability 321
MacDonell, W. R., mortality 39
McGregor, I. A., mortality 45
McKenzie, Richard,
  government 649
McKeown, Thomas, disease
  51, 54–5, 57, 58
McNeill, William H., disease 52
Maddison, A.
  economic growth 136, *139*,
    141
  productivity *168*
magnesium
  availability *317*
  prices 311, *312*
  recycling 315
malaria, DDT and 589,
  602
Malaysia
  deforestation *340*
  nutrition *400*, *402*
  rice production *374*
  smoking *109*, *112*
malnutrition
  body mass index and
    mortality risk 65–9, 70–1
  causes and consequences 63
  energy requirements and 63–5
  US poverty levels and 250,
    251
  see also nutrition
Malthus, T. R., demographics
  367
Manabe, S., climate models *549*
manganese 311–12, *317*, *318*,
  319
Manley, S. L., CFC estimates
  537
Mann, Charles C., species loss
  355, 360
Mann, Patrick, water demand
  431
Manne, Alan S., greenhouse
  effect 566
Manthy, Robert, wood
  resources 330
Manton, K. G., health 79
manufacturing industry,
  productivity 272, *273*
Marcucci, M. A., ozone layer 539
Margo, R. A.
  African-Americans' living
    standards 178–87
  unemployment 198
marine accidents, fatalities *101*
market mechanisms
  agricultural productivity and
    367–9, 386, 406–9

market mechanisms (*Cont.*)
  for progress 24–6
  risk reduction policies 633–4
  socialist economies and,
    pollution trends compared
    503–20
  water supply and demand
    429–32
Marshall Institute, greenhouse
  effect 565, 566
Martin, A., air pollution *479*
Mathers, C. D., health 77
Mayr, Ernst, species 348
Meadows, D. H., mineral
  resources 313, 319
measles 52, 53, 57, 90
meat consumption 248, 249, *250*
meat production 388–9, 398
media
  news bias 4–5, 6, 208–9,
    218–22, 654–5, 659
  newspaper readership 22
  public opinion and 612,
    624–6
  use of term 'ecology' 637
Medicaid and Medicare 234,
  252
Meehl, G. A., climate model
  550, *551*, 553
Mellanby, Kenneth, natural
  ecology 637–40
Melograni, P., education *217*,
  218
mental illness, homelessness
  and *263–5*, 266
Merck-INBio arrangements 343
mercury
  availability *318*, 320
  prices 304, *306*, 320
metals
  availability and price 303–12
  substitutability 313–22
methane
  Antarctic ozone hole 538
  CFCs and 536
  as greenhouse gas 488, 545,
    560
Mexico
  agricultural labor force 128,
    129, 133–4[8]
  deforestation *340*
  education 218
  homicide rates 92
  maize production 395
  nutrition *143*, *250*, *400*, *402*
  smoking *109*, *110*
Michaels, Patrick J., greenhouse
  effect and global change
  544–61
microwave ovens, radiation
  exposure 602
migration
  living standards and 650
  population growth and 622–3
military defense
  incidence of war 654
  US: land use 439; spending
    275
Miller, Kenton R., species loss
  351, 352
Mills, A. L., lead pollution 459
Mims, F. M., ozone layer 539
minerals (dietary) 251

minerals (fuel) *see* coal; oil;
  wood fuel
minerals (non-fuel)
  availability and price 303–12,
    645
  substitutability 313–22
mining accidents, fatalities *102*
Mitchell, agricultural labor
  force *11*
Mitchell, B. R.
  beer output *141*
  cotton consumption *142*
  economic growth *139*
  mortality rates *41*, *45*, *140*
mobility 21
  agriculture and 368, 386
Mokyr, Joel, living standards
  135–45
molybdenum, availability *318*,
  319, *321*
monks, mortality rates 31
monuments, environmental
  damage 446–7, 531
Moore, Stephen, non-fuel
  minerals 303–12
morbidity *see* disease
Morse, Gary, homelessness
  264–5
mortality 8–10, 14, 30–4
  African-Americans 179, *180*,
    181
  causes of death: accidents
    98–105; disease 52–7;
    lynchings *186*; murder and
    suicide 91–7
  health and disease and 51–9:
    control of disease 57–8;
    US trends explained 78–9,
    80, 81–2
  infants *see* infant and child
    mortality
  nutrition and 61–9, 70–1
  risk predictors, body mass
    index (BMI) 63, 65–9,
    70–1
  smoking and 107, 110
  USSR 85–9
motor vehicles
  exhaust emissions: acid rain
    523, 524; economic system
    and *504*, 505, 511, *515*,
    516–17; UK 486–8; USA
    491, 496–7, 511, *515*,
    516–17, 524
  top speed *166*
  US accident rates 98, 100
  US ownership and spending
    on *159*, *160*, 245, *246*, 247
  water used in production of
    431
Mowbray, C. V., homelessness
  263
Mulhall, Michael, sugar
  consumption *143*
Mulkern, V., homelessness *263*
multinational firms, government
  role and 649
murder rates 91–2, 96
  lynchings *186*
Murozumi, aerosol pollutants
  459
Muth, Richard D., housing
  203–7

Myers, John G., non-fuel
  minerals 303–12
Myers, Norman
  deforestation 338, *340–2*
  species loss 346, 348–51, 354
Nader, Ralph, nuclear power
  576
NAPAP (National Acid
  Precipitation Assessment
  Program) 523, 524, *526*
  buildings 531
  cost benefits of reducing acid
    rain *533*, 534
  forests 528–9, 530
  health 531
  lakes *533*
NAS (National Academy of
  Sciences)
  forests 338
  global cooling 646
  greenhouse gases 565, 566
NASA
  ozone layer 537, 538–9
  urban heat islands 551–2
natural ecology 637–40
natural gas
  fossil fuel pollution 342
  petrochemical substitutability
    319–20
  price and consumption *283*
natural resources 10–13
  energy *see* energy resources
  forecasts 645, 647
  forests 328–43
  mechanisms for progress 24–6
  non-fuel minerals 303–12,
    313–22, 645
  outdoor recreation 323–7,
    650
  species loss 15, 343, 346–60,
    602: tropical deforestation
    and 339, 351, 352, 354–5,
    356, 359–60
Natural Resources Defense
  Council (NRDC), Alar
  595–6
Nelson, Robert H., outdoor
  recreation 323–7
Nemeskeri, J., life expectancy
  30, 38
Nepal
  deforestation *340*
  rice production *374*
  smoking *109*
Netherlands
  agricultural labor force 124,
    125, 126, 133[3]
  agricultural productivity 365,
    366–7, 368
  air pollution *504*, *512*, *516–17*
  child mortality 42
  education *219*, *220*, 222
  future challenges 657
  income *139*, *194*
  labor productivity *168*
  pre-industrial living
    standards 137
  slavery in Dutch colonies 174
  smoking *109*
  water pollution *509*, *518*
New York City Harbor,
  pollution 451–4

New York Times
  environmental issues 610,
    611, 612, *613*, 617
  population growth references
    624–5, 626
New Zealand
  education *222*
  forests 335
  infectious diseases *55*, 56, 58
  smoking *109*, *112*
  water pollution *518*
Newacheck, P. W., health 77
newspapers
  news bias 4–5, 6, 208–9,
    218–22, 654–5, 659
  public opinion and 612,
    624–6
  readership 22
  use of term 'ecology' 637
Nicholson, E. M., ecology 637
nickel
  availability *317*, *318*
  carcinogenic effects 582, 592
  prices 304, *306*
Niemi, R. G., energy resources
    615
Nigeria
  nutrition *143*, *402*
  smoking *109*, *111*
Nile river, pollution 461–4, *465*
niobium, availability *318*
nitrogen
  consumption and availability
    316
  leguminous crops 367, 381
  nitrate fertilizers 364–5
  soil runoff 419
nitrogen oxides
  acid rain 483, *484*, 485–6:
    causation 523, 525;
    distribution and trends
    525, *527*; effects 528,
    530–1
  CFCs and 536
  economic growth, and,
    economic systems
    compared 508, 511,
    512–13, *514*
  as greenhouse gas
    545
  ozone concentrations and 487
  UK levels 487–8
  USA *17*, 455, *474*, 497,
    498–500, *502*
Nordhaus, William D.,
    greenhouse effect 566, 568
North America *see* Canada; USA
Norway
  air pollution *504*, *512*, *516–17*
  education *219*, *220*, *222*
  height trends *144*
  housing *244*
  income *139*
  labor productivity *168*
  mortality risk, body mass
    index (BMI) and 65, *66–7*,
    68
  poverty *238*, 239
  smoking *109*, *111*
  water pollution *518*
NRDC (Natural Resources
    Defense Council), Alar
    595–6

nuclear power
  costs 294–301: plant design
    298–301; regulation and
    297–9, 300
  forecasts 645, 647
  hazards 576–85: radiation
    risks 576–7, 583–5;
    radioactive waste 580–3;
    reactor accidents 578–80;
    routine emissions 577
nutrition
  availability of food resources
    11, *12*, 13, 394–403:
    Africa 401, 403; Asia 401,
    403; behavioral evidence of
    dietary adequacy 398–400;
    quantification problems
    395–6; vulnerable groups
    401; *see also* agricultural
    productivity
  effect on economic growth
    136
  energy requirements 63–5,
    142–4
  morality and 61–9, 70–1
  sea food 411–15
  US living standards 150,
    154–6: daily per capita
    calories *142*; low-income
    Americans 247–51; slaves
    178–9; sugar consumption
    143, *155*

Oates, Wallace E.,
    environmental quality
    444–75
oceans, global warming and:
    cloud brightening 558;
    ocean dynamics 546–7,
    548, *550*, 553, *554*; sea
    level rise 557
oil 11
  development of 25–6
  doomsayers' forecasts 22, 23
  fossil fuel pollution 342
  petrochemical substitutability
    319–20
  prices and supply 287–93:
    public opinion and
    615–16; USA *283*, *284*,
    287–91, *293*
  spills 414–15, 633
  US whale oil crisis 25
  water used in refining *431*
oilseed production 378, 382
Olson, Storrs L., species loss
    357
Omran, Abdel R., disease and
    mortality 51, 52, 59
OPEC, oil prices 288
opinion polls *see* public opinion
organic fertilizers, crop yields
    and 364–5
Osler, William, medical care
    153
outdoor recreation 323–7, 650
Owens, William, education 651
oxygen, consumption and
    availability *316*
ozone concentrations
  UK 486–7
  USA *17*, *474*, *494*, 496,
    498–500, *502*

ozone layer 17–18, 536–41,
    542, *543*
  Antarctic ozone hole 537–8
  future and 645–6
  global ozone decrease and
    538–40
  ozone photochemistry theory
    536–7
  skin cancer 540–1

Paddock brothers, *Famine 1975*
    22
Pakistan
  infant mortality 401
  nutrition *143*, *402*, 403
  rice production 374
  smoking *109*, *111*
parks, outdoor recreation 323–6
particles, suspended *see*
    suspended particles
PCBs
  Great Lakes *472*
  health effects 599–600
  in human tissue and milk
    510, 600
Pebley, A. R., child mortality 46
Pellekan, J. V., hunger 394–5
Perrenoud, Alfred, disease and
    mortality 51, 53
Persson, Reidar, forests 338
pesticides 385–6
  Great Lakes *470*, *471*, *472*
  health effects 588–94, 602–3
  residue levels 508, *510*
  soil runoff 419, 519
Peto, Richard
  cancer 589
  smoking 110
petrochemicals, substitutability
    319–20
petrol *see* oil
Petty, William, the human
    condition 3–4
pH scale, acid rain 483, *524*
Philippines
  agricultural labor force
    129–30, *131*
  forests 335, *340*
  homicide rates 92
  nutrition 399, *400*, *402*
  rice production 374
  smoking *109*
Phillips, J. C., substitutability
    321
philosophers, role of 658–9
phones, USA *159*, *160*, 246
phosphorus
  consumption and availability
    316, *318*, 320–1
  Great Lakes levels of *470*
  substitutability 319
photochemical pollution
  UK 486–8
  USA 491–2, 496, 524
photochemistry theory, ozone
    536–7
Piliavin, Irving, homelessness
    263
Pitts, J. N., Jr, CFC estimates
    537
plague 51–2, 53, 57
Planned Parenthood,
    population growth 620

plantation forestry 328, 330, 334–6
platinum 307, *308*, 312, *317*
Poe, G. S., health surveys 73
Poland
  agricultural productivity 377, 382
  forest damage 484
  lead pollution 460
  nutrition *400*
  pollution and economic growth 511, *512–13*, 516–17
  smoking *109*, *112*
Poleman, Thomas T., food and nutrition 394–403
polio, USSR 90
pollution 14–15, *16*, 17–19
  acid rain 17–18, 483–6, 523–34, 638–9
  data issues 492–4: evidence collection 444–8, 461–4, *465*
  doomsayers' forecasts 22
  greenhouse effect 17–18, 342, 488, 544–61, 565–74
  long-term trends 444–75
  market and socialist systems compared 503–20
  mechanisms for progress 24–5
  nuclear power compared with coal 579–80, 582, 583, 585
  oil spills 414–15, 633
  ozone layer 17–18, 536–41, 542, *543*
  photochemical 486–8, 491–2, 496, 524
  public opinion about 612, 614, 616, 633
  soil acidification 484, 529–30
  soil contamination: nuclear power 580–3; toxic waste 603–4
  soil runoff 418–19, 422, 432
  vehicle exhaust emissions *see* motor vehicles, exhaust emissions
  see also air pollution; water pollution
population growth 16–17
  agricultural productivity 367–8, 379, 388–9, 403
  child mortality in prehistory 38
  economic growth and 16–17, 24–6, 657–8
  estimates of 35–6
  fertility forecasts 649, 657–8
  infectious disease and 52
  public opinion about 619–26
  scarcity trends 13
Portugal
  consumption per capita *159*
  education *219*, *220*, *222*
  housing *244*
  nutrition *400*
  slavery 173, 175
  smoking *109*, *112*
  water pollution *509*
Postel, Sandra, forests 336
potassium
  consumption and availability *316*

Great Lakes 449, 450, 451
Potter, Neal, wood resources 330
poverty
  nutrition and 247–51, 396, 398–9, *400*, 401, *402*, 403
  USA 151, 231–40, 241–52
  see also living standards
PRA (Probabilistic Risk Analysis), nuclear power 578–80
Pramaggiore, M., US health 80
Pratt, W. P., mineral resources 319
precipitation *see* rainfall
press reporting *see* journalism
Preston, Samuel H., mortality 9, 30–4, 54, 55–6, 57, 58
Probabilistic Risk Analysis (PRA), nuclear power 578–80
productivity 161–70
  agricultural *see* agricultural productivity
  manufacturing industry 272, *273*
  USA *161*, *168–70*, 271, *272*
psychiatric disorder, homelessness and *263–5*; 266
public health
  disease control 57, 58–9
  urbanization and 32
public opinion
  environment and energy 610–17, 629, 633
  perceptions and beliefs 654–6
  population growth 619–26
  press reporting and 6, 612, 624–6
Puerto Rico
  slavery 174
  species loss 351, 354
Puranen, Bi., tuberculosis 56
purchasing power 13
  international comparisons 147–8

Quinn, James F., species loss 356

radiation
  food irradiation 597
  from microwave ovens 602
  nuclear power 576–7, 583–5: radioactive waste 580–3; reactor accidents 578–80; routine emissions 577
railroads
  top speed *166*
  US fatalities *102*
rainfall
  acid rain 17–18, 483–6, 523–34, 638–9
  climate predictions 548
  soil erosion by 417, 418–19, *421*, 422, 423
Rakhimov, E. D., Aral Sea pollution *519*
Ramanathan, V.
  Antarctic ozone hole 538
  climate models 548
Raper, A. F., lynchings *186*

rapeseed oil 382
Raven, Peter, species loss 357, 358
raw materials *see* natural resources
reading, literacy rates 214, *216*
recreation
  leisure time 21
  outdoor 323–7, 650
Rector, Robert, US poverty 241–52
recycling
  of garbage *475*
  of minerals 315
Reed, M. H., CFC estimates 537
Reher, David, disease and mortality 51, 58
Reid, W. V., species loss 351, 352, 353
Reilly, William K., dioxin risks 599
religious orders, mortality rates 30–1
Repetto, Robert, deforestation 329
resources, natural *see* energy resources; natural resources
"Restricted Activity Days", health indicators 73
Reutlinger, Shlomo, hunger 394–5
Reynolds, Alan, US economy 271–7
rhenium, availability *318*, *321*
Ribaudo, Marc O., soil erosion 419, 420, 423
rice, starchy staple ratios 398
rice yields 378, *379*, 405, 406
  acid tolerant strains 383
  Africa 383, 387
  Asia *374*, *383*, *384*, 388
Richels, Richard G., greenhouse effect 566
Ricossa, S., education *217*, *218*
Riley, J. C., health 51, 77, 79, 80
Rinsland, C. P., CFC estimates 537
risk analysis, nuclear power accidents 578–80
risk management 628–34
  estimation mistakes 629–30
  informational approach 631
  market references 633–4
  misplaced conservatism 632
  regulatory efforts 632
  relative gains and losses 632–3
  responsibility assignment 634
  valuation distortions 630–1
Roberts, James S., alcohol 114–20
Robinson, John P., free time 224–30
Rodhe, H., air pollution *18*, 478, *480*
Roemer, Ruth, smoking 112
Roman Empire
  agricultural productivity 366
  life expectancy 30
  plagues and epidemics 51, 52
  slavery 172–3

Romer, Christina, unemployment 201[3]
Roper Reports
environmental opinions 611, 612, *613*, 614–15, 616, 617
population growth 622, *623*
Rorabaugh, W. J., alcohol *114*, 115–16
Rose, W. I., CFC estimates 537
Rosnow, M. T., homelessness 263
Rossi, Peter H., homelessness 259, 263, 267[1]
Roth, Dee, homelessness 264
Rothman, Stanley, environment and energy 610–17
Rowland, F. S., Antarctic ozone hole 537, 538
rubbish *see* waste
rubidium *317*
Russia *see* USSR
Rwanda *143*, *340*, *402*
Ryan, John, poverty 232

Salas, L. J., CFC estimates 537
Sansom, J., temperature patterns *556*
Saudi Arabia, smoking *109*, *111*
Sayer, J. A., species loss 346, 351–3
scandium *318*
scarlet fever 57
Schama, Simon, living standards 137
Schneider, living standards 6
Schneider, David M., homelessness 260
Schneider, S. H., climate models 548, 565
Schofield, R. S., life expectancy 31, *140*
children 40, *140*
disease and 51, 58
mortality crises 61–3, 64
schooling *see* education
Schrenk, H. H., air pollution 491
Schultz, Theodore W., soil erosion 416–23
Schutt, Russell K. homelessness 263
Schwar, M. J. R., air pollution 484
Science, global cooling 646
Science Digest, global cooling 646
Scotland *see* UK
Scotto, J., UV–B radiation 541
Scrimshaw, Nevin S., mortality and nutrition 64
sea food 411–15, 428, 639
Sedjo, Roger A., forests 328–43
selenium
availability *318*, 319, *321*
health effects 592
soil runoff pollution 432
Selowsky, Marcelo, hunger 394–5
Sen, Amartya, living standards 7, 135
Senegal *109*, *111*, *402*
serfdom 171, 173

sewage discharges, economic system and 518
Seymour, living standards 6
Shammas, Carole, nutrition *143*
Sharples, Jerry A., grain production 408–9
shipping accidents, fatalities *101*
silicon *317*, 319
silver 307, *318*, 321
Simberloff, D., species loss 352
Simon, Herbert A., agricultural labor force 133[3]
Simon, J. L.
agricultural productivity *373–5*
air pollution *480*
body mass and longevity 70–1
education 208–23
forecasts 642–59
living standards 158–60, 184–7
non-fuel minerals 303–12
oil production *292, 293*
population estimates 35–6
poverty 240
productivity 165–70
species loss 346–60
Simon, Rita J., population growth 619–26
Singapore
nutrition *400*
smoking *109*, 110, *112*
Singer, S. Fred, ozone layer 536–41
Singh, H. B., CFC estimates 537
Sivonen, S., forest products 331
skin cancer, ozone layer and 540–1
slavery 171–5
slave diet 178–9
wealth concentration and 190
Slayton, Robert A., homelessness 260–1
smallpox 52, 53, 57
Smeeding, Timothy, poverty *238*
Smith, Adam, slavery 172
Smith, James P., African-Americans 181
Smith, Robert Angus, acid rain 483
smoke control, UK 476–81
smoking 106–13, *158*, 631
Smolensky, Eugene, poverty 232–3
Snooks, Graeme, economic growth 137
Snow, C. E., life expectancy 38
social class, unemployment and 200
social service programs, US 234–5, 238, 239, *249*, 249, 251, 252
socialist economies, pollution trends 503–20
sodium
consumption and availability *316*
Great Lakes 449, 450, 451
soil
acidification 484, 529–30

contamination: nuclear power and 580–3; toxic waste 603–4
erosion 416–23: conservation tillage 381; erodibility index (EI) 416–17, 420; extent of in USA 416–18; as a social problem 418–22
fertility 367–8
solar cycle, ozone concentrations and 539–40
solar electricity
forecast 645
nuclear power compared 582, *583*
solid waste *see* waste
Somalia *340*, 396
South Africa
forests 335
mineral resources 320
slavery 175
smoking *109*
South America
agricultural productivity *377*, *381*, *385*, 388–9
forests *329, 330, 333, 334, 335–9, 340*, 352
lead pollution in Peru 460
nutrition 398–9, *400*, *402*
smoking *109*, *110*
see also named countries
Soviet Union *see* USSR
space heating, USA 150–1, 152–3, 160
space travel, productivity growth 165–6
Spain
agricultural labor force 127, *128*
consumption per capita *159*
education 218, 219, 220, 222
housing *244*
slavery 173
smoking *109*, *112*
sugar consumption *143*
Sparhawk, W. N., forests 336, *337*, 338
species loss 15, 346–60
biologists' goals 356–7
commercialization of biodiversity 343
DDT and 602
defining species 347–8
loss estimates 348–55
tropical deforestation and *339*, 351, 352, 354–5, 356, 359–60
value of species-saving 357–8
spirits, alcoholic, consumption of 114–20
Sri Lanka
child mortality 47
DDT use 589, 602
deforestation *340*
nutrition 398, 399, *400*, *402*
rice production *374*
smoking *109*, *111*
SST (supersonic transport) program, skin cancer and 540
standards of living *see* living standards
Steckel, Richard H., slaves 179

steel
  acid rain's effects on 531
  availability and price 319
  water used in production of
    431
Stein, Robert, sugar
    consumption 143
Stiles, R. E., CFC estimates
    537
stocks, US economic
    performance and 276–7
stratospheric ozone 17–18,
    536–41, 542, 543
streams, acid rain's effects on
    532–3
strontium 317
Stuart, S. N., species loss 352,
    353
substance abuse, homelessness
    and 263–5, 266
Sudan
  famine 396
  smoking 109, 111
  sorghum hybrids 382
sugar consumption 143, 155
suicide rates 92–4, 97
sulfur
  consumption and availability
    316
  water pollution by 449, 450,
    451, 518
sulfur dioxide
  acid rain: causation 523, 524;
    distribution and trends
    525, 527; effects 528, 531,
    534, 638
  albedo effects 557–8, 559,
    560
  market and socialist
    economies compared 508,
    511, 512–13, 514
  UK 476, 477–86, 638
  USA 17: early concerns 491;
    long-term trends 454–5,
    468²⁷, 494, 495, 498–500,
    502; photochemical smogs
    491–2
Sullivan, Richard J., agricultural
    labor force 124–34
Summers, L. H.,
    unemployment 201⁵
sunlight, skin cancer 540–1
sunspot activity, ozone
    concentrations 539–40
supersonic transport (SST)
    program 540
Surinam
  deforestation 340
  smoking 109
suspended particles
  albedo effects 557–8, 559,
    560
  market and socialist
    economies compared 508,
    511, 512–13, 514
  USA 17, 494, 495, 498–500,
    502
Swanson, soil erosion 418, 419
Sweden
  air pollution 504, 516–17
  body height 67, 68, 144
  education 218, 219, 220, 222
  homicide rates 91

income 139, 194
  labor productivity 168, 169
  mortality and life expectancy
    9, 31, 32: accident
    fatalities 102; children 39,
    40, 41–2; height and 67,
    68; suicide rates 93, 94;
    tuberculosis death rates 56
  poverty 238, 239
  smoking 109, 112
  sugar consumption 143
  water pollution 445–6, 518
Switzerland
  child mortality 39–40
  education 222
  income 139
  labor productivity 168
  poverty 238, 239
  smoking 109, 112
Symonds, R. B., CFC
    estimates 537

Taiwan
  education 218
  life expectancy 33, 47
  rice production 374
Tanner, James, height trends
    144
tantalum, availability 318
Tanzania
  agricultural productivity 387
  nutrition 402
  smoking 109
taxation
  census data and poverty
    estimates 251–2
  home ownership and 205
  multinational firms,
    government role and 649
  US economic performance
    and 274–5
TB (tuberculosis) 54–7, 85, 89
technological change 25–6
  continued progress 646–9
  journalistic bias and 655
  productivity and 161–70
teenagers
  education 210, 214, 215,
    217: dropout rates 208–9,
    211, 212
  unemployment 199–200
telephones, USA 159, 160, 246
tellurium 318, 321
temperature changes, global
    warming see greenhouse
    effect
Terry, L., smoking 107
Thailand
  agricultural labor force
    129–30, 131, 133⁷
  deforestation 340
  homicide rates 92
  nutrition 400, 402
  slavery 175
  smoking 109, 111
thallium 318, 321
Third World countries see
    developing countries;
    named countries
Thompson, M. A., air
    pollution 480
Three Mile Island, reactor
    accident 579

timber
  net volumes 332, 333
  price trends 330–1
time-diaries, leisure time 226–9
tin
  availability 318, 319
  prices 309, 312
titanium 317, 318
tobacco smoking 106–13, 158,
    631
tornadoes, fatalities 101
total suspended particles
    (TSPs) see suspended
    particles
Tousson, Prince Omar, Nile
    river 462, 463, 464
Toutain, J., calorie
    consumption 64
transport
  agriculture and 368, 386
  productivity growth 165–6,
    167
  supersonic transport program
    (SST) 540
  US fatal accidents 101, 102
  see also motor vehicles
Tranter, Neil, mortality 140
trees see forests; tropical forests;
    wood
tropical forests 328–30, 335–42
  species loss 339, 351, 352,
    354–5, 356, 359–60
TSPs (total suspended
    particles) see suspended
    particles
tuberculosis see TB
Tucker, William, homelessness
    259, 268¹
tungsten 311, 312, 318
Turkey
  agricultural labor force 128,
    129, 130
  consumption per capita 159
  irrigation 386
  nutrition 400, 402
  smoking 109, 112
typhoid 90, 150

UCS (Union of Concerned
    Scientists), nuclear power
    576, 578–9, 583, 585
UK
  agricultural labor force
    124–5, 126, 133³
  agricultural productivity 365,
    366–7, 639: estimates 369,
    370, 371; population
    growth 367–8
  air pollution 18, 455–6,
    476–89: acid deposition
    481–6, 483–6, 638–9;
    economic growth and 504,
    512, 516–17; greenhouse
    gases 488; photochemical
    pollution and vehicle
    exhaust emissions 486–8;
    pre-20th century 447, 448,
    464³, 466³, 476–8; smoke
    and sulfur dioxide decline
    in twentieth century
    478–83, 485–6
  alcohol consumption 115,
    117, 118–20

UK(*Cont.*)
  coal 25, 280–2
  consumption per capita *159*
  economic growth and living
    standards 137–9, *140*: air
    pollution *504*, *512*,
    516–17; beer output *141*;
    cotton consumption *142*,
    145; daily per capita
    calories *142*; height trends
    144; housing *244*; income
    139; income distribution
    194; international
    comparisons *147*; labor
    productivity *168*, *169*;
    meat consumption *250*;
    poverty *238*, 239, 240;
    sugar consumption *143*;
    water pollution 508, *509*
  education 218, 219, 220, 222
  forests 484, 638–9
  health 77: disease control
    57, 58
  mortality and life expectancy
    9, 31–2: accident fatalities
    103; Benedictine monks
    31; calorie consumption
    64–5; causes of death *86*;
    children 39, 40, *43*, 44,
    140; crisis mortality 61–3;
    economic growth 136,
    139, *140*; height and *67*,
    68; infectious diseases *43*,
    44, 54–6; violent deaths
    91, 93, 94–7
  Pesticide residues in human
    tissue and milk *510*
  slavery 172, 173, 174
  smoking 108, *109*, *112*
  water pollution 447–8,
    456–7, 467[13], 508, *509*
Ulrich, Alice H., wood
  resources 330
ultraviolet light, skin cancer
  and 540–1
unemployment, trends in
  196–202
  see also labor force
UNESCO, education *215–16*
Union of Concerned Scientists
  (UCS), nuclear power
  576, 578–9, 583, 585
United Nations
  Food and Agriculture
    Organization *see* FAO
  life expectancy 33–4: health
    and disease and 51, 52,
    53; infant mortality rates
    41, *45*, 46, 401
uranium, radon problems and
  582, 583
urbanization
  agricultural labor force 124,
    133[6]
  agricultural productivity
    368
  heat islands, greenhouse
    effect 551–2
  life expectancy 32, 56, 58–9:
    children 38–9, 41; slaves
    179
  ozone concentrations 487
US land use trends 434–5

Uruguay
  deforestation *340*
  housing space 243
  smoking *109*, *110*
USA
  acid rain 523: cost benefits of
    reduction *533*, 534; crop
    yields and 528;
    distribution and trends
    525, *526*; forests 528–30;
    health 531; hydrogen
    peroxide 524; lakes
    532–3; pH scale *524*;
    visibility 534
  agriculture 380: acid rain
    528; Acreage Reduction
    Programs (ARPs) 406–8;
    Conservation Reserve
    Program (CRP) 420–2;
    corn *373*, *375*, 378, *392*,
    406–8; cotton *375*, *392*,
    436–7; labor force 126;
    land use trends 434,
    435–7, 440; population
    growth 388, 389; soil
    erosion 416–23; starchy
    staple ratio 398; total and
    per capita *397*; wheat *12*,
    375, *392*, 395, 406–8
  air pollution *15–17*, 491–502,
    504: data issues 492–4;
    economic growth and *504*,
    505–8, 511, *512–13*,
    514–17; general trends
    494–7, 498–502; long-term
    trends 446–7, 454–5, 457,
    458, 460, *474*; *see also*
    USA, acid rain
  alcohol consumption *114*,
    115–16, 118–20, *158*
  Civil Rights movement 181,
    182
  economic performance
    271–7: global warming
    566, *568*; income 191–2,
    193–4, 272–4; pollution
    504, 505–8, *509*, 511,
    512–13, 514–17, *518*, 519;
    poverty 235
  education 208–23:
    African-Americans 182,
    209, *213–14*; dropout rates
    208–9, 211, *212*;
    expenditure and
    attainments 217–23
  employment 196–202,
    271–2: African- Americans
    181–2, *185*
  energy resources: from
    1800–1989 282–4; coal
    280–1, 282, *283*;
    consumption rates *159*,
    506, 507–8, 520; nuclear
    power costs 294–301;
    nuclear power hazards
    576–85; oil prices and
    supply *283*, 284, 287–91,
    293; public opinion about
    610, 614–17; whale oil
    crisis 25; wood fuel 282,
    283, 284
  forests 331–2, 333, 334: acid
    rain 528–30; forest

    products 330, 331, *332*;
    land use 434, *435*, 436;
    species loss 352
  health 72–82: acid rain 531;
    cancer 540, 589, 593;
    indicators 73, 82; measles
    90; medical care 153, *159*,
    234; National Health
    Interview Survey 72–3, 77;
    trends explained 77–81;
    tuberculosis and
    diphtheria 85; *see also*
    USA, mortality and life
    expectancy
  housing 151, 160:
    African-Americans 180,
    181, 182; cost and quality
    203–7; economic
    performance and 276–7;
    homelessness 237,
    257–68; poverty levels and
    241–5; subsidies 234
  land use 434–40
  living standards 149–60:
    African-Americans
    178–83, 184–7, 236–7,
    250; consumer spending
    156, *158–60*; food *142*,
    143, 150, 154–6, 247–51;
    free time 224–30; heat
    150–1, 152–3, 160; height
    144; housework and
    household appliances
    152–3, 155, *159*, 160,
    245–7; housing *see* USA,
    housing; international
    comparisons 147–8;
    medical care 153, *159*,
    234; outdoor recreation
    323–7; per capita income,
    growth rates *139*; poverty
    151, 231–40, 241–52;
    smoking 106–8, *109*, *110*,
    158; water 149–50, *160*;
    wealth and income *139*,
    179–83, *186*, 188–94
  mortality and life expectancy
    9, 32; causes of death *86*;
    fatal accidents 98–105;
    height and body mass
    index 68–9, 71; homicide
    rates 91–2; infectious
    diseases 55, 56, 58;
    lynchings *186*; by race
    4–5, 179, *180*, 181; suicide
    rates 93, 95
  pollution 14–15, *16–17*:
    pesticide residues *510*;
    solid waste 460–1, *475*; *see
    also* USA, air pollution;
    USA, water pollution
  population growth, public
    opinion 619–26
  productivity 161, *168–70*,
    271, 272, *273*: *see also*
    USA, agriculture
  slavery 173–4
  water, soil erosion by 417,
    418–19, *421*, 422, 423, 519
  water management 425–32
  water pollution: drinking
    water *473*; economic
    growth and 508, *509*, *518*,

USA (*Cont.*)
519; Great Lakes 449–51, *452*, 467[18], *470–2*; New York City Harbor 451–4
USSR
agricultural productivity *384*: cereal production *378*; corn 382; cotton 428–9, 519; per capita grain production *380*; total and per capita *397*
fish landings 413
forecasts for 654
forests 333, *334*, 484
housing space *242*, 243
international consumption comparisons 158, *159*, 160
meat consumption *250*
morbidity 89–90
mortality *33*, 85–9
pollution and economic growth 511, *512–13*, 514, *515–19*
serfdom in Russia 173
smoking *109*, *112*
sugar consumption *143*
water pollution 517–19; Aral Sea 428–9, 519
UV light, skin cancer and 540–1

Vallin, Jacques, disease and mortality 51
Valroff, J., ecology 637
vanadium *318*, 319
Vedder, R., unemployment *185*
vehicles *see* motor vehicles
Venezuela
forests 335
meat consumption *250*
smoking *109*, *110*
Verbrugge, L. M., US health 77, 78, 79, 80
Vickland, Katherine H., greenhouse effect 566
Viscusi, W. Kip, risk management 628–34
visibility, acid rain's effect on 534
vitamins, dietary deficiency 251
von Mises, Ludwig, ideas 658, 659
voting
African-Americans 181
trends 22

Waaler, Hans Th., height and mortality *66*, *67*
wages *see* income
Wahl, Richard W., water values *428*, 432
Waidmann, T., health 77, 80
Wales *see* UK
Wall Street Journal, population growth 624–5, 626
war
confidence destroying effect 656
predictions about 654
warning labels 631
Washington Post
population growth 624–6
reporting slant 4, 23, 208–9, 218–22

Washington, W. M., climate model *550*, *551*, 553
waste
accumulation and disposal, USA 460–1, *475*
economic system and *518*
health effects 603–4
radioactive 580–3
water 425–32
acidification effects 483–4: drinking water 531; lakes and streams 532–3
drought: African food production and 386–7; soil erosion and 418
flood fatalities *101*
fluoridation of 601
global warming: cloud brightening 558; ocean dynamics 546–7, 548, *550*, 553, *554*; riverine site bias 553; sea level rise 557
irrigation 386, 427: Africa 387; Aral Sea and 428–9, 519; water management policy 427, 428–9, 432
radioactive waste and 581–2
soil erosion by 417, 418, *421*, 423: soil runoff pollution 418–19, 422, 519
supply and demand 425–6: market-based approach 429–32; political control 426–9; USA 149–50, *160*, 425–32
US land use trends, artificial reservoirs 440
water pollution 14, *15*
Baltic Sea 445–6, 466[9]
doomsayers' forecasts 22
drinking water *473*
economic systems compared 508, *509*, 517–19
Great Lakes 449–51, *452*, 467[18], *470–2*
London 447–8, 456–7, 467[13]
markets and 431–2
New York City Harbor 451–4
Nile river 461–4, *465*
oil spills 414–15, 633
public opinion about 614, 633
soil runoff 418–19, 422, 519
Watt Committee, acid rain 638
Wattenberg, B., housing *185*
wealth and income distribution, USA 188–94
African-Americans 179–83, *186*, 236–7
poverty 231–7, 241–52
Weinberg, A. M., mineral resources 313, 321
Weir, D. R., life expectancy 61, *62*, 64
welfare programs, USA 234–5, 238, 239, 252
food consumption 248, *249*, 251
Welfeld, I., crop yields *375*
West Germany *see* Germany
West Indies
forests *329*, *340*
nutrition *402*

slavery 174
smoking *109*
Western, David, species loss 355, 356
Wetherald, R. T., climate models *549*
whale oil crisis 25
wheat
population growth and 388–9
starchy staple ratios 398
US prices *12*, 406, *407*, 408
winter 382
world stocks *407*
yields 378, *379*, *405*: China *383*, *384*, 388 USA *375*, *392*, 395
Whelan, Elizabeth M., cancer scares 595–605
Whitmore, T. C., species loss 346, 351–3
Wigley, T. M. L., temperature trends *547*
Wilcoxin, Peter J., greenhouse effect 566
Wildavsky, A., species loss 346–60
wilderness areas, outdoor recreation 326, 650
wildlife
game hunting 325–6
species loss 15, 346–60; commercialization of biodiversity 343; DDT and 602; defining species 347–8; tropical deforestation and 339, 351, 352, 354–5, 356, 359–60; value of species-saving 357–8
US National Wildlife Refuge system 323
Wilkins, R., health 77
Williams, G. P., alcohol *117*
Williamson, Jeffrey G.
economic growth 139
poverty rates *232*
US wealth distribution 188–95
Wilson, Edward O., species loss 347, 355
Wilson, G. B., alcohol *117*
Wilson, R., US health 80
wind erosion, of soil 417, 418, 422–3
wine consumption 114–20
Winkler, E. M. environmental deterioration 447
Wise, John P., sea food 411–15
Wolfe, B., health 77, 80
Wolff, E. N., productivity *169*
women
mortality rates 9
in workforce 224: free time 227, 228; USA 200, 271–2; working years 225
wood (fuel), price and availability 280–1, 282, *283*, 284
wood (lumber)
net volumes *332*, 333
price trends 330–1
Woodin, S. J., air pollution 484

Wooten, H. H., land use
    434–5, *435, 439*
work *see* labor
Working, Holbrook, wheat
    yields 395
World Bank
    forests 335
    hunger quantification 394–5,
        402
    longevity *140*
    nutrition *143, 400*
World Food Council,
    starvation 394
World Health Organisation
    smoking *109,* 110
    suicide rates *94*
World Resources Institute
    (WRI), forests 336,
        339
World Wildlife Fund 346, 356
Worldwatch, forests 336
"Worried Sick" syndrome 80
WRI *see* World Resources
    Institute

Wrigley, E. A.
    agricultural labor force 124,
        132–3[2,3]
    English economy 282
    life expectancy 31, 40, 61–3,
        64, *140*
Wuebbles, D. J., supersonic
    transport 540

yams *380*
Yasuba, Yasukichi, Japanese
    living standards 136
Ycas, M. A., health 77
Young, Arthur, slavery 172
yttrium *318*
Yugoslavia
    pollution 511, *512–13*
    smoking *109, 112*

Zaire
    cobalt crisis 320
    deforestation *340*
    nutrition *402*
    smoking *109*

Zambia
    cobalt crisis 320
    deforestation *340*
    smoking *109*
Zanzibar, slavery 175
Zaychenko, A. S., US housing
    242
Zeckhauser, Richard J., risk
    management 628–34
Zimbabwe
    agricultural productivity 387
    smoking *109*
zinc
    availability *318,* 321
    prices 307, *308*
    recycling 315
    substitutability 314–15, 319
zirconium *318, 321*
Zon, R., forests 336, *337,*
    338
Zucker, A., mineral resources
    313, 316, 321
Zweig, Stefan, belief in
    progress 656